16th International Symposium on Nuclei in the Cosmos (NIC-XVI)

EPJ Web of Conferences Volume 260 (2022)

Chengdu, China
21 - 25 September 2021

Editors:

Weiping Liu
Youbao Wang
Bing Guo
Xiaodong Tang
Sheng Zeng

ISBN: 978-1-7138-4456-3

Printed from e-media with permission by:

Curran Associates, Inc.
57 Morehouse Lane
Red Hook, NY 12571

Some format issues inherent in the e-media version may also appear in this print version.

For additional information, please contact EDP Sciences – Web of Conferences
at the address below.

EDP Sciences – Web of Conferences
17, Avenue du Hoggar
Parc d'Activité de Courtabœuf
BP 112
F-91944 Les Ulis Cedex A
France

Phone: +33 (0) 1 69 18 75 75
Fax: +33 (0) 1 69 28 84 91

contact-edps@webofconferences.org

Additional copies of this publication are available from:

Curran Associates, Inc.
57 Morehouse Lane
Red Hook, NY 12571 USA
Phone: 845-758-0400
Fax: 845-758-2633
Email: curran@proceedings.com
Web: www.proceedings.com

TABLE OF CONTENTS

edp sciences

EPJ Web of Conferences

Statement of Peer review

In submitting conference proceedings to EPJ Web of Conferences, I certify to the Publisher that I adhere to the **Policy on Publishing Integrity** of the journal in order to safeguard good scientific practice in publishing.

1. All articles have been subjected to peer review administered by the proceedings editors.
2. Reviews have been conducted by expert referees, who have been requested to provide unbiased and constructive comments aimed, whenever possible, at improving the work.
3. Proceedings editors have taken all reasonable steps to ensure the quality of the materials they publish and their decision to accept or reject a paper for publication has been based only on the merits of the work and the relevance to the journal.

Title, date and place of the conference

The 16th International Symposium on Nuclei in the Cosmos (NIC-XVI), September 21-25, 2021, Beijing

Proceedings editor(s):

Weiping Liu, Youbao Wang, Bing Guo, Xiaodong Tang, Sheng Zeng

Date and editor's signature

Feb. 17, 2022

EPJ Web of Conferences **260**, 01001 (2022) https://doi.org/10.1051/epjconf/202226001001
NIC-XVI

An indirect technique in nuclear astrophysics: alpha-cluster transfer reaction

Yang-Ping Shen[1,*], *Bing* Guo[1], and *Wei-Ping* Liu[1]

[1]China Institute of Atomic Energy, P.O. Box 275 (10), Beijing 102413, China

Abstract. Helium (^{4}He, or α) is the second most abundant element in the observable Universe. The α-particle induced reactions such as (α, γ), (α, n) and (α, p) play a crucial role in nuclear astrophysics, especially for understanding stellar helium burning. Because of the strong Coulomb repulsion, it is greatly hindered to directly measure the cross sections for these α-capture reactions at stellar energies. Alpha-cluster transfer reaction is a powerful tool for investigation of astrophysical (α, γ), (α, n) and (α, p) reactions since it can preferentially populate the natural-parity states with an α-cluster structure which dominantly contribute to these astrophysical α-capture reactions during stellar helium burning. In this paper, we review the theoretical scheme, the experimental technique, astrophysical applications and the future perspectives of such approach based on α-cluster transfer reactions.

1 Introduction

Nuclear astrophysics research is currently a frontier in the quest to understand how the elements in Universe were created and how stars evolve over time. Thousands of nuclear processes are responsible for the synthesis of the elements and drive the evolution of stars. These nuclear reactions are mostly triggered by hydrogen, helium, and heavy ions. Apart from the major so-called Big Bang Nucleosynthesis (BBN) [1, 2], the ^{4}He nuclide (namely, α cluster or α particle) can also be created during the hydrogen-burning phase within stars [3], which did not occur until about 100 million years after the Big Bang. Because ^{4}He is the second most abundant element in the observable Universe after hydrogen, α-cluster induced reactions like (α, γ), (α, n), and (α, p) play a crucial role in nuclear astrophysics, especially for understanding stellar helium burning, which is a critical stage during the evolution of stars.

Because the energies corresponding to typical temperatures in stars are significantly below the high Coulomb barrier, the direct measurements of α-particle induced reactions at stellar energies (so-called Gamow window) are greatly hindered due to the vanishing cross section resulting from the small Coulomb penetrability at energies of astrophysical interest. The ^{12}C(α, γ)^{16}O cross section, for example, is expected to be on the order of 10^{-17} b at 300 keV corresponding to the average temperature of helium burning. By far, this is more than five orders of magnitude lower than the maximum sensitivity achieved by the most advanced measurements. Generally the estimations of the cross sections at stellar energies are extrapolated from experimental data with a much higher energy range. However, because unknown

*e-mail: fermi09@foxmail.com

resonant states offen exist in the energy range of astrophysical interest, and free parameters in extrapolations are not able to be properly constrained by data at high energies. Therefore, such extrapolations may lead to considerable uncertainty. As a result, indirect approaches (for example, α-cluster transfer reactions) are particularly useful for reactions that are difficult, if not impossible, to measure directly. These techniques can be used to derive level parameters (e.g., energies, asymptotic normalization coefficients (ANC) or spectroscopic factors (SFs), lifetimes) that can then be used in the analysis of the R-matrix or other reaction model [4].

Several indirect techniques for studying astrophysical α-capture reactions have been developed and implemented. In present article, we focuse on a specific and complementary method, α-cluster transfer reactions, which specially aims at determining the cross sections and the stellar rates of α-particle induced reactions in the stellar helium burning phase, which are more difficult directly to measure at the Gamow window than proton induced reactions since they have higher Coulomb barrier. Because the α transfer reaction is most likely to occur by α-cluster transfer mechanism, it can not only be used to study α-particle induced astrophysical reactions [5], but also to evaluate the nuclear structure (e.g., α-decay widths [6]) and nuclear reaction mechanisms [7].

2 DWBA theory

In reactions of type $A(a, b)B$, the nuclei A and a usually start in their ground states. In general, direct reaction and compound-nucleus reaction mechanisms are used to describe transfer reactions. Only a few nucleons or the nucleus as a whole on the surface of the nucleus are involved in the reactions that proceed most quickly. They are called direct reactions, and usually occur at high incident energies. This is because such reactions finish more quickly, leading to fewer internal collisions. Because the final nuclei's directions are significantly more impacted by the initial direction in these quick reactions, they often exhibit high cross sections at forward angles.

The DWBA theory, which assumes a one-step transition between the initial and final scattering states, is widely applied to model the direct reactions. This theory proves very useful in description of such reactions. Of course, the compound-nucleus reaction process is still viable at lower energies or at backward angles. However, this mechanism produces isotropic angular distributions which can be easily distinguished from forward-peaked cross sections. It can also be evaluated and, if necessary, subtracted. In addition one-step assumption may be improved by including two and higher-order steps, as in a perturbation series. Coupled-channels approaches must be employed in this scenario. More details about the Higher-order effects analysis can be found in Ref. [8].

In the DWBA theory, the most important equation is the relation between the reduced DWBA and experimental differencial cross sections. Considering the α-cluster transfer reaction $A + a \to b + B$, the relation is given to be

$$\left(\frac{d\sigma}{d\Omega}\right)_{\text{exp}} = \sum_{l_B j_B l_a j_a} S_{A\alpha l_B j_B} S_{b\alpha l_a j_a} \sigma^{\text{DWBA}}_{l_B j_B l_a j_a}, \tag{1}$$

where $\sigma^{\text{DWBA}}_{l_B j_B l_a j_a}$ denotes the reduced DWBA cross section. In addition the asymptotic normalization coefficient (ANC) can be related to the spectroscopic factor by

$$(C_{\beta\gamma l_\varphi j_\varphi})^2 = S_{\beta\gamma l_\varphi j_\varphi} \cdot b^2_{\beta\gamma l_\varphi j_\varphi}. \tag{2}$$

$\sigma^{\text{DWBA}}_{l_B j_B l_a j_a}$ depend on the optical model potential (OMP) parameters for the initial $(a + A)$ and final $(b + B)$ channels and the real binding potentials for the initial (a) and final nuclei

(B). Usually the OMP parameters of the initial and final channels can be fixed by fitting the experimental angular distributions of elastic scattering for the initial and final channels. The geometrical parameters for the binding potentials can be constrained by a minimum-χ^2 fitting to the experimental data of transfer reaction angular distributions. Another typical method to constrain the binding potential is to reproduce the root-mean-square (rms) radius of the α-cluster wave function (e.g., for the $B = (A\alpha)$ system) using the formula [9]

$$\left\langle r_B^2 \right\rangle = \frac{m_{\mathrm{He}}}{m_B} \left\langle r_{\mathrm{He}}^2 \right\rangle + \frac{m_A}{m_B} \left\langle r_A^2 \right\rangle + \frac{m_{\mathrm{He}} m_A}{m_B^2} \left\langle r^2 \right\rangle . \tag{3}$$

Here $\left\langle r_B^2 \right\rangle$, $\left\langle r_{\mathrm{He}}^2 \right\rangle$, and $\left\langle r_A^2 \right\rangle$ are the the rms radii of the compound nucleus B, the valence cluster He, and the core A, which can be determined experimentally or theoretically. $\left\langle r^2 \right\rangle$ is the root-mean-square (rms) radius of the α-cluster wave function.

3 Experimental techniques

From the perspective of nuclear astrophysics, α transfer reactions with large cross sections provide an alternative method for extracting level parameters such as the α spectroscopic factor (S_α), spectroscopic amplitude (SA), ANCs, or reduced α widths for the subthreshold resonances crucially involved in determining the astrophysical S-factor of the challenging α-capture reactions. Thus, in studies involving α-transfer processes, the selection of the α-transfer systems is the highest priority to consider. The most common transfer systems are the (^{6}Li, d) and (^{7}Li, t) reactions. The (^{11}B, ^{7}Li) reaction has also been proposed as a suitable transfer system for studying cluster structures and astrophysical reactions in recent years.

In the DWBA calculations, the S_α or the ANC of the α provider in the transfer systems, such as ^{6}Li in the (^{6}Li, d) system, ^{7}Li in the (^{7}Li, t) system and ^{11}B in the (^{11}B, ^{7}Li) system, is necessarily required to derive the desirable level parameters. The detailed table of S_α/SA and ANC results of the ^{6,7}Li and ^{11}B ground states for the (^{6}Li, d), (^{7}Li, t) and (^{11}B, ^{7}Li) transfer systems can be found in Ref. [10].

In addition to (^{6}Li, d), (^{7}Li, t) and (^{11}B, ^{7}Li), other transfer systems, such as (^{16}O, ^{12}C) and (^{20}Ne, ^{16}O), also have potential for investigating α-cluster structure and α induced reactions of astrophysical interest. To date the ANC of the ^{16}O GS has been extensively investigated. There is a large discrepancy of up to a factor 240 between these reported experimental values of the GS ANC ranging from 13.9 ± 2.4 fm$^{-1/2}$ to 3390 fm$^{-1/2}$ [11–15]. Although the most recent measurement by Shen et al. [15] gives a more precise value, further high-precision measurements of the ^{16}O GS ANC are still desirable. This is because the ^{16}O GS ANC is crucial for restricting the GS external capture in the ^{12}C(α, γ)^{16}O reaction, as well as for studying astrophysical reactions as a powerful tool. As for the ^{20}Ne GS ANC, although Motobayashi et al. [16] presented the first experimental determination through the elastic transfer reaction ^{20}Ne(^{16}O, ^{20}Ne)^{16}O using ΔE-E silicon detector telescope, it is still necessary to perform high-precision measurement of the ^{20}Ne GS ANC because of its potential application to astrophysical reactions.

With the transfer systems introduced above, here we discuss three kinds of typical experiments, including sub-Coulomb measurement of α-transfer reactions, high-energy measurement of α-transfer reactions, and high-resolution coincidence measurement of absolute α-decay widths.

Performing transfer reaction measurements at sub-Coulomb energies is challenging experimentally, because the cross sections become rather small and also the energies of the recoil nuclei are low. However, such measurements are crucial for reducing the dependence of the results on optical potential parameters and the influence of compound-nuclear contributions. This is because, at sub-Coulomb energies, DWBA calculations are primarily governed by Coulomb potentials for both the exit and entrance channels, and seldom depend on nuclear potential parameters. The DWBA cross sections are thus essentially model independent. The pioneering sub-Coulomb measurements of the ^{12}C(^{6}Li, d)^{16}O and ^{12}C(^{7}Li, t)^{16}O cross sections were performed by Brune et al. [17] to investigate the ^{12}C(α, γ)^{16}O S-factor by determining the reduced α widths of the 2^+ and 1^- subthreshold states in ^{16}O. The normalized DWBA calculations well reproduced the experimental excitation functions for both ^{12}C(^{6}Li, d)^{16}O and ^{12}C(^{7}Li, t)^{16}O leading to the 2^+ and 1^- states of ^{16}O.

An alternative technique is sub-Coulomb measurement of transfer-reaction differential cross sections as a function of angle, namely, angular distributions. For example, the astrophysical S-factor of the ^{13}C(α, n)^{16}O reaction was investigated by determining the ANC for the 6356 keV $1/2^+$ state of ^{17}O using the ^{13}C(^{6}Li, d)^{17}O reaction in inverse kinematics, performed at two sub-Coulomb energies [18, 19]. It should be noted that the technique based on excitation function measurement is not suitable for a high-energy measurement because in that case compound-nucleus contribution and higher-order effects can no longer be ignored, causing DWBA calculations to fail to reasonably reproduce experimental total cross sections.

Comparing with Sub-Coulomb transfer measurement, the high-energy transfer measurement allows one to use a high-precision magnetic spectrograph for resolving the closely spaced levels which cannot be achieved only by silicon detectors in sub-Coulomb transfer measurement. Therefore, these two techniques can be used complementarily to study α-cluster structures and then astrophysical nuclear reactions. In Fig. 1 we schematically show a typical experimental setup for high-energy transfer measurement performed using the Q3D magnetic spectrograph at HI-13 tandem accelerator of China Institute of Atomic Energy (CIAE). The reaction products were separated and focused by the magnetic spectrograph and detected by two-dimensional position sensitive silicon detector array fixed at the focal plane. The two-dimensional position information enables the products emitted into the acceptable solid angle to be completely recorded, and the energy information was used to remove the impurities with the same magnetic rigidity, as discussed in Refs. [9, 15, 20]. Typically the Q3D magnetic spectrograph has an energy resolution of 0.02% and an angular resolution of 0.1 degree, which makes it possible to resolve the closely spaced states which cannot be achieved only by silicon detectors, and to obtain high-precision data of angular distributions at forward angles where direct mechanism dominates. High-precision data of the angular distributions at forward angles are helpful for constraining the binding potential parameters with the minimum χ^2 fitting.

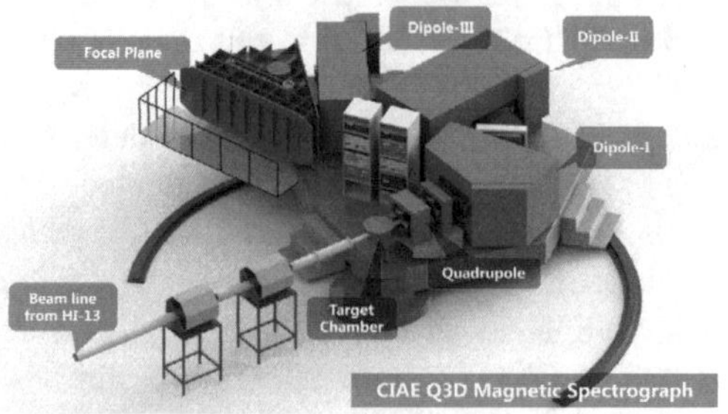

Figure 1. Typical experimental setup for high-energy transfer measurement performed using the Q3D magnetic spectrograph at HI-13 tandem accelerator of CIAE. The Q3D magnetic spectrograph consists of a target chamber, a quadrupole, three dipoles, and detector arrays at the focal plane.

As for the resonances far above the α-decay threshold, large α-decay widths of these resonances allow us to perform high-resolution coincidence measurement of absolute α-decay widths. For example, Wheldon et al. [21] unambiguously measured the absolute α-decay widths from the excited states in ^{16}O within the energy range 13.85 to 15.87 MeV through the ^{12}C(^{6}Li, d)^{16}O reaction by using a large-acceptance position-sensitive silicon detector array placed near the target position in coincidence with the high-resolution Q3D magnetic spectrograph. The deuteron ejectiles were analyzed by the Q3D spectrograph, while the recoil or the breakup products were recorded by the silicon array comprised of four double-sided silicon-strip detectors. Such a coincidence measurement technique is useful to investigate the resonances with considerable α widths. Similar application of this technique has been achieved in other astrophysical reactions such as the ^{15}O(α, γ)^{19}Ne reaction (see, e.g., Ref. [22]).

4 ^{12}C(α, γ)^{16}O, a example for the application of the transfer reaction method

The ^{12}C(α, γ)^{16}O reaction, which is one of the most important reactions in nuclear astrophysics, is dominated by multiple supra- and sub-threshold resonance to ^{16}O excited states. The complicated reaction mechanism in ^{12}C(α, γ)^{16}O provides an excellent case for the application of indirect techniques, especially the transfer reaction method. The resonance to the ^{16}O 6917 keV 2^+ and 7117 keV 1^- states are two dominant subthreshold resonances and the external capture to the ^{16}O ground state is also expected to be unneglectable. Many works with (^{6}Li, d), (^{7}Li, t) and (^{11}B, ^{7}Li) transfer reactions are reported. We recently measured the ANC of the ^{16}O 6917 keV 2^+ and ground states with ^{12}C(^{11}B, ^{7}Li)^{16}O transfer reaction [15, 23]. A comparison of $S_{E2}(300)$ is shown in Fig. 2.

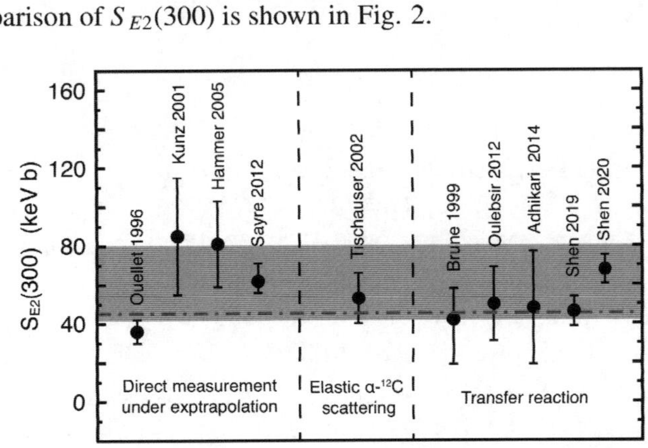

Figure 2. The $S_{E2}(300)$ comparison [11, 15, 17, 23–29]. The grey shadow represents the compilation value of NACRE II (2013) [30]. The blue dot-dashed line is the value in deBoer et al. (2017) [4].

It's seen that the S_{E2} factors given by the transfer reaction method are in good agreement with the values given by direct measurements and other indirect methods like elastic scattering. And the results with different transfer systems are also consistent with each other, which proves the systematic uncertainty from the diversity of different transfer systems is under control. One thing that should be mentioned is that the S_{E2} factor given by Shen et al. (2020) [15] is significantly higher than others. This is because the contribution from the external capture to the ^{16}O ground state is considered and found to be large in Shen et al. (2020) [15].

5 Discussion and outlook

The alpha-cluster transfer reaction has long been a powerful tool for determining the level parameters of the unknown resonant states with an α-cluster structure and further investigating astrophysical (α, γ), (α, n) and (α, p) reactions during helium burning phase of stars, and it continues to be so in the latest state-of-the-art analyses. To date this technique has been unambiguously verified by reproducing the well-known width, and has been extensively used to investigate astrophysical α-particle induced reactions on stable nuclei such as the main neutron source reaction ^{13}C(α, n)^{16}O of the *s*-process nucleosynthesis and the so-called "holy grail" reaction ^{12}C(α, γ)^{16}O, as well as reactions on long-lived unstable nuclei, such as the ^{14}C$(\alpha, \gamma)^{18}$O reaction.

As an indirect technique, it turns out to be one of the most useful methods in determining the contributions from the subthreshold states that are extremely difficult to constrain using the high-energy cross sections data from direct measurements, since the data at high energies have no effective constraints on such contributions while low-energy measurements are greatly hindered by the Coulomb barrier. Therefore, indirect techniques will continue to be highly desirable in the future, in addition to development of novel or underground direct measurements.

Acknowledgments

We would like to thank our colleagues of nuclear astrophysics team, without whom the results reviewed in this article would not have been achieved. This work is supported by the National Natural Science Foundation of China under Grants No. 12125509, 12005304, 11490561, U1867214, 11961141003.

References

[1] B.D. Fields, Annu. Rev. Nucl. Part. Sci. **61**, 47 (2011)
[2] R.H. Cyburt, et al., Rev. Mod. Phys. **88**, 015004 (2016)
[3] E.M. Burbidge, et al., Rev. Mod. Phys. **29**, 547 (1957)
[4] R.J. deBoer, et al., Rev. Mod. Phys. **89**, 035007 (2017)
[5] L. Buchmann, et al., Astrophys. J. **324**, 953 (1988)
[6] D.F. Jackson, M. Rhoades-Brown, Nature **267**, 593 (1977)
[7] P. Hodgson, E. Beták, Phys. Rep. **374**, 1 (2003)
[8] I.J. Thompson, F.M. Nunes, Cambridge University Press, 2009 (2009)
[9] B. Guo, et al., Astrophys. J. **756**, 193 (2012)
[10] Y.P. Shen, B. Guo, W.P. Liu, Prog. Part. Nucl. Phys. **119**, 103857 (2021)
[11] D.B. Sayre, et al., Phys. Rev. Lett. **109**, 142501 (2012)
[12] S. Adhikari, C. Basu, Phys. Lett. B **682**, 216 (2009)
[13] S. Adhikari, et al., J. Phys. G **44**, 015102 (2017)
[14] M.C. Morais, R. Lichtenthäler, Nucl. Phys. A **857**, 1 (2011)
[15] Y.P. Shen, et al., Phys. Rev. Lett. **124**, 162701 (2020)
[16] T. Motobayashi, et al., Nucl. Phys. A **331**, 193 (1979)
[17] C.R. Brune, et al., Phys. Rev. Lett. **83**, 4025 (1999)
[18] E.D. Johnson, et al., Phys. Rev. Lett. **97**, 192701 (2006)
[19] M.L. Avila, et al., Phys. Rev. C **91**, 048801 (2015)
[20] Y.P. Shen, et al., Phys. Lett. B **797**, 134820 (2019)

[21] C. Wheldon, et al., Phys. Rev. C **83**, 064324 (2011)
[22] W.P. Tan, et al., Phys. Rev. Lett. **98**, 242503 (2007)
[23] Y.P. Shen, et al., Phys. Rev. C **99**, 025805 (2019)
[24] J.M.L. Ouellet, et al., Phys. Rev. C **54**, 1982 (1996)
[25] R. Kunz, et al., Phys. Rev. Lett. **86**, 3244 (2001)
[26] J.W. Hammer, et al., Nucl. Phys. A **752**, 514 (2005)
[27] P. Tischhauser, et al., Phys. Rev. Lett. **88**, 072501 (2002)
[28] N. Oulebsir, et al., Phys. Rev. C **85**, 035804 (2012)
[29] S. Adhikari, et al., Phys. Rev. C **89**, 044618 (2014)
[30] Y. Xu, et al., Nucl. Phys. A **918**, 61 (2013)

The ^{12}C+^{12}C fusion reaction at stellar energies

Xiaodong Tang[1,2,3,*] and *Longhui* Ru[1,2,**]

[1]Institute of modern physics, Chinese Academy of Sciences, Lanzhou, 730000, China
[2]School of Nuclear Science and Technology, University of Chinese Academy of Sciences, Beijing 100049, China
[3]Joint department of nuclear physics, Lanzhou University and institute of modern physics, Lanzhou, 730000, China

Abstract. The carbon fusion reaction is crucial in stellar evolution. Despite six decades of studies, there is still a large uncertainty in the reaction rate which limits our understanding of various stellar objects, such as massive stars, type Ia supernovae, and superbursts. In this paper, we review the experimental and theoretical studies of the carbon fusion reaction at sub-barrier energies. An outlook for future studies is also presented.

1 Introduction

Stars with a mass of more than 8 solar masses and less than 10 solar masses can ignite the ^{12}C+^{12}C fusion reaction and proceed with carbon burning inside of their cores. These stars end up their lives as Ne/O white dwarfs. More massive stars will continue with Ne-, O- and Si-burnings in their cores and shells and eventually become supernovae. During these hydrostatic and explosive burning processes, ^{12}C+^{12}C is one of the primary reactions of the burning processes which shape the stellar evolution and the final nucleosynthesis. The typical temperature ranges span from 0.6 to 2.5 GK [1, 2].

The ^{12}C+^{12}C fusion reaction is considered to be the ignition reaction of type Ia supernova and superburst. In type Ia supernova, the ignition happens in the core of white dwarf typically at T$\sim$ 0.15-0.7 GK. In superburst, the heating sources in the crust of neutron star raise the temperature of the ash and eventually start carbon ignition at a temperature of $\sim$0.5 GK[3]. The ignition conditions mentioned above strongly depend on the ^{12}C+^{12}C reaction rate as well as the treatment of reaction rate estimation in dense matter[4–6].

2 Experimental progress

At sub-barrier energies, the main products of ^{12}C+^{12}C are n, p and α. To obtain the fusion reaction cross section at such energies, two different techniques had been used. One is the particle spectroscopy and the other is the characteristic γ-ray spectroscopy. Patterson[7], Mazarakis[8], Becker[9] and Zickefoose[19] extended the measurements below 2.7 MeV by counting the protons and/or alphas from the fusion reaction process. As energy approaches

*e-mail: xtang@impcas.ac.cn
**e-mail: rulonghui@impcas.ac.cn

the astrophysical region, backgrounds arising from the target impurity and other sources produce a significant background. Some proton and alpha channels could not be identified, leading to underestimation of the total fusion cross sections. Kettner[10], Aguilera[11] and Spillane[12] measured the cross sections of the characteristic γ-rays of the fusion residues, ^{23}Na, ^{20}Ne and ^{23}Mg. Since some decays of the fusion residues bypass the decay of the characteristic γ-rays, the sum of the characteristic γ-ray cross sections reflects only a fraction of the total fusion cross section. To get a cleaner background, the particle-γ coincidence technique has been used to measure the cross section of the p_1 and α_1 channels [13–15]. But the statistics was limited by the lower detection efficiency and the available beam current. The total fusion cross section can only be obtained by the theoretical branching ratios which introduce systematic uncertainties as shown later in this paper. A recent direct measurement of ^{12}C(^{12}C,n)^{23}Mg has been performed at stellar energies, providing a reliable rate[16]. The ^{12}C(^{12}C,^{8}Be)^{16}O is an open channel that has been overlooked due to the lack of experimental work[17]. The results of some measurements are shown in Fig. 1 as the modified S factor, $S^*(E_{c.m.}) = \sigma(E_{c.m.})E_{c.m.}exp(\frac{87.21}{\sqrt{E_{c.m.}}} + 0.46E_{c.m.})$.

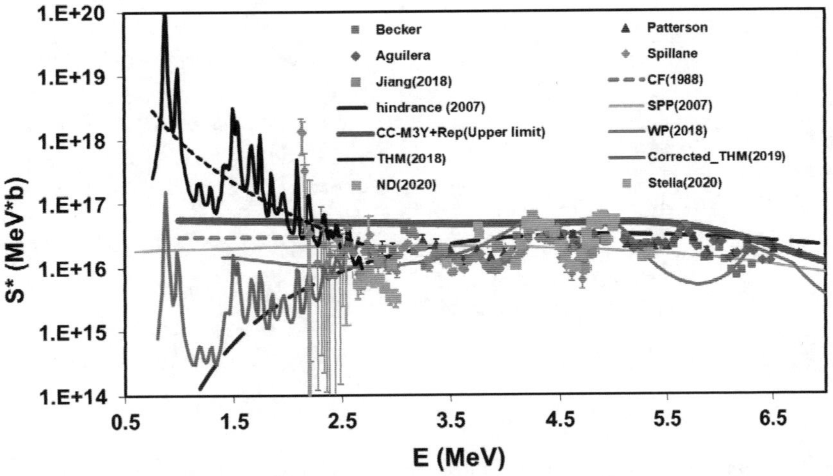

Figure 1. (Color online) S* factors of ^{12}C+^{12}C. The ^{12}C+^{12}C data from Ref. [7], [9], [11], [12], [14] and [15] shown as blue, red, green diamond, magenta, blue, green square and orange points, respectively. Model calculations, CC-M3Y+Rep (thick dark red), SPP(oragen), TDWP(light blue) and hindrance(blue dashed) are also shown, respectively. The recommended averaged S* factor by CF88 is shown as red dashed line. The THM result and the fit are shown as black dashed and solid lines, respectively.

3 Data compilation

Here we discuss three important factors in the compilation of the S* factor of the ^{12}C+^{12}C fusion reaction: energy calibration, background evaluation and the branching ratios of the observed/missing channels.

An accurate and precision energy calibration is the basic to compare the data set. Due to the Coulomb barrier tunneling effect, the reaction cross section has a strong dependence on the reaction energy. For example, at $E_{\rm c.m.}$=3 MeV, a 0.2% change in $E_{\rm c.m.}$ results in a 5% change in the reaction cross section. The measurement of Mazarakis found an upturning trend of the S* factor at the lower energies. It was interpreted as absorption under the

Coulomb barrier. By aligning the resonances at higher energies with latter measurements, an energy shift of +100 keV was found for the data of Mazarakis and the upturning trend disappeared[10, 20].

Barrón-Palos *et al.* reported an upturning trend in their measured S* factor. As they used a smoothing procedure to wash out the resonance structure, one can not check their energy calibration with the reported S* factor. We compared the thick target yields of the 1634-keV transition measured by Spillane *et al.* and Barrón-Palos *et al.*. The result is shown in Fig.2. Although the resonant peaks in the two measurements seem to be aligned with each other, the measurement of Barrón-Palos *et al.* is generally higher than the measurement of Spillane *et al.* except the agreement around the resonance at $E_{c.m.}$=3.2 MeV. For example, the thick target yield of Barrón-Palos *et al.* at $E_{c.m.}$=2.4 MeV is more than a factor of 10 higher than the measurement of Spillane *et al.*, resulting in the upturning trend of the S* factor at astrophysical energies. Such a discrepancy may arise from the unknown backgrounds. The strong resonance reported by Spillane *et al.* has not been confirmed by any experiment yet. It will be useful to re-do the measurement, clarify the discrepancy and draw conclusion on the resonances at stellar energies.

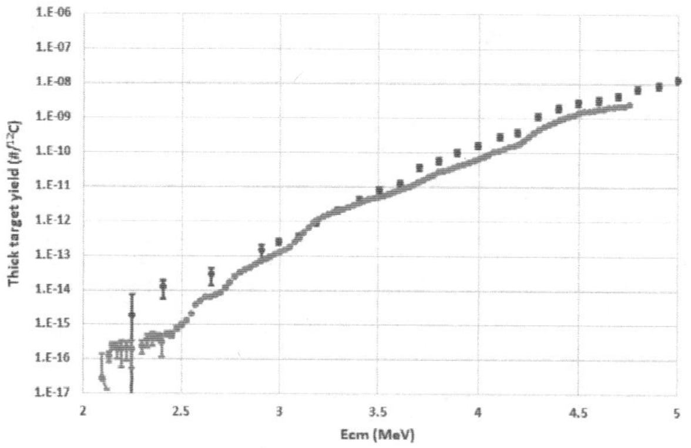

Figure 2. (Color online) Comparison of the thick target yields of the 1634-keV transition measured by Barrón-Palos *et al.* (blue points) and Spillane *et al.* (orange points).

Limited by the detection capability, some channels have to be missed in every experiment. Lacking systematic studies, the correction of the missing channels has been done in various ways. For example, it was thought that the p_0 and α_0 channels were the only missing channels in the γ-spectroscopy measurement[11]. We have performed a systematic analysis to investigate the fractions of missing channels based on a statistical model [21]. In this study, the spin population of the entrance channel was taken from the coupled channel calculation. The decay branching ratio of each channel was evaluated with Talys and calibrated using the experimental data of Becker [9]. Our study shows that the branching ratios of p_1/p_{tot}, γ_{440}/p_{tot}, α_1/γ_{1634}, and $\gamma_{1634}/\alpha_{tot}$ are not constants and the average trends of these branching ratios can be well described with the calibrated statical model prediction. Here p_{tot} and α_{tot} refer to the summed cross sections of the proton and α channels, respectively. We also found the fluctuations around these predictions decrease as the branching ratios increase. Some of the results are shown in Fig. 3. Taking the proton channel as an example, at $E_{c.m.}$=5 MeV, the ratio of p_1/p_{tot} is round 0.2 with a relative fluctuation of 42%(σ) around the prediction. By including more observable channels, the ratio of $\sum_{i=0}^{5} p_i/p_{tot}$ increases to about 0.5 while the corresponding relative fluctuation decreases to 12%(σ). And this relative fluctuation of this ratio decrease as this ratio increases towards lower energies. Therefore, future experiments are needed to measure as many channels as possible to reduce this systematical uncertainty.

After applying corrections based on branching ratios predicted by the statistical model, the compilation of the experimental S* factors are shown in Fig. 4. An agreement within ±30% is achieved among the total S* factors obtained using the experimental data sets at $E_{c.m.} > 4$ MeV, while some discrepancies remain at lower energies. For example, the recent measurement of α_1 by Tan [14] is about one tenth of the result of Fruet [15] at $E_{\text{c.m.}} < 3$ MeV. Such kind of discrepancies hinder the development of reliable models and should be resolved by better measurements in the near future.

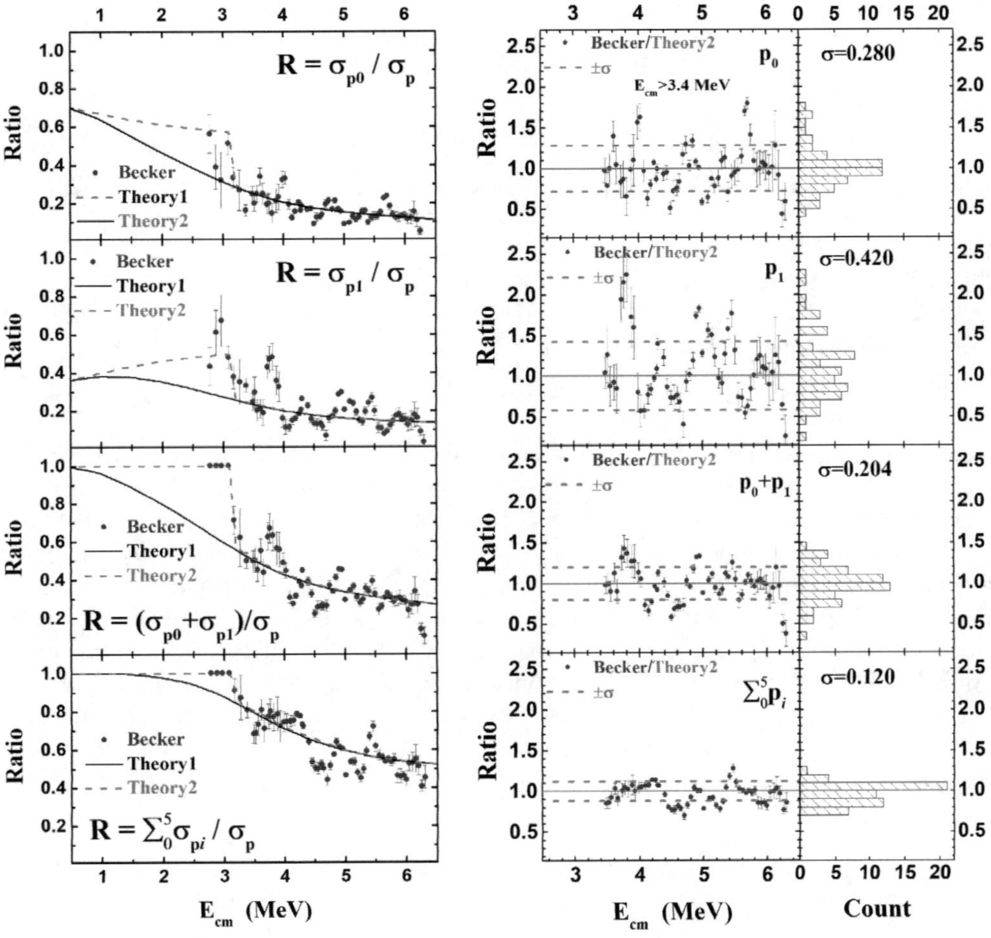

Figure 3. The comparison of the branching ratios calculated with experimental data [9] for the p-channel. (left) The calculated σ_{p0}/σ_p, σ_{p1}/σ_p, $(\sigma_{p0}+\sigma_{p1})/\sigma_p$, $\sum_{i=0}^{5}\sigma_{p_i}/\sigma_p$ ratios are displayed with experimental data of Becker *et al.* [9]. **Theory1** is the present calculation by Talys, **Theory2** takes the condition of the cut-off exist in the experiment of Becker *et al.* [9] into consideration. (right) The values of Becker/Theory2 for the ratios shown in (left) are calculated and displayed. The statistics are also achieved with average value (Mean=1) and standard deviation (σ). The standard deviation (σ) according to Mean=1.0 is given together with the Becker/Theory2.

Another systematic uncertainty often overlooked is the angular distribution of the detected particles or γ-rays. At the deep sub-barrier energies, as the dominated angular momenta of

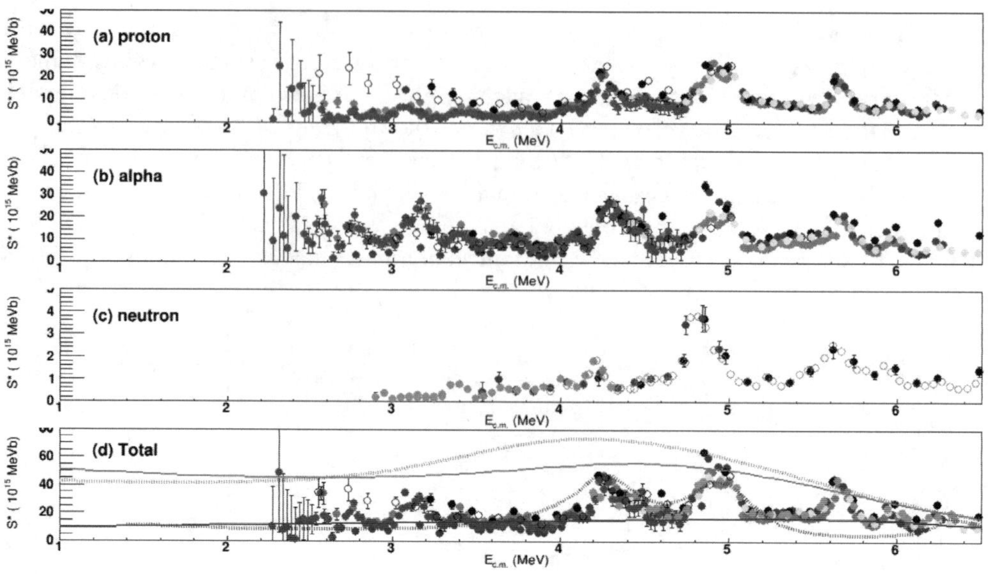

Figure 4. (Color online) The modified S factors of the proton, alpha, neutron channels and their sum are shown as (a), (b), (c) and (d), respectively. The branching ratio corrections have been applied. The explaination of legends can be found in Ref. [21].

the entrance channel are L_i =0 and 2, the angular distribution of the emitted charged particles, p and α, can be anisotropic. The allowed orders of the angular distribution coefficients are 0, 2 and 4. The angular distributions of the exit channels require a complete measurement using a charged particle array and good statistics. They have only been measured by the limited amount of experiments at sub-barrier energies. When the angular distribution coefficient of the fourth order can be ignored, one may approximate the averaged differential cross section by using the differential cross section at 55 or 125 degrees in the center of mass frame. An isotropic angular distribution was assumed to get the integrated cross section. Measurements performed at other angles often use this assumption to determine the integrated cross section, resulting in an unknown systematic uncertainty. More experimental and theoretical investigations are needed to quantify this uncertainty.

4 Test of extrapolating model and the establishment of the upper limit

Modeling $^{12}C+^{12}C$ heavy ion fusion cross sections at deep sub-barrier energies has been a long standing challenge for nuclear reaction theory [20]. The Gamow energy window for most astrophysical applications typically spans 500 keV or more, and with no reliable method for predicting the resonances at lower energies, the standard reaction rate (CF88) was established by modeling the averaged S* factor,$S^*(E) = S(E)exp(0.46E)$, with a constant value corresponding to a S(E) with a rising trend towards lower energies [7, 22]. The rising trend of S-factor towards lower energies is supported by various phenomenological and microscopic models, such as density-constrained time dependent Hartree-Fock method (DC-TDHF) [23], wave-packet dynamics (TDWP) [24], barrier penetration model based on the global São Paulo potentials (SPP) [5] or the Krappe-Nix-Sierk potential (KNS) [11], and coupled channel calculations [25, 26] such as CC-M3Y+Rep (Fig. 1(a)). However, the hindrance model, a global phenomenological model based on the systematics observed in systems with 64≳A≳30, predicts that the $^{12}C+^{12}C$ S-factor reaches its maximum at $E_{c.m.}$=3.68±0.38 MeV [13, 27–29]. At lower energies, this model predicts a rapid drop in the S-factor leading to a reduced reaction rate that is many orders of magnitude smaller than the standard rate used for astrophysical modeling. However, the complicated resonant structure makes it difficult to test the predictive power of the extrapolating models. As we demonstrated in Fig.5 and Table 1, all the models yield similarly poor reduced-χ^2s of about 5–9.5. Recently, the cross sections were determined indirectly for $E_{c.m.}$= 0.8 MeV to 2.7 MeV using the Trojan Horse Method [30], recommending a new S-factor with a slope rising faster than any models presented in Fig. 1, resulting in a new rate which is 1 or 2 orders of magnitude higher than the standard one. The large uncertainty of the $^{12}C+^{12}C$ rate drastically impacts a number of models such as late-time massive star evolution, the ignition dynamics of type Ia supernovae and x-ray superbursts [5, 6].

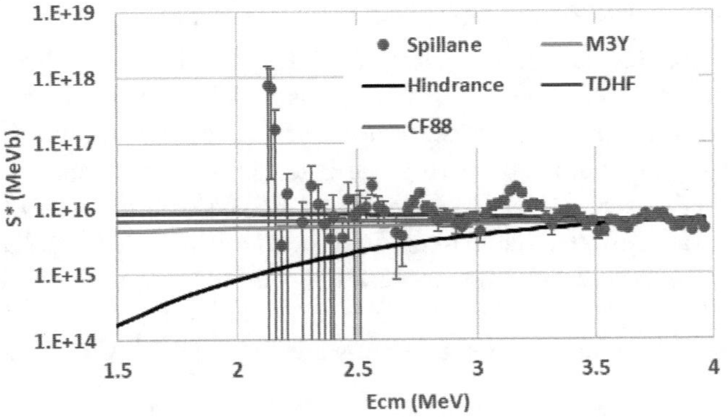

Figure 5. (Color online) The best fits of various models to the experimental S* factor of Spillane in the range of $E_{c.m.}$= 2.68 to 3.97 MeV. The resulting reduced-χ^2 for the models, M3Y-Rep, Hindrance, TDHF, and CF88, are listed in Table 1.

Table 1. The reduced-χ^2 for different models (scenario 1: No normalization; scenario 2: with renormalization;)

Models	M3Y-Rep	Hind	TDHF	CF88
scenario 1	1888	680	3232	915
scenario 2	6.69	9.47	5.03	6.01

While the complicated resonance-like structure in $^{12}C+^{12}C$ and the lack of reliable measurements at lower energies prevent us from drawing a clear conclusion [13], the isotope fusion reaction $^{12}C+^{13}C$ offers an ideal opportunity to constrain the $^{12}C+^{12}C$ S-factor. It has been observed at energies below and above the Coulomb barrier that $^{12}C+^{13}C$ and $^{13}C+^{13}C$ cross sections are upper bounds of the $^{12}C+^{12}C$ cross sections, and match the maxima of the resonance-like structure seen in $^{12}C+^{12}C$ in a wide range from 10^{-8} b to 1 b [31]. This strong correlation among the three systems has been well explained by coupled channel calculations and the significantly different level densities of the compound states [32]. At sub-barrier energies, the valence neutron(s) in $^{12}C+^{13}C$ and $^{13}C+^{13}C$ increase the level densities of their compound states by at least one order of magnitude in comparison to $^{12}C+^{12}C$ and result in smooth cross sections. According to the following equation [32],

$$\sigma = \sum_J \sigma_{cc}^J \left(1 - exp(1 - 2\pi\overline{\Gamma}_J/D_J)\right), \tag{1}$$

the upper limit of the ^{12}C+^{12}C cross section is reached using the high level density limit and can be modeled consistently with the other C+C isotope systems [32]. The average of the ^{12}C+^{12}C fusion cross section is also predicted by modulating the upper limit with the averaged ratio of the level width <Γ> and the level spacing <D> of ^{24}Mg [32]. Since the effect of $<\Gamma/D>$ is not sensitive to the energy, the shape of the averaged cross section is essentially determined by the model used for the upper limit prediction. Therefore, a model well constrained by ^{12}C+^{13}C is crucial for us to set a reliable upper limit and constrain the shape of the averaged cross section for ^{12}C+^{12}C.

We performed an activity measurement in SLANIC mine for the ^{12}C+^{13}C system with energies down to $E_{\mathrm{c.m.}}$=2.323 MeV, at which the ^{12}C(^{13}C,p)^{24}Na cross section is found to be 0.22(7) nb. The ^{12}C+^{13}C fusion cross section is derived with a statistical model calibrated using experimental data. Our result of the ^{12}C+^{13}C fusion cross section provided the first decisive evidence in the carbon isotope systems which rules out the existence of the astrophysical S-factor maximum predicted by the phenomenological hindrance model, while confirming the rising trend of the S-factor towards lower energies predicted by other models, such as CC-M3Y+Rep, DC-TDHF, KNS, SPP and ESW. After normalizing the model predictions with our data, a more reliable upper limit is established for the ^{12}C+^{12}C fusion cross sections at stellar energies [33]. However, this upper limit is lower than the strong resonance observed by Spillane at 2.14 MeV and the indirect THM measurement based on the PWIA assumption. Experimental and theoretical efforts are needed to resolve the issues.

5 Extrapolation towards stellar energies and Reaction Rate

We proposed a new extrapolation based on the lower and upper limits for the S* factor of ^{12}C+^{12}C in parallel to the other existing extrapolation [21]. The KNS extrapolation based on the non-resonant component of the existing S* at higher energies is one of the lower limits we considered. The current TDWP approach does not include the cluster effect which may account for the resonances at deep sub-barrier energies. Therefore it can provide the other lower limit. It is interesting to note that the TDWP calculation agrees with the empirical lower limit (KNS) with a deviation below 33% at energies below 3 MeV. Combining our upper limits with the lower limit based on the prediction of TDWP, the ^{12}C+^{12}C S* factors are better constrained despite the unknown resonances within the unmeasured energy range as shown in Fig. 4(d). The reaction rate of ^{12}C+^{12}C was calculated with the measured S* factors at E≥2.7 MeV, where a reasonable agreement among the experimental total S* factors exists, and using the random S* factor limited by the lower and upper limits for <2.7 MeV. The average of these two limits is used to estimate the mean value of the S* factor. The resulting reaction rates are shown in Fig. 6 together with the rates obtained with the hindrance model and the THM indirect measurement.

6 New approaches

The precise measurement of the tiny reaction cross section demands new experimental techniques with lower background, higher detection efficiency and higher luminosity. The measurement of the channels with the γ-ray emissions is being planned at the LUNA underground laboratory. Jingpin Underground Nuclear Astrophysics collaboration (JUNA) is also pushing

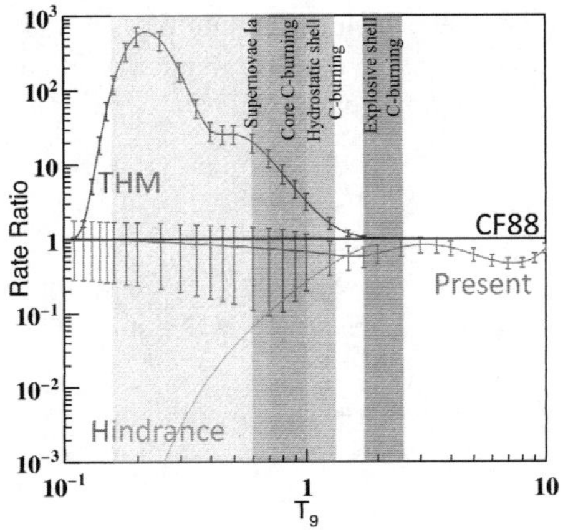

Figure 6. The relative reaction rate obtained in the current work (Red) together with the rates based on the THM measurement [30] (Blue) and hindrance model [28] (Green) to the CF88 rate [22]. Temperatures for core carbon burning (T = 0.6-1.0 GK), hydrostatic shell carbon burning (T = 1.0-1.2 GK), and explosive shell carbon burning (T = 1.8-2.5 GK) are marked by colored bands.

the project of upgrading their accelerator to study the heavy ion fusion reactions at stellar energies. As a complementary approach, we are developing a charged-particle measurement based on the Low Energy Accelerator Facility (LEAF) at IMP [34]. Using the state of art LINAC and powerful ECR sources, LEAF has delivered 50-pμA carbon beam on target, the highest beam intensity ever realized in the carbon fusion experiments. We plan to reach 100 pμA by optimizing the transmission efficiency. To suppress the beam induced backgrounds and the natural background, we developed a new experimental technique using the Time Projection Chamber. With the supreme tracking capability and particle identification, the preliminary thick target yield of ^{12}C(^{12}C,α_0)^{20}Ne has been determined to 1.4(1.4)E-17 evt/^{12}C at $E_{\text{c.m.}}$=2.22 MeV. We are planning to check the strong resonance reported at $E_{\text{c.m.}}$=2.14 MeV in the near future.

The indirect measurement like THM offers a way of extrapolation. But this approach needs to be validated with the direct measurement before applying it at the deep sub-barrier energies [35]. Lacking reliable information of the resonances at deep sub-barrier energies, extrapolating models are not able to provide more precise extrapolation. It has been pointed out theoretically that the molecular resonances in the channel of ^{12}C+^{12}C can be populated via the isoscalar transitions. Therefore, the ^{24}Mg(α,α')^{24}Mg* offers another opportunity to access the resonances at the deep sub-barriers which is not accessible by direct measurement. Such measurements carried out at RCNP [36] and iThemba will provide the important nuclear structure information for the extrapolation at stellar energies.

7 Summary

The ^{12}C+^{12}C reaction is one of the most important reactions in stellar evolution. Despite some progress being made in recent years, there is no satisfactory answer yet for the reaction rate to be used in the astrophysical models. Reliable measurement of the major reaction channels of ^{12}C+^{12}C at energies below 3 MeV, complimentary indirect measurements and standardized data compilation are essential to resolve the existing discrepancies and guide the theory towards the right direction. New experimental and theoretical approaches are being developed to push the measurement further at deep sub-barrier energies and model the complicated resonance beyond the reach of measurement. With all these efforts, one expects to achieve a more reliable reaction rate for the interest of astrophysics.

8 Acknowledgment

This work was supported in part by the National Key Research and Development Program of China (No. 2016YFA0400501), the Strategic Priority Research Program of Chinese Academy of Sciences (No. XDB34020200).

References

[1] S. Woosley, A. Heger, and T. Weaver, Rev. Mod. Phys. **74**, p. 1015 (2002)
[2] C. Iliadis, *Nuclear Physics of Stars* (WILEY-VCH Verlag GmbH & Co. KGaA, 2007)
[3] R. Cooper, A. Steiner, and E. Brown, Apj. **702**, p. 660 (2009)
[4] L. R. Gasques, A. V. Afanasjev, E. F. Aguilera, M. Beard, L. C. Chamon, P. Ring, M. Wiescher, and D. G. Yakovlev, Phys. Rev. C **72**, p. 025806 (2005)
[5] L. R. Gasques, Phys. Rev. C **76**, p. 035802 (2007)
[6] K. Mori, M. A. Famiano, T. Kajino, M. Kusakabe, and X.D. Tang, Monthly Notices of the Royal Astronomical Society **482**, Pages L70–L74(2019)
[7] J. R. Patterson, H. Winkler, and C. S. Zaidins, Ap.J. **157**, p. 367(1969)
[8] M.G.Mazarakis and W.E.Stephens, Phys.Rev. C **7**, 1280 (1973)
[9] H.W. Becker, K. U. Kettner, C. Rolfs, and H. P. Trautvetter, Zeitschrift für Physik A **303**, 305–312 (1981)
[10] K. Kettner, H. Lorenz-Wirzba, and C. Rolls, Z. Phys. A **298**, p. 65 (1980)
[11] E. F. Aguilera, P. Rosales, E. Martinez-Quiroz, G. Murillo, M. Fernndez, H. Berdejo, D. Lizcano, A. Gmez- Camacho, R. Policroniades, A. Varela, E. Moreno, E. Chvez, M. E. Ortz, A. Huerta, T. Belyaeva, and M. Wiescher, Phys. Rev. C **73**, p. 064601 (2006)
[12] T. Spillane *et al.*, Phys. Rev. Lett. **98**, p. 122501 (2007)
[13] C. L. Jiang *et al.*, Phys. Rev. C **97**, 012801(R) (2018)
[14] W.P. Tan *et al.*, Phys. Rev. Lett. **124**,192702 (2020)
[15] G. Fruet *et al.*, Phys. Rev. Lett. **124**, 192701 (2020)
[16] B. Bucher *etal.*, Phys. Rev. Lett. **114**, p. 251102 (2015)
[17] B.Cujec, I.Hunyadi, and I.M.Szoghy, Phys.Rev. C **39**, 1326 (1989)
[18] L. Barrón-Palos *etal.*, Nucl. Phys. A **779**, P.318 (2006)
[19] J.Zickefoose *et al.*, Phys.Rev. C **97**, 065806 (2018)
[20] C. A. Barnes, S. Trentalange, and S. C. Wu, "Treatise on heavy-ion science," (Plenum Preess, New York, 1985) Chap. 1,, pp. 3–60.
[21] Y. J. Li, X. Fang, B. Bucher, K. A. Li, L. H. Ru and X. D. Tang, Chin. Phys. C **44**, 115001 (2020)
[22] G. R. Caughlan and W. A. Fowler, Atomic Data And Nuclear Data Tables **40**, p. 283 (1988)
[23] K.Godbey, C.Simenel, and A.S.Umar, Phys.Rev. C **100**, 024619 (2019)
[24] A.Diaz-Torres and M.Wiescher, Phys. Rev. C **97**, 055802 (2018)
[25] M. Assuncao and P. Descouvemont, Phys. Lett. B **723**, 355–359 (2013)
[26] H. Esbensen, X. Tang and C.L. Jiang, Phys. Rev. C **84**, 064613 (2011)
[27] B. B. Back, H. Esbensen, C. L. Jiang, and K. E. Rehm, Rev. Mod. Phys. **86**, 317 (2014)

[28] C. L. Jiang, K. E. Rehm, B. B. Back, and R. V. F. Janssens, Phys. Rev. C **75**, 015803 (2007)
[29] C. L. Jiang, B. B. Back, K. E. Rehm, K. Hagino, G. Montagnoli and A. M. Stefanini, Eur. Phys. J. A **57**, 235 (2021)
[30] A. Tumino *et al.*, Nature **557**, 687 (2018)
[31] M. Notani *et al.*, Phys. Rev. C **85** (2012) 014607
[32] C.L. Jiang *et al.*, Phys. Rev. Lett. **110**, 072701 (2013)
[33] N.T. Zhang *et al.*, Phys. Lett. B **801**, 135170(2020)
[34] Y. Yang *et al.*, J. Phys.: Conf. Ser. **1401**, 01201(2020)
[35] A.M. Mukhamedzhanov, A.S. Kadyrov and D.Y. Pang, Eur. Phys. J. A **56**, 233 (2020)
[36] T. Kawabata *et al.*, Few-Body Systems 54, 93 (2013)

Proton partial widths evaluation through the ^{30}Si(^{3}He,d)^{31}P transfer reaction for understanding abundance anomalies in Globular Clusters

Djamila-Sarah Harrouz[1,*], *Nicolas* de Séréville[1,**], *Philip* Adsley[2], *Faïrouz* Hammache[1], *Richard* Longland[3,4], *Beyhan* Bastin[5], *Thomas* Faestermann[6], *Ralf* Hertenberger[7], *Marco* La Cognata[8], *Livio* Lamia[8,9], *Anne* Meyer[1], *Sara* Palmerini[10,11], *Rosario Gianluca* Pizzone[8], *Stefano* Romano[8,9,12], *Aurora* Tumino[8,13], and *Hans-Friedrich* Wirth[7]

[1]Université Paris-Saclay, CNRS/IN2P3, IJCLab, 91405 Orsay, France
[2]Cyclotron Institute and Department of Physics & Astronomy, Texas A&M University, College Station, Texas 77843, USA
[3]North Carolina State University, Raleigh, NC 27695
[4]Triangle Universities Nuclear Laboratory, Durham, NC 27708
[5]Grand Accélérateur National d'ions Lourds (GANIL) CEA/DRF-CNRS/IN2P3 Bd. Henry Bequerel, 14076 Caen, France
[6]Physik Department E12, Technische Universität München, D-85748 Garching, Germany
[7]Fakultät für Physik, Ludwig-Maximilians-Universität München, D-85748 Garching, Germany
[8]Laboratori Nazionali del Sud - Istituto Nazionale di Fisica Nucleare, Via Santa Sofia 62, 95123 Catania, Italy
[9]Dipartimento di Fisica e Astronomia E. Majorana, Univ. di Catania, Catania, Italy
[10]Dipartimento di Fisica e Geologia, Università degli Studi di Perugia, Perugia, Italy
[11]Istituto Nazionale di Fisica Nucleare, Sezione di Perugia, Perugia, Italy
[12]Centro Siciliano di Fisica Nucleare e Struttura della Materia-CSFNSM, Catania, Italy
[13]Facoltà di Ingegneria e Architettura, Università degli Studi di Enna, Italy

Abstract. Some observed abundances in globular clusters have suggested the existence of multiple generations of stars within the clusters as the observations require temperature ranges higher than current stars. The ^{30}Si(p,γ)^{31}P reaction plays a key role in the synthesis of the observed abundances. The study of the ^{30}Si(^{3}He,d)^{31}P transfer reaction is a tool for constraining the strengths of low-lying resonances, and the proton partial widths are the main ingredients for calculating those strengths. We present the method used for estimating the proton partial widths and their associated uncertainties.

1 Introduction

Globular clusters are among the oldest objects in the Universe, which makes them important sites for constraining the formation and early evolution models of galaxies. Globular clusters are mainly populated by low-mass stars in the main sequence, with some of them exhibiting anti-correlation of abundances between pairs of light elements such as C-N, O-Na, and

*e-mail: djamila-sarah.harrouz@ijclab.in2p3.fr
**e-mail: nicolas.de-sereville@ijclab.in2p3.fr

Mg-Al [1]. These observed abundance cannot originate from the clusters stars because the temperature is too low to make the observed anti-correlations, indicating that they have been partially inherited from previous generation of stars called polluters[2]. The nature of these polluters is still uncertain.

The NGC 2419 globular clusters is an interesting case since it exhibits an additional anticorrelation between Mg-K which requires a higher temperature for hydrogen burning when compared to other anticorrelations [3]. Monte Carlo network calculations suggest that the temperature for polluter candidates should range between 120 - 200 MK [4]. This range is sensitive to few proton-capture reactions, and mainly to the ^{30}Si(p,γ)^{31}P reaction [5]. For these stellar temperature, the ^{30}Si(p,γ)^{31}P reaction rate depends on resonances of energies $E_r^{c.m.} = 100 - 500$ keV, where the strengths are directly proportional to the proton partial widths since the latter are much smaller than γ-ray partial widths.

In this proceeding we report on the analysis of the associated one-proton ^{30}Si(^{3}He,d)^{31}P transfer reaction and the determination of proton widths in the framework of DWBA analysis.

2 Data Analysis

The ^{30}Si(^{3}He,d)^{31}P reaction was performed at Maier-Leibnitz-Laboratorium (MLL) Germany. Ejectiles were momentum analysed using the Q3D spectrometer for different laboratory angles. Deuterons were selected using the focal-plane detectors and experimental details can be found in ref.[6] where excited states between $E_x = 6.8$ keV MeV and $E_x = 8.1$ keV MeV have been studied. Figure 1 shows the deuteron focal plane position spectrum obtained at an angle of $\theta_{lab} = 12°$, containing both bound and unbound levels (Sp=7296 keV) of ^{31}P which were fitted with skewed Gaussians with same resolution. The red curve shows the best least-square fit of the deuteron spectrum. The experimental resolution achieved was 6.2 keV (FWHM) at $\theta_{lab} = 12°$, three times better then a previous study of the same reaction [7]. The high sensitivity allowed the observation of weakly populated states at $E_x = 7446$ keV and 7470 keV which are of interest for temperature range of polluter candidates in Globular clusters.

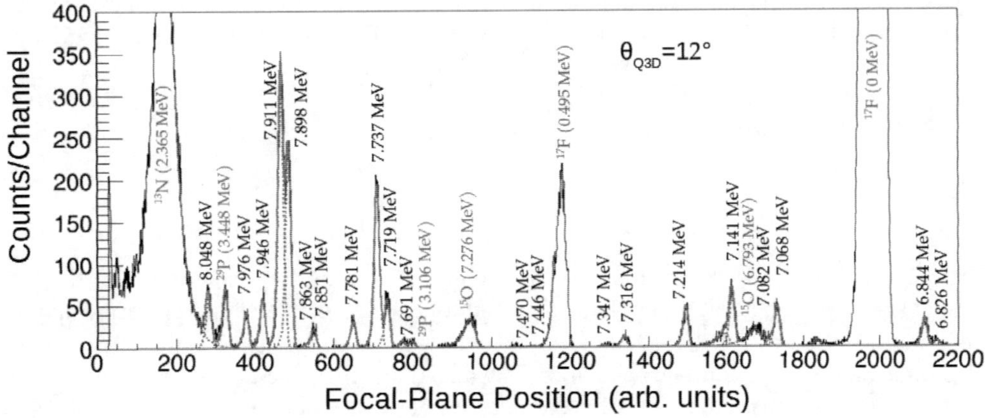

Figure 1. (Color online) Deuteron magnetic rigidity spectrum at $\theta_{lab} = 12°$. Excitation energies in ^{31}P between $E_x = 6.8$ MeV and 8.1 MeV are covered and are reported for each state. The best fit is shown as a solid line, while the individual contributions are in red/green dashed lines for ^{31}P states and contaminant peaks, respectively.

3 Angular distribution and DWBA analysis

The extracted deuteron yield at each spectrometer angle was used, after normalisation, to calculate the differential cross sections for the ^{31}P populated states. The latter are compared to theoretical differential cross sections computed in the Finite-Range DWBA framework, using the FRESCO code[8].

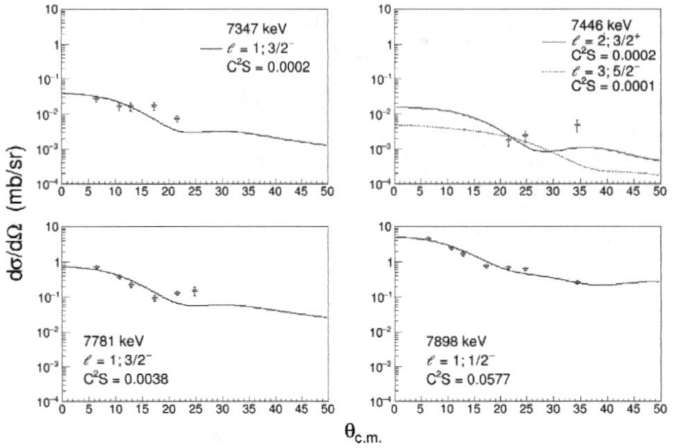

Figure 2. (Color online) Angular distributions of ^{31}P levels that dominate the thermonuclear reaction rate in the ^{30}Si(p,γ)^{31}P reaction. Blue curves are finite-range DWBA calculations normalized to experimental data points in red.

Figure 2 shows the experimental angular distributions and the best fit with FR-DWBA calculation for the resonances that contribute the most to the ^{30}Si(p,γ)^{31}P reaction rates, namely $E_r^{c.m.} = 50, 149, 485$ and 602 keV corresponding to excited states at $E_x = 7347, 7446, 7781$ and 7898 keV, respectively. The optical potential parameters for the entrance channel were obtained from a previous experimental study of the elastic scattering ^{30}Si(p,p)^{30}Si with the same beam energy as the present work [9], and the parameters for exit channel are adapted from global deuteron potentials, set F of ref. [10]. The overlap function $\langle ^3\text{He}|\text{d}\rangle$ for the Finite-Range calculations was taken from ref. [11]. A Woods-Saxon potential was used to describe the ^{30}Si+p wave function, with the depth adjusted to match the binding energy of each calculation. The spectroscopic factor C^2S_p represents the normalisation factor obtained when fitting the theoretical angular distribution to the experimental data points. The spectroscopic factor is used to calculate the proton partial width, which is crucial for the resonance strengths and reaction rate calculations. The proton width is defined as the product $\Gamma_p = C^2S_p \times \Gamma_{s.p}$ where $\Gamma_{s.p}$ is the single-particle proton width calculated as [12]:

$$\Gamma_{s.p} = \frac{\hbar^2 s}{2\mu} P_\ell(E_r, s)|R(s)|^2, \tag{1}$$

where μ is the reduced mass of the ^{30}Si+p system, $P_\ell(E_r, s)$ is the penetrability of the centrifugal and Coulomb barrier, calculated for a radius s and associated to a transferred angular momentum ℓ, and $|R(s)|$ is the radial part of the wave function, estimated at the same radius s where it achieves an asymptotic behaviour. The uncertainties on the proton width come from (i) the optical potential parameters for entrance and exit channels (ii) the binding potential for the relative ^{30}Si+p wave function, and (iii) experimental uncertainties associated with the differential cross sections. The geometry of the binding potential is poorly constrained, thus the associated uncertainty has been investigated through a Monte Carlo study where the radius and the diffuseness of the Wood-Saxon well have been varied according to a Gaussian distribution, with a full width half maximum of 25% and 35% with respect to the nominal values of $r = 1.25$fm and $a = 0.65$fm, respectively. A sample of 100000 calculations were performed using the DWUCK4 code since it uses a straightforward method for solving the

https://doi.org/10.1051/epjconf/202226001003

Schrödinger equation for unbound levels, in a Zero-Range approximation. The spectroscopic factor and single-particle widths obtained follow a log-normal distributions shown in Figure 3, both with an associated uncertainty of $\approx 30\%$. A strong correlation between C^2S_p and $\Gamma_{s.p}$ is found and the product gives rise to an uncertainty on Γ_p smaller than 1%. The proton width distribution is not well described by any analytical probability distribution function (PDF), and thus the cumulative distribution function is used to extract the 1σ uncertainty as the difference between the 0.84 and the 0.16 quantiles. The final uncertainty on the proton width is then dominated by the optical potential parameters and the statistical fit errors.

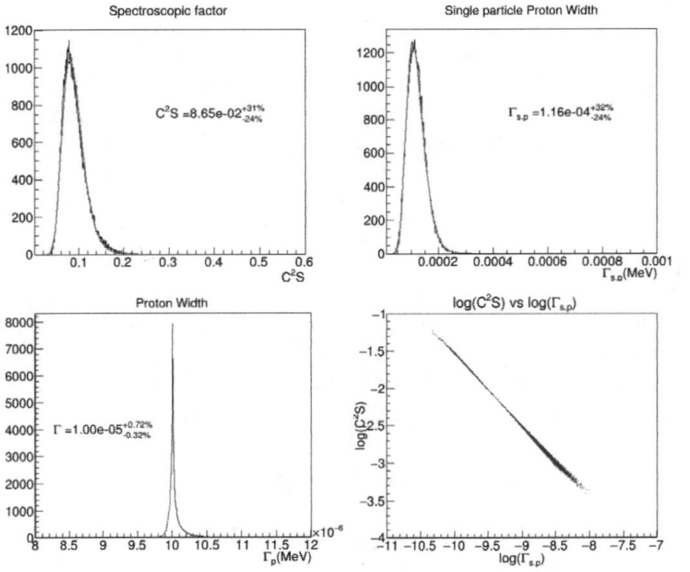

Figure 3. (Color online) Statistical distributions for: (top left) the spectroscopic factor, (top right) the single particle partial width, and (bottom left) the proton partial width. The red curves are the log-normal distributions fitted to the data. The bottom right panel shows the correlation between the spectroscopic factor and the single particle widths, in log scales.

4 Conclusions

Proton partial widths in the compound nucleus ^{31}P have been extracted and the associated uncertainties estimated with a statistical method. These proton widths, obtained with the ^{30}Si(^{3}He,d)^{31}P transfer reaction are key ingredients for the evaluation of the proton-capture rate ^{30}Si(p,γ)^{31}P which is crucial for understanding the abundance anomalies in Globular Clusters. More details on the evaluation the evaluation of the ^{30}Si(p,γ)^{31}P reaction rate can be found in ref. [6].

References

[1] Pancino et al., A&A **601**, A112 (2017)
[2] D. C. Powell et al., Nucl. Phys. A **660**, 349 (1999)
[3] N. Prantzos, C. Charbonnel, and C. Iliadis, A&A **470**,179 (2007)
[4] C. Iliadis et al., ApJ **818**, 98 (2016)
[5] J. R. Dermigny and C. Iliadis, ApJ **848**, 14 (2017)
[6] D. S. Harrouz et al., Phys. Rev, C (submitted 2021)
[7] J. Vernotte et al., Phys. Rev. C **41**, 1956 (1990)
[8] I. J. Thompson, Computer Physics Report **7**, 167 (1988)
[9] J. Vernotte et al., Nucl. Phys. A **390**, 285 (1982)
[10] W. W. Daehnick, J. D. Childs, and Z. Vrcelj, Phys. Rev. C **21**, 2253 (1980)
[11] I. Brida, S. C. Pieper, and R. B. Wiringa, Phys. Rev. C **84**, 024319 (2011)
[12] C. Iliadis, *Nuclear Physics of Stars* (Wiley-VCH, 2015)

Direct Measurement of Carbon Fusion at Astrophysical Energies with Gamma-Particle Coincidences

M. Heine[1,*], *G.* Fruet[1], *S.* Courtin[1,2], *D.G.* Jenkins[3], *P.* Adsley[4], *A.* Brown[3], *R.* Canavan[5,6], *W.N.* Catford[5], *E.* Charon[7], *D.* Curien[1], *S.* Della Negra[4], *J.* Duprat[8], *F.* Hammache[4], *J.* Lesrel[4], *G.* Lotay[5], *A.* Meyer[4], *E.* Monpribat[1], *D.* Montanari[1,2], *L.* Morris[3], *M.* Moukaddam[1], *J.* Nippert[1], *Zs.* Podolyák[5], *P.H.* Regan[5,6], *I.* Ribaud[4], *M.* Richer[1], *M.* Rudigier[5], *R.* Shearman[5,6], *N.* de Séréville[4], and *C.* Stodel[9]

[1]Université de Strasbourg, CNRS, IPHC UMR 7178, F-67000 Strasbourg, France
[2]USIAS/Université de Strasbourg, Strasbourg, F-67083, France
[3]University of York, York, YO10 5DD, UK
[4]IPN, CNRS/IN2P3, Université Paris-Sud, Université Paris-Saclay, F-91406 Orsay Cedex, France
[5]Department of Physics, University of Surrey, Guildford, GU2 7XH, UK
[6]National Physical Laboratory, Teddington, Middlesex, TW110 LW, UK
[7]NIMBE, CEA, CNRS, Université Paris-Saclay CEA Saclay F-91191 Gif sur Yvette, France
[8]CSNSM, Université Paris Sud, UMR 8609-CNRS/IN2P3, F-91405 Orsay, France
[9]GANIL, CEA/DSM-CNRS/IN2P3, Caen, F-14076, France

Abstract. We present $^{12}C+^{12}C$ direct fusion measurements with STELLA UK-FATIMA, that reach into the region of astrophysics interest relevant to massive stars ($M_{\odot} \approx 25$) using self-supporting thin rotating carbon foils [1]. We demonstrate that detecting gammas and light charged particles in coincidence with nanosecond timing is key for effective background reduction achieving reliable measurements in the sub-nanobarn range. We give details about core developments of the detection apparatus as well as the coincidence-analysis procedure of low count statistics. The present data largely follows the phenomenological hindrance interpolation and shows indication for resonant behaviour at the lowest energy explored.

Introduction

Fusion reactions play an essential role in the energy production, the nucleosynthesis of chemical elements and the evolution of massive stars. Among these reactions, carbon burning is a crucial ingredient to understand the late stages of massive stars [2] essentially driven by the $^{12}C+^{12}C$ reaction. It presents prominent resonances at energies ranging from a few MeV/nucleon down to sub-Coulomb barrier energies, possibly due to molecular ^{12}C-^{12}C configurations of ^{24}Mg [3, 4]. The possible persistence of these resonances to deep sub-Coulomb barrier relative energies causes colossal uncertainties of extrapolations of $^{12}C+^{12}C$ fusion cross section towards the experimentally challenging region of astrophysics interest with extremely small cross sections. At deep sub-Coulomb barrier energies, fusion hindrance

*e-mail: marcel.heine@iphc.cnrs.fr

has been evidenced for heavier systems [5], in present microscopic calculations being interpreted as Pauli blocking of fusion states [6], globally reducing the probability of reaction even further.

Direct Measurements with Thin Targets

The direct measurement of key fusion reactions at stellar energies offers an unbiased and evident experimental access, but calls for innovative measures for efficient background reduction. In the present approach, with relative energies E_{rel} = 2.2 to 5.4 MeV, for identifying

$$^{12}\text{C} + ^{12}\text{C} \rightarrow ^{20}\text{Ne}^* + \alpha \text{ and } ^{12}\text{C} + ^{12}\text{C} \rightarrow ^{23}\text{Na}^* + \text{p}$$

reactions by means of detecting the associated de-excitation gammas ($LaBr_3$(Ce) detectors) of light charged evaporated particles (annular silicon detectors), experiments were performed with the STELLA (STELar LAboratory) apparatus [7] in synchronized trigger-less time-stamped data recording mode with UK-FATIMA (FAst TIMing Array) [8] at Andromède at IJC Lab, Orsay (France). The $C^{2+,3+}$ beam intensity was increased from 30 pnA around the Coulomb barrrier at 6.6 MeV to 2 pμA for the astrophysically relevant region.

The accuracy of data-taking is guaranteed on the one hand by time-walk corrections using simulations of the self activity of the $LaBr_3$(Ce) detectors [7], that is associated with ^{138}La. This background component leads, combined with the decay of members of the ^{227}Ac chain, to a counting rate of 1 to 10 Bq/cm^3, and is efficiently removed from the physics spectra by demanding associated charged particles coincidences. On the other hand, the well isolated background from elastic scattering on deuterium is traced in a combined analysis with repeated alpha-source calibrations of the silicon detectors to ensure reliable operation.

The rotating target mechanism supports large diameter (~5 cm) thin (20 to 70 μg/cm^2) carbon foils that spin with up to 1000 rpm for efficient heat dissipation in order to prevent target deterioration. The online target-monitoring routine of comparing the in-beam intensity with off-beam elastic scattering rates was backed by off-line foil-thickness and homogeneity measurements after irradiation confirming that the effect of possible carbon buildup in the beam spot position is prevented by the ultra-high vacuum environment ($\approx 10^{-8}$ mbar) and below 1%. In addition, Raman spectrometry analysis of irradiated and non-irradiated areas of selected target foils concludes no significant change of structure of the carbon material under beam exposure.

STELLA UK-FATIMA Fast Timing Coincidence Data

A particle spectrum in coincidence with a gamma signal is displayed in Figure 1 as a function of the gamma-particle time-difference Δt, where the $LaBr_3$(Ce) act as a "start" detector. Evaporation particles populating the i-th excited states with alpha emission are indicated in black (α_i) and with proton emission in red (p_i), respectively, to the left. The synchronization of the 125 MHz STELLA acquisition with the 1 GHz UK-FATIMA system enables timing gates of around 15 ns and clear separation for exclusive cross section measurements. The timing difference between alphas and protons is caused by different interaction times in the silicon detector substrate. Horizontal bands are caused by the internal background of the $LaBr_3$(Ce) crystals, that is virtually removed by gating on the associated gamma energy as demonstrated to the right for the α_1 and p_1 transition. Note that all characteristic proton energies with gamma cascades through the first exited states of ^{23}Na populated by p_1 are also visible.

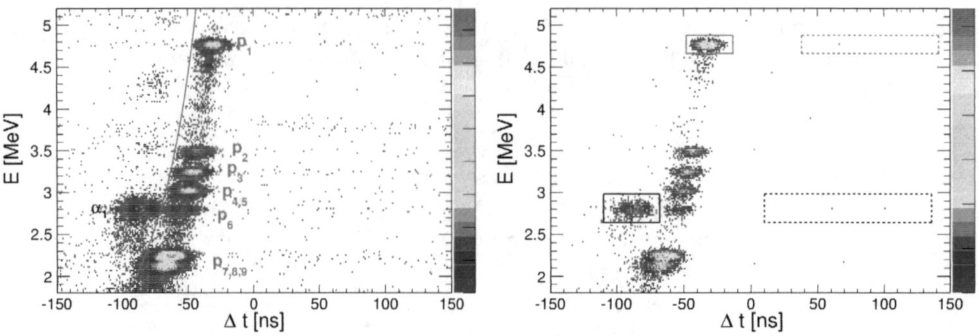

Figure 1. Correlation of charged particle energies with the gamma-particle time-difference. (Left) Evaporation particles with gamma coincidences populating the i-th excited states for $^{12}C+^{12}C\rightarrow$ $^{20}Ne^{*}+\alpha_i$ are indicated in black and for $^{12}C+^{12}C\rightarrow$ $^{23}Na^{*}+p_i$ in red. (Right) The event selection gates for observations (solid boxes) and background events (dashed boxes) are displayed for the α_1 and p_1 transition, after selection of the associated gamma energy.

The event selection gates for coincident observation (solid boxes) and non-coincident background events (dashed boxes) choosing a wider range for higher statistics are displayed for the α_1 and p_1 transition in Figure 1. The signal $\mu = n - b$ is derived by subtracting the background b from the coincident observation n. Towards the lowest relative energies, with exponentially dropping cross sections, a statistical framework adapted to low counting rates was applied [9]. The analysis routine assures that reliable confidence intervals can be attributed *e.g.* to signals $\mu = 0$ with non-zero observation and background measurements, assigning accurate limits depending on the background level.

Carbon Burning S-Factors

The gamma-particle coincidence technique combined with the merit of nanosecond-timing measurements have been used to minimize background contributions for $^{12}C+^{12}C$ cross-section determination at the highest precision reached so far in the region of astrophysics interest [10–12]. This has allowed to obtain astrophysical S-factors, which are presented in Fruet *et al.* [1], supplemented by beam and target relevant details of the campaigns as well as the numerical cross-section values.

The results identify three distinct sub-Coulomb energy-regimes: (i) In the relative-energy range E_{rel} = 4.5 to 5.5 MeV the cross sections are typically of the order of millibarns. The initial data are in excellent agreement with earlier measurements [13, 14] giving confidence in the present methodology and experimental accuracy of measuring the excitation function. (ii) In the deep-sub-barrier regime from E_{rel} = 2.5 to 4 MeV, where the S-factor plateau is located according to the phenomenological hindrance model [15], the Fowler standard extrapolation systematically overestimates the results and hindrance is observed. The present data show remarkable agreement with earlier coincidence measurements utilizing a Germanium-silicon detector array [11] (iii) below E_{rel} = 2.5 MeV, well inside the 25 $M_{\odot}$ Gamow window, the S-factor rises up that may indicate a change in the fusion mechanism. The situation may indicate either the presence of a resonance and/or might be interpreted as the consequence of the low level density of states in the highly symmetrical compound nucleus ^{24}Mg at these excitation energies [16].

Following the discussion in Fruet *et al.*, the observation of hindrance behavior over a wide energy range necessitates resonant doorway configurations of ^{24}Mg for interpreting the measurement at the lowest explored energy. In this picture, fusing is decomposed artificially into two steps, *i.e.* population of resonances at high excitation energies with subsequent cascading into the rotational bands of ^{24}Mg. A more natural explanation might be based on limited phase space in the highly symmetrical even-even nucleus ^{24}Mg with large spacing of narrow levels, so that only a few final states are available for the fusion process to occur.

In the context of quiescent carbon burning in massive stars, the described fusion-hindrance bahaviour will lead to elevated carbon-burning ignition temperatures [17, 18], as compared to calculations with the Fowler extrapolation. In comparison to this model, less heat is generated by ^{12}C fusion while heat dissipation by neutrinos is basically temperature and density-dependent. Sensitivity studies with plain variations of the S-factor across the entire energy range showed furthermore, that the efficiency of the ^{22}Ne(α, n)^{25}Mg and the ^{13}C(α, n)^{16}O reaction in neutron-seed generation for the *s*-process can change drastically [17]. The present results can be used in further investigations of the advanced burning phases in massive stars, to refine the understanding of their development and the produced neutrino flux of neutrinos. This knowledge will improve the understanding of the probability of type Ia supernovae [18].

Acknowledgments

STELLA was funded by the University of Strasbourg IdEX program and CNRS Strasbourg, the Andromède facility by ANR-10-EQPX-23. This work was partially supported by the UK grant No. ST/L005743/1, No. ST/P003885/1, and No. ST/P005314/1. P.H.R. and R.S. acknowledge support from the UK Department BEIS. D.G.J. and S.C. acknowledge support from their Fellowships USIAS. J.N. acknowledges support from QMat by grants ANR 10 IDEX 0002 and ANR 20 SFRI 0012.

References

[1] G. Fruet *et al.*, Physical Review Letters **124**, 192701 (2020)
[2] C.E. Rolfs and W.S. Rodney, *Cauldrons in the Cosmos* (Univ of Chicago Press, 1988)
[3] D. Jenkins and S. Courtin, Journal Physics G **42**, 034010 (2015)
[4] Y. Chiba and M. Kimura, Physical Review C **91**, 061302(R) (2015)
[5] C.L. Jiang *et al.*, Physical Review Letters **89**, 052701 (2002)
[6] C. Simenel *et al.*, Physical Review C **95**, 031601(R) (2017)
[7] M. Heine *et al.*, Nuclear Instruments and Methods A **903**, 1 (2018)
[8] M. Rudigier *et al.*, Nuclear Instruments and Methods A **969**, 163967 (2020)
[9] G.J. Feldman and R.D. Cousins, Phys Rev D **57**, 3873 (1998)
[10] T. Spillane *et al.*, Physical Review Letters **98**, 122501 (2007)
[11] C.L. Jiang *et al.*, Physical Review C **97**, 012801(R) (2018)
[12] W.P. Tan *et al.*, Physical Review Letters **124**, 192702 (2020)
[13] H.W. Becker *et al.*, Zeitschrift für Physik A **303**, 305 (1981)
[14] E.F. Aguilera *et al.*, Physical Review C **73**, 064601 (2006)
[15] G. Montagnoli *et al.*, Physical Review C **97**, 024610 (2018)
[16] C.L. Jiang *et al.*, Physical Review Letters **110**, 072701 (2013)
[17] M. Pignatari *et al.*, Astrophysical Journal **762**, 31 (2013)
[18] O. Straniero *et al.*, Journal of Physics: Conference Series **665**, 012008 (2016)

Oxygen Formation in the Light of Gamma-Beams: [Precision Measurements of the ^{12}C(α,γ) Reaction with Gamma-Beams and a Time Projection Chamber]

Moshe Gai[1,*]

[1]LNS at Avery Point, University of Connecticut, 1084 Shennecossett Rd., Groton, CT 06340, USA

Abstract. The carbon/oxygen (C/O) ratio at the end of stellar helium burning is not known with sufficient accuracy due to the large uncertainty in the cross-section of the ^{12}C$(\alpha,\gamma)^{16}O$ reaction. We developed a new method to measure this reaction that is significantly different from the experimental efforts of the past four decades. Data were measured with vanishingly small background, inside one detector, that also serves as an active target (AT-TPC). Angular distributions of the ^{12}C$(\alpha,\gamma)^{16}O$ reaction were obtained by measuring the inverse ^{16}O$(\gamma,\alpha)^{12}C$ reaction with gamma-beams (from the HIγS facility) and an optical readout time projection chamber (O-TPC) detector. We agree with current world data on the measured total reaction cross-section. We further evidence the strength of our method with angular distributions measured over the 1^-resonance at $E_{cm} \sim 2.4$ MeV. We extract the *E1*-*E2* mixing phase angle (ϕ_{12}) over this resonance and obtained values that follow the trend of prediction based on unitarity of the scattering amplitude. Our technique promises to yield results that will surpass the quality of the currently available data. We continue these measurements to lower energies at the HIγS facility of TUNL at Duke University, and anticipate measurements down to $E_{cm} = 1.1$ MeV at the ELI-NP, the EU facility in Romania..

1 Introduction

The ^{12}C$(\alpha,\gamma)^{16}O$ reaction was measured in terrestrial laboratories at energies near or below 1.5 MeV. We refer the reader to [1] for a thorough review of all measurements performed over the last 47 years, including the most recent data measured by the Stuttgart group with the EUROGAM array [2]. These data were used extensively to extrapolate to stellar conditions [1].

In stellar conditions, two partial waves, $\ell = 1$ and 2, contribute. They are denoted by the spectroscopic *E1* and *E2* amplitudes. The challenges in this field are measurements of angular distributions of the ^{12}C$(\alpha,\gamma)^{16}O$ reaction from which the *E1* and *E2* cross sections are extracted, for accurate extrapolations to stellar conditions.

The *E1*-*E2* interference phase angle (ϕ_{12}) was calculated at first in the frame of R-Matrix theory [3], in term of the elastic phase shifts: $\phi_{12} = \delta_2 - \delta_1 + atan(\eta/2)$ where η is the Sommerfeld parameter. However, more recently, Brune [4] and Gai [5]

*e-mail: moshe.gai@uconn.edu. Supported by USDOE grant No. DE-FG02-94ER40870.

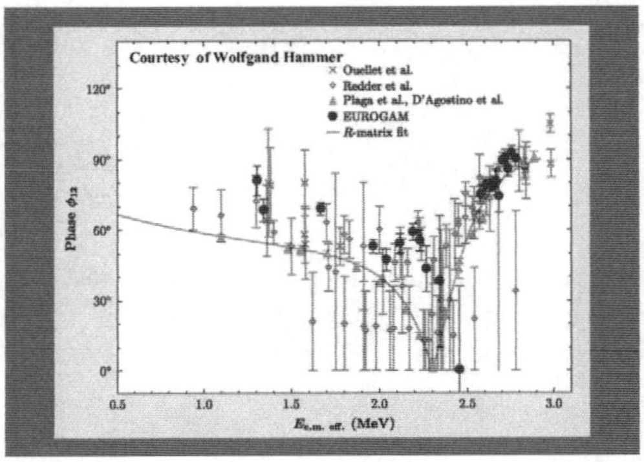

Figure 1. The world data on ϕ_{12} compiled by Professor Wolfgang Hammer. The Stuttgart EUROGAM data [2] are shown in blue. Figure courtesy of Professor Hammar.

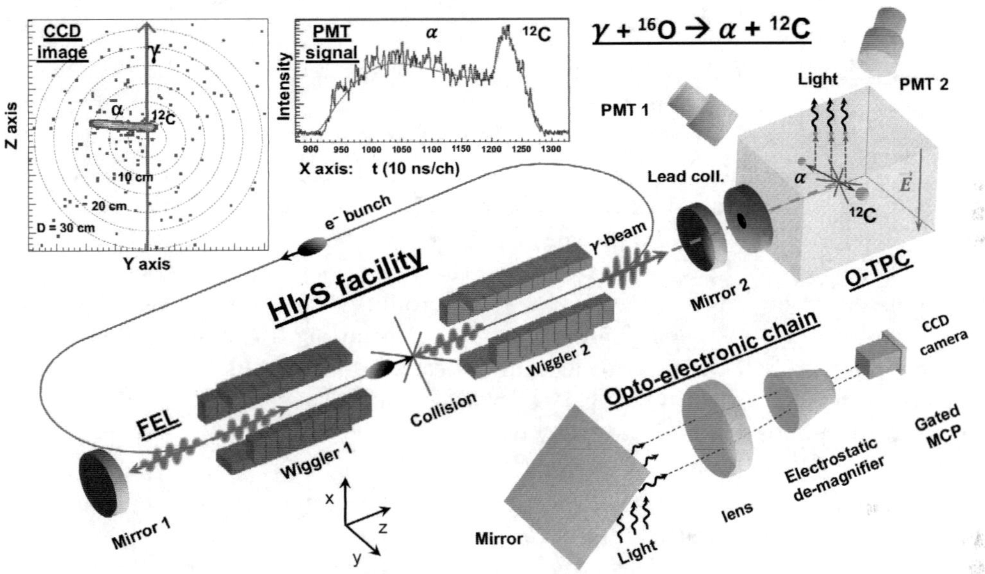

Figure 2. Schematic diagram of the HIγS beam and O-TPC. CCD image and PMT signal are shown on top left. Figure reproduction from [6] with permission from Springer Nature. Open Access Creative Commons license, creativecommons.org/ licenses/by/4.0/

noted that the theoretical prediction for ϕ_{12} is a consequence of the Watson theorem, which is derived assuming the unitarity of the scattering matrix. Hence the predicted ϕ_{12} is an elementary consequence of quantum mechanics. It must be observed in all measured angular distributions.

But as seen in Fig. 1, so far, no available measurement of ϕ_{12} over the broad 1^- resonance of ^{16}O (at E_{cm} = 2.4 MeV) show agreement with this fundamental prediction of quantum theory, even though at these energies the cross sections are large. Most disturbing is the disagreement of the modern measurements of the Stuttgart group [2], since these data were used for extrapolation [1]. As we show in Fig. 1, all 9 low energy data points at 1.3 - 2.3 MeV, that were measured with small error bars, disagree with the predicted ϕ_{12}, suggesting an ill understood systematic error(s), that were already noted before [5].

2 Measurement of the ^{16}O(γ, α) at HIγS

We chose to measure the ^{12}C$(\alpha, \gamma)^{16}O$ reaction by measuring the time reversed ^{16}O$(\gamma, \alpha)^{12}C$ reaction, from which the direct reaction can be deduced using detailed

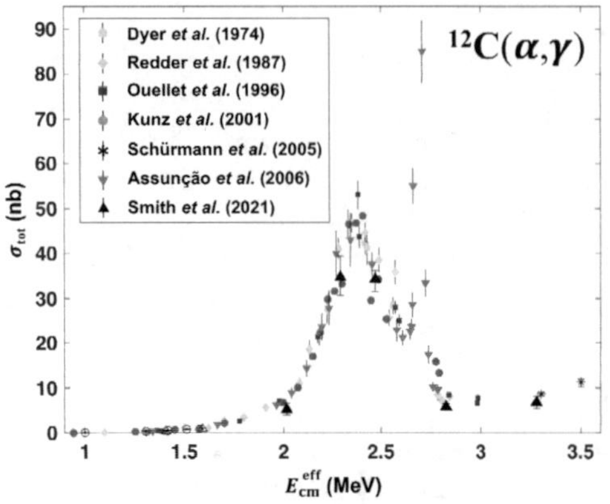

Figure 3. The world data on the total reaction cross section compared to our result [6]. Figure reproduction from [6] with permission from Springer Nature. Open Access Creative Commons license, creativecommons.org/ licenses/by/4.0/

balance. The setup of our measurement, data analyses, reduction and results are described in great details in our recently published Nature Communications paper [6]. But this discussion is beyond the scope of the current short contribution.

Briefly, the quasi mono-energetic gamma-beam ($\Delta E/E \sim 3\%$) was produced by the back-angle scattering of FEL photons with approximately 1 GeV electron beams, as discussed in [7]. The gamma beams incident on the Optical readout Time Projection Chamber (O-TPC) constructed at the University of Connecticut [8]. The photo-dissociation of oxygen in the CO_2 gas produces the α+^{12}C ions. These ions travel through the gas-filled chamber and the ionization tracks are amplified and recorded by the CCD camera through an optical readout system. A time projection of tracks is also recorded by using a set of Photo Multiplier Tubes. Again, we refer the reader to [6] for a complete review of our measurement.

The in plane angle (α) is measured ($\pm 2°$) from the CCD image, and the out of plane angle (β) is measured with the time projection signal recorded by the PMTs. The scattering and azimuthal angles (θ and ϕ) are deduced: $\tan\phi = \tan\beta/\sin\alpha$ and $\cos\theta = \cos\beta$ x $\cos\alpha$. Complete (0° - 180°) angular distributions were measured at nominal beam energies of $E_\gamma = 9.08$, 9.38, 9.48, 9.58, 9.78, 10.1 and 10.4 MeV.

The measured angular distributions were fitted with the standard $E1 + E2$ amplitude [9] :

$$W(\theta) = (3|A_{E1}|^2 + 5|A_{E2}|^2)P_0(\cos\theta) + \left(\frac{25}{7}|A_{E2}|^2 - 3|A_{E1}|^2\right)P_2(\cos\theta)$$
$$- \frac{60}{7}|A_{E2}|^2 P_4(\cos\theta) + 6\sqrt{3}|A_{E1}||A_{E2}|\cos\phi_{12}\,[P_1(\cos\theta) - P_3(\cos\theta)]$$

2.1 Results

The total reaction cross section measured in our HIγS experiment is shown in Fig. 3 taken from [6]. The agreement with the world data on the total reaction cross section serves as "proof of principle" of our new method. The extracted $E1 - E2$ mixing phase angle (ϕ_{12}) is shown in Fig. 4. The current results are from "data on tape" taken in 2012 (and analyzed during the COVID-19 pandemic). These data are with low statistics and we intend to remeasure it with higher statistics. Never-the-less the $E1 - E2$ mixing phase angle (ϕ_{12}) extracted by us is in agreement with the trend of the theoretical prediction. It serves as an impetus to repeat this measurement with higher statistics and continue to low energies.

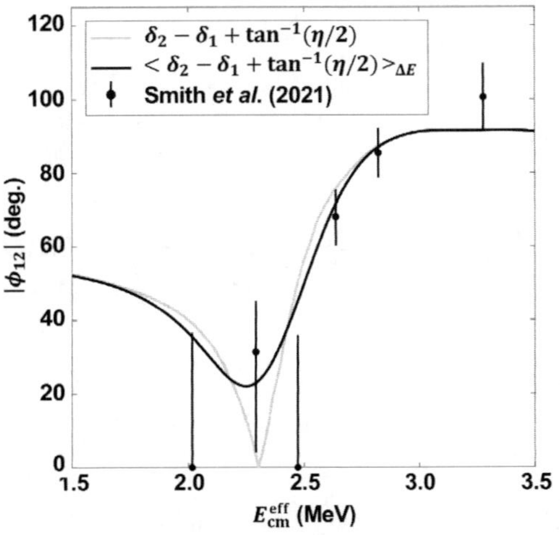

Figure 4. The measured $E1 - E2$ mixing phase angle (ϕ_{12}). Figure reproduction from [6] with permission from Springer Nature. Open Access Creative Commons license, creativecommons.org/ licenses/by/4.0/

3 Future Prospects at the HIγS and the ELI-NP

The current results were obtained with an optical readout TPC (O-TPC). A considerably better electronic readout TPC [10, 11] was constructed at the University of Warsaw (the ELITPC) with considerably better performance characteristics. We intend to repeat the measurement reported here with the O-TPC at the HIγS using the ELITPC, and continue to lower energies approaching ~1.7 MeV. After completing the measurements at the HIγS we plan to continue these measurements down to 1.1 MeV at the newly constructed ELI-NP facility in Romania [12] using the VEGA beam.

References

[1] R. J. deBoer, J. Goerres, M. Wiescher, R. E. Azuma, A. Best, C. R. Brune, C. E. Fields, S. Jones, M. Pignatari, D. Sayre, K. Smith, F. X. Timmes, Rev. Mod. Phys. **89**, 035007 (2017).

[2] M. Assuncao *et al.* Phys. Rev. C **73**, 055801 (2006).

[3] F.C. Barker, and T. Kajino, Aust. J. Phys. **44**, 369 (1991).

[4] C.R. Brune, Phys. Rev. C **64**, 055803 (2001).

[5] M. Gai, Phys. Rev. C **88**, 062801(R) (2013).

[6] R. Smith , M. Gai, S. R. Stern, D. K. Schweitzer and M. W. Ahmed Nature Communications **12**, 5920 (2021).

[7] H.R. Weller, M.W. Ahmed, H.Gao, W. Tornow, Y. Wu, M. Gai, R. Miskimen, Prog. Part. Nucl. Phys. **62**, 257 (2009).

[8] M. Gai, M.W. Ahmed, S.C. Stave, W.R. Zimmerman, A. Breskin, B. Bromberger, R. Chechik, V. Dangendorf, Th. Delbar, R.H. France III, S.S. Henshaw, T.J. Kading, P.P. Martel, J.E.R. McDonald, P.-N. Seo, K. Tittelmeier, H.R. Weller, and A.H. Young., Jour. Instr. **5**, 12004 (2010).

[9] P. Dyer, C.A. Barnes, Nuc. Phys. A **233**, 495 (1974).

[10] M. Gai *et al.* Nucl. Instr. Meth. A **954**, 161770 (2020).

[11] M. Cwiok, M. Bieda, J.S. Bihalowicz, W. Dominik, Z. Janas, Janiak, J. Manczak, T. Matulewicz, C. Mazzocchi, M. Pfuetzner, P. Podlaski, S. Sharma, M. Zaremba, D. Balabanski, A. Bey, D.G. Ghita, O. Tesileanu, M. Gai, Act. Phys. Pol. B **49**, 1001 (2018).

[12] D. Filipescu *et al.*, Eur. Phys. J. A **51**, 185 (2015).

Direct measurements of the ^{12}C(^{12}C,p)^{23}Na and ^{12}C(^{12}C,α)^{20}Ne reactions at low energies for Nuclear Astrophysics

Lizeth Morales-Gallegos[1,2,3,*], *Marialuisa* Aliotta[2,3], *Andreas* Best[4], *Carlo G.* Bruno[2], *Raffaele* Buompane[3], *Thomas* Davinson[2], *Mario* De Cesare[5], *Antonino* Di Leva[4], *Antonio* D'Onofrio[3], *Jeremias G.* Duarte[3,1], *Leandro* Gasques[6], *Lucio* Gialanella[3], *Gianluca* Imbriani[4], *Giuseppe* Porzio[3], *David* Rapagnani[4], *Mauro* Romoli[1], and *Filippo* Terrasi[3]

[1]INFN, Sezione di Napoli, Napoli, Italy
[2]SUPA, School of Physics and Astronomy, University of Edinburgh, Edinburgh, UK
[3]Dipartimento di Matematica e Fisica, Università degli Studi della Campania "Luigi Vanvitelli", Caserta, Italy
[4]Dipartimento di Fisica "E. Pancini", Università degli Studi di Napoli "Federico II", Napoli, Italy
[5]Centro Italiano di Ricerche Aerospaziali, Capua, Italy
[6]Departamento de Física Nuclear, Instituto de Física da Universidade de São Paulo, São Paulo, Brazil

Abstract. ^{12}C+^{12}C reactions are crucial in the evolution of massive stars and explosive scenarios. The measurement of these reactions at astrophysical energies is very challenging due to their extremely small cross sections, and the presence of beam induced background originated by the natural ^{1,2}H contaminants in the C targets. In addition, the many discrepancies between different data sets and the complicated resonant structure of the cross sections make the extrapolation to low energies very uncertain. Recently, we performed a direct measurement of the ^{12}C+^{12}C reactions at the CIRCE Laboratory in Italy. Results from a study on target contamination were used, allowing us to measure cross sections at $E_{c.m.}$=2.51 – 4.36 MeV with 10-25 keV energy steps. Two stage ΔE-Erest detectors were used for unambiguous particle identification. Branching ratios of individual particle groups were found to vary significantly with energy and angular distributions were also found to be anisotropic, which could be a potential explanation for the discrepancies observed among different data sets.

1 Introduction

Carbon fusion reactions determine the destiny of massive stars and the properties of a star before a supernova explosion. They are also essential to model X-ray bursts and explosions on the surface of neutron stars [1–10]. With the present knowledge of carbon burning, the Gamow window of these reactions corresponds to $E_{\mathrm{c.m.}} = 1.5 \pm 0.3$ MeV [11]. At these energies, the ^{12}C+^{12}C reactions proceed through the ^{23}Na+p and ^{20}Ne+α channels. Carbon-fusion cross sections are extremely small ($\ll 10^{-9}$ b) and thus difficult to measure in the laboratory [12, 13]. In addition, the extrapolation from high energy data to astrophysical

*e-mail: moralesgallegos@na.infn.it

energies is uncertain by the presence of several resonances and beam induced background due to $^{1,2}H$ impurities in the target [14–19]. Several attempts have been made over the past five decades to determine the $^{12}C+^{12}C$ reactions cross-sections [12, 20–34]. However, data still carry large uncertainties and show significant discrepancies between different data sets. Furthermore, no direct measurement has been possible at energies below $E_{c.m.} = 2.14$ MeV and indirect measurements [35] incited an intense debate [36, 37]. For these reasons, further direct experimental investigations are required.

2 Experimental setup

Measurements of the $^{12}C+^{12}C$ reactions were performed at the 3 MV Pelletron Tandem Accelerator of the CIRCE Laboratory, Department of Mathematics and Physics of the University of Campania "Luigi Vanvitelli" in Caserta, Italy. Thick (1 mm) HOPG targets were mounted on a water-cooled target ladder surrounded by a sphere kept at −300 V for electron suppression, allowing for beam-current reading directly on target. The detection system consisted of four telescope detectors called GASTLY (GAs Silicon Two-Layer sYstem), each comprising an ionisation chamber (IC, ΔE stage) and a large area ($\simeq$ 25 cm^2) silicon strip detector (SSD, E_{rest} stage). Further details on the full detector array and its commissioning are reported in [38]. For the present study, the silicon detector was used as a single pad. Three detectors were mounted on a vertical plane at 121° (D121) and 156° (above and below the beam axis; D156), and one on the horizontal plane at an angle 143° (D143) to the beam axis, as shown in figure 1. See [39] for a full experimental setup description.

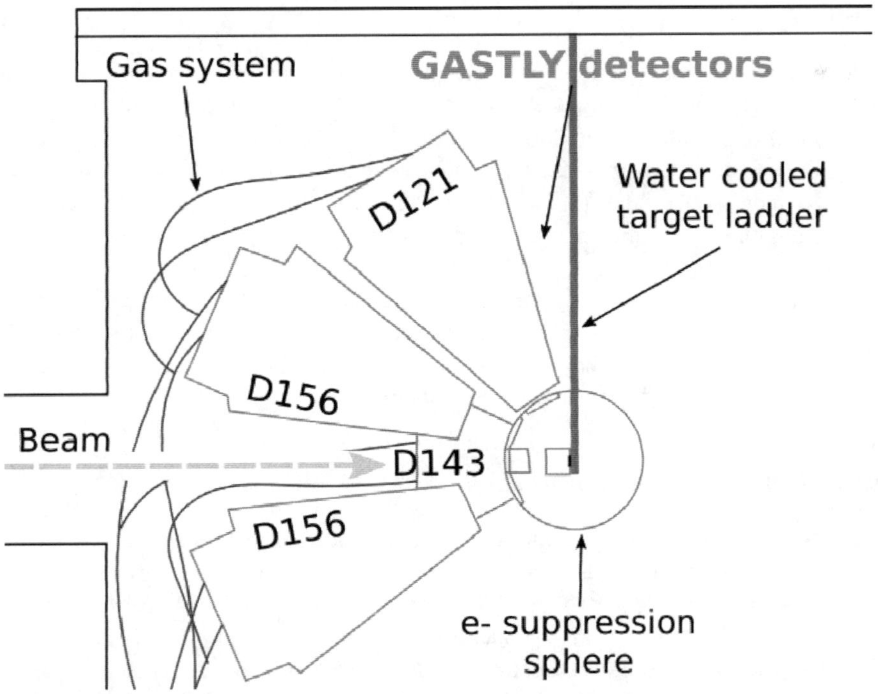

Figure 1. Sketch of the experimental configuration used in this work. Shown are four GASTLY detectors, target ladder, sphere for electron suppression and the beam direction.

Data were taken at energy intervals of 20-50 keV in the laboratory system. Target temperature was constantly monitored with a thermo-camera and maintained to at least 400 °C (using intense beams) to reduce deuterium contamination as a previous study recommends [40]. The scattering chamber was surrounded by an enclosure filled with dry nitrogen.

2.1 Data analysis

For the analysis of each particle channel, background runs of several days were taken, time-normalized and subtracted from the relevant proton and α-particle spectra at each beam energy. Subsequently, proton and α-particle peaks from the $^{12}C+^{12}C$ reactions were identified through kinematic reconstruction and comparing with simulations [41]. As many particle peaks overlap, the number of events within each was extracted using the maximum likelihood method from a combined fit of skewed Gaussian functions. Given that all analysed protons at the energies studied here arrive to the SSD, only its spectra were used in the proton analysis. Some deuterium-induced peaks were still visible (despite its minimization) in the proton spectra. In most cases, it was possible to disentangle this beam-induced contribution from the peaks of interest. Otherwise, the affected proton peaks were discarded from further analysis. Unlike for the proton channel, data analysis for the α-particles channel was performed on reconstructed total energy spectra, (E_{tot} = IC+SSD). Thick-target yields were calculated from the net number of events at each beam energy, then differentiated at two consecutive beam energies to finally extract the differential cross sections. Each cross section was later associated to an effective energy expressed in the centre of mass system and finally converted into $\tilde{S}$-factors. See [39] for a complete description of the data analysis.

3 Results

Differential $\tilde{S}$-factors for individual particle groups were obtained from each detector. In Morales-Gallegos et al. [39] detailed results are reported. Deuterium contamination of the target was identified at all times and thus excluded from the data points. Our data confirm the presence of resonance-like structures across the entire energy region explored in this work, as also reported in previous studies [22, 25, 29, 30, 42]. Given that many working groups assume an isotropic angular distribution and constant branching ratios, we investigated whether such assumptions are correct or not. Even though an angular distribution measurement was not possible due to the use of only three (large) detection angles, our results reveal the presence of anisotropies over all energies. Also branching ratios show significant variations as a function of energy which indicates that they should not be considered constant. Similar results for all proton and α particle groups and angles were obtained.

4 Summary and conclusions

The $^{12}C+^{12}C$ fusion reactions are among the most important in astrophysics as they govern the evolution and fate of massive stars. In this work, we reported on the measurement of the $^{12}C(^{12}C,p)^{23}Ne$ and $^{12}C(^{12}C,\alpha)^{23}Na$ reactions at $E_{c.m.}$ = 2.51 – 4.36 MeV with energy steps < 25 keV in the centre of mass. Background-subtracted proton- and α-particle spectra were fitted with skewed Gaussian functions to obtain the number of events under all particle peaks ($p_0 - p_6$ and $\alpha_0 - \alpha_1$). Cross sections were obtained by differentiating thick-target yields measured at consecutive beam energies and converted into differential $\tilde{S}$-factors. Our results confirm the presence of resonance-like structures as revealed by previous measurements. However, non-constant branching ratios and anisotropic angular distributions were

observed for all particle groups. We note that existing discrepancies among various data sets available in the literature may arise from incorrect assumptions on branching ratios and angular distributions. Improvements to our setup are currently underway: seven GASTLY detectors were produced and mounted at different angles and the SSDs have all their strips fully functioning allowing for an angular distribution measurement.
A thorough study of the intrinsic alpha-emitters contamination of the detectors is being conducted at INFN underground site of the Laboratori Nazionali del Gran Sasso (LNGS) as a joint project between the ERNA and the LUNA collaborations. This in view of a possible use of the GASTLY detectors for the direct measurement of the $^{12}C+^{12}C$ reactions into the Gamow window using the intense carbon beam that will be provided by the upcoming 3.5 MV underground accelerator facility at LNGS (INFN).

Acknowledgments

This work was supported by the Italian National Institute of Nuclear Physics (INFN), V:ALERE and the Royal Society. MA acknowledges the support of a Visiting Professorship from the Università degli Studi della Campania "Luigi Vanvitelli". LMG thanks CoNaCYT (scholarship 311745), the University of Edinburgh and COST Action ChETEC: CA16117 (STSM number 38819) for their financial support.

References

[1] S. Cristallo *et al. Journal of Physics (NPA6)*, **665**, 012019 (2016)
[2] M. Limongi, M., *Proceedings of Science (**ENAS 6**)*, 010 (2011)
[3] E. Bravo *et al.*, *Astronomy and Astrophysics* **535**, A114 (2011)
[4] R. L. Cooper *et al.*, *Astrophysical Journal* **702**, 660 (2009)
[5] M. E. Bennett *et al.*, *Monthly Notices of the Royal Astronomical Society* **420**, 2737–2755 (2012)
[6] B. Bucher *et al.*, *EPJ Web of Conferences* **93**, 03009 (2015)
[7] A. Cumming and L. Bildsten, *The Astrophysical Journal* **559**, L127-L130 (2001)
[8] X. Tang *et al.*, *Nuclear Physics and Atomic Energy* **14**, 224-232 (2013)
[9] Y. Sato *et al.*, *The Astrophysical Journal* **807**, 105 (2015)
[10] K. Mori *et al.*, *Monthly Notices of the Royal Astronomical Society: Letters* **482**, L70-L74 (2019)
[11] F. Strieder and C. Rolfs, *Progress in Particle and Nuclear Physics* **59**, 562-578 (2007)
[12] L. Barrón-Palos *et al.*, *Nuclear Physics A* **779**, 318-332 (2006)
[13] C. Angulo, *Lecture Notes in Physics* **764**, 253-282 (2009)
[14] B. R. Fulton, *Contemporary Physics* **40**, 299-311 (1999)
[15] M. Freer and A. C. Merchant, *Journal of Physics G: Nuclear and Particle Physics* **23**, 261 (1997)
[16] D. A. Bromley *et al.*, *Physical Review Letters* **4**, 365 (1960)
[17] L. J. Satkowiak *et al.*, *Physical Review C* **26**, 2027-2034 (1982)
[18] C. L. Jiang *et al.*, *Physical Review Letters* **110**, 072701 (2013)
[19] M. E. Bennett *et al.*, *Proceedings of Science (**NIC XI**)*, 182 (2010)
[20] J.R. Patterson *et al.*, *The Astrophysical Journal* **157**, 367-373 (1969)
[21] M. G. Mazarakis and Stephens, *Physical Review C* **7**, 1280-1287 (1973)
[22] H. W. Becker *et al.*, *Zeitschrift für Physik A, Atoms and Nuclei* **303**, 305-312 (1981)
[23] J. Zickefoose *et al.*, *Physical Review C* **97**, 065806 (2018)

[24] M. D. High and B. Ĉujec, *Nuclear Physics A* **282**, 181 (1977)

[25] K. U. Kettner *et al.*, *Zeitschrift für Physik A, Atoms and Nuclei* **298**, 65-75 (1980)

[26] P. Rosales and E. F. Aguilera, *Revista Mexicana de Física S4* **49**, 88-91 (2003)

[27] L. Barrón-Palos *et al.*, *Revista Mexicana de Física S2* **50**, 18-23 (2004)

[28] L. Barrón-Palos *et al.*, *European Physical Journal A* **645-646**, 25 (2005)

[29] E. Aguilera *et al.*, *Physical Review C* **73**, 064601 (2006)

[30] T. Spillane *et al.*, *Physical Review Letters* **98**, 122501 (2007)

[31] C. L. Jiang *et al.*, *Nuclear Instruments and Methods in Physics Research Section A* **682**, 12-15 (2012)

[32] C.L. Jiang *et al.*, *Physical Review C* **97**, 012801 (2018)

[33] G. Fruet *et al.*, *Physical Review Letters* **124**, 192701 (2020)

[34] W.P. Tan *et al.*, *Physical Review Letters* **124**, 192702 (2020)

[35] A. Tumino *et al.*, *Nature* **557**, 687-690 (2018)

[36] A. M. Mukhamedzhanov *et al.*, *arXiv*:1806.05921v1 (2018)

[37] A. Tumino *et al.*, *arXiv*:1807.06148 (2018)

[38] M. Romoli, L. Morales-Gallegos *et al.*, *European Physical Journal A* **54**, 142 (2018)

[39] L. Morales-Gallegos *et al.*, *European Physical Journal A* **submitted**, (2022)

[40] L. Morales-Gallegos *et al.*, *European Physical Journal A* **54**, 132 (2018)

[41] L. Zhang, *private communication* (2016)

[42] J. Zickefoose, $^{12}C + ^{12}C$ *Fusion: Measurement and Advances Toward the Gamow Energy*, University of Conneticut Ph. D. Thesis (2010) https://www.proquest.com/docview/908637546

Supernova Nucleosynthesis, Radioactive Nuclear Reactions and Neutrino-Mass Hierarchy

Xingqun Yao[1,2,*], *Motohiko* Kusakabe[1,2,**], *Toshitaka* Kajino[1,2,3,***], *Silvio* Cherubini[4,****], *Seiya* Hayakawa[5,†], and *Hidetoshi* Yamaguchi[5,‡]

[1]School of Physics & International Research Center for Big-Bang Cosmology and Element Genesis, Beihang University, Beijing 100083, China
[2]National Astronomical Observatory of Japan, Mitaka, Tokyo 181-8588, Japan
[3]Graduate School of Science, The University of Tokyo, Bunkyo-ku, Tokyo 113-0033, Japan
[4]Department of Physics and Astronomy, "Ettore Majorana" University of Catania, Catania, Italy
[5]Center for Nuclear Study, The University of Tokyo, 2-1 Hirosawa, Wako, Saitama 351-0198, Japan

Abstract. The ν-process nucleosynthesis in core-collapse supernovae is a sensitive probe of unknown neutrino mass hierarchy through the MSW effect. We carefully studied the uncertainties of almost one hundred ν-induced and nuclear reactions associated with the nucleosynthesis and found that the ν-^{16}O and ^{11}C(α,p)^{14}N reactions among them have the biggest effect on the final ^{7}Li/^{11}B isotopic abundance ratio. The neutrino mass hierarchy is constrained in our nucleosynthetic method with measured ^{7}Li/^{11}B value in SiC-X presolar grains. The inverted hierarchy is statistically more favored at the 2-σ C.L. [1].

1 Introduction

Core-collapse supernova (CCSN) ejects a huge number of three-flavour neutrinos and anti-neutrinos which provide observational signals through the Mikheyev–Smirnov–Wolfenstein (MSW) matter effect [2, 3]. Neutrinos propagate through the inner iron core and several layers and eventually reach the He layer. When the electron number density satisfies the specific MSW resonance condition, the flavor conversion occurs between $\bar{\nu}_e$ and $\bar{\nu}_{\mu,\tau}$ for the inverted hierarchy and ν_e and $\nu_{\mu,\tau}$ for the normal hierarchy, respectively. Thus, the energy spectra of three-flavor neutrinos vary as a function of radius depending on the mass hierarchy.

These neutrinos interact with abundant nuclei in each layer, and various isotopes such as ^{7}Li, ^{11}B, ^{19}F, ^{92}Nb, ^{98}Tc, ^{138}La, ^{180}Ta, etc. are synthesized [4–7]. The neutrino oscillation effect that depends on the mass hierarchy considerably changes the final yields of these nuclei via the charged current interactions. The isotopic abundances in CCSNe are the very unique and sensitive probe of the neutrino mass hierarchy. Both ^{7}Li and ^{11}B were discovered [8] in presolar silicon-carbide X (Si-C X) grains which form from the supernova (SN) ejecta. The measured ^{7}Li/^{11}B ratio could be used to constrain the neutrino mass hierarchy.

*e-mail: sternyao@buaa.edu.cn
**e-mail: kusakabe@buaa.edu.cn
***e-mail: kajino@buaa.edu.cn
****e-mail: cherubini@lns.infn.it
†e-mail: hayakawa@cns.s.u-tokyo.ac.jp
‡e-mail: yamag@cns.s.u-tokyo.ac.jp

^{7}Li is synthesized mostly in the He layer, while ^{11}B is synthesized over the entire region from O-Ne-Mg to He layers [1]. In order to provide nuclear abundances of ^{7}Li and ^{11}B, we need to study how reliable the ν-nucleus cross sections are. We also need to study the nuclear reactions which secondarily destroy or produce them. The important ν-nucleus cross sections include those for the production of unstable nuclei ^{7}Be which decay to ^{7}Li in 53.22 days by electron capture and ^{11}C and ^{11}Be which decay to ^{11}B in 20.34 min and 13.76 sec by $\beta^{+/-}$-decays, respectively. We studied the sensitivity of the final yields of the $A = 7$ and 11 nuclear systems to all ν-induced reactions on ^{4}He, ^{12}C and ^{16}O. The secondary nuclear reactions include both radioactive and stable nuclear reactions such as ^{11}C(α,p)^{14}N and ^{11}B(α,n)^{14}N. We studied the sensitivity of the final yields to the 91 nuclear reactions.

In this paper we report the result of our theoretical studies of the sensitivity of the A = 7 and 11 abundances to the ν-^{16}O and ^{11}C(α,p)^{14}N reactions because these two reactions have the biggest effect on changing the ^{7}Li/^{11}B isotopic abundance ratio among all reactions which we studied. We then discuss how to constrain the neutrino-mass hierarchy by comparing our theoretical prediction with the measured ^{7}Li/^{11}B ratio in the presolar Si-C X grains [1].

2 Supernova Model

We adopt the pre-SN model for SN 1987 calculated [9] for metallicity of Large Magellanic Cloud. The initial mass of the progenitor star is $20M_{\odot}$, and the star evolves to a helium core of $6M_{\odot}$. The hydrodynamic evolution of exploding materials and nucleosynthesis were calculated in the same method adopted in [7]. The total neutrino energy is 3×10^{53} erg, the exponential decay timescale of neutrino luminosity is 3 s, and the neutrino temperatures are set to be 3.2 MeV for ν_e, 5.0 MeV for $\bar{\nu}_e$, and 6.0 MeV for ν_x, where ν_x stands for ν_μ, $\bar{\nu}_\mu$, ν_τ and $\bar{\nu}_\tau$ [6]. Details are reported in [1].

3 Result and Discussions

3.1 ν-process nucleosynthesis of ^{7}Li, ^{7}Be, ^{11}B and ^{11}C

As displayed in Fig. 1, both ^{7}Li+^{7}Be and ^{11}B+^{11}C abundances depend strongly on neutrino mass hierarchies. The integrated yields in normal hierarchy (blue curves) are higher than those calculated in inverted hierarchy (red curves). Since the original temperatures are different from one another $T_{\nu_e} < T_{\bar{\nu}_e} < T_{\nu_x}$ at the neutrino-spheres, the MSW-effect makes more efficient swapping of the neutrino energy spectra between T_{ν_e} and T_{ν_x} than the swapping between $T_{\bar{\nu}_e}$ and T_{ν_x}. As a result, ν_e has the largest enhancement of the high energy component after high-density MSW resonance occurs. This leads to stronger charged current reactions via ν_e than $\bar{\nu}_e$ to enhance the production through, for example ^{4}He$(\nu_e, e^{-}$p$)^{3}$He$(\alpha,\gamma)^{7}$Be and ^{12}C$(\nu_e, e^{-}$p$)^{11}$C, etc., in the normal hierarchy (blue curves).

Stellar interior is divided into four layers, i.e. O-Ne-Mg layer (region I), O-C layer (II), C-He layer (III) and He layer (IV) in Fig. 1. Most of ^{7}Li, ^{7}Be, ^{11}B and ^{11}C are respectively produced in the regions of Lagrangian masses of 4.5–6.0 $M_{\odot}$ in IV, 3.5–5.5 $M_{\odot}$ in II–IV, 1.8–5.5 $M_{\odot}$ in I–IV, and 1.8–5.7 $M_{\odot}$ in I–IV. Among them, ^{11}B and ^{11}C are of particular interest. In the ^{16}O-rich layers I and II, both nuclei are predominantly produced by the ^{16}O+ν reaction. Suzuki et al. [10] have recently calculated ^{16}O+ν reaction cross sections precisely by using modern shell-model Hamiltonian which includes spin-isospin flip interaction with tensor force. This shell model calculation describes very well several nuclear properties such as magnetic-moments of p-shell nuclei, Gamow-Tellar strength for ^{12}C→^{12}N and ^{14}C→^{14}N, ^{12}C+ν reaction cross sections via both charged and neutral current interactions, and exotic nuclear properties of neutron-rich unstable nuclei. Our nucleosynthesis calculation using

Suzuki's new ^{16}O+ν rates gives one order of magnitude larger ^{11}B+^{11}C abundances (solid curves) than those using Hoffman & Woosley's rate [11] (dashed curves) in region I.

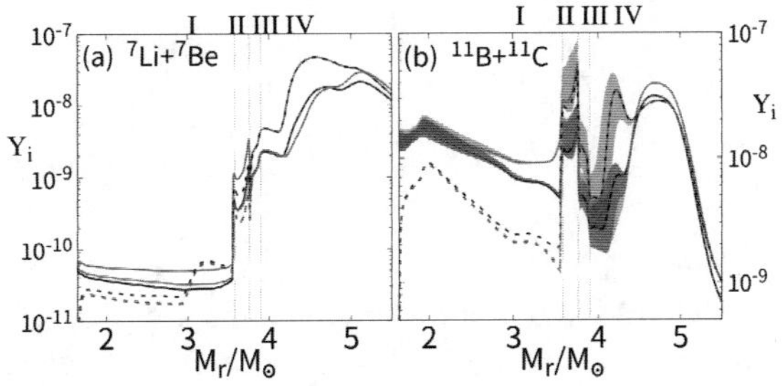

Figure 1. The calculated abundances of (a) ^{7}Li+^{7}Be and (b) ^{11}B+^{11}C as a function of Lagrangian mass coordinate in units of $M_{\odot}$. Red and blue lines correspond to inverted and normal hierarchies, respectively. Colored bands reflect the uncertainty of the ^{11}C$(\alpha, p)^{14}$N reaction rate. Solid and dashed lines are the results using Suzuki et al. [10] and Hoffman & Woosley [11] rates for ^{16}O+ν, respectively.

3.2 Sensitivity to the nuclear reactions

We calculated the sensitivity of nuclear abundances to all ν-^{4}He, ^{12}C and ^{16}O reaction rates whose uncertainties are estimated to be ±20% and ±10% for charged and neutral current reactions, respectively [1]. These uncertainties result in changes in the final yields by less than 20% and 10% because of the non-linear response made by the secondary nuclear reactions.

The reaction rates for ^{11}C(α,p)^{14}N, ^{11}C(n,p)^{11}B and ^{11}B(α,n)^{14}N among 91 nuclear reactions have rather large uncertainties. In particular, the first one was found to affect strongly the final abundances of ^{11}C and ^{11}B. Although some of the other reactions also have error bars in the measured reaction cross sections or theoretical estimates, those uncertainties result in less than a few % change in the final yields [1].

^{11}C is produced at $T_9 = T/(10^9\ \mathrm{K}) \approx 0.2 - 1.0$ in SN ν-process. It subsequently encounters the secondary destruction process ^{11}C$(\alpha, p)^{14}$N whose most effective temperature corresponds to the Gamow window energy ~ 0.3 MeV. Hayakawa et al. [12] carried out the direct measurement of this reaction cross sections for the first time, and Ingalls et al. [13] studied reverse reaction. Although these experiments cover 0.6 MeV $< E_{\mathrm{CM}} <$ 1.3 MeV, they did not reach E_{CM} = 0.3 MeV. There still exists five resonances at $E_{\mathrm{CM}} <$ 0.6 MeV. We therefore estimated the contribution of these resonances by assuming the Wigner limit in Breit-Wigner formula, which serves an upper limit on the total cross section, in addition to the non-resonant contribution as a lower limit. We then found that the newly estimated upper limit could be factors of ~ 4 and 40 larger at T_9 ~ 0.3 and 0.15, respectively, than the standard rate of CF88 [14] in JINA reaclib database [15]. The ν-process nucleosynthesis of ^{11}B and ^{11}C is thus subject to large uncertainties as shown in Fig. 1.

3.3 Constraining the ν-mass hierarchy from meteoritic ^{7}Li/^{11}B

Because the lifetimes of ^{11}C and ^{7}Be are negligible compared with the age of SN remnant, they finally decay to ^{11}B and ^{7}Li. Therefore, the calculated ratio (^{7}Li+^{7}Be)/(^{11}B+^{11}C) provides the key information.

The pre-solar grains form from the dusts which condensate during the cooling process of SN ejecta. The efficiency of the SN ejecta from each layer is not known very well because

Table 1. Calculated abundance ratios (^{7}Li+^{7}Be)/(^{11}B+^{11}C) in our nucleosynthesis model. Associated errors represent uncertainties of the final yields as shown in Fig. 1 which arise from the uncertainty (2-σ C.L.) of the nuclear reaction rate for ^{11}C(α, p)^{14}N. See text for the four cases.

ejected layers	I+II+III+IV	II+III+IV	III+IV	IV
inverted hierarchy	$0.559^{+0.030}_{-0.060}$	$0.961^{+0.039}_{-0.077}$	$1.045^{+0.031}_{-0.038}$	$1.077^{+0.025}_{-0.025}$
normal hierarchy	$0.798^{+0.119}_{-0.167}$	$1.353^{+0.292}_{-0.347}$	$1.612^{+0.319}_{-0.295}$	$1.726^{+0.312}_{-0.273}$

of the complicated hydrodynamic processes. We assume therefore the four cases in which the nucleosynthesis products are ejected in different manner as tabulated in Table 1. We also assume that the materials of SN ejecta are well mixed during the ejection process. Mathews et al. [5] inferred the constraint ^{7}Li/^{11}B < 0.56 at the 2-σ C.L. from the measured abundances of ^{7}Li and ^{11}B discovered in presolar SiC-X grains [8]. Only the inverted hierarchy in case I+II+III+IV of Table 1 meets with this constraint.

4 Conclusion

Still unknown neutrino mass hierarchy affects the final abundances of ^{7}Li, ^{7}Be, ^{11}B and ^{11}C in the SN ν-process nucleosynthesis. The final ^{11}C abundance among them was found to be subject to uncertainties of ^{11}C(α, p)^{14}N reaction rate. Although we assumed a single set of neutrino energy spectra and decay constant of their luminosities in a single SN progenitor model, the calculated abundance ratio ^{7}Li/^{11}B does not depend strongly on these model parameters as far as $T_{\nu_e} < T_{\bar{\nu}_e} < T_{\nu_x}$ is satisfied. This is because of the self-regulating reaction mechanism that balances between production and destruction of these nuclei in a narrow temperature range [6, 16]. The inverted hierarchy is suggested to be statistically more favored at the 2-σ C.L. We however need more theoretical calculations in many different models. It is also highly desirable to measure the ^{11}C(α,p)^{14}N reaction cross sections at lower energies and to find more samples of presolar SiC-X grains which include both ^{7}Li and ^{11}B.

References

[1] X. Yao, M. Kusakabe, T. Kajino, S. Cherubini, S. Hayakawa and H. Yamaguchi , to be submitted (2021).
[2] L. Wolfenstein, Phys. Rev. D **17**, 2369-2374 (1978)
[3] S. P. Mikheyev and A. Y. Smirnov, Sov. J. Nucl. Phys. **42**, 913-917 (1985)
[4] T. Yoshida, T. Kajino, H. Yokomakura *et al.*, Phys. Rev. Lett. **96**, 091101 (2006)
[5] G. J. Mathews, T. Kajino, W. Aoki, *et al.*, Phys. Rev. D **85**, 105023 (2012)
[6] T. Kajino, G. J. Mathews and T. Hayakawa, J. Phys. G **41**, 044007 (2014)
[7] M. Kusakabe, M. K. Cheoun, K. S. Kim *et al.*, Astrophys. J. **872**, 164 (2019)
[8] W. Fujiya, P. Hoppe and U. Ott, Astrophys. J. Lett. **730**, L7 (2011)
[9] Y. Kikuchi, M. a. Hashimoto, M. Ono and R. Fukuda, PTEP **2015**, no.6, 063E01 (2015)
[10] T. Suzuki, S. Chiba, T. Yoshida *et al.*, Phys. Rev. C **98**, no.3, 034613 (2018)
[11] https://dbserv.pnpi.spb.ru/elbib/tablisot/toi98/www/astro/hw92_1.htm
[12] S. Hayakawa, S. Kubono, D. Kahl *et al.*, Phys. Rev. C **93**, no.6, 065802 (2016)
[13] P. D. Ingalls, *et al.*, Phys. Rev. C **13**, 524-527 (1976)
[14] G. R. Caughlan and W. A. Fowler, Atom. Data Nucl. Data Tabl. **40**, 283-334 (1988)
[15] Richard H. Cyburt, *et al.*, Astrophys. J. Suppl. **189**, 240 (2010)
[16] H. Ko, *et al.*, Astrophys. J. Lett. **891**, L24 (2020)

Nuclear cosmochronometers for supernova neutrino-process

Takehito Hayakawa[1,2,*], *Heamin* Ko[3], *Myung-ki* Cheoun[3], *Motohiko* Kusakabe[4], *Toshitaka* Kajino[4,5,6], *Satoshi* Chiba[7], *Ken'ichi* Nomoto[8], *Masa-aki* Hashimoto[9], *Masaomi* Ono[10], *Toshihiko* Kawano[11], and *Grant J.* Mathews[12]

[1]National Institutes for Quantum Science and Technology, Ibaraki 319-1106, Japan
[2]Institute of Laser Engineering, Osaka University, Osaka 567-0871, Japan
[3]Soongsil University, Seoul 156-743, Korea
[4]Beihang University, Beijing 100083, China
[5]National Astronomical Observatory of Japan, Mitaka, Tokyo 181-8588, Japan
[6]The University of Tokyo, Tokyo 113-0033, Japan
[7]Tokyo Institute of Technology, Tokyo 113-0033, Japan
[8]Kavli Institute for the Physics and Mathematics of the Universe (WPI), The University of Tokyo, Chiba 277-8583, Japan
[9]Kyushu University, Fukuoka 812-8581, Japan
[10]RIKEN, Saitama 351-0198, Japan
[11]Theoretical Division, Los Alamos National Laboratory, Los Alamos, NM 87545, USA
[12]University of Notre Dame, Notre Dame, IN 46556, USA

Abstract. The short-lived unstable isotopes with half-lives of 0.1−10 My have been used as nuclear cosmochronometers to evaluate from an astrophysical event such as supernova (SN) explosion or AGB s-process to the solar system formation. We have proposed shorted-lived radioisotopes of ^{92}Nb and ^{98}Tc as the nuclear cosmochronometers for supernova neutrino-process

1 Introduction

A huge number of neutrinos emitted in core-collapse supernova (SN) explosions (ν process) [1] play an important role in stellar nucleosyntheses of rare some nuclides such as ^{7}Li, ^{11}B, ^{19}F, ^{138}La, and ^{180}Ta [1–4]. When the high-energy neutrinos pass through the outer layers of the star they can induce nuclear reactions on pre-existing nuclei. Many nuclides are, in principle, generated by the ν process in SNe but the produced abundances are smaller than production by other major processes such as the s or r process by a few orders of magnitude. Thus, the ν process can only play a significant role in the synthesis of a rare isotope when the isotope is not produced by the major processes.

Short-lived unstable isotopes with half-lives of 10^6−10^8 y have been used as nuclear cosmochronometers to evaluate the time from an astrophysical event such as an AGB s-process or a SN explosion to the solar system formation (SSF) [5–8]. The unstable isotope ^{92}Nb decays to the daughter nucleus ^{92}Zr by β decay with a half-life of 3.47×10^7 y. Although ^{92}Nb does not naturally exist at the present solar system, its existence at the SSF has been

*e-mail: hayakawa.takehito@qst.go.jp

found by analysis of primitive meteorites [9–13]. Thus, ^{92}Nb has the potential to be used as a nuclear cosmochronometer for a nucleosynthesis episode which produces ^{92}Nb. However, the astrophysical origin of ^{92}Nb has not been established. Hayakawa et al. [14] have proposed the ν process origin for ^{92}Nb. Furthermore, the radioisotope ^{98}Tc ($T_{1/2} = 4.2\times10^6$ y) is another candidate for the ν-process cosmochronometer [15], although only an upper limit of ^{98}Tc/^{98}Ru $< 6\times10^{-5}$ has been reported [16] for the ^{98}Tc initial abundance at the SSF.

2 Supernova ν-process calculation

There are six species of neutrinos: electron neutrinos, muon neutrinos, tau neutrinos and their anti-neutrinos. The neutrino-induced reactions can be classified into three groups: the neutral current (NC) reaction with all six neutrinos, the charged current (CC) reaction with electron neutrinos, and the CC reaction with electron anti-neutrinos [15]. Previous studies for ^{92}Nb [14], ^{138}La, and ^{180}Ta [1, 2] have shown that individual ν-process isotopes are predominantly synthesized by the CC reaction with ν_e and the NC reaction. Figure 1 shows a partial nuclear

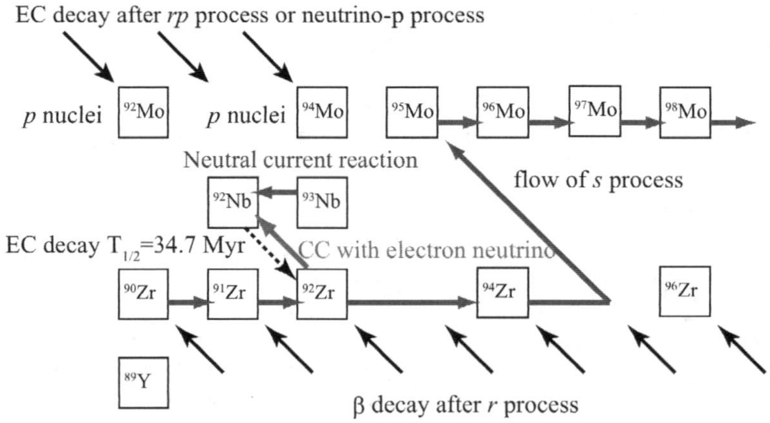

Figure 1. Nucleosynthesis flow and key nuclear reactions around ^{92}Nb

chart and nucleosynthesis flows around ^{92}Nb. ^{92}Nb is predominantly generated by the CC reaction with ν_e on ^{92}Zr and it is also produced by the NC reaction on ^{93}Nb. Figure 2 shows nucleosynthesis flows for ^{98}Tc. Among the CC reactions with ν_e, the ^{98}Mo(ν_e, e$^-$)^{98}Tc reaction is the dominant reaction. There are two NC reactions: ^{99}Ru(ν, ν'p)^{98}Tc and ^{99}Tc(ν, ν'n)^{98}Tc. One of the remarkable features for ^{98}Tc production is that ^{98}Tc is also produced by the CC reaction with $\overline{\nu_e}$ though the ^{99}Ru($\overline{\nu}_e$, e$^+$n)^{98}Tc and ^{100}Ru($\overline{\nu}_e$, e$^+$2n)^{98}Tc reactions. This suggests that the ^{98}Tc abundance may be sensitive to the average energy of the electron anti-neutrinos. We have performed calculations of the neutrino-induced reaction cross sections using a QRPA model [17] and the branching ratios are calculated using a Hauser-Feshbach calculation with a CCONE nuclear reaction calculation code [18]. We have calculated ν-process production rates using a core-collapse SN model for SN 1987A with an kinetic energy of 10^{51} erg [19]. We have used a 20 $M_\odot$ progenitor with a 6 $M_\odot$ He core with a metallicity of $Z_\odot/4$. Because the neutron-induced reaction cross sections in the proton rich-side have not been well studied, we have calculated the neutron capture cross sections in this mass region [21]. We have calculated evolution of the progenitor star including the weak *s*-processes [20] with the calculated neutron capture reactions. The neutrino flux decays exponentially with a time constant of 3 s. The six neutrino species can be treated as three groups: electron neutrino, electron

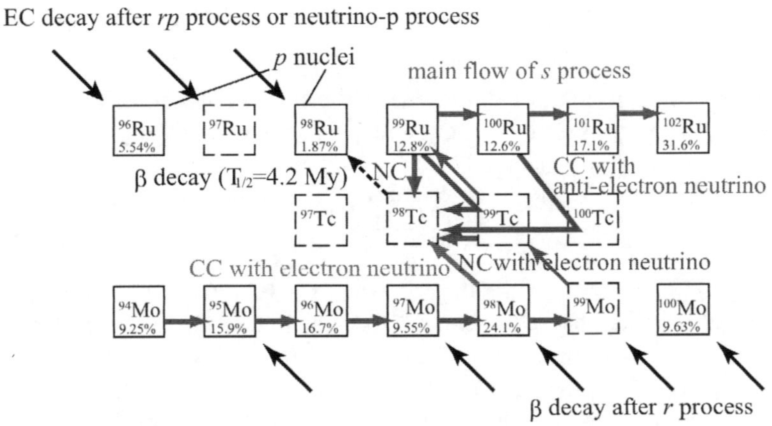

Figure 2. Nucleosynthesis flow and key nuclear reactions around ^{98}Tc

anti-neutrino, and the other four neutrinos. Previous studies [22] for the energy spectra of the neutrinos have suggested the following energy hierarchy: $\langle\nu_e\rangle < \langle\overline{\nu}_e\rangle < \langle\nu_{\mu,\tau}, \overline{\nu}_{\mu,\tau}\rangle$. In the present calculation, we adopt average energies of kT = 3.2, 5.0, 6.0 MeV for $\langle\nu_e\rangle$, $\langle\overline{\nu}_e\rangle$, and $\langle\nu_{\mu,\tau}, \overline{\nu}_{\mu,\tau}\rangle$, respectively.

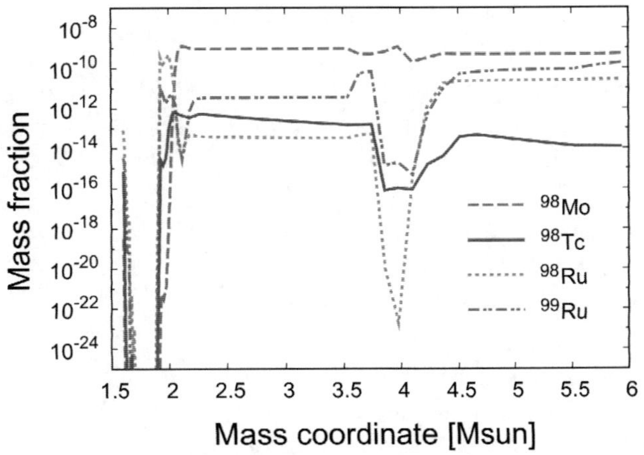

Figure 3. Calculated abundances. The solid line indicates the abundance of ^{98}Tc. The dashed, dotted, dashed double-dotted lines are ^{98}Mo, ^{98}Ru, and ^{99}Ru, respectively.

Figure 3 shows the calculated abundances. Integrating the layers within the mass range of $1.8 < M < 3.7$, we obtain masses of 5.1×10^{-13} and 3.4×10^{-11} $M_\odot$ for ^{98}Tc and ^{98}Ru, respectively. The contribution from the CC reactions with electron anti-neutrinos is relatively large compared to that of other heavy ν-process isotopes. The integrated mass fraction of ^{98}Tc decreases by approximately 20% compared to one with all six neutrino spices without the CC reactions with electron anti-neutrinos. ^{98}Tc is the most sensitive to the temperature of the electron anti-neutrinos among heavy elements because the contribution of the CC reaction with electron anti-neutrinos to ^{92}Nb, ^{138}La, and ^{180}Ta was considered to be negligibly small in the previous studies.

3 Age from the last SN to SFF

It is assumed that short-lived unstable isotopes are produced by a nearby SN before the SSF and subsequently they are mixed with the collapsing protosolar cloud. The isotopic abundance ratio at the time of SSF can then be expressed as

$$\left[\frac{^{98}\mathrm{Tc}}{^{98}\mathrm{Ru}}\right]_{SSF} = \frac{fN(^{98}\mathrm{Tc})_{SN}e^{-\Delta/\tau 98}}{N(^{98}\mathrm{Ru})_{\odot} + fN(^{98}\mathrm{Ru})_{SN}}\ , \tag{1}$$

where $N(^{98}\mathrm{Tc})_{SN}$ and $N(^{98}\mathrm{Ru})_{SN}$ are the numbers of ^{98}Tc and ^{98}Ru, respectively, in the SN ejecta, $N(^{98}\mathrm{Ru})_{\odot}$ is the number of the initial ^{98}Ru nuclei in the collapsing cloud, Δ is the time from the SN to SSF, and f is the dilution fraction. The timescales Δ in the range of $3\times10^{7}-10^{8}$ y have been previously estimated from several short-lived radioisotopes [7]. The dilution factor has been estimated to the values from 7×10^{5} to 2×10^{3}. The initial solar abundance of ^{92}Nb has been reproduced using the SN ν-process model with the parameters of $\Delta = 10^{6}$ (or 3×10^{7} y) and $f = 3\times10^{-3}$ [14]. The ^{98}Tc/^{98}Ru ratios calculated using the ν-process model and $f = 3\times10^{-3}$ are ^{98}Tc/^{98}Ru $= 1.3\times10^{-5}$ and 1.1×10^{-7} for $\Delta = 10^{6}$ and 3×10^{7} y, respectively. These calculated ratios are lower than the measured upper limit of ^{98}Tc/^{98}Ru $< 6\times10^{-5}$ [16]. Thus, it is possible to explain both the initial abundances of ^{92}Nb and ^{98}Tc by the contribution of a single SN ν-process.

References

[1] S. E. Woosley, *et al* Astrophys. J. **356**, 272 (1990).
[2] A. Heger, *et al*, Phys. Lett. B **606**, 258 (2005).
[3] T. Yoshida, T. Kajino, D. H. Hartmann, Phys. Rev. Lett. **94**, 231101 (2005).
[4] S. M. Austin, A. Heger, and C. Tur, Phys. Rev. Lett. **106**, 152501 (2011).
[5] G. J. Wasseerburg, M. Busso, R. Gallino, Astrophys. J. **466**, L109 (1996).
[6] B. S. Meyer and D. D. Clayton, Space Sci. Rev. **92**, 133 (2000).
[7] N. Dauphas, T. Rauscher, B. Marty, L. Reisberg, Nucl. Phys. A **719**, 287c (2003).
[8] M. Lugaro, *et al*, PNAS, **113**, 907 (2016).
[9] C. L. Jr., Harper, Astrophys. J. **466**, 437 (1996).
[10] Q. Z., Yin, *et al.* Astrophys. J. **536**, L49 (2000).
[11] M. Schönbächler, *et al.* Science **295**, 1705 (2002).
[12] T. Iizuka, *et al.* Earth Plan. Sci. Lett. **439**, 172 (2016).
[13] M.L. Haba, *et al.* PNAS, **118 (8)**, e2017750118 (2021).
[14] T. Hayakawa, *et al.* Astrophys. J. **779**, L9 (2013).
[15] T. Hayakawa, *et al.* Phys. Rev. Lett. **121**, 102701 (2018).
[16] H. Becker, R. J. Walker, Chem. Geol. **196**, 43 (2003).
[17] M. -K. Cheoun, *et al.* Phys. Rev. C, **85**, 065807 (2012).
[18] O. Iwamoto, *et al.* Nucl. Data sheets, **131**, 259 (2016).
[19] T. Shigeyama, K. Nomoto, M. Hashimoto, Astrophys. Astronm. **196**, 141 (1988).
[20] Y. Kikuchi, *et al.* Prog. Theor. Exp. Phys. **2015(6)**, 063E01 (2015).
[21] T. Kawano, *et al.* J. Nucl. Sci. Technol., **47**, 462 (2010).
[22] T. Fischer, *et al*, Phys. Rev. D **85**, 083003 (2012).

Analysis of the impact of the ^{204}Tl neutron capture cross section on the *s*-process only isotope ^{204}Pb

Adrià Casanovas-Hoste[1,2,*], *César* Domingo-Pardo[2], *Carlos* Guerrero[3], *Ariel* Tarifeño-Saldivia[1], *Francisco* Calviño[1], and *Jorge* Lerendegui-Marco[3]

[1]Institut de Tècniques Energètiques, Universitat Politècnica de Catalunya, Spain
[2]Institut de Física Corpuscular, CSIC-Universitat de València, Spain
[3]Universidad de Sevilla, Spain

Abstract. In the study of the slow (s) process of nucleosynthesis branching nuclei become of particular interest. These nuclei have half-lives of the order of 1-100 years, and in the stellar environment their decay rate can compete with the neutron capture rate, which effectively leads to a split of the *s*-process path. Due to the dependence of the decay and the neutron capture rates on the the physical conditions -temperature and neutron density- of the nucleosynthesis environment, variations in these conditions lead to a change in the abundances of the immediately following nuclei. A very interesting branching point is the *s*-process only ^{204}Tl, which decays to ^{204}Pb. In this work we show how the capture cross section of ^{204}Tl is a key nuclear input which, in addition to being crucial in fixing the ultimate ^{204}Pb *s*-process abundance, makes the latter sensible to the temperature and neutron density of the stellar environment where the *s*-process takes place.

1 Introduction

Approximately half of the solar elemental abundances heavier than iron are produced by the so-called slow (s) process of stellar nucleosynthesis. This process is characterized by successive neutron capture and beta decay reactions, following the path of the valley of nuclear stability. Elements with $A > 90$ are mainly produced by the *main* *s*-process, which takes place in the Asymptotic Giant Branch (AGB) phase of low mass ($1.5 - 3\ M_{\odot}$), thermally pulsing stars. During the AGB phase, *s*-process happens in two main stages, characterized by different physical conditions. Between two Thermal Pulse (TP) episodes, H is burned radiatively in a shell at the bottom of the convective envelope of the star. In the ashes of the H-burning shell, conditions are such that a ^{13}C pocket forms, and neutrons are released by the $^{13}C(\alpha,n)$ reaction. When temperature approaches $\sim 0.9 \cdot 10^8$ K (corresponding to a thermal energy of $kT \sim 8$ keV), neutron densities of $\sim 10^7$ cm^{-3} trigger the *s*-process chain. After some 40,000 years, the partially degenerated helium that has accumulated in a thin He intershell finally ignites fusion, leading to a short and intense increase in star luminosity –hence the TP name. During the TP, temperature in the H and He shells can rise up to $\sim 3 \cdot 10^8$ K (corresponding to $kT \sim 26$ keV) and a large convection zone is created. At the bottom of this convective zone the increase in temperature leads to a release of neutrons by the partial activation of the

*e-mail: adria.casanovas@cern.ch

^{22}Ne(α,n) reaction. Although these events last for a few years, neutron densities can reach up to $\sim 10^{10}$ cm^{-3}, changing the isotopic patterns around some particular isotopes. An up to date review of the *s*-process, including a more detailed analysis of the stellar scenarios, can be found in [1].

Along the *s*-process path, some unstable nuclei are produced, with half-lives from a few years to thousands of years. At these isotopes the neutron capture reactions may compete with beta-decays. In some cases, the high temperature reached during the TP can change considerably the capture-to-decay ratio. When this occurs, the usual path of the nucleosynthesis process is changed, thus leaving its fingerprint on the abundances of the neighbouring stable nuclei. Hence, by affecting the local isotopic ratios, these branching-point nuclei become probes of the stellar conditions where they are produced. For a quantitative study of the branching nuclei one needs to know both their β-decay half-lives and the neutron capture cross sections. While the former are quite straight-forward to measure, the determination of the cross section is very challenging. This is due to difficulties in producing a sufficiently large amount of sample material with a significant enrichment on the isotope of interest, and also to the background induced in most cases by the decay of the sample material itself during the experiment.

2 The ^{204}Tl branching point

^{204}Tl is a relevant branching point because it crucially determines the abundance of its daughter isotope, ^{204}Pb, and of the radioactive ^{205}Pb. Both isotopes are sheltered from any contribution from the *r*-process, and thus are largely (>90%) produced by the *s*-process [2][3]. ^{205}Pb, with a half-life of 17.3 My, has several cosmochemical applications [4]. In addition it has the potential to be used as a chronometer of the last *s*-process events that contributed to the Solar System composition[5][6]. The strength of a branching point like ^{204}Tl can be quantified by the *branching factor* f_β,

$$f_\beta = \frac{\lambda_\beta}{\lambda_n + \lambda_\beta}, \tag{1}$$

where $\lambda_\beta = \ln(2)/t_{1/2}$ is the beta decay rate, and $\lambda_n = n_n \langle \sigma_A \rangle v_T$ is the neutron capture rate. The quantity $\langle \sigma_A \rangle$ is defined as the capture cross section averaged by the Maxwellian spectrum of the neutrons with thermal velocity v_T, or MACS. For values close to 1 of f_β, most of ^{204}Tl will decay to ^{204}Pb whereas for low f_B values neutron capture will dominate, leading to a decrease in the production of ^{204}Pb. Hence, the abundance of the latter is directly related to the strength of the branching ratio. More interestingly, the branching ratio itself is very sensitive to the physical conditions, that is, temperature and neutron density, that characterize the *s*-process. In the first place, the half-life of the decay of ^{204}Tl is predicted to decrease dramatically as temperature rises [7]. On the other hand, the strength of the neutron sources, especially the ^{22}Ne(α,n) reaction, is also strongly dependent on temperature. However, in order to extract information on the *s*-process physical conditions, it is necessary to know as accurately as possible the nuclear inputs associated to the decay and capture rates, that is, the stellar half-life of ^{204}Tl and the MACS.

In order to understand the *s*-process behaviour at the ^{204}Tl branching, we carried out a sensitivity study using previously available theoretical estimates of the ^{204}Tl cross section. In this first step we could study the dynamics of the ^{204}Tl branching –and the ^{204}Pb resulting abundance– as a function of the ^{204}Tl cross section during the two stages of the *s*-process. For that, we have employed the post-processing nucleosynthesis tools and AGB models provided by the NuGrid Collaboration [8]. The calculation has been performed for a full sequence of a ^{13}C pocket and the following TP events. Concerning the input data for the ^{204}Tl cross section

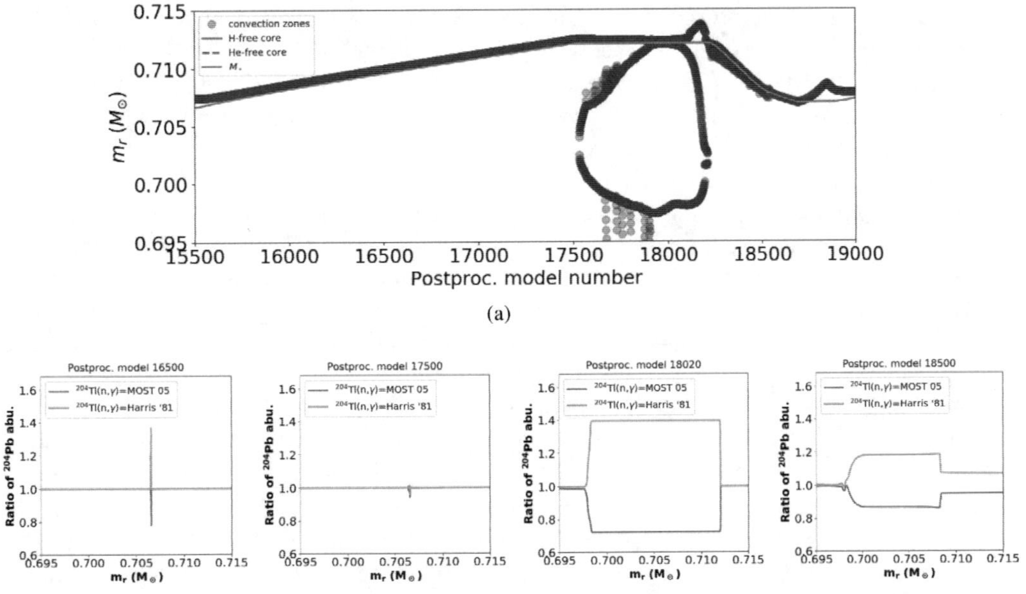

Figure 1: (a) Kippenhahn diagram of the evolution of the structure of the $3M_\odot$, $Z = 0.006$ star model employed in the analysis of the ^{204}Tl branching provided in the text. The bottom panels show the ^{204}Pb abundance profile of the star for the two extreme cases of the ^{204}Tl(n, γ) cross section mentioned in the text, which have been normalized to the Kadonis reference value. The profiles are taken at four specific moments of the star evolution: (b) during the ^{13}C-pocket (cycle 16500), (c) at the end of it and prior to the TP event (cycle 17500), (d) at the time of the highest neutron irradiation in the TP (cycle 18020), and (e) during the Third Dredge-up (cycle 18500).

for the calculations, theoretical values were employed since there was no experimental data before the results reported later in this contribution. As reported in the Kadonis database [9], the theoretical calculations of the cross section differ by more than a factor of three, spanning between 97 mb [10] and 328 mb [11]. This situation is in contrast with the ~18% relative uncertainty ascribed to this cross section in Kadonis, and most probably indicates a large underestimation of the true uncertainty. We have selected these values as extreme cases in our nucleosynthesis calculations, with the recommended value by Kadonis as the reference value. Results of a nucleosynthesis calculation of a full ^{13}C-pocket and the following TP event are depicted in Figure 1 for a stellar model with 3 $M_\odot$ and a metallicity of $Z = 0.006$, approximately equivalent to half solar. In particular, the plot in figure 1a, known as Kippenhahn diagram, shows the time evolution of the structure of the star; the y-axis values correspond to its radius, expressed by the mass that they enclose (in units of solar masses), and the values of the x-axis correspond to post-processing cycles. Figures 1b to 1e show the stellar profile of the abundance of ^{204}Pb at different times during the ^{13}C-pocket event and the subsequent TP for the different ^{204}Tl cross sections mentioned before, and normalized to the reference value of Kadonis. During the first stages of the ^{13}C-pocket event (Figure 1b), important variations in the ^{204}Pb abundance due to the different cross section are observed. A lower ^{204}Tl(n, γ) cross section leads to a ^{204}Tl buildup and consequently a higher ^{204}Pb abundance. However, the long time span and low neutron exposure during the ^{13}C-pocket event ensures that the equilibrium of nucleosynthesis flow is maintained, and any excess –or defect– of the ^{204}Pb abundance tends to be counterbalanced (Figure 1c). On the other hand, in the TP *s*-process episodes, and despite the predicted reduction in the ^{204}Tl half-life, the neutron density is high enough to lead to values of $f_\beta \sim 0.5$. The high neutron capture rate makes the abundance of ^{204}Pb during these episodes especially sensitive to the capture cross section of ^{204}Tl, as can be seen in Figure 1d. Finally, when the TP is over and the subsequent third dredge-up occurs, the material synthesized during the preceding TP is directly injected into the stellar envelope (Figure 1e). Since the AGB phase is characterized by several ^{13}C leading to a differentiation of the ^{204}Pb abundance as a function of the ^{204}Tl capture cross section. At the end of the AGB phase the cumulative effect of the contributions of the successive ^{13}Cplus TP events (10 for the model employed in this work) leads to a differentiation of around 30%. for the two extreme cases of the ^{204}Tl cross section analyzed here.

The same analysis has been carried out for a star model of the same metallicity, but a lower mass of 1.65 $M_\odot$. The differentiation in the abundance of ^{204}Pb obtained is lower, which we ascribe not only to the possible different number of TP events, but also to the much higher intensity of the thermal pulses –in terms of temperatures and neutron density– in the more massive star. A more detailed analysis of the differentiation of the ^{204}Pb abundance as a function of the stellar model is being currently performed, and is expected to be published in the near future. In any case, the crucial role of the ^{204}Tl neutron capture cross section in determining the abundance of ^{204}Pb has been highlighted. Up to now, a more accurate and precise determination of the *s*-process dominant contribution to the ^{204}Pb abundance has been hampered by the unknown ^{204}Tl neutron capture cross section. The first experimental measurement of the ^{204}Tl(n, γ) cross section was performed at the neutron Time-of-flight facility, n_TOF, at CERN, in the year 2015. Details of the sample production, the experimental methods and the analysis procedure can be found in [12] and in [13]. This experiment has provided the first ever direct measurement of neutron capture cross resonances on ^{204}Tl. The first experimental determination of the MACS, based on these new results, is currently under way. The new MACS will be employed to constrain the *s*-process contribution to the primordial ^{204}Pb abundance. Current nucleosynthesis models predict a deficiency in the *s*-process abundance of ^{204}Pb [14]. If this deficiency is confirmed, it could be an indication of additional processes contributing to the observed abundance of ^{204}Pb, such as fractionation effects [15] and/or contributions from the γ-process [16].

References

[1] F. Käppeler, R. Gallino, S. Bisterzo, W. Aoki, Rev. Mod. Phys. **83**, 157 (2011)

[2] C. Arlandini, F. Kappeler, K. Wisshak, R. Gallino, M. Lugaro, M. Busso, O. Straniero, Astrophys. J. **525**, 886 (1999)

[3] C. Travaglio, R. Gallino, M. Busso, R. Gratton, Astrophys. J. **549**, 346 (2001)

[4] C. Palk, R. Andreasen, M. Rehkämper, A. Stunt, K. Kreissig, B. Coles, M. Schönbächler, C. Smith, Meteorit. Planet. Sci. **53**, 167 (2018), `https://onlinelibrary.wiley.com/doi/pdf/10.1111/maps.12989`

[5] K. Yokoi, K. Takahashi, M. Arnould, Astron. Astrophys. **145**, 339 (1985)

[6] R. Baker, M. Schönbächler, M. Rehkämper, H. Williams, A. Halliday, Earth Planet. Sci. Lett. **291**, 39 (2010)

[7] K. Takahashi, K. Yokoi, At. Data Nucl. Data Tables **36**, 375 (1987)

[8] C. Ritter, F. Herwig, S. Jones, M. Pignatari, C. Fryer, R. Hirschi, Mon. Not. R. Astron. Soc. **480**, 538 (2018)

[9] I. Dillmann, M. Heil, F. Käppeler, R. Plag, T. Rauscher, F. Thielemann, AIP Conference Proceedings **819**, 123 (2006), http://www.kadonis.org

[10] M.J. Harris, Astrophys. Space Sci. **77**, 357 (1981)

[11] S. Goriely, *Hauser-feshbach rates for neutron capture reactions (version 9/12/2002)*

[12] A. Casanovas, A.E. Tarifeño-Saldivia, C. Domingo-Pardo, F. Calviño, E. Maugeri, C. Guerrero, J. Lerendegui-Marco, R. Dressler, S. Heinitz, D. Schumann et al., J. Phys. Conf. Ser. **1668**, 012005 (2020)

[13] A. Casanovas-Hoste, Ph.D. thesis (2020), presented 27 Oct 2020, https://cds.cern.ch/record/2746344

[14] S. Bisterzo, R. Gallino, F. Käppeler, M. Wiescher, G. Imbriani, O. Straniero, S. Cristallo, J. Görres, R.J. deBoer, Mon. Not. R. Astron. Soc. **449**, 506 (2015), 1507.06798

[15] G. Gonzalez, Mon. Not. R. Astron. Soc. Lett. **443**, L99 (2014), https://academic.oup.com/mnrasl/article-pdf/443/1/L99/9420491/slu083.pdf

[16] F. Käppeler, H. Beer, K. Wisshak, Rep. Prog. Phys. **52**, 945 (1989)

The impact of isospin dependence of pairing on fission barriers in the fission cycling regions

A. V. Afanasjev[1,*] and *A.* Taninah[1]

[1]Department of Physics and Astronomy, Mississippi State University, MS 39762

Abstract. A systematic analysis of the ground state and fission properties of actinides and superheavy nuclei important for the *r* process modeling has been performed within the framework of covariant density functional theory for the first time in Ref. [1]. A brief review of the results related to the heights of primary fission barriers and systematic uncertainties in their prediction is presented. In addition, new results on the potential impact of the isospin dependence of pairing on fission barriers in fission cycling regions is provided for the first time.

1 Introduction

The *r* process is responsible for the synthesis of approximately half of the nuclei in nature beyond Fe [2] and it is the only process which leads to the creation of nuclei heavier than Bi [3] . Fission becomes important in the *r* process simulations for the neutron-to-seed ratios which are large enough to produce fissioning nuclei [4, 5]. The *r* process can reach the region beyond neutron shell closure at $N = 184$ for these ratios exceeding 100: the fission plays a dominant role in this region. Fission leads to the termination of the hot *r* process by means of *fission cycling* which returns matter to lighter nuclei [4, 5]. It also determines the strength of fission cycling, the ratio of the actinides to light and medium mass *r* process nuclei, and thus the shape of the final element abundance pattern. In addition, it defines the possibility of the formation of neutron-rich superheavy nuclei in the *r* process [6].

The outcome of the r-process modeling sensitively depends on the quality of employed theoretical frameworks and associated theoretical uncertainties and their propagation on going to neutron-rich nuclei in the situation when experimental data are not known. So far only non-relativistic frameworks have been used in such modeling and in the analysis of fission cycling (see review in the introduction of Ref. [1]). The first attempt to produce nuclear input required for the *r* process modeling within the relativistic framework (covariant density functional theory (CDFT) [7]) has been carried out in Ref. [1]. The goal of this contribution is to briefly review the results of this study with major focus on fission properties and to analyze potential impact of isovector dependence of pairing on the fission barriers of very neutron-rich nuclei.

2 Fission barriers and related systematic uncertainties

The distributions of primary fission barrier (PFB) heights in the (Z, N) plane obtained in the axial relativistic Hartree-Bogoliubov (RHB) calculations are shown in Fig. 1. The calcu-

*e-mail: aa242@msstate.edu

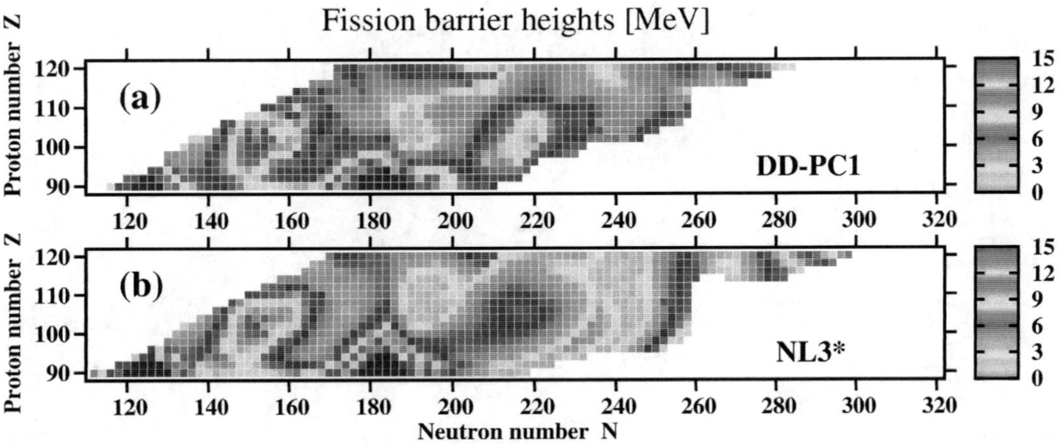

Figure 1. The heights E^B (in MeV) of PFBs obtained in axial RHB calculations as a function of proton and neutron numbers for nuclei located between two-proton and two-neutron drip lines. Based on Fig. 15 of Ref. [1].

lations have been performed with four state-of-the-art covariant energy density functionals (CEDFs) DD-PC1, DD-ME2, NL3* and PC-PK1 in order to evaluate systematic theoretical uncertainties in the predictions of fission barriers (see Ref. [1] for details). There is a large similarity of the results obtained with DD-PC1 and DD-ME2 on the one hand and NL3* and PC-PK1 on the other: thus only those obtained with DD-PC1 and NL3* are shown. In general, the topologies of the fission barrier maps shown in Fig. 1 are similar. However, the fission barriers obtained with DD-PC1 are on average higher by approximately 2 MeV as compared with those calculated with NL3*. This is a consequence of different nuclear matter properties of these two functionals (see discussion in Ref. [1]). These two functionals also differ with respect of the predictions of the formation of superheavy elements with $N > 240$ in the r process. This is because the band of the nuclei around $N \approx 240$ has extremely low fission barriers with heights of around 2 MeV in the NL3* CEDF (see Fig. 1(b)). The nuclear flow during most of the neutron irradiation step of the r process follows the neutron drip line and this flow will most likely be terminated at $N \approx 240$ nuclei because of these low fission barriers in the calculations with NL3*.

Theoretical systematic uncertainties in the heights of PFBs given by the spreads ΔE^B are shown as a function of proton and neutron numbers in Fig. 2. They are relatively modest in some regions but are enhanced near the $N = 184$ and $N = 258$ shell closures, for the $Z \approx 90$ nuclei with $N = 166 - 184$ and in the wide band of nuclei parallel to the two-neutron drip line. The analysis of these spreads allows us to identify two major sources of theoretical uncertainties in the predictions of the heights of PFBs (see Ref. [1] for details). These are underlying single-particle structure (especially the one in the vicinity of shell closures) and nuclear matter properties of employed CEDFs. The former mostly affects the predictions for the ground states (and thus for the heights of PFBs) in the first two regions and the latter the predictions for PFBs of the nuclei located in the vicinity of the neutron drip line.

3 The impact of isospin dependence of pairing on fission barriers

The systematic calculations of Ref. [1], the results of which are presented in Figs. 1 and 2, have been performed with separable pairing of Ref. [9] the scaling factors f_i of which for the

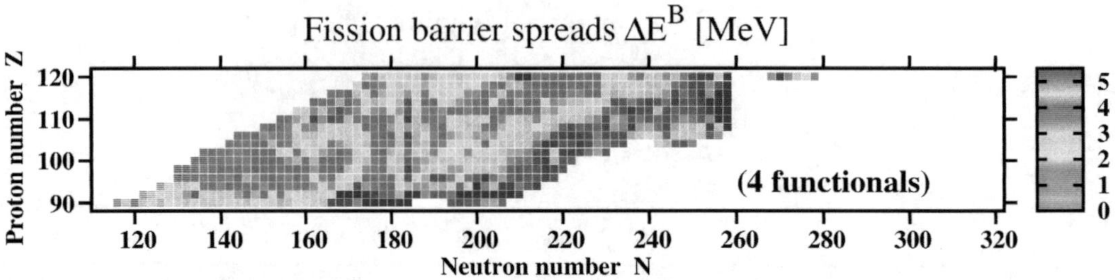

Figure 2. The spreads ΔE^B of the heights of PFBs as a function of proton and neutron numbers. $\Delta E^B(Z,N) = |E^B_{max}(Z,N) - E^B_{min}(Z,N)|$, where, for given Z and N values, $E^B_{max}(Z,N)$ and $E^B_{min}(Z,N)$ are the largest and smallest heights of PFBs obtained with the employed set of four functionals. Based on Fig. 15 of Ref. [1].

Table 1. Different versions (v1, v2 and v3) of separable pairing interaction as defined by the particle number dependencies of their scaling factors f_i ($i = \pi$ or ν). The constants C_i and α_i are taken from Table 4 of Ref. [8].

subsystem	v1	v2	v3
proton	$f_\pi = 1.0$	$f_\pi = C_\pi * (N+Z)^{\alpha_\pi}$	$f_\pi = C_\pi * e^{\alpha_\pi \lvert N-Z \rvert}$
neutron	$f_\nu = 1.0$	$f_\nu = C_\nu e^{\alpha_\nu \frac{\lvert N-Z \rvert}{N+Z}}$	$f_\nu = C_\nu * \lvert N-Z \rvert^{\alpha_\nu}$
		$C_\pi = 1.877$, $\alpha_\pi = -0.1072$	$C_\pi = 1.178$, $\alpha_\pi = -0.0026$
		$C_\nu = 1.208$, $\alpha_\nu = -0.674$	$C_\nu = 1.264$, $\alpha_\nu = -0.0495$

nuclei with $Z > 88$ are set to $f_\pi = f_\nu = 1.0$ (see Ref. [10] for definition of pairing strength). This pairing is labeled as "v1" below, see Table 1. However, in general more complicated particle number dependencies of scaling factors f_i are allowed (see Refs. [8, 10]). Indeed, recent systematic analysis of Ref. [8] clearly reveals isospin dependence of scaling factors f_ν of neutron pairing. However, the situation is less certain in the proton subsystem since similar accuracy of the description of pairing indicators can be achieved both with isospin-dependent and mass-dependent scaling factors f_π.

The fission barrier heights sensitively depends on the strength of pairing interaction (see Ref. [11]). Thus, it is important to understand by how much and in which direction the fission barriers can be affected by these particle number dependencies of pairing interaction. For that the primary fission barriers of very neutron-rich ^{316}Fm and ^{324}Rf nuclei have been calculated with three versions (v1, v2 and v3) of scaling factors for separable pairing (see Table 1). These nuclei are located at the neutron-rich side of expected fission cycling region (see Fig. 1 in Ref. [1]). The versions v2 and v3 are the combinations of proton and neutron scaling factors favored by the analysis of Sec. IV of Ref. [8]. One can see in Table 2 that in all cases the inclusion of isospin dependence of pairing increases the heights of primary fission barriers but the magnitude of the increase depends both on the CEDF and nucleus. The latter feature is mostly due to the differences between the functionals in underlying single-particle structure. The magnitude of the increase varies from $E^B_{v2} - E^B_{v1} = 0.25$ MeV to $E^B_{v3} - E^B_{v1} = 1.57$ MeV (see DD-ME2 results in ^{324}Rf). This large difference in ^{324}Rf is caused by substantially weakened proton pairing in the calculations with the v3 version of separable pairing which according to Ref. [11] leads to a significant increase of fission barrier height. However, in most of the cases the difference between $E^B_{v2} - E^B_{v1}$ and $E^B_{v3} - E^B_{v1}$ is relatively small being typically around 0.2 MeV and the magnitude of these two quantities is around 1 MeV. Another observation is that the v3 version of separable pairing leads to somewhat

Table 2. Fission barriers heights E^B [in MeV] for the indicated nuclei, functionals and versions of separable pairing. The columns 6 and 7 show the increases of fission barrier heights for two isospin dependent pairing interactions (v2 and v3) as compared to the one obtained with pairing of constant strength (v1).

Nucleus	CEDF	E^B_{v1}	E^B_{v2}	E^B_{v3}	$E^B_{v2} - E^B_{v1}$	$E^B_{v3} - E^B_{v1}$
1	2	3	4	5	6	7
^{316}Fm	NL3*	5.71	6.49	6.38	0.78	0.756
	DD-PC1	8.83	9.11	9.45	0.28	0.62
	DD-ME2	9.77	10.36	10.53	0.59	0.76
^{324}Rf	NL3*	6.39	7.06	7.23	0.67	0.84
	DD-PC1	9.08	10.44	10.58	1.36	1.50
	DD-ME2	10.44	10.69	12.01	0.25	1.57

higher fission barriers as compared with the ones obtained with v2 version. Note that these nuclei are extremely neutron-rich with N/Z ratio exceeding 2.0. Thus, it is reasonable to expect that the magnitude of the increase of fission barriers due to isospin dependence of pairing will be lower in the nuclei located closer to the β-stability line.

4 Conclusions

The detailed analysis of the ground state and fission properties of actinides and superheavy nuclei important for the r process modeling has been performed within the framework of CDFT for the first time (see Ref. [1]). In particular, it allowed to establish the systematic uncertainties in the heights of primary fission barriers and their sources. In addition, the present investigation reveals isospin dependence of pairing as an additional factor affecting fission barriers in fission cycling regions. Its inclusion leads to a substantial increase of fission barriers in very neutron rich nuclei. Available covariant [8] and non-relativistic Skyrme [12, 13] DFT investigations strongly point to the existence of isospin dependence of effective pairing interaction. However, its details and accurate form are still under debate.

This material is based upon work supported by the U.S. Department of Energy, Office of Science, Office of Nuclear Physics under Grant No. DE-SC0013037.

References

[1] A. Taninah, S. E. Agbemava and A. V. Afanasjev, Phys. Rev. C 102, 054330 (2020).
[2] G. Martínez-Pinedo, Eur. Phys. J.: Spec. Top. 156, 123 (2008).
[3] V. Liccardo *et al*, Eur. Phys. J. A54, 221 (2018).
[4] G. Martínez-Pinedo *et al*, Prog. Part. Nucl. Phys. 59, 199 (2007).
[5] S. Goriely, Eur. Phys. J. A51, 22 (2015).
[6] I. Petermann *et al*, Eur. Phys. J. A 48, 122 (2012).
[7] D. Vretenar, A. V. Afanasjev, G. A. Lalazissis, and P. Ring, Phys. Rep. 409, 101 (2005).
[8] S. Teeti and A. V. Afanasjev, Phys. Rev. C 103, 034310 (2021).
[9] Y. Tian, Z. Y. Ma, and P. Ring, Phys. Lett. B 676, 44 (2009).
[10] S. E. Agbemava, A. V. Afanasjev, D. Ray, and P. Ring, Phys. Rev. C 89, 054320 (2014).
[11] S. Karatzikos *et al*, Phys. Lett. B 689, 72 (2010).
[12] C. A. Bertulani, H. Liu, and H. Sagawa, Phys. Rev. C 85, 014321 (2012).
[13] M. Yamagami *et al*, Phys. Rev. C 86, 034333 (2012).

Beyond-mean-field calculations of allowed and first-forbidden β^- decays of r-process waiting-point nuclei

Caroline Robin[1,2,*], *Elena* Litvinova[3,4], and *Gabriel* Martínez-Pinedo[1,5,6]

[1]GSI Helmholtzzentrum für Schwerionenforschung, Planckstraße 1, 64291 Darmstadt, Germany
[2]Fakultät für Physik, Universität Bielefeld, D-33615, Bielefeld, Germany
[3]Department of Physics, Western Michigan University, Kalamazoo, Michigan 49008, USA
[4]National Superconducting Cyclotron Laboratory, Michigan State University, East Lansing, Michigan 48824, USA
[5]Institut für Kernphysik (Theoriezentrum), Fachbereich Physik, Technische Universität Darmstadt, Schlossgartenstraße 2, 64298 Darmstadt, Germany
[6]Helmholtz Forschungsakademie Hessen für FAIR, GSI Helmholtzzentrum für Schwerionenforschung, Planckstraße 1, 64291 Darmstadt, Germany

Abstract. β-decay rates of neutron-rich nuclei, in particular those located at neutron shell closures, play a central role in simulations of the heavy-element nucleosynthesis and resulting abundance distributions. We present β-decay half-lives of even-even $N = 82$ and $N = 126$ r-process waiting-point nuclei calculated in the approach based on relativistic quasiparticle random phase approximation with quasiparticle-vibration coupling. The calculations include both allowed and first-forbidden transitions. In the $N = 82$ chain, the quasiparticle-vibration coupling has an important impact close to stability, as it increases the contribution of Gamow-Teller modes and improves the agreement with the available data. In the $N = 126$ chain, we find the decay to proceed dominantly via first-forbidden transitions, even when the coupling to vibrations is included.

1 Introduction

Modeling the r-process nucleosynthesis requires a huge amount of nuclear physics input, and in particular, the knowledge of β-decay rates for thousands of neutron-rich nuclei. Due to their very short lifetimes, most of these nuclei are not accessible experimentally and astrophysical simulations thus heavily rely on theoretical methods, which should be as universal and reliable as possible.

As of today the only approach able to provide global sets of β-decay rates is known as Quasiparticle Random Phase Approximation (QRPA). Such method, which corresponds to the small amplitude limit of the time-dependent mean-field approach, has the advantage to be applicable to a wide range of masses at relatively low numerical cost. QRPA, however, suffers well-known limitations related to a simplified treatment of nuclear correlations which restrict its reliability and predictive power, in particular in unexplored regions of the nuclear chart. Currently, three global sets of β-decay rates are available, which have been calculated within different QRPA frameworks. The earlier set based on the finite range droplet model and

*e-mail: c.robin@gsi.de

schematic interaction [1], and two more recent self-consistent and fully microscopic QRPA calculations based on a relativistic Langrangian [2] and non-relativistic Skyrme functional [3]. Overall the predictions of these three QRPA frameworks disagree to some degree, in particular on the contribution of the FF transitions in different mass regions, and on the order of magnitude of the total half-lives above $N > 126$ which are predicted to be substantially lower by Ref. [2] than by the other two approaches. This large spread in the prediction of β-decay half-lives is an important issue as it can yield large uncertainties in the predictions of elemental abundances [4]. For example, it was found in Refs. [2, 5] that the shortenings of half-lives in the heavy region yields a broadening of the third r-process abundance peak towards lower masses.

While further studies are needed to fully understand the origin of the discrepancies, it is desirable to improve the reliability of β-decay rates by developing theoretical methods which incorporate a more precise description of nucleonic correlations, while remaining applicable to a wide range of masses. In this paper we use the approach based on charge-exchange relativistic QRPA (RQRPA) that is extended to account for the coupling between nucleons and collective nuclear vibrations, as first developed in Ref. [6], and apply it to the description of allowed and first-forbidden β decays of $N = 82$ and $N = 126$ r-process waiting-point nuclei.

2 Quasiparticle-vibration coupling effect on allowed and first-forbidden β^- decay of $N = 82$ and $N = 126$ nuclei.

In this work we calculate the β-decay rates λ including both allowed Gamow-Teller (GT) and first-forbidden (FF) transitions according to Ref. [2, 7]:

$$\lambda = \frac{\ln 2}{K} \sum_{n, \Omega_n < \Delta M_{nH}} f(\Omega_n, Z+1) \, . \tag{1}$$

In Eq. 1, $K = 1644 \pm 2$ s [8], the index n denotes the states of the daughter nucleus $(Z+1, N-1)$ accessible via β-decay, Ω_n is the excitation energy of these states with respect to the parent ground state, $\Delta M_{nH} \simeq 0.78$ MeV is the neutron-Hydrogen mass difference and $f(\Omega_n, Z+1)$ is the so-called integrated shape function which takes the form

$$f(\Omega_n, Z+1) = \int_1^{W_n} dW W \sqrt{W^2-1} (W_n - W)^2 F(Z+1, W) \left[C_{allowed}(W) + C_{FF}(W) \right] \, , \tag{2}$$

where $W = E_e / m_e c^2$ denotes the electron energy in unit of the electron mass, $W_n \equiv (M_i - M_n)/(m_e c^2) = [(m_n - m_p) - \Omega_n]/(m_e c^2)$ is the mass difference between initial and final states, $F(Z+1, W)$ is the Fermi function [7], and $C_X(W)$ are the allowed and FF shape factors. In the case of allowed transitions $C_{allowed}(W)$ coincides with the GT reduced transition probabilities, while $C_{FF}(W)$ takes a complicated expression which depends on energy and transition amplitudes of $0^-, 1^-$ and 2^- FF modes (for more details see [2, 9]). In this work we adopt the bare value of the weak axial coupling constant $g_A \simeq -1.27641$ [10] which appears in the definition of $C_X(W)$.

The nuclear transition amplitudes are obtained within the approach based on the relativistic QRPA which is extended to account for the coupling between single quasiparticles and collective nuclear vibrations. For details on the formalism and numerical scheme, see Refs. [6, 9]. In this work we use the NL3 parametrization of the meson-nucleon Lagrangian [11] for the particle-hole channel of the static interaction and a monopole-monopole force for the isovector pairing channel [12]. The isoscalar pairing is only included dynamically via quasiparticle-vibration coupling and is not present in the static interaction [9].

We show in Fig. 1 the β-decay half-lives $T_{1/2} = \ln 2/\lambda$ of $N = 82$ and $N = 126$ neutron-rich nuclei obtained without and with quasiparticle-vibration coupling (QVC). The empty symbols show the half-lives obtained in the GT allowed approximation while the full symbols are the half-lives obtained with the contribution of FF transitions.

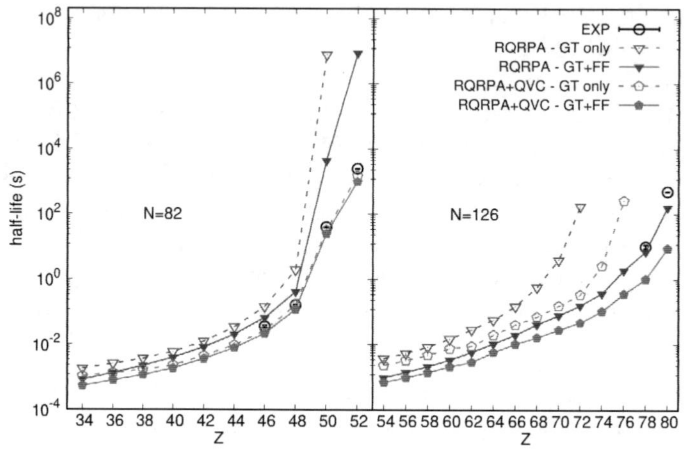

Figure 1. β-decay half-lives of $N = 82$ (left) and $N = 126$ (right) nuclei. We show the results obtained within RQRPA (green triangles) and RQRPA+QVC (red diamond) in the allowed GT approximation (empty symbols) and including FF transitions (full symbols). When no symbol is shown, the half-life is predicted infinite. The experimental data is taken from [13].

In both isotonic chains, the QVC correlations, which fragment the transition strength distributions [6], lead to a decrease of the half-lives which can be of several orders of magnitude compared to RQRPA. The effect is typically stronger close to stability, in particular in the $N = 82$ chain, as the decay Q value becomes small and is populated by very few states. In that case, the presence of correlations is crucial and leads to a much better agreement with the experimental data, although the decrease can be slightly too strong in some cases. Far from stability, however, the Q value is large so that the details of the transition strength distributions matter less and the effect of correlations on the half-lives becomes small. In the $N = 126$ chain the decrease of the half-lives due to QVC is of one order of magnitude at most when the FF transitions are included. In fact, in this case, RQRPA is closer to the data for $Z = 78$ and $Z = 80$. The reader should remember, however, that we did not include any quenching of the GT or FF transition matrix elements.

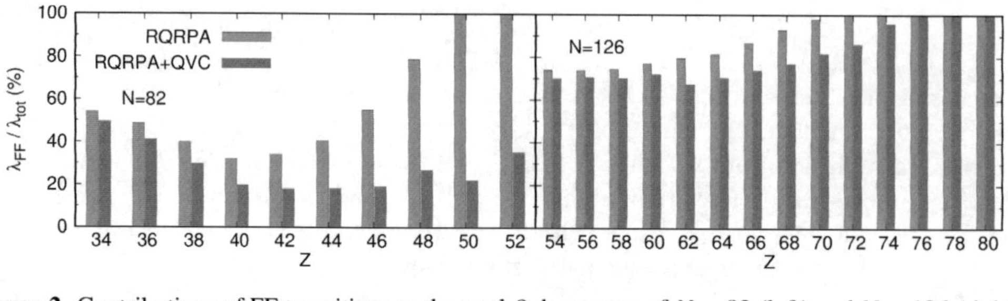

Figure 2. Contributions of FF transitions to the total β-decay rate of $N = 82$ (left) and $N = 126$ (right) nuclei, obtained within RQRPA (green boxes) and RQRPA+QVC (red boxes).

In Fig. 2 we show the contribution of FF transitions to the total decay rate (in percent). In the $N = 82$ isotonic chain, the contribution of FF modes increases far from stability from $Z = 40$ to $Z = 34$ where it reaches about 50 – 55% within both RQRPA and RQTBA. Examining the results in more details, we find that these FF transitions mainly occur via

0^- and 1^- modes. Going towards stability, the weight of FF transitions within RQRPA increases again until it reaches $\simeq$ 100% in $Z = 50 - 52$. In these nuclei the proton $1g_{9/2}$ is fully occupied and thus, the main low-energy GT transition $1g_{7/2} \rightarrow 1g_{9/2}$ is blocked. The decay then proceeds solely via 2^- FF transitions (mostly $1h_{11/2} \rightarrow 1g_{7/2}$). With QVC the contribution of FF modes remains below $\simeq$ 35%, which is of the same order of what was found in Shell-Model calculations [14]. Although the proton $1g_{9/2}$ is also filled for $Z \geq 50$, GT modes caused by transitions to other levels are lowered in energy due to the correlations, and contribute significantly to the decay due to the phase space. In the $N = 126$ chain, FF transitions dominate compared to GT modes. In RQRPA we find a contribution of $\simeq$ 75% for $Z = 54 - 58$ which then increases to 100% close to stability as the proton $1h_{11/2}$ sub-shell is filled. The trend remains similar when QVC correlations are included, apart from some variations. In this chain, Shell-Model studies have found a steeper increase of the FF contributions, from $\simeq 20 - 30\%$ to $\simeq 60 - 75\%$ in the range $Z = 66 - 72$ [14, 15]. Looking closer at the different forbidden transitions, we find that 0^- operators are responsible for the FF decay by > 75% along the chain, within RQRPA+QVC. Among those, the relativistic component (γ_5 operator) appear to be more important than the spin-dipole component.

3 Conclusion

We have calculated β^--decay rates of $N = 82$ and $N = 126$ isotonic chains within an extension of RQRPA accounting for QVC, including GT and FF transitions. We find the effect of QVC to be important near stability, in particular in the $N = 82$ chain, where it increases the contribution of GT modes. In $N = 126$ isotones, the FF decay remains dominant, even when QVC is included. In a future work we will examine the role of FF transitions in other magic isotonic and isotopic chains, and compare to different theoretical frameworks [9].

This work is supported by ERC-885281-KILONOVA Advanced Grant and US-NSF Career Grant PHY-1654379, and used resources of the National Energy Research Scientific Computing Center, a DOE Office of Science User Facility supported by the Office of Science of the U.S. Department of Energy under Contract No. DE-AC02-05CH11231.

References

[1] P. Möller, B. Pfeiffer, K.L. Kratz, Phys. Rev. C **67**, 055802 (2003)
[2] T. Marketin, L. Huther, G. Martínez-Pinedo, Phys. Rev. C **93**, 025805 (2016)
[3] E.M. Ney, J. Engel, T. Li, N. Schunck, Phys. Rev. C **102**, 034326 (2020)
[4] T. Shafer *et al.*, Phys. Rev. C **94**, 055802 (2016)
[5] M. Eichler *et al.*, The Astrophysical Journal **808**, 30 (2015)
[6] C. Robin, E. Litvinova, Eur. Phys. J. A **52**, 205 (2016)
[7] H. Behrens, W. Buehring, Nuclear Physics A **162**, 111 (1971)
[8] J.C. Hardy, I.S. Towner, Phys. Rev. C **79**, 055502 (2009)
[9] C. Robin, G. Martínez-Pinedo, E. Litvinova, in preparation
[10] B. Märkisch *et al.*, Phys. Rev. Lett. **122**, 242501 (2019)
[11] G.A. Lalazissis, J. König, P. Ring, Phys. Rev. C **55**, 540 (1997)
[12] E. Litvinova, P. Ring, V. Tselyaev, Phys. Rev. C **78**, 014312 (2008)
[13] IAEA LiveChart, https://www-nds.iaea.org/relnsd/vcharthtml/VChartHTML.html
[14] Q. Zhi *et al.*, Phys. Rev. C **87**, 025803 (2013)
[15] T. Suzuki, T. Yoshida, T. Kajino, T. Otsuka, Phys. Rev. C **85**, 015802 (2012)

Experimental studies on astrophysical reactions at the low-energy RI beam separator CRIB

H. Yamaguchi[1,*], *S.* Hayakawa[1], *N.R.* Ma[1], *H.* Shimizu[1], *K.* Okawa[1], *L.* Yang[1,2], *D.* Kahl[1,3,4], *M.* La Cognata[5], *L.* Lamia[5,6,7], *K.* Abe[1], *O.* Beliuskina[1,8], *S.M.* Cha[9,10], *K.Y.* Chae[9], *S.* Cherubini[5,6], *P.* Figuera[5], *Z.* Ge[11,8], *M.* Gulino[5,12], *J.* Hu[13], *A.* Inoue[14], *N.* Iwasa[15], *A.* Kim[16,17], *D.* Kim[10,16], *G.* Kiss[11,18], *S.* Kubono[11,13], *M.* La Commara[19,20], *M.* Lattuada[5,6], *E.J.* Lee[9], *J.Y.* Moon[21], *S.* Palmerini[22,23], *C.* Parascandolo[20], *S.Y.* Park[16], *V. H.* Phong[11,24], *D.* Pierroutsakou[20], *R.G.* Pizzone[5], *G.G.* Rapisarda[5], *S.* Romano[5,6,7], *C.* Spitaleri[5,6], *X.D.* Tang[13], *O.* Trippella[22,23], *A.* Tumino[5,12], *N.T.* Zhang[13], *Y.H.* Lam[13,25], *A.* Heger[26], *A.M.* Jacobs[27,28], *S.W.* Xu[13], *S.B.* Ma[13], *L.H.* Ru[13], *E.Q.* Liu[13], *T.* Liu[13], *C.B.* Hamill[3], *A.* St J. Murphy[3], *J.* Su[29], *X.* Fang[30], *M.S.* Kwag[9], *N.N.* Duy[9], *N.K.* Uyen[9], *D.H.* Kim[9], *J.* Liang[31], *A.* Psaltis[31,32], *M.* Sferrazza[33], *Z.* Johnston[27,28], and *Y.Y.* Li[34]

[1] Center for Nuclear Study (CNS), University of Tokyo, Wako, Japan
[2] China Institute of Atomic Energy, Beijing, China
[3] School of Physics and Astronomy, the University of Edinburgh, Edinburgh, UK
[4] Extreme Light Infrastructure – Nuclear Physics, Horia Hulubei National Institute for R&D in Physics and Nuclear Engineering (IFIN-HH), Bucharest-Măgurele, Romania
[5] Istituto Nazionale di Fisica Nucleare - Laboratori Nazionali del Sud, Catania, Italy
[6] Department of Physics and Astronomy, University of Catania, Catania, Italy
[7] Centro Siciliano di Fisica Nucleare e Struttura della Materia, CSFNSM, Catania, Italy
[8] Department of Physics, University of Jyväskylä, Jyväskylä, Finland
[9] Department of Physics, Sungkyunkwan University, Suwon, Korea
[10] Center for Exotic Nuclear Studies, Institute for Basic Science, Daejeon, Korea
[11] The Institute of Physical and Chemical Research (RIKEN), Wako, Japan
[12] Faculty of Engineering and Architecture, Kore University of Enna, Enna, Italy
[13] Institute of Modern Physics, Chinese Academy of Sciences, Lanzhou, China
[14] Research Center for Nuclear Physics (RCNP), Osaka University, Osaka, Japan
[15] Department of Physics, Tohoku University, Sendai, Japan
[16] Department of Physics, Ewha Womans University, Seoul, Korea
[17] Korea University, Seoul, Korea
[18] Institute for Nuclear Research (Atomki), Debrecen, Hungary
[19] Department of Physics, University of Naples Federico II, Naples, Italy
[20] Istituto Nazionale di Fisica Nucleare - Section of Naples, Naples, Italy
[21] Rare Isotope Science Project, Institute for Basic Science, Daejeon, Korea
[22] Istituto Nazionale di Fisica Nucleare - Section of Perugia, Perugia, Italy
[23] Department of Physics and Geology, University of Perugia, Perugia, Italy
[24] Faculty of Physics, VNU University of Science, Hanoi, Vietnam
[25] School of Nuclear Science and Technology, University of Chinese Academy of Sciences, Beijing, China
[26] School of Physics and Astronomy, Monash University, Victoria, Australia
[27] The Joint Institute for Nuclear Astrophysics, Michigan State University, East Lansing, MI, USA
[28] Department of Physics and Astronomy, Michigan State University, East Lansing, MI, USA

*e-mail: yamag@cns.s.u-tokyo.ac.jp

[29] College of Nuclear Science and Technology, Beijing Normal University, Beijing, China
[30] Sino-French Institute of Nuclear Engineering and Technology, Sun Yat-Sen University, Zhuhai, China
[31] Department of Physics & Astronomy, McMaster University, Ontario, Canada
[32] Technische Universität Darmstadt, Darmstadt, Germany
[33] Département de Physique, Université Libre de Bruxelles, Bruxelles, Belgium
[34] School of Nuclear Science and Technology, Lanzhou University, Lanzhou, China

Abstract. Experimental studies on astrophysical reactions involving radioactive isotopes (RI) often accompany technical challenges. Studies on such nuclear reactions have been conducted at the low-energy RI beam separator CRIB, operated by Center for Nuclear Study, the University of Tokyo. We discuss two cases of astrophysical reaction studies at CRIB; one is for the ^{7}Be+n reactions which may affect the primordial ^{7}Li abundance in the Big-Bang nucleosynthesis, and the other is for the ^{22}Mg(α, p) reaction relevant in X-ray bursts.

1 Introduction

Astrophysical reactions involving radioactive isotopes (RI) often play an important role in explosive stellar environments. Although the RI are seldom seen on the earth due to the finite lifetime, they do exist in stars, and contribute to the evolution and thermal dynamics of stellar objects. Experimental efforts have been made for the studies on such RI-involving reactions. In a normal experimental condition, short-lived RI or neutrons can only be used as the beam, not as the target. However, reaction measurements with RI beams often suffer from the limitation of beam intensities, which are typically as small as 10^5 particles per second (pps) or less, while $> 10^{14}$ pps is available for light-ion beams. This great difference in the beam intensity is fundamental for the feasibility of the measurement.

Here we discuss possible approaches to study RI-involving reactions in spite of the technical limitation of the RI beam, introducing recent representative results from the low-energy RI beam facility CRIB [1–3] of the University of Tokyo. CRIB is an RI beam separator operated by Center for Nuclear Study (CNS), the University of Tokyo, and located at the RIBF facility of RIKEN Nishina Center. CRIB can produce low-energy ($<$ 10 MeV/u) RI beams by the in-flight technique, using primary heavy-ion beams accelerated at the AVF cyclotron of RIKEN (K=70). Most of the RI beams are produced via 2-body reactions such as (p, n), (d, p) and (^{3}He, n), taking place at an 8-cm-long gas target with a maximum pressure of 760 Torr. A cryogenic target system, in which the target gas can be cooled down to about 90 K, is presently available, and an intense ^{7}Be beam of 2×10^8 pps was produced with the system [4]. A diagram to show the RI beams ever produced at CRIB is found in [3]. In 2021, a development of ^{6}He beam was carried out, where we successfully produced a ^{6}He beam at 8.0 MeV/u with an intensity greater than 2×10^5 pps. To further increase the RI beam intensity, a high heat-proof target is under development. Using molybdenum foils as the sealing windows of the gas target, the heat capacity is doubled.

We introduce below two latest works at CRIB [5, 6], in which RI-involving astrophysical reactions were measured at astrophysical energies in indirect ways.

2 ^{7}Be+n reaction measurement with the Trojan horse method

The primordial production of light nuclides is well described with the standard model of Big Bang nucleosynthesis (BBN), however, the ^{7}Li/H abundance remains overestimated by a factor of 3–4, known as the cosmological lithium problem. We have studied the ^{7}Be(n, p_0)^{7}Li, ^{7}Be(n, p_1)^{7}Li*and ^{7}Be(n, α)^{4}He reactions, which may affect the primordial ^{7}Li abundance [5].

To study the reactions of two unstable particles (^{7}Be and n), we applied the Trojan horse method (THM, see References in [5, 7]), which is an indirect method to study a 2-body reaction via a 3-body reaction measurement under a quasi-free kinematical condition. The ^{7}Be$(n, \alpha)^4$He reaction had been measured with the THM in a previous work [7], and our new work expanded the sensitivity also to the ^{7}Be$(n, p_0)^7$Li and ^{7}Be$(n, p_1)^7$Li* reactions, by achieving a sufficient energy resolution to separate the contributions of these two reactions.

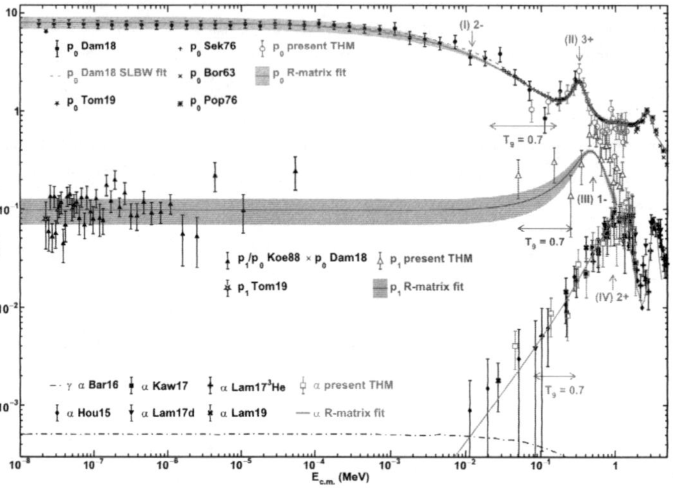

Figure 1. The cross sections multiplied by $\sqrt{E_{\rm cm}}$ of the ^{7}Be$(n, p_0)^7$Li, ^{7}Be$(n, p_1)^7$Li*and ^{7}Be$(n, \alpha)^4$He reactions in [5] are shown as open red circles, blue triangles, green squares, respectively. The previous experimental data are also shown for comparison. See [5] for details.

The cross sections multiplied by $\sqrt{E_{\rm cm}}$ obtained in this work are shown in Fig.1. The ^{7}Be$(n, p_0)^7$Li and ^{7}Be$(n, \alpha)^4$He rates were basically consistent with the previous works, but we assigned a smaller uncertainty on the ^{7}Be$(n, p_0)^7$Li reaction rate compared to the previous work [8]. The ^{7}Be$(n, p_1)^7$Li* reaction was measured at the BBN energy for the first time in this work. With the simulation using `PRIMAT` code, we demonstrated the new reaction rate, including the ^{7}Be$(n, p_1)^7$Li* contribution, could reduce the primordial ^{7}Li abundance by $\sim 10\%$.

3 ^{25}Al+p elastic resonant scattering for the ^{22}Mg(α, p) reaction

^{22}Mg(α, p) reaction is known to be one of the relevant reactions in X-ray bursts, greatly affecting to its light curve. Recently, the first direct measurement on this reaction has been performed [9], but only at energies much higher than the Gamow energy of X-ray burst, due to the limited intensity (900 pps) of the ^{22}Mg RI beam. The reaction is dominated by several resonances above the α-threshold energy, which may imply a simple extrapolation with a statistical-model calculation down to the Gamow energy could be unreliable. Therefore, we studied this reaction by measuring the resonances in the compound ^{26}Si nucleus with resonant scattering of ^{25}Al+p.

The measurement was performed at CRIB with an ^{25}Al RI beam at an intensity of 2×10^5 pps with essentially the same method as previous experiments [10, 11]. Fig. 2 shows the obtained excitation functions of the elastic and inelastic scatterings. We obtained resonance parameters just above the α particle threshold, as indicated in the figure, with an R-matrix analysis. The parameters were employed in a new X-ray burst model calculation, and we obtained a better agreement with the observed light curves of GS 1826–24 clocked burster and SAX J1808:4–3658 PRE burster compared to the previous works.

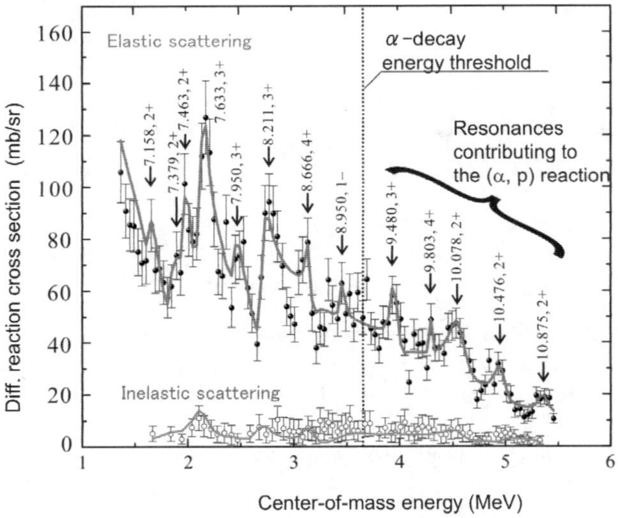

Figure 2. Excitation function of ^{25}Al+α scatterings with an R-matrix analysis. See [6] for details.

4 Summary

Astrophysical reactions involving RI are often difficult to study experimentally, mainly due to the limitation of the present RI beam technique. In the above two example, we demonstrated that such astrophysical reactions can be studied even with a low-intensity RI beam, by employing an indirect method (THM) or studying resonance parameters.

The experiments were performed at RI Beam Factory operated by RIKEN Nishina Center and CNS, the University of Tokyo. This work was partly supported by JSPS KAKENHI (No. 18H01218, and 19K03883) from the Ministry of Education, Culture, Sports, Science and Technology (MEXT) of Japan.

References

[1] S. Kubono, Y. Yanagisawa et al., Eur. Phys. J. **A13**, 217 (2002)
[2] Y. Yanagisawa, S. Kubono et al., Nucl. Instrum. Meth. Phys. Res., Sect. A **539**, 74 (2005)
[3] H. Yamaguchi, D. Kahl et al., Nucl. Phys. News **30**, 21 (2020)
[4] H. Yamaguchi, Y. Wakabayashi et al., Nucl. Instrum. Meth. Phys. Res., Sect. A **589**, 150 (2008)
[5] S. Hayakawa, M.L. Cognata et al., **915**, L13 (2021)
[6] J. Hu, H. Yamaguchi et al., Phys. Rev. Lett. **127**, 172701 (2021)
[7] L. Lamia, M. Mazzocco et al., ApJ **879**, 23 (2019)
[8] L. Damone, M. Barbagallo et al., PhRvL **121**, 042701 (2018)
[9] J.S. Randhawa, Y. Ayyad et al., Phys. Rev. Lett. **125**, 202701 (2020)
[10] H. Yamaguchi, Y. Wakabayashi et al., Phys. Lett. B **672**, 230 (2009)
[11] J.J. He, L.Y. Zhang et al., Phys. Rev. C **88**, 012801 (2013)

The Impact of Nuclear Physics Uncertainties on Interpreting Kilonova Light Curves

Yonglin Zhu[1,2,*], *Jennifer* Barnes[3], *K. A.* Lund[1,4], *T. M* Sprouse[4], *N.* Vassh[5], *G. C* McLaughlin[1,2], *M. R* Mumpower[4,6], and *R.* Surman[2,5]

[1]Dept. of Physics, North Carolina State University, Raleigh NC 27695 USA,

[2]Joint Institute for Nuclear Astrophysics - Center for the Evolution of the Elements, USA,

[3]Dept. of Physics and Columbia Astrophysics Laboratory, Columbia University, NY 10027 USA,

[4]Theoretical Division, Los Alamos National Laboratory, Los Alamos NM 87545 USA

[5]Dept. of Physics, University of Notre Dame, Notre Dame IN 46556 USA,

[6]Center for Theoretical Astrophysics, Los Alamos National Laboratory, Los Alamos, NM, 87545, USA

Abstract. Recently, analysis of the optical counterpart AT2017gfo to the gravitational wave-detected neutron star merger GW170817 has suggested a promising resolution of a long-standing debate on neutron star mergers as a source of some of the heaviest elements. However, making quantitative progress in these areas requires an accounting of the uncertainties in different aspects of physics, which is input into simulations of merging compact objects and their associated phenomena, remarkably rapid neutron capture nucleosynthesis. We investigate the uncertainties from the nuclear inputs to rapid neutron capture nucleosynthesis calculations combining different theoretical nuclear mass models, spontaneous fission rates, and fission daughter product distributions on top of the experimental nuclear data. We report that such nuclear physics uncertainties typically generate at least one order of magnitude uncertainty in the nuclear heating, which leads to uncertainties in the bolometric luminosity and the inferred mass of r-process material from the kilonova light curve.

Introduction

The detailed observations of GW170817 provides a definitive answer to the question of heavy element origin: binary neutron star mergers (NSMs) are a source of elements heavier than iron. Current simulations could fit optical counterpart to GW170817 qualitatively well; meanwhile, quantitative differences in simulations could be by orders of magnitude (see discussion in [1]). In order to accurately interpret kilonova emission and understand the evolution of r-process elements in the Universe, recent studies has attempted to explore the complexity from astrophysics[2–4], nuclear physics[5–7] and atomic physics[8].

*e-mail: zhuygln@gmail.com

Objective

We investigate uncertainties from the nuclear inputs to rapid neutron capture nucleosynthesis calculations with a focus on nuclear heating and the impact on kilonova light curves. We demonstrate how these nuclear physics uncertainties propagate to nuclear heating and inferred astrophysical quantities such as ejecta mass.

Method

We consider lanthanide-rich outflows to emphasize contributions from species with A $\geq$ 120 in the light curve, following setup and treatments from [5]. These lanthanide-rich outflows will be useful for elucidating neutron-rich ejecta from the mergers of binary neutron stars, as well as black hole-neutron star systems. We start with a base astrophysical trajectory of a standard parameterized wind from [9] and use the electron fraction Y_e as a proxy for all astrophysical inputs. We generate seed nuclear abundance wtih SFHo equation of state (Steiner et al. 2012) and perform the nucleosynthesis simulations with the Portable Routines for Integrated nucleoSynthesis Modeling(PRISM). We explore the uncertainties from theoretical nuclear inputs in regions where experimental measured mass from AME2016 [10] and experimental data from NUBASE2016 [11] are not available. We adopt the treatment in thermalization efficiencies for effective heating from [12] and opacity from [13] as input into a semi-analytical light curve model. We selected 13 highlighted simulations as in Table 1 to cover reasonable electron fractions and popular nuclear physics inputs(nuclear mass models, fission prescriptions). In our kilonvoa light curve calculations, we use a single-component (red) model of a range of given mass and velocity of $v_{ej} = 0.15c$.

Table 1. Highlighted Simulations Setup (for details see [5])

Index	Mass Model	Y_e	Yield Distribution	Spontaneous Fission rates
1	FRDM2012	0.28	Symmetric	Karpov/Zagrebaev
2	FRDM2012	0.16	Symmetric	Karpov/Zagrebaev
3	HFB22	0.16	Symmetric	Karpov/Zagrebaev
4	HFB27	0.16	Kodama	Karpov/Zagrebaev
5	DZ33	0.16	Symmetric	Karpov/Zagrebaev
6	UNEDF1	0.16	Kodama	Karpov/Zagrebaev
7	UNEDF1	0.16	Kodama	Xu & Ren
8	UNEDF1	0.24	Symmetric	Karpov/Zagrebaev
9	SLY4	0.18	Symmetric	Karpov/Zagrebaev
10	SLY4	0.21	Symmetric	Karpov/Zagrebaev
11	TF+D3C*	0.16	Symmetric	Karpov/Zagrebaev
12	DZ33	mixture[a]	Kodama	Karpov/Zagrebaev
13	UNEDF1	mixture[b]	Symmetric	Karpov/Zagrebaev

[a]

[b]Linear combinations of trajectories with different Y_e values that fit the solar r-process abundance.

Results

As can be seen in Figure 7 in [5], the final abundances from the selected simulations differ in the abundance of r-process elements by two to three orders of magnitude. Relevant to the opacity calculations used in kilonova light curve simulations, the mass fraction of lanthanides and actinides at one day after merger differs from 10^{-2} to more than 10^{-1} as shown in sub-barplot of Figure 1. We use this to explore the light curve with a grid of ejecta mass values for these 13 models as colored and grey solid lines in Figure 1, comparing with the observed values (diamond markers). We highlight one light curve for each of the 13 simulations that best matches the late time observation from GW170817 with color and find that their ejecta masses range from $0.01M_{\odot}$ to $0.08M_{\odot}$. We different power law fits to different light curves - for simulations with more contribution from fission and/or α-decay, such as simulation 3 in blue line, the light curves follow a slower decay rate like $L \propto t^{-1}$; while simulations that are not neutron-rich enough to make heavy nuclei that undergo fission and α-decay, such as simulation 1, follow a faster decay rate as $L \propto t^{-2.7}$.

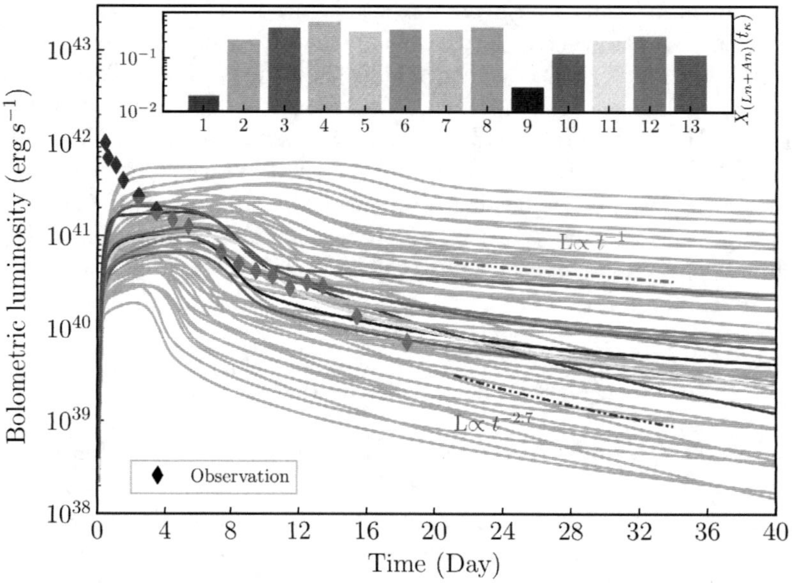

Figure 1. Bolometric luminosity of selected simulations with different eject masses. The subplot in the top is mass fractions of lanthanides (*Ln*) and actinides (*An*), $X_{\mathrm{Ln+An}}$ at one day after merger. Blue and red diamond markers are observed bolometric luminosity. Dash-doted lines with labels of $L \propto t^{-\alpha}$ show power fit to the bolometric luminosity. The colored lines show light curve simulations that are closest to the AT2017gfo observation of the inferred ejecta masses grid. The inferred ejecta masses used for each colored lines are $0.05M_{\odot}$, $0.02M_{\odot}$, $0.01M_{\odot}$, $0.01M_{\odot}$, $0.01M_{\odot}$, $0.02M_{\odot}$, $0.02M_{\odot}$, $0.05M_{\odot}$, $0.08M_{\odot}$, $0.08M_{\odot}$, $0.02M_{\odot}$, $0.05M_{\odot}$ by order. The grey lines shows the light curves with the rest of ejecta mass grid for each simulations. This figure is made from same calculations as in figure 8 and 12 from [5].

Conclusion

For lanthanide-rich outflows, nuclear inputs introduce uncertainty in lanthanide and actinide production to at least one order of magnitude around one day after merger. We show that the uncertainty in effective nuclear heating propagates to bolometric luminosity with the same order of magnitude from days to hundreds of days after merger. We find that power fit may be useful to estimate light curve at a different times with consideration of dominating reaction channels. We demonstrate that the inferred quantities like mass of ejecta could be significantly impacted by the choice of nuclear physics inputs in the simulations.

References

[1] A. Aprahamian, R. Surman, A. Frebel, G. C. McLaughlin, A. Arcones, A. B. Balantekin, J. Barnes, T. C. Beers, E. M. Holmbeck and J. Yoon, *et al.* [arXiv:1809.00703 [astro-ph.HE]].

[2] O. Korobkin, R. Wollaeger, C. Fryer, A. L. Hungerford, S. Rosswog, C. Fontes, M. Mumpower, E. Chase, W. Even and J. M. Miller, *et al.* Astrophys. J. **910**, no.2, 116 (2021) doi:10.3847/1538-4357/abe1b5 [arXiv:2004.00102 [astro-ph.HE]].

[3] F. Foucart, P. Moesta, T. Ramirez, A. J. Wright, S. Darbha and D. Kasen, [arXiv:2109.00565 [astro-ph.HE]].

[4] R. T. Wollaeger, C. L. Fryer, E. A. Chase, C. J. Fontes, M. Ristic, A. L. Hungerford, O. Korobkin, R. O'Shaughnessy and A. M. Herring, Astrophys. J. **918**, no.1, 10 (2021) doi:10.3847/1538-4357/ac0d03 [arXiv:2105.11543 [astro-ph.HE]].

[5] Y. L. Zhu, K. Lund, J. Barnes, T. M. Sprouse, N. Vassh, G. C. McLaughlin, M. R. Mumpower and R. Surman, Astrophys. J. **906** (2021) no.2, 94 doi:10.3847/1538-4357/abc69e [arXiv:2010.03668 [astro-ph.HE]].

[6] J. Barnes, Y. L. Zhu, K. A. Lund, T. M. Sprouse, N. Vassh, G. C. McLaughlin, M. R. Mumpower and R. Surman, Astrophys. J. **918**, no.2, 44 (2021) doi:10.3847/1538-4357/ac0aec [arXiv:2010.11182 [astro-ph.HE]].

[7] Y. L. Zhu, T. Sprouse, M. R. Mumpower, N. Vassh, R. Surman and G. C. McLaughlin, Springer Proc. Phys. **219**, 469-472 (2019) doi:10.1007/978-3-030-13876-9_92

[8] K. Hotokezaka, M. Tanaka, D. Kato and G. Gaigalas, doi:10.1093/mnras/stab1975 [arXiv:2102.07879 [astro-ph.HE]].

[9] I. V. Panov and H. T. Janka, Astron. Astrophys. **494**, 829 (2009) doi:10.1051/0004-6361:200810292 [arXiv:0805.1848 [astro-ph]].

[10] M. Wang, W. J. Huang, F. G. Kondev, G. Audi and S. Naimi, Chin. Phys. C **45**, no.3, 030003 (2021) doi:10.1088/1674-1137/abddaf

[11] G. Audi, F. G. Kondev, M. Wang, W. J. Huang and S. Naimi, Chin. Phys. C **41**, no.3, 030001 (2017) doi:10.1088/1674-1137/41/3/030001

[12] D. Kasen and J. Barnes, Astrophys. J. **876**, no.2, 128 (2019) doi:10.3847/1538-4357/ab06c2 [arXiv:1807.03319 [astro-ph.HE]].

[13] D. Kasen, N. R. Badnell and J. Barnes, Astrophys. J. **774**, 25 (2013) doi:10.1088/0004-637X/774/1/25 [arXiv:1303.5788 [astro-ph.HE]].

EPJ Web of Conferences **260**, 03005 (2022) https://doi.org/10.1051/epjconf/202226003005
NIC-XVI

Competition between allowed and first-forbidden β decays and the r-process

Zsolt Podolyák[1,*]

[1]Department of Physics, University of Surrey, Guildford, GU2 7XH, UK

Abstract. β^- decay lifetimes are essential ingredients for r-process yield calculations. In N≈126 r-process waiting point nuclei first-forbidden and allowed β decays are expected to compete. Recent experiments performed at CERN/ISOLDE showed that ^{207,208}Hg decay predominantly via first-forbidden decays. In addition, following on a high statistics study of the β^+/EC decay of ^{208}At, it is suggested that the Z>82, N<126 nuclei provide an excellent testing ground for global calculations addressing the competition between first-forbidden and allowed β decays.

1 Introduction

Half of the nuclei heavier than iron were synthesised in the r process. This process is based on a succession of neutron captures and β^- decays. Since heavy (A>150) r-process path nuclei still cannot be produced in laboratories, yield calculations have to rely on β decay half-lives predicted by theoretical calculations. In contrast to lighter mass regions, around neutron-number N=126 there is competition between (parity conserving) allowed and (parity changing) first-forbidden β decays [1, 2]. First-forbidden (FF) transitions can be dominant, with profound implications on their half-lives and therefore on the r-process, specifically on the third r-process peak at A~195 [3]. However, FF transitions are notoriously difficult to calculate. Here we present results obtained in β-decay studies on nuclei around ^{208}Pb in experiments performed at the ISOLDE Decay Station at CERN.

2 First-forbidden β decays

An ideal atomic nucleus to study the competition between allowed and forbidden β decays should have a small number of both positive and negative parity levels which could be populated in β decay, and they should have simple and well-understood wave functions. Quantum excitations in nuclei in the vicinity of doubly magic numbers have well defined wave functions. In addition, the ^{208}Tl nucleus (with one proton-hole and one neutron-particle outside the closed shells) is expected to have a small number of both positive and negative parity low-energy states. Indeed, shell model calculations predict two 0^+, five 1^+, as well as one 0^- and three 1^- states at excitation energies below the Q_β=3.48 MeV value [4] (note, that in addition, a number of collective octupole states with negative parity are also expected above 2.6 MeV, however these could not be calculated within the used model space). Therefore, the

*e-mail: z.podolyak@surrey.ac.uk

Table 1. Experimentally determined lifetime and first-forbidden beta decay contribution for the β decays of ^{208}Hg and ^{207}Hg, compared with theoretical predictions.

Nucleus		Exp	Theory					
			[6]	[7]	[8]	[9]	[2]	[1]
^{208}Hg	$T_{1/2}$ (s)	135(10) [5]	168.9	8.1	12.1	5.3	0.9	70
	FF/total	≈1 [5]				0.040	0.87	1
^{207}Hg	$T_{1/2}$ (s)	174(12) [11]	40.1	9.1		313.2	4.0	61
	FF/total	≈1 [10]				0.986	0.77	1

decay of ^{208}Hg into ^{208}Tl provides an ideal testing ground for the study of the competition between allowed and FF β decays. This was studied at ISOLDE at CERN [5]. Three negative parity excited states, one 0^- and two 1^-, were populated directly in β decay. The FF decay probabilities with $\log ft$ values in the range of 5.2-6.0 are in line with systematics in this region, however at present these cannot be reliably calculated. In contrast none of the positive parity states were populated. This latter can be understood by considering the properties of the single proton and neutrons involved. The half-life of the ^{208}Hg ground-state was measured as $T_{1/2}$=135(10) s [5]. There are several published global calculations covering the region of interest for the *r*-process. Since shell-model calculations are not feasible far from closed shells, all global calculations use mean field approaches, and all recent ones include first-forbidden decays. These are based on FRDM+QRPA [6], KTUY5 [7], DF3+cQRPA [8], FRDM+QRPA [9], RHB+RQRPA [2] and FAM [1] methods, and their results are compared with the experimental value in table 1. From a similar, but much higher yield experiment it was established that ^{207}Hg (Q_β=4.55 MeV [4]) also decays predominantly (several observed states have no firm parity assignment) via FF β decays [10]. The experimental and theoretical values for the ^{207}Hg decay are also listed in table 1.

None of the existing calculations can reproduce the experimental lifetimes of N≈126 nuclei in the region [12–16]. This is not surprising, as even such state-of-the-art calculations are not expected to provide accurate single-particle energies, essential for nuclei close to closed shells where the β-decay strength is dominated by few transitions. There are some shell-model calculations available for N=126 nuclei [17, 18], however these do not consider neutrons above the N=126 shell closure, therefore cannot provide a reliable lifetime estimate for ^{207,208}Hg.

It is difficult to populate heavy neutron-rich nuclei in the laboratory. It is much easier to produce proton rich ones. Recently, the β^+/EC decay of ^{208}At into ^{208}Po was studied within a high statistics experiment [19]. A roughly equal role of allowed and FF decays were found. This can be understood considering the relevant shell model orbitals in the region. Since the FF decays populate mainly excited states at high energies, some older experiments suffered from the pandemonium effect. The role of the FF β decays in the region is illustrated in figure 1. Note that some nuclei have very low Q_{EC} values and consequently there are few excited states in their daughter nuclei within the available energy window. The decays of ^{208}Po and ^{209}Po are peculiar; all β decays proceed via larger degree of forbiddenness due to the lack of daughter states with spin-parities consistent with FF and allowed β decays. The indicated small experimentally determined FF contribution for some Rn and At isotopes with larger Q_{EC} values are most likely due to the pandemonium effect, with the real value expected to be much larger.

It is suggested that the neutron-deficient region "north-west" of ^{208}Pb provides an excellent testing ground for the competition between allowed and first forbidden β-decay calculations [19]. However, presently very few global calculations for β^+/EC decays exists on this

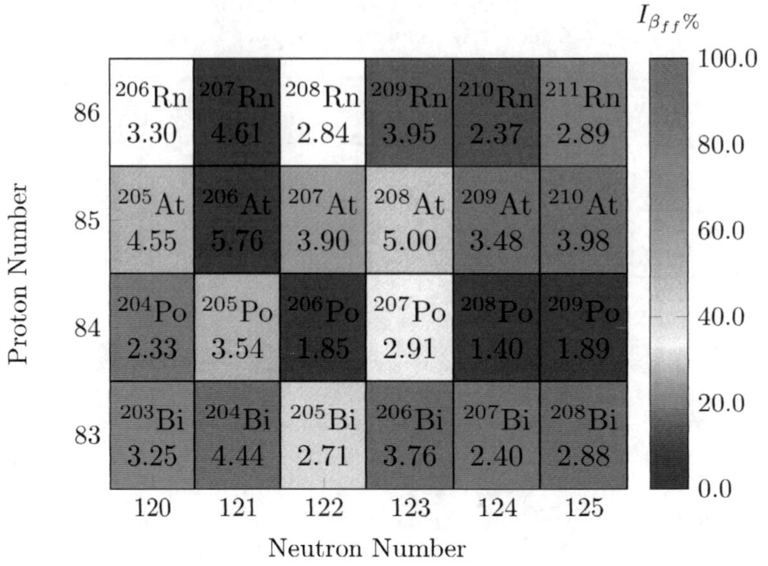

Figure 1. Role of first forbidden decays for nuclei 'north-west' of ^{208}Pb. The colour code refers to the $I_{\beta\ FF}/I_{\beta\ TOTAL}$ value. The parent nuclei are indicated. Q_{EC} values, in MeV, are also given. Figure modified from [19].

part of the nuclide chart. These predict half-lives of 105.7 s [7] and 2412 s [20], as compared to the experimental $T_{1/2}$=5900(1100) s [21]. Note that [20] considers only allowed decays.

3 The Δn=0 selection rule

The well-known selection rules for allowed β decay are: angular momentum change ΔI=0,1 and no parity change. However, there is one additional selection rule: the number of nodes in the radial wave function, *n*, of the parent and daughter states must be equal. The validity of this less known Δn=0 selection rule in Gamow-Teller β decay was recently investigated [22] by studying the β decay of ^{207}Hg into ^{207}Tl. The level of forbiddenness of the Δn=1 $\nu 1g_{9/2} \rightarrow \pi 0g_{7/2}$ transition has been determined. From the non-observation of this decay at the level of <3.9×10^{-3}%, corresponding to logft>8.8 (95% confidence limit), one can conclude that the selection rule holds. This is the most stringent test of the Δn=0 selection rule to date. Without this rule the lifetime of ^{207}Hg would be approximatively 7% shorter, while in the case of ^{209}Pb the effect is about ≈20%. This selection rule has little importance for nuclei close to stability, but is essential for the Z<82, N>126 *r*-process waiting point nuclei, lengthening their lifetimes.

4 Conclusions and outlook

In conclusion, recent ISOLDE results addressing the competition between allowed and first-forbidden β decays were reported. In particular it was found that both ^{207}Hg and ^{208}Hg decay via first-forbidden β decays into negative parity daughter states, despite the availability of positive-parity states. These were understood by examining the properties of individual neutrons and protons within the shell model approach. Following the study of the β^+/EC decay of ^{208}At, it is suggested that the Z>82, N<126 nuclei, which can be studied in high yield experiments provide an excellent testing ground of the competition of FF and allowed decays. This region would merit modern global β decay calculations.The presented results on β decays are important for the detailed understanding of the nucleosynthesis of heavy elements produced in the rapid neutron-capture process.

References

[1] E.M. Ney, J. Engel, T. Li, N. Schunck, Phys.Rev. C **102**, 034326 (2020).
[2] T. Marketin, L. Huther, G. Martinez-Pinedo, Phys. Rev. C **93**, 025805 (2016).
[3] N. Nishimura, Zs. Podolyák, D.-L. Fang, T. Suzuki, Phys.Lett. B **756**, 273 (2016).
[4] Meng Wang *et al.*, Chinese Phys. C **45**, 030003 (2021).
[5] R.J. Carrol *et al.*, Phys. Rev. Lett. **125**, 192501 (2020).
[6] P. Moller, B.Pfeiffer, K.-L.Kratz, Phys. Rev. C **67**, 055802 (2003).
[7] H. Koura, T. Tachibana, M. Uno, M. Yamada, Prog. Theor. Phys. **113**, 305 (2005).
[8] I. Borzov, Nucl. Phys. A **777**, 645 (2006).
[9] D.L. Fang, B.A. Brown, T. Suzuki, Phys. Rev. C.**88**, 034304 (2013).
[10] T.A. Berry *et al.*, Phys. Rev. C. **101**, 054311 (2020).
[11] B.Jonson, O.B.Nielsen, L.Westgaard, J.Zylicz, Proc.Int.Conf.Nuclei Far from Stability, Helsingor, Denmark, Vol.2, p.640 (1981); CERN-81-09 (1981).
[12] G. Benzoni *et al.*, Phys. Lett. B **715**, 293 (2014).
[13] A.I. Morales *et al.*, Phys.Rev.Lett. **113**, 022702 (2014).
[14] A.I. Morales *et al.*, Europhys.Lett. **111**, 52001 (2015).
[15] R. Caballero-Folch *et al.*, Phys. Rev. Lett. **117**, 012501 (2016).
[16] R. Caballero-Folch *et al.*, Phys. Rev. C **95**, 064322 (2017).
[17] T. Suzuki, T. Yoshida, T. Kajino, T. Otsuka, Phys. Rev. C **85**, 015802 (2012).
[18] Q. Zhi, E. Caurier, J.J. Cuenca-Garcia, K. Langanke, G. Martinez-Pinedo, K. Sieja, Phys. Rev. C **87**, 025803 (2013).
[19] M. Brunett et al., Phys. Rev. C **103**, 054327 (2021).
[20] P. Moller, J. R. Nix, and K.-L. Kratz, Atomic Data Nucl. Data Tables **66**, 131 (1997).
[21] M.J. Martin, Nucl. Data Sheets **108**,1583 (2007).
[22] T.A. Berry et al., Phys. Lett. B **793**, 271 (2019).

New capability for indirect neutron capture measurements: The DICER instrument at LANSCE

A. Stamatopoulos[1,*], *P.* Koehler[1], *A.* Matyskin[2], *E. M.* Bond[2], *T.A.* Bredeweg[2], *A.* Couture[1], *B.* Di Giovine[1], *M.E.* Fassbender[2], *A. C.* Hayes-Sterbenz[3], *A. L.* Keksis[2], *K.* Parsons[4], *G.* Rusev[2], *J.* Ullmann[1], and *C.* Vermeulen[2]

[1]Physics division, Los Alamos National Laboratory, 87545, NM, USA
[2]Chemistry Division, Los Alamos National Laboratory, 87545, NM, USA
[3]Theoretical Division, Los Alamos National Laboratory, 87545, NM, USA
[4]X-Computational Physics Division, Los Alamos National Laboratory, 87545, NM, USA

Abstract. The Device for Indirect Capture on Radionuclides (DICER) implements a new indirect technique for (n,γ) studies in which the neutron capture rate is determined from analysis of resonance neutron transmission data. The DICER instrument and associated radionuclide production at the Isotope Production Facility (IPF), both at the Los Alamos Neutron Science Center (LANSCE), as well radioactive sample fabrication, have been under development in the last few years. First measurements on a radioactive sample (^{88}Zr , $t_{1/2}$=83.4 days), which was recently reported to have an extremely large thermal neutron capture cross section and resonance integral [1, 2], are planned for the winter of 2021. A performance overview, brief details on the ^{88}Zr fabrication and proof of good operation results will be presented.

1 Introduction

The accurate quantification of radiative neutron capture cross sections is essential in various applications such as radiochemical diagnostics, nuclear forensics and nuclear astrophysics. Although several studies have been performed on stable nuclei and a few on long-lived radionuclides [3] using direct techniques, the measurement of (n, γ) cross sections on short-lived radionuclides is far more challenging due to backgrounds from the decay.

Hence, a number of indirect methods have been pursued and substantial effort has been devoted to quantify systematic errors associated with these techniques. The most popular indirect techniques include the surrogate method [4], the γ-ray strength function method [5, 6], the Oslo method [7] and the β-Oslo method [8].

Total cross section measurements are less affected by the decay background due to the long sample-to-detector distances, which are typically of the order of tens of meters. From neutron transmission measurements, (n, γ) cross sections can be tightly constrained and even accurately calculated through the Nuclear Statistical Model (NSM) [9].

*e-mail: athanasios.stamatopoulos@lanl.gov

Figure 1. The DICER instrument is located at flight path 13, at Manuel Lujan Jr. Neutron Scattering Center at LANSCE. The collimation components along with the 31 m detection station are shown.

The technique is presented in detail in Ref. [10]. Briefly, neutron transmission spectra provide directly the level spacing (D_0, distance between transmission dips), the total (Γ, width of the transmission dip) and neutron resonance widths (Γ_n, depth of the transmission dip). There are resonances where the radiation width (Γ_γ) is not significantly smaller than Γ_n and at the same time other reaction channels are suppressed. In those cases, the Γ_γ can be easily calculated: $\Gamma = \Gamma_n + \Gamma_\gamma$. Γ_γ distributions are known to have small fluctuations, therefore only a small number of resonances is essentially needed to calculate the average radiation width ($\langle\Gamma_\gamma\rangle$) with a sufficient accuracy.

The design of the Device for Indirect Capture Experiments on Radionuclides (DICER)[9, 11, 12], was based on the aforementioned technique and is being developed at the Los Alamos Neutron Science Center (LANSCE) where a high neutron flux is delivered. A plethora of radionuclides relevant to radiochemical diagnostics, nuclear forensics, nuclear astrophysics and nuclear data in general [9, 11, 13] are expected to be studied at DICER. The instrument is designed to study small samples (tens of μg, 0.1-1 mm in diameter) with half-lives of the order of tens of days or higher and level spacings of the order of tens of eV or smaller.

2 The DICER instrument

DICER is located at the Manuel Lujan Jr. Neutron Scattering Center (flight path 13) at LANSCE. Source-to-detector distances of 31 and a 64 m are available, however, only the 31 m station, shown in Fig. 1, will be discussed. A detailed description of the first DICER generation is provided in Ref. [14].

2.1 Neutron source

The neutron beam delivered at DICER is produced through spallation from LANSCE's 800 MeV proton beam, pulsed at 20 Hz[15]. The beam is impinging on a split, 10 cm in diameter, cylindrical tungsten target. Each proton pulse has a 125 ns FWHM and average intensity of $\sim$ 100 μA. The spallation neutrons are moderated in a liquid hydrogen moderator, a process which results in a neutron spectrum that spans from meV-MeV in energy regime, as shown in Fig. 2.

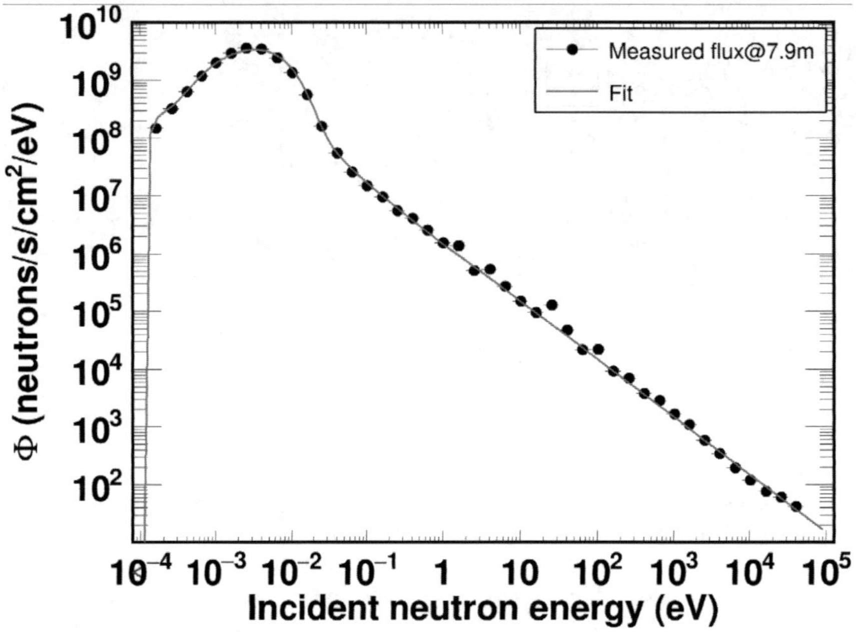

Figure 2. Measured and analytical description of DICER's neutron spectrum at 7.9 m from the spallation target.

2.2 Collimation system

The collimation system allows DICER to perform simultaneously sample in and sample out measurements, unlike the traditional neutron transmission measurements where periodic sample insertion/removal is required. Periodically inserting and removing the sample from the beam, involves positioning errors, therefore the samples used in traditional measurements are significantly larger than the beam size. In summary, the DICER approach allows the measurement of small samples and reduces the measuring time by a factor of two by utilising two non-parallel neutron beam lines, which converge at the same spot on the liquid hydrogen moderator. A brief description on how this is achieved, will be provided below.

A cylindrical shell of brass (Rotating Beam Blocker, Fig. 1), 30.5 cm in length and 21.5 cm in diameter, is installed at a 14.35 m distance from the spallation target. A rotating cylindrical piece of brass, 10 cm in diameter, is inserted in the cylindrical shell and is equipped with three 0.8 cm diameter holes, which provide four different beam line configurations: (a) two beam lines, (b) no beam lines, (c) a beam line to the left and (d) a beam line to the right.

At a 14.85 m distance from the exit of the neutron source, a right rectangular prism made of brass (Binocular Collimator, Fig. 1), 30 cm long and 15 cm wide, is installed. This component serves both as a collimator and sample holder and provides two well defined and narrow beam lines, 1 mm in diameter, which point to the same area on the moderator, as illustrated in Fig. 3. The binocular collimator can house a cylindrical sample envelope of 1.5 cm in length and 1 cm in diameter. The positioning of the sample canisters with respect to the binocular collimator are pictured in Fig. 3.

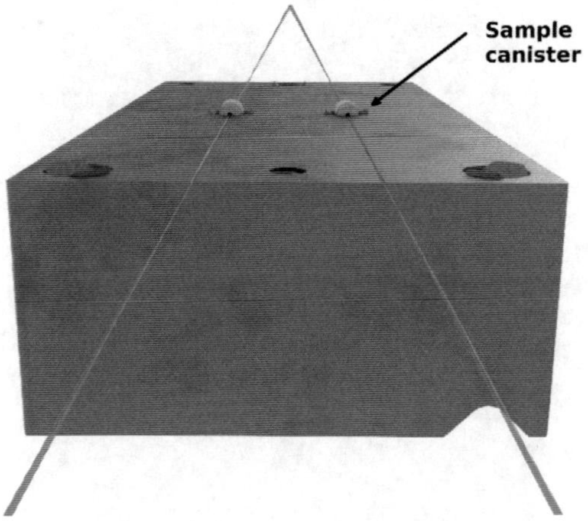

Figure 3. The binocular collimator of DICER allows to perform simultaneous sample in and out measurements, while it also serves as a sample holder.

Finally, the last shaping of the neutron beam takes place at 18.5 m from the neutron moderator. At that position, a rectangle brass collimator (aperture stop, Fig. 1) 30 cm long and 15 wide, is located. The aperture stop cleans up the beam penumbra from the sample collimator and ensures two well defined beam spots at the detector position.

2.3 Detectors

The standard detection system of DICER consists of two ^{6}Li-glass disks, 10 cm in diameter and various thicknesses (1, 2, 4, 6.3, 12.7 mm) coupled to dual photomultipliers (PMTs). The PMTs are perpendicular with respect to the neutron beam propagation, hence not interacting with neutrons. This helps minimizing backgrounds. The two detectors are installed 31 m from the neutron source and are shown in Fig. 1.

3 Production of radioactive material at LANSCE and neutronic considerations: The ^{88}Zr case

The DICER approach relies on the synergy between two LANSCE facilities: the Manuel Lujan Jr. neutron scattering center and the Isotope Production Facility (IPF)[16, 17]. A demonstration of this collaboration is the production of the ^{88}Zr radionuclide. Briefly, 29 g of a cylindrical yttrium metallic target, 2.90 mm in length

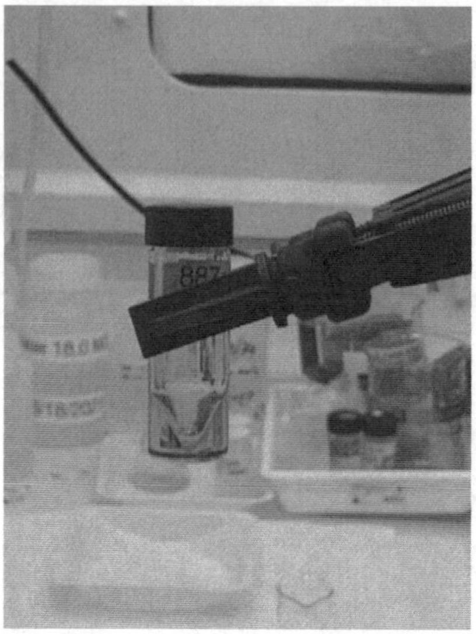

Figure 4. About 200 mCi of ^{88}Zr were produced at IPF at LANSCE. The radioactive material is diluted in 1.9 mol/L DCl. The image was taken inside a hot cell.

and 46 mm in diameter, encapsulated in an aluminum holder, was irradiated at the IPF.

The target was irradiated for a total of $\sim$ 9 hours at an average beam current of 96.1 μA. After cooling the irradiated material for 82 days, the target was disassembled and dissolved in a hot cell environment in 150 mL of 6 mol/L HCl by slow addition of 5-10 mL portions. About 200 mCi/8 GBq of ^{88}Zr were separated and diluted in 1.9 mol/L DCl solution. Finally, the solution was evaporated almost to dryness and washed with 10 mL of 6 mol/L HCl. This solution was transferred into a 10 mL v-vial. The sample is shown in Fig. 4.

The successful yttrium irradiation and ^{88}Zr recovery was confirmed through the means of gamma-ray spectroscopy, using a High Purity Germanium (HPGe) detector and a small aliquot (25 μL) of the separated material. The aliquot was diluted in 10 mL of 1 mol/L HCl and was then placed in front of an HPGe detector. The recorded spectrum shown in Fig. 5, illustrates the successful production and separation of ^{88}Zr : The 392.7 keV photo-peak from ^{88}Zr and the 898.7 and 1836.8 keV from the ^{88}Y daughter are dominating the spectrum.

Since DICER is an instrument that measures all neutron reactions that take place during a measurement, it has to be ensured that the chemical format/solution of the sample is fairly transparent to neutrons. In other words, the minimum amount of neutron interactions with materials other than the one of interest (^{88}Zr in this example) is preferred. Hydrogen has a significant high probability to scatter neutrons, therefore is appreciably not transparent to neutrons. To mitigate that effect, HCl was replaced with DCl. The ^{88}Zr +HCl sample was dried once again and the same procedure described above was used to produce ^{88}Zr in DCl solution. As shown in Fig. 6, the transmission through DCl, for a 1.4 mol/L, 1 cm long and 1.2 mm in diameter cylindrical sample, ranges between 85 – 95% in the region where a ^{88}Zr resonance is expected. For this calculation, (n,tot) cross sections from the ENDF/B-VIII.0 library[18] were used for ^{2}H, ^{35}Cl and ^{37}Cl.

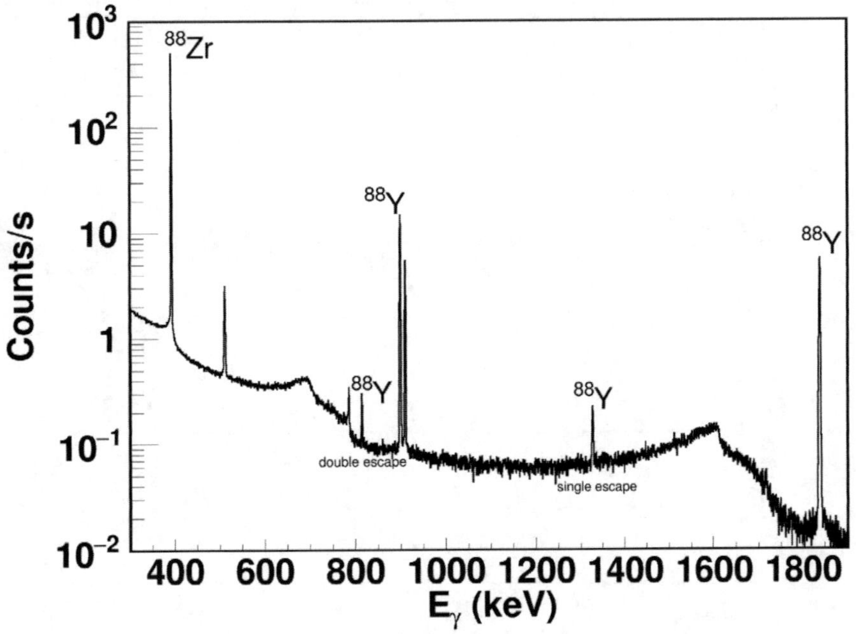

Figure 5. Gamma ray spectrum of 25μL of the sample taken with an HPGe detector. Only the ^{88}Zr and ^{88}Y gamma rays along with their escape peaks are labeled.

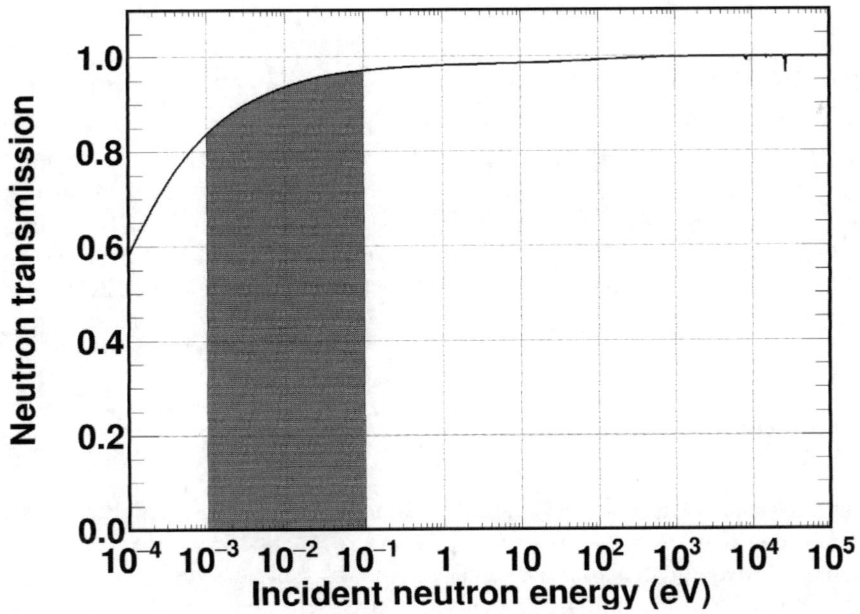

Figure 6. Neutron transmission through a 1.4 mol/L, 1 cm long and 1.2 mm in diameter cylindrical sample made of DCl. The grey shaded area corresponds to the main area of interest where a ^{88}Zr resonance is expected. ENDF/B-VIII.0 cross sections for ^{2}H, ^{35}Cl and ^{37}Cl were used.

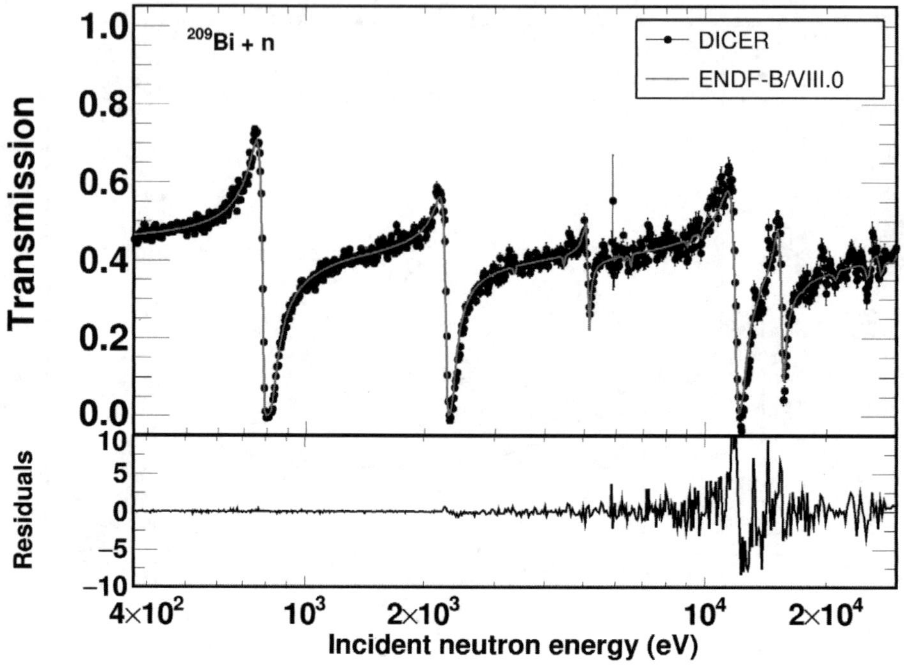

Figure 7. DICER data in comparison with ENDF/B-VIII.0. A satisfactory reproduction was achieved.

4 Performance overview

DICER is being commissioned since the autumn of 2019 and although the commissioning phase will finish at the end of 2022, many stable nuclei have already been studied such as ^{147,149}Sm, ^{191,193}Ir, ^{95}Mo, ^{209}Bi, ^{197}Au, ^{nat}Cd and ^{nat}Gd. To demonstrate the good understanding of the new device and the reproduction of well-known reactions, Fig. 7 shows a satisfactory reproduction of DICER data using ENDF/B-VIII.0 resonance parameters.

5 Conclusion

A new instrument at flight path 13 at the Manuel Lujan Jr. Neutron Scattering Center of LANSCE is being developed to study indirect neutron capture on short-lived radionuclides, through neutron transmission measurements and resonance analysis. The DICER concept is based on the synergy between experimental facilities at LANSCE, such as production of radioactive samples at IPF and their neutron irradiation at the Manuel Lujan Jr. Neutron Scattering Center. Great effort has been made to design and precisely align the instrument as well as to develop an efficient and reliable data reduction scheme. Both efforts led to a deep understanding of the instrument's performance, which was demonstrated by the satisfactory reproduction of DICER data from evaluated resonance parameters [14]. Finally, DICER is getting ready to perform its first measurements on radioactive samples on ^{88}Zr and ^{88}Y.

Acknowledgements

This work benefited from the use of the LANSCE accelerator facility and was supported by the US Department of Energy through the Los Alamos National Laboratory. Los Alamos National Laboratory is operated by Triad National Security, LLC, for the National Nuclear Security Administration of US Department of Energy (Contract No. 89233218CNA000001). Research presented in this article was supported by the Laboratory Directed Research and Development program of Los Alamos National Laboratory under project number 20200108DR.

References

[1] J.A. Shusterman, N.D. Scielzo, K.J. Thomas, E.B. Norman, S.E. Lapi, C.S. Loveless, N.J. Peters, J.D. Robertson, D.A. Shaughnessy, A.P. Tonchev, Nature **565**, 328 (2019)

[2] J.A. Shusterman, N.D. Scielzo, E.P. Abel, H.K. Clause, N.D. Dronchi, W.D. Frey, N. Gharibyan, J.A. Hart, C.S. Loveless, S.R. McGuinness et al., Phys. Rev. C **103**, 024614 (2021)

[3] A. Couture, R. Reifarth, At. Data and Nucl. Data Tables **93**, 807 (2007)

[4] J.E. Escher, J.T. Burke, R.O. Hughes, N.D. Scielzo, R.J. Casperson, S. Ota, H.I. Park, A. Saastamoinen, T.J. Ross, Phys. Rev. Lett. **121**, 052501 (2018)

[5] H. Utsunomiya, S. Goriely, H. Akimune, H. Harada, F. Kitatani, S. Goko, H. Toyokawa, K. Yamada, T. Kondo, O. Itoh et al., Phys. Rev. C **82**, 064610 (2010)

[6] H. Utsunomiya, S. Goriely, T. Kondo, C. Iwamoto, H. Akimune, T. Yamagata, H. Toyokawa, H. Harada, F. Kitatani, Y.W. Lui et al., Phys. Rev. C **88**, 015805 (2013)

[7] M. Guttormsen, T. Ramsøy, J. Rekstad, Nucl. Instrum. Meth. A **255**, 518 (1987)

[8] A. Spyrou, S.N. Liddick, A.C. Larsen, M. Guttormsen, K. Cooper, A.C. Dombos, D.J. Morrissey, F. Naqvi, G. Perdikakis, S.J. Quinn et al., Phys. Rev. Lett. **113**, 232502 (2014)

[9] P. Koehler, Tech. Rep. LA-UR-14-21466, Los Alamos National Laboratory (2014), `https://permalink.lanl.gov/object/tr?what=info:lanl-repo/lareport/LA-UR-14-21466`

[10] P.E. Koehler, K.H. Guber, Phys. Rev. C **88**, 035802 (2013)

[11] P. Koehler, et al., Tech. Rep. LA-UR-14-21656, Los Alamos National Laboratory (2014), `https://permalink.lanl.gov/object/tr?what=info:lanl-repo/lareport/LA-UR-14-21656`

[12] P. Koehler, et al., Tech. Rep. LA-UR-18-22995, Los Alamos National Laboratory (2018)

[13] G. Keksis, et al., Tech. Rep. LA-UR-21-23034, Los Alamos National Laboratory (2021), `DRAFT`

[14] A. Stamatopoulos, P. Koehler, A. Couture, B. DiGiovine, G. Rusev, J. Ullmann, Nucl. Instrum. Meth. A **1025** (2022)

[15] P.W. Lisowski, K.F. Schoenberg, Nucl. Instrum. Meth. A **562**, 910 (2006)

[16] K.F. Johnson, H.W. Alvestad, W.C. Barkley, D.B. Barlow, D.S. Barr, L.S. Bennett, L.J. Bitteker, E.A. Bjorklund, W. Boedeker, M.J. Borden et al. (2004)

[17] LANSCE, *Ipf*, `https://lansce.lanl.gov/facilities/ipf/index.php`

[18] D. Brown, M. Chadwick, R. Capote, A. Kahler, A. Trkov, M. Herman, A. Sonzogni, Y. Danon, A. Carlson, M. Dunn et al., Nucl. Data Sheets **148**, 1 (2018)

EPJ Web of Conferences **260**, 04001 (2022) https://doi.org/10.1051/epjconf/202226004001
NIC-XVI

Can we distinguish quark stars from neutron stars with measurements of global properties?

Ang Li[1,*]

[1]Department of Astronomy, Xiamen University, Xiamen, Fujian 361005, China

Abstract. The phase state of the dense stellar matter is an exciting topic in the area of nuclear astrophysics. It may be probed by observed properties of neutron stars from, for example, the currently operating satellites (NICER, Neutron star Interior Composition Explorer) and the gravitational-wave laser interferometers (Advanced LIGO, Virgo, and KAGRA). Based on our recent constrained parameter spaces of the equation of states of neutron stars and quark stars from LIGO/Virgo and NICER, we discuss the important role of an even-accurate determination of the stellar radius for distinguishing possible quark stars from neutron stars and our understanding of the QCD phase transition at finite density.

1 Introduction

Neutron stars are the remnants of the gravitational collapse of very massive stars during the supernova event. They may have masses in the range of $M \sim$ 1-3 $M_{\odot}$ and radii of $R \sim$ 10–20 km. So the matter inside the star is quite dense, with an average density exceeding the ordinary nuclear density. They are the densest and smallest stars observed in the Universe, composed of mainly nuclear matter plus a small fraction of electrons and muons. A neutron star is bound by gravity, and it is kept in hydrostatic equilibrium only by the pressure produced by the compressed nuclear matter. The hadronic matter might undergo a phase transition to quark matter inside the high-density cores of neutron stars, forming hybrid stars. In a hypothetic scenario [1, 2], neutron stars might even be self-bound with deconfined quarks constituting the entire star, i.e., quark stars. The equation of state (EOS), mainly pressure as a function of the (energy) density $p(\varepsilon)$ is informative of the composition and the inner structure of a neutron star; And experimental nuclear physics provides critical data needed to benchmark theories of dense matter EOS residing in neutron stars [3].

However, such extreme conditions make it impossible to attain the dense matter EOS by experimental methods only; several recent observational developments provide powerful constraints on the EOS and the internal composition. Among these are the measurable global properties of neutron stars, including the mass (PSR J0348+0432 [4], PSR J2215+5135 [5], PSR J0740+6620 [6]), the radius (PSR J0030+0451 [7, 8], PSR J0740+6620 [9, 10]), and the tidal deformability (GW170817) [11]. Fig. 1 demonstrates the effective constraining on the mass-radius relations from these data (all results are shown to the 90% confidence level if otherwise specified). In particular, the not-too-high tidal deformability of GW170817 can

*e-mail: liang@xmu.edu.cn

exclude very stiff EOSs, the relatively large radii of PSR J0030+0451 (with a mass around the typical mass $1.4\,M_\odot$) disfavors very soft EOSs. The simultaneous measurement of the masses and radii of 2-solar-mass pulsars (like the presently available PSR J0740+6620) can exclude the mass-radius parameter space from both the stiff and the soft sides, in this sense serving as an ideal probe for the EOS study of dense matter.

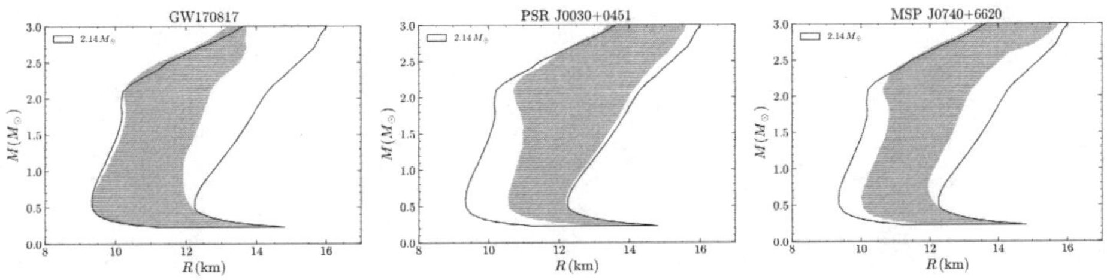

Figure 1. Posterior probability distributions (pink regions) of the mass-radius relations from the tidal deformability of GW170817 (left) and the simultaneous measure of the neutron star mass and radius of low-mass PSR J0030+0451 (middle) and high-mass PSR J0740+6620 (right), compared to the black area in each panel, obtained based on the EOSs consistent with the lower limit on the maximum mass $M_{\rm TOV}$ from the mass distribution measured for PSR J0740+6620.

The discussions in the present contribution follow the limited references of our recent works [12–18], and we mainly focus on one of the critical problems in the EOS study: Is it possible to distinguish gravity-bound (quark hybrid) neutron stars from self-bound quark stars with measurements of global properties? Since the strange quark matter in hybrid stars and quark stars are described under different stability conditions, and both of them can't coexist stably [19], a confirmation (or exclusion) of either of them can shed light on the remaining-unsolved QCD at finite baryon density [30]. We mention here that [12] made use of several nuclear many-body frameworks for various modeling of dense matter and found that the "effective" stiffness of the EOS, including the sound velocity $c_s = \sqrt{dP/d\varepsilon}$ and the adiabatic index $\Gamma = (\varepsilon + P)(dP/d\varepsilon)/P$, can not effectively signify the matter's composition in intermediate (nonperturbative) densities relevant to compact stars.

2 Models and Analysis

Model calculations showed that massive stars (like heaviest PSR J0740+6620) could be hybrid stars with a stiff quark-matter core [20]. The parameter space for neutron star EOSs could be extended significantly through the introduction of a hadron-quark phase transition [14], while maintaining consistency with the observational data from tidal deformability of the GW170817 and GW190425 binary neutron star merger as detected by LIGO/Virgo and the mass and radius of PSR J0030+0451 and PSR J0740+6620 as detected by *NICER*. The EOSs for hybrid stars [16, 17] constrained within the Bayesian statistical approach are shown in the left panel Fig. 2. For the analysis, we have assumed that the hadron-quark phase transition is of first order, where the low-density hadronic matter described in a unified manner by the QMF EOS [21] transforms into a high-density phase of quark matter modeled by the generic "Constant-sound-speed" (CSS) parameterization [22]. The CSS parameterization is a general parametrization suitable for expressing experimental constraints in a model-independent way. The value of the sound speed is varied from $1/\sqrt{3}$ (the perturbative QCD value) to 1 (the casual limit). The parameter spaces of the hybrid star EOSs and mass-radius relations are found not sensitive to the low-density hadronic EOS adopted [16, 17].

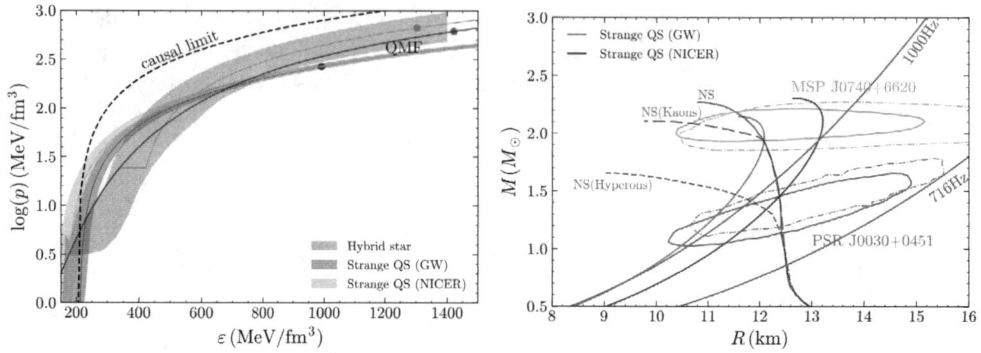

Figure 2. Left: Posterior probability distributions of the EOSs for (hybrid) neutron stars (pink region) [16, 17] from combined data of LIGO/Virgo and NICER, in comparison with those of quark stars from LIGO/Virgo (purple region) [15] and from NICER (blue region) [18]. Three dots indicate the densities when the maximum masses of the stars are reached in the most-preferred cases for three analyses (shown in solid lines). The extreme EOS in the causal limit is also shown in the black dashed line. Right: Most preferred mass-radius relations for both (hybrid) neutron stars (black-solid curve) and quark stars (solid red and blue curves) from the EOSs shown in the left panel, along with the microscopic calculations on possible exotic EOSs, like the ones including hyperons (short-dashed curve) [25] and K^- condensation (dashed curve) [26]. The green lines define the lower boundaries of the mass-radius parameter spaces if certain spin frequencies (716 Hz [24], 1000 Hz) are discovered from pulsar-like objects. Also shown are the mass-radius constraints of the NICER mission for PSR J0030+0451 [7, 8] and PSR J0740+6620 [9, 10].

In the left panel of Fig. 2, the posterior distributions of hybrid star EOSs are compared to those of quark stars [15, 18], resulting from an analysis based on the widely-used bag-model EOS parametrization incorporating the same data from LIGO/Virgo and NICER. From their comparison, we see that the quark star EOSs become stiff earlier than neutron star ones at low densities and are softer than the neutron star ones at high densities. Consequently, a heavy compact star does not necessarily demand a superconformal sound velocity in the quark star case, and it resulted in a lower (static) maximum mass for quark stars than that of hybrid stars [15]. Note that the maximum mass of neutron star may be lowered due to the appearance of strange baryons (i.e., hyperons, kaons), which competes with that of deconfined quarks in the high-density cores of neutron stars [23].

In the right panel of Fig. 2 the most preferred mass-radius relations for both hybrid stars and quark stars are compared. The dashed and short-dashed curves represent two EOS examples including other types of strangeness phase transitions: The EOS modeling with the inclusion of hyperons [25] are performed in the framework of the Brueckner-Bethe-Goldstone many-body approach with three-body forces and the NSC89 nucleon-hyperon interactions, and the Brueckner theory is combined with a standard chiral model for kaon condensation for the EOS with K^- [26]. A general comment is that different types of compact stars are not distinguishable from the available data [27]. A possible discovery of sub-millisecond pulsars (represented by the green curve labeling 1000 Hz) might not help either, since various theoretical neutron star and quark star models can reconcile with the observation. We expect a more precise radius measurement is possible (at least $\sim$ 1 km accuracy) in the near future to change the situation.

In Fig. 3, we include another possible global property possibly to be measured, i.e., the moment of inertia of PSR J0737-3039 A [28], the 1.338 $M_{\odot}$ primary component of the first

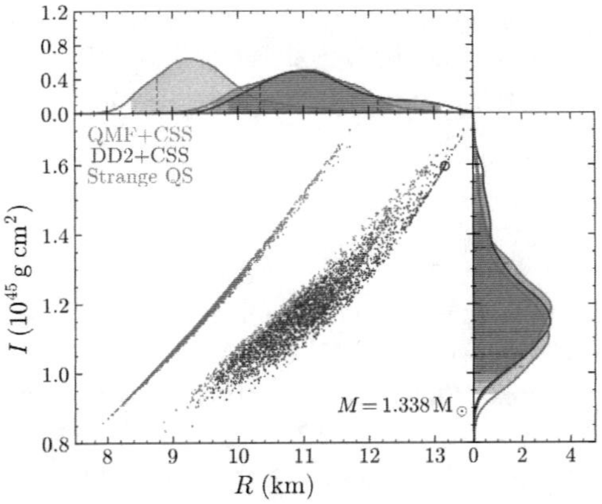

Figure 3. Posterior distributions of I vs. R for a $M = 1.338\,M_\odot$ star, transformed from the EOS parameters posteriors. The probability density of R and I are shown in the upper and right panel, respectively, with the 68% credible interval in dashed lines and the 95% credible interval in shade regions. Two markers denote the two results for pure neutron stars. Besides the results of the soft QMF EOS [21], we also include those of the stiff DD2 EOS [29]. Figure taken from [16].

double pulsar system PSR J0737-3039. The moment of inertia of PSR J0737-3039 A is expected to be measured with ~ 10% accuracy within the next decade. The posterior distributions of I as functions of R for both hybrid stars and quark stars are shown in Fig. 3. For the analysis of hybrid stars, both the soft QMF EOS [21] and the stiff DD2 EOS [29] are employed for the low-density hadronic EOS. One observes that the two I vs. R distributions are located in separate regions and become more and more separated with each other when increasing R, stressing the idea that it can be possible to distinguish them from simultaneous measurement of the stellar radius and the moment of inertia, provided that the accuracy of the radius measurement is at least ~ 1 km. In particular, we calculate the moment of inertia of PSR J0737-3039 A to be $I_A = 1.27^{+0.18}_{-0.14} \times 10^{45}\,\mathrm{g\,cm^2}$ ($1.29^{+0.26}_{-0.15} \times 10^{45}\,\mathrm{g\,cm^2}$) within QMF+CSS (DD2+CSS) as a hybrid star, and a smaller value of $1.13^{+0.14}_{-0.13} \times 10^{45}\,\mathrm{g\,cm^2}$ as a quark star.

3 Summary

The EOS of dense stellar matter is an exciting topic in the area of nuclear astrophysics. Although it is known that the degree of freedom is hadron around nuclear saturation density, the QCD phase state for cold, dense matter for intermediate densities is a very complex problem [30]. A great deal of effort is undergoing in the communities of nuclear physics, particle physics, and astrophysics due to its crucial importance related to the stellar birth and evolution, the merger remnants of neutron star binaries, as well as many puzzles like the central engine of short gamma-ray bursts and the plausible mass gap between neutron stars and black holes. For example, see discussions in our recent review [13].

One key issue of the EOS is still unclear: Does the matter go through a phase transition from hadron matter to quark matter at some intermediate densities, or is quark matter the absolute ground state of strongly interacting matter (i.e., the above-mentioned Witten' conjecture)? After decades of speculation, quark stars still serve as a viable alternative physical model for compact stars. For example, besides the usual binary neutron star merger scenario, binary quark stars have been proposed to be the possible origin for the GW170817 event [31]; Also, it is found that a binary quark star merger for some binary configurations could eject a comparable amount of matter (to the binary neutron star case) [32, 33], to account for the electromagnetic observation in the optical/infrared/UV bands (i.e., kilonova); And a magnetar with quark star EOS is preferred as the post-merger remnant to explain some groups of short gamma-ray burst observations [34, 35].

In the present contribution, we review our recent works on the possibility of distinguishing neutron stars from quark stars from their observed global properties. We mainly find that it is possible to distinguish between the two through high-accuracy radius measurement possible in the near future with joint efforts of the multimessenger missions like aLIGO/Virgo, NICER, SKA. Other possible methods include e.g., the post-merger gravitational wave signals with the help of merger simulation, see discussions in e.g., [32, 33, 36–39].

Acknowledgements

I appreciate discussions with Sophia Han, Jinliang Jiang, Baoyuan Sun, Shaopeng Tang, Chengjun Xia, Enping Zhou, Xia Zhou, Zhenyu Zhu, and the XMU neutron star group. I am also thankful to the latter for assistance with figures. The work is supported by the National SKA Program of China (No. 2020SKA0120300), the National Natural Science Foundation of China (grant No. 11873040), the science research grants from the China Manned Space Project (No. CMS-CSST-2021-B11), and the Youth Innovation Fund of Xiamen (No. 3502Z20206061).

References

[1] Bodmer A. R., Physical Review D **4**, 1601 (1971)
[2] Witten E., Physical Review D **30**, 272 (1984)
[3] Arcones A., et al., arXiv e-prints **ar**Xiv:1603.02213 (2016)
[4] Antoniadis J., et al., Science **340**, 448 (2013)
[5] Linares M., Shahbaz T., Casares J., The Astrophysical Journal **859**, 54 (2018)
[6] Cromartie H. T., et al., Nature Astronomy **4**, 72 (2020)
[7] Miller M. C., et al., The Astrophysical Journal **887**, L24 (2019)
[8] Riley T. E., et al., The Astrophysical Journal **887**, L21 (2019)
[9] Miller M. C., et al., The Astrophysical Journal **918**, L28 (2021)
[10] Riley T. E., et al., The Astrophysical Journal **918**, L27 (2021)
[11] Abbott B. P., et al., Physical Review Letters **121**, 161101 (2018)
[12] Xia C., Zhu Z., Zhou X., Li A., Chinese Physics C **45**, 055104 (2021)
[13] Li A., et al., Journal of High Energy Astrophysics **28**, 19 (2020)
[14] Miao Z., Li A., Zhu Z., Han S., The Astrophysical Journal **904**, 103 (2020)
[15] Miao Z., Jiang J.-L., Li A., Chen L.-W., The Astrophysical Journal **917**, L22 (2021)
[16] Miao Z., Li A., Dai Z.-G., arXiv e-prints **ar**Xiv:2107.07979 (2021)
[17] Li A., Miao Z., Han S., Zhang B., The Astrophysical Journal **913**, 27 (2021)
[18] Li A., Miao Z.-Q., Jiang J.-L., Tang S.-P., Xu R.-X., Monthly Notices of the Royal Astronomical Society **506**, 5916 (2021)
[19] Weber F., Progress in Particle and Nuclear Physics **54**, 193 (2005)
[20] Li A., Zuo W., Peng G. X., Physical Review C **91**, 035803 (2015);
[21] Zhu Z.-Y., Zhou E.-P., Li A., The Astrophysical Journal **862**, 98 (2018)
[22] Alford M. G., Han S., Prakash M., Physical Review D **88**, 083013 (2013)
[23] Li A., Burgio G. F., Lombardo U., Peng G. X., International Journal of Modern Physics E **17**, 1635 (2008)
[24] Hessels J. W. T., et al., Science **311**, 1901 (2006)
[25] Burgio G. F., Schulze H.-J., Li A., Physical Review C **83**, 025804 (2011)
[26] Li A., Zhou X. R., Burgio G. F., Schulze H.-J., Physical Review C **81**, 025806 (2010)
[27] Li A., Peng G.-X., Lu J.-F., Research in Astronomy and Astrophysics **11**, 482 (2011)
[28] Lyne A. G., et al., Science **303**, 1153 (2004)

[29] Fortin M., et al., Physical Review C **94**, 035804 (2016)
[30] Gorda T., Kurkela A., Paatelainen R., Säppi S., Vuorinen A., Physical Review Letters **127**, 162003 (2021)
[31] Zhou E.-P., Zhou X., Li A., Physical Review D **97**, 083015 (2018)
[32] Bauswein A., et al., Physical Review Letters **103**, 011101 (2009)
[33] Zhu Z., Rezzolla L., Physical Review D **104**, 083004 (2021)
[34] Li A., et al., Physical Review D **94**, 083010 (2016)
[35] Li A., Zhu Z.-Y., Zhou X., The Astrophysical Journal **844**, 41 (2017)
[36] Most E. R., Weih L. R., Rezzolla L., Schaffner-Bielich J., Physical Review Letters **120**, 261103 (2018)
[37] Montaña G., Tolós L., Hanauske M., Rezzolla L., Physical Review D **99**, 103009 (2019)
[38] Chatziioannou K., Han S., Physical Review D **101**, 044019 (2020)
[39] Zhou E., Kiuchi K., Shibata M., Tsokaros A., Uryu K., arXiv e-prints **ar**Xiv:2111.00958 (2021)

First measurement of ^{25}Al+p resonant scattering relevant to the astrophysical reaction ^{22}Mg(α,p)^{25}Al

J. Hu[1,2,*], *H.* Yamaguchi[3,4], *Y.H.* Lam[1,2], *A.* Heger[5,6,7,8], *D.* Kahl[9,10], *A.M.* Jacobs[8,11], *Z.* Johnston[8,11], *S.W.* Xu[1], *N.T.* Zhang[1], *S.B.* Ma[1], *L.H.* Ru[1], *E.Q.* Liu[1], *T.* Liu[1], *S.* Hayakawa[3], *L.* Yang[3], *H.* Shimizu[3], *C.B.* Hamill[10], *A. St J.* Murphy[10], *J.* Su[12], *X.* Fang[13], *K.Y.* Chae[14], *M.S.* Kwag[14], *S.M.* Cha[14], *N.N.* Duy[14,15], *N.K.* Uyen[14], *D.H.* Kim[14], *R.G.* Pizzone[16], *M.* La Cognata[16], *S.* Cherubini[16], *S.* Romano[16,17,18], *A.* Tumino[16,19], *J.* Liang[20], *A.* Psaltis[20], *M.* Sferrazza[21], *D.* Kim[22], *Y.Y.* Li[1,2], and *S.* Kubono[3,23]

[1]Institute of Modern Physics, Chinese Academy of Sciences, Lanzhou 730000, China.
[2]School of Nuclear Science and Technology, University of Chinese Academy of Sciences, Beijing 100049, China
[3]Center for Nuclear Study(CNS), University of Tokyo, RIKEN campus, 2-1 Hirosawa, Wako, Saitama 351-0198, Japan
[4]National Astronomical Observatory of Japan, 2-21-1 Osawa, Mitaka, Tokyo 181-8588, Japan
[5]School of Physics and Astronomy, Monash University, Victoria 3800, Australia
[6]OzGrav-Monash – Monash Centre for Astrophysics, School of Physics and Astronomy, Monash University, VIC 3800, Australia
[7]Center of Excellence for Astrophysics in Three Dimensions (ASTRO-3D), Australia
[8]The Joint Institute for Nuclear Astrophysics, Michigan State University, East Lansing, MI 48824, USA
[9]Extreme Light Infrastructure - Nuclear Physics, IFIN-HH, 077125 Bucharest-Măgurele, Romania
[10]SUPA, School of Physics & Astronomy, University of Edinburgh, Edinburgh EH9 3FD, United Kingdom
[11]Department of Physics and Astronomy, Michigan State University, East Lansing, MI 48824, USA
[12]College of Nuclear Science and Technology, Beijing Normal University, Beijing 100875, China
[13]Sino-French Institute of Nuclear Engineering and Technology, Sun Yat-Sen University, Zhuhai 519082, Guangdong, China
[14]Department of Physics, Sungkyunkwan University, Suwon 16419, Korea
[15]Institute of Research and Development, Duy Tan University, Da Nang 550000, Vietnam
[16]Laboratori Nazionali del Sud-INFN, Via S. Sofia 62, Catania 95123, Italy
[17]Dipartimento di Fisica e Astronomia "Ettore Majorana" - Università degli Studi di Catania
[18]Centro Siciliano di Fisica Nucleare e Struttura della Materia (CSFNSM)
[19]Facoltà di Ingegneria e Architettura, Università degli Studi di Enna "Kore", Enna 94100, Italy
[20]Department of Physics & Astronomy, McMaster University, Ontario L8S 4M1, Canada
[21]Département de Physique, Université Libre de Bruxelles, Bruxelles B-1050, Belgium
[22]Department of Physics, Ewha Womans University, Seoul 03760, Korea
[23]RIKEN Nishina Center, 2-1 Hirosawa, Wako, Saitama 351-0198, Japan

Abstract. Type I X-ray bursts (XRBs) are the most frequently observed thermonuclear explosions in nature. The ^{22}Mg(α,p)^{25}Al reaction plays a critical role in XRB models. However, experimental information is insufficient to deduce a precise ^{22}Mg(α,p)^{25}Al reaction rate for the respective XRB temperature range. A new measurement of ^{25}Al+p resonant scattring was performed up to the astrophysically interested energy region of ^{22}Mg(α,p)^{25}Al. Several resonances

*e-mail: hujunbaggio@impcas.ac.cn

were observed in the excitation functions, and their level properties have been determined based on an R-matrix analysis. In particular, proton widths and spin-parities of four natural-parity resonances above the α threshold of ^{26}Si, which can contribute the reaction rate of ^{22}Mg(α,p)^{25}Al, were first experimentally determined.

1 Introduction

Type I X-ray bursts are the most frequently observed thermonuclear explosions in nature [1–3]. These astrophysical phenomena occur in the envelope of accreting neutron star in low-mass X-ray binary systems [4, 5]. The investigation of X-ray bursts is significant to the understanding of the neutron star's properties and the underlying physics [6]. The bursts are driven by the tripe-α reaction, the αp-process [7] and the rp-process [8, 9]. After the breakout from the hot CNO cycle, the nucleosynthesis path is characterized by the αp-process which is the dominant process in sd-shell nuclei. The αp-process is a sequence of α- and proton-induced reactions that rapidly transport nuclear material, along the neutron deficient side of the stable nuclei, from the CNO cycle toward heavier masses region.

The X-ray light curve is the only direct observable of X-ray bursts, which is affected significantly by the αp-process. According to a recent sensitivity study by Cyburt *et al.* [10], the ^{22}Mg(α, p)^{25}Al reaction is thought to be the most sensitive one within the αp-process according to the calculation of multi-zone burst model (KEPLER code) and has a prominent impact on the burst light curve. However, the reaction rates were calculated using the statistical models due to the missing of important experimental information.

2 Experimental method

A resonant scattering measurement of ^{25}Al+p has been performed to experimentally examine the ^{22}Mg(α, p)^{25}Al reaction rates. The experiment was carried out using the CNS radioactive ion beam separator (CRIB) [11], installed by the Center for Nuclear Study (CNS), University of Tokyo, in the RIKEN Accelerator Research Facility. A primary beam of ^{24}Mg^{8+} was accelerated up to 8.0 MeV/u by the AVF cyclotron (K = 70) with an average intensity of 1 eμA. The primary beam bombarded a liquid-nitrogen-cooled D_2 gas target [12] where a secondary beam of ^{25}Al was produced via the ^{24}Mg(d, n)^{25}Al reaction in inverse kinematics. The D_2 gas at 200 Torr and 90 K was confined in a small cell with a length of 80 mm. The entrance and exit windows were made of 2.5 μm thick Havar foils. The ^{25}Al beam was separated by the CRIB separator using the in-flight method. The ^{25}Al beam, with an energy of 142 ± 1 MeV and an average intensity of 2.0×10^5 pps, was then delivered to F3 experimental chamber and bombarded a 150-μm-thick $(CH_2)_n$ target in which the beam was stopped. The setup at F3 experimental chamber is shown in figure 1, which is quite similar to that used in Ref. [13]

A PPAC (Parallel Plate Avalanche Counter) [14] and a MCP (Micron Channel Plate) [15] were used for measuring time and position information of the beam particles. The beam particles were identified in an event-by-event mode using the abscissa of MCP, and the ToF between MCP and the RF signal provided by the cyclotron. After passing through a Wien-Filter, the ^{25}Al beam purity was typically 70%. Figure 2(a) shows the beam particle identification before the secondary target.

The recoiling light particles were measured using three sets of Si telescopes at average angles of θ_{lab} = 0°, 20° and 23°, respectively. Each telescope consisted of a 65-μm-thick

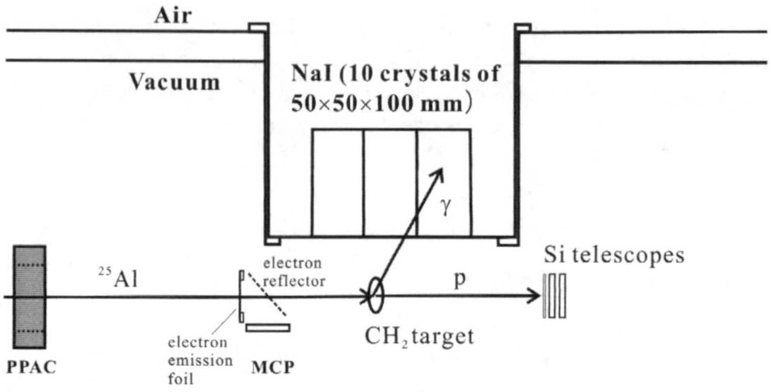

Figure 1. Schematic diagram (side view) of the experimental setup at F3 chamber.

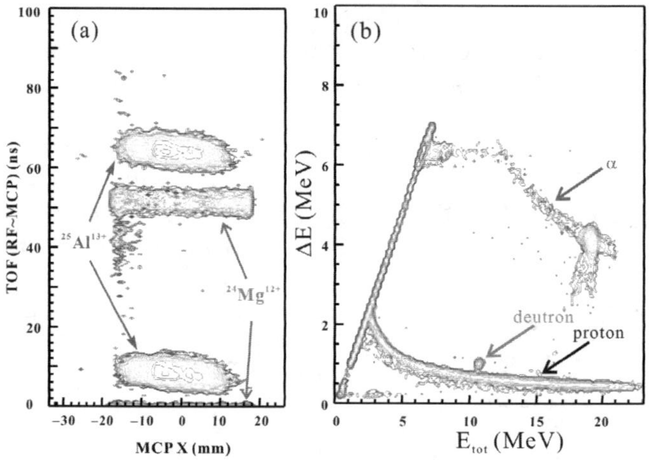

Figure 2. (a) Identification plot for the beams via the ToF technique. (b) Identification plot for the recoiling particles via the ΔE-E technique.

double-side-strip (16×16 strips) silicon detector and two 1500-μm-thick pad detectors. The recoiling particles were clearly identified by using the ΔE-E method as shown in figure 2(b). An array of ten NaI detectors was mounted directly above the target and used to detect the γ rays from the decay of the excited states in ^{25}Al. Each NaI detector is with a geometry of 50 × 50 × 100 mm, covering 20% of the total solid angle altogether. In addition, an 80-μm-thick carbon target was used in a separate run for evaluating the background contribution.

The de-excitation γ rays in the inelastic events have been measured with the NaI detectors. These detectors have an average energy resolution of 13.5% (FWHM) for 662-keV γ rays. The 450-keV photopeak detection efficiency ϵ was measured as 5%, using γ-ray sources placed at the target position. The γ-ray enery spectrum of proton-γ coincident events showed a peak at 450 keV and the contribution to the excitation function by the inelastic scattering events was successfully deduced. The inelastic contribution, less than 12% of the elastic scattering, was subtracted from the total excitation function.

$E_{c.m.}$ resolution of the excitation function was 30-90 keV in full width at half maximum (FWHM) for the Si telescope at $\theta_{lab} = 0°$. The uncertainty was mostly from energy straggling of the particles in the thick target, along with the energy resolution of the silicon detectors. At larger angles, the angular resolution of the recoiling proton produced large energy uncertainty and the resulting energy resolution was 75-200 keV at $\theta_{lab} \sim 20°$.

3 Exprimental results and R-matrix analysis

The excitation function of ^{25}Al+p elastic scatering has been reconstructed using the procedure decribed previously [13, 16–18]. The excitation function of elastic scattering for the proton target was deduced by subtracting the carbon and inelastic scattering contributions. The excitation function around $\theta_{lab} = 0°$ is shown in figure 3; data from the other two telescopes were not included in the analysis due to their awful energy resolution. Several resonant structures were clearly observed in the spectrum. In order to determine the resonant parameters of observed resonances, R-matrix calculations have been performed with the R-matrix code AZURE2 [19]. A channel radius of $R = 1.4 \times (1 + 25^{1/3}) \approx 5.5$ fm appropriate for the ^{25}Al+p system has been utilized in the calculation.

For the discussions on the level properties obtained by R-matrix analysis and the astrophysical implications on the X-ray bursts, please refer to our published paper [20].

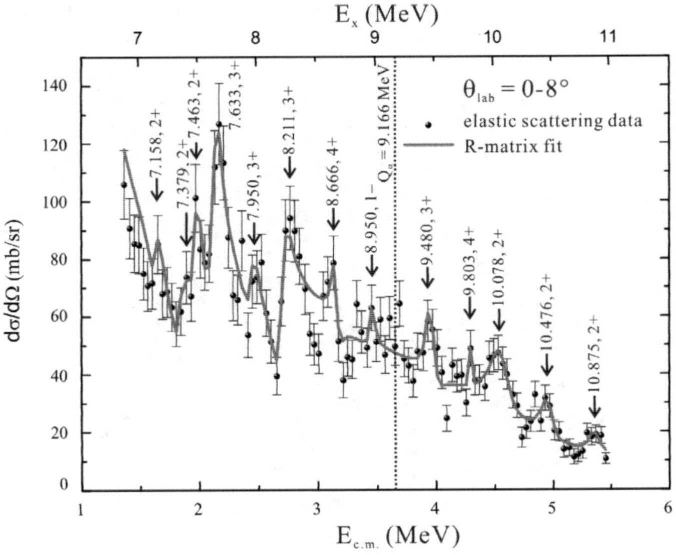

Figure 3. Excitation function of ^{25}Al+p elastic scattering at θ_{lab} = 0-8°. The red curve represents the best overall R-matrix fit.

References

[1] W.H.G. Lewin et al., Space Sci. Rev. **62**, 223 (1993)
[2] H. Schatz et al., Nucl. Phys. **A 777**, 601 (2006)
[3] A. Parikh et al., Prog. Part. Nucl. Phys. **69**, 225 (2013)
[4] S.E. Woosley, R.E. Taam, Nature **263**, 101 (1976)
[5] P.C. Joss, Nature **270**, 310 (1977)
[6] A.W. Steiner et al., Astrophys. J. **722**, 33 (2010)
[7] S.E. Woosley, T.A. Weaver, in *High Energy Transients in Astrophysics, AIP Conf. Proc., vol. 115*, edited by S.E. Woosley (AIP, New York, 1984), p. 273

[8] R.K. Wallace et al., Astrophys. J. Suppl. **45**, 389 (1981)
[9] M. Wiescher et al., Astrophys. J. **316**, 162 (1987)
[10] R. Cyburt et al., Astrophys. J. **830**, 55 (2016)
[11] Y. Yanagisawa et al., Nucl. Instr. Meth. **A 539**, 74 (2005)
[12] H. Yamaguchi et al., Nucl. Instr. Meth. **A 589**, 150 (2008)
[13] H. Yamaguchi et al., Phys. Lett. **B 672**, 230 (2009)
[14] H. Kumagai et al., Nucl. Instr. Meth. **A 470**, 562 (2001)
[15] S. Hayakawa, M. Sc. thesis, Department of Physics, University of Tokyo (2008)
[16] T. Teranishi et al., Phys. Lett. **B 556**, 27 (2003)
[17] T. Teranishi et al., Phys. Lett. **B 650**, 129 (2007)
[18] J.J. He et al., Phys. Rev. **C 76**, 055802 (2007)
[19] R.E. Azuma et al., Phys. Rev. **C 81**, 045805 (2010)
[20] J. Hu et al., Phys. Rev. Lett. **127**, 172701 (2021)

Effects of Nuclear Equation of State on Type-I X-ray Bursts and Implication for *Clocked Burster* GS 1826–24

Akira Dohi[1,2,*], *Nobuya* Nishimura[3,4], *Masa-aki* Hashimoto[1], *Yasuhide* Matsuo[1], *Tsuneo* Noda[5], and *Shigehiro* Nagataki[3]

[1]Department of Physics, Kyushu University, Fukuoka 819-0395, Japan
[2]Interdisciplinary Theoretical and Mathematical Sciences Program (iTHEMS), RIKEN, Wako, Saitama 351-0198, Japan
[3]Astrophysical Big Bang Laboratory, CPR, RIKEN, Wako, Saitama 351-0198, Japan
[4]RIKEN Nishina Center for Accelerator-Based Science, Wako, Saitama 351-0198, Japan
[5]Department of Education and Creation Engineering, Kurume Institute of Technology, Kurume, Fukuoka 830-0052, Japan

Abstract. We study the effects of nuclear equation of state (EOS) on Type-I X-ray bursts, focusing on the *Clocked burster* GS 1826–24. According to the *shell-flush* model, the surface gravity of the neutron stars (NSs) strengthens burst activities, where the recurrence time and peak luminosity are higher for larger-radius EOSs. We show that such a prediction matches with the *full* multi-zone model, in which whole NS regions are considered in burst calculation. We also show that the large-radius EOSs are not preferred to account for the observed light curves of GS 1826–24.

1 Introduction

Type-I X-ray bursts, which are caused by unstable H or He burning onto a neutron star (NS), are valuable phenomena to understand the various properties of X-ray binaries such as the accretion rate, compositions of accreted matter, and nuclear reaction rates relevant to the rp process. Few observed bursters show the constant shape of light curves in a burst sequence. They are called clocked bursters, which are useful for constraining the model parameters of low-mass X-ray binaries. Several works have been done for modeling of a representative clocked burster, GS 1826–24, and finally constrained model parameters. These studies, however, mostly focused on the outer layers of the NS (e.g., [1]).

The physical properties inside the NS, such as the nuclear equation of state (EOS), crustal heating, and neutrino cooling processes, can affect the X-ray-burst light curves. Thanks to the recent developments of NS observations, the nuclear EOS has been rigidly constrained. In fact, the observation of the highest NS mass beyond $2M_\odot$ [2–4] ruled several out softer EOSs. For the NS radius, several experimental and observational constraints have been established (for a recent review, see [5]), but there remains significant uncertainties.

In this study, we investigate the dependence of light curves for X-ray bursts on the NS EOS, focusing on the uncertainties of the NS radius. We compare our calculated burst models with the observations of GS 1826–24.

*e-mail: dohi@phys.kyushu-u.ac.jp

2 The methods of multi-zone X-ray burst calculations

Table 1. Physical quantities at the nuclear saturation density for adopted EOSs

EOS	ρ_0 [10^{14} g cm^{-3}]	w_0 [MeV]	K_0 [MeV]	$E_{\rm sym}$ [MeV]	L [MeV]
Togashi	2.66	−16.0	245	30.0	35.0
TM1	2.41	−16.3	281	36.9	111
TM1e	2.41	−16.3	281	31.4	40.0
LS220	2.57	−16.0	220	28.6	73.8

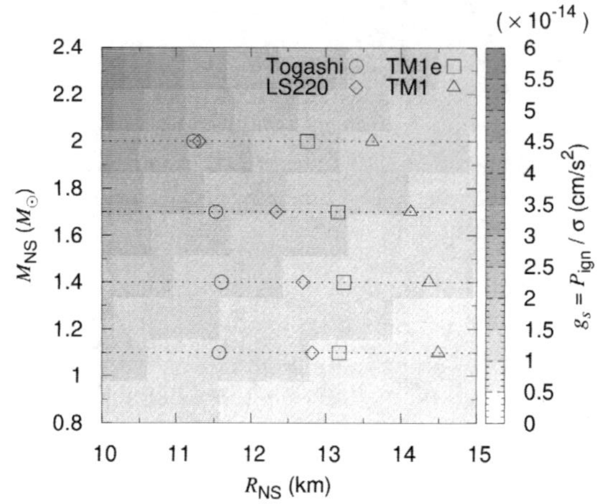

Figure 1. The surface gravity g_s on the mass–radius plane based on the *shell-flash* model. The symbols indicate the values with our adopted EOSs.

To calculate X-ray-burst light curves considering the central NS, we use a one-dimensional general-relativistic stellar-evolution code [6] (hereinafter called *full* multi-zone model). Namely, we follow the thermal evolution of accreting NSs in quasi-hydrostatic equilibrium. To reduce the costs of numerical calculations, we adopt a reduced nuclear-reaction network with 88 nuclei for H or He mixed burning (APRX3 in [7]). For time evolution of H and He mass fractions and energy generation rates, APRX3 reproduces the resulting energy generation rates with a large reaction network with 897 nuclei within the 40% error.

We implement sets of realistic nuclear EOSs of Togashi [8], LS220 [9], TM1e [10], and TM1 [11]. The information of NS EOS such as the pressure and energy density comes from the energy per nucleon $w(\rho_{\rm B}, Y_p)$, where $\rho_{\rm B}$ is the baryon density and Y_p is the proton fraction, which can be expanded around the saturation density ρ_0:

$$w(\rho_{\rm B}, Y_p) = w_0 + \frac{K_0}{18\rho_0^2}(\rho_{\rm B} - \rho_0)^2 + \cdots + \left[E_{\rm sym} + \frac{L}{3\rho_0}(\rho_{\rm B} - \rho_0) + \cdots\right]\left(1 - 2Y_p\right)^2, \quad (1)$$

where w_0, K_0, $E_{\rm sym}$, and L are, respectively, energy per nucleon, incompressibility, symmetry energy, and its slope at ρ_0. Thus, any EOS can be mostly characterised by the five parameters. For our adopted EOSs, we show their values in Table 1. The important parameter for the NS radius is L, being still uncertain (e.g., [12]). In our EOSs, the TM1 has the largest L value and radius of $R_{1.4}$ = 14.3 km, while the Togashi smallest L value and $R_{1.4}$ = 11.6 km, where

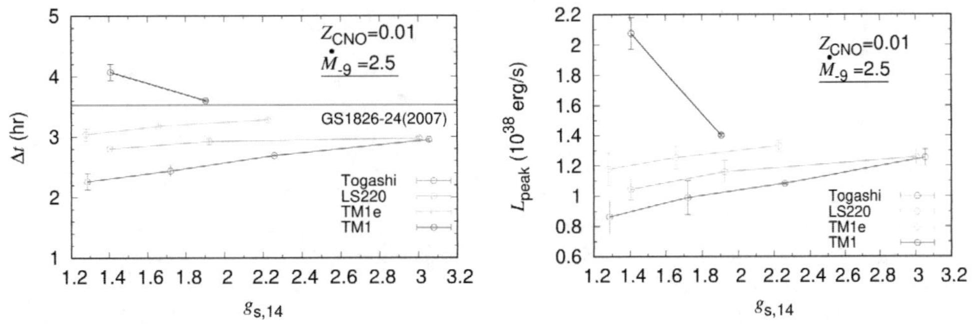

Figure 2. Δt and $L_{\rm peak}$ plotted by $g_{\rm s,14}$ with different EOSs. The horizontal solid line corresponds to the observed value of GS 1826–24 in 2007 [17]. Some burst models are omitted if hydrostatic equilibrium conditions are not achieved.

$R_{1.4}$ is the radius at $M_{\rm NS} = 1.4\ M_\odot$. Thus, the radius is highly different in the range of about 11–14 km, depending on the symmetry energy.

The relation between the mass–radius and burst behavior is roughly expressed by the shell-flush model, assuming the constant pressure during a burning onto the NS [13]:

$$P_{\rm ign} = g_{\rm s}\sigma, \tag{2}$$

where σ is the column depth and $g_{\rm s}$ is the surface gravity, which considers the general relativistic effect. We show the value of g_s in Figure 1. Suppose that $P_{\rm ign}$ is independent of the NS mass and radius, σ is higher with a large-radius EOS. This implies that the observable recurrence time Δt and the peak luminosity $L_{\rm peak}$ should also be higher. The EOS uncertainties affect burst behavior, which should be clarified in full multi-zone X-ray burst models.

Since $P_{\rm ign}$ comes from the heating and cooling processes inside a NS, it should not be a free parameter, unlike the shell-flush model. Many previous studies with multi-zone X-ray-burst models, which cover the only accreted layers, introduce a similar artificial heating parameter Q_b to impose the envelope and crust boundary conditions (e.g., [1]). To treat the internal energy of NS consistently, we consider the conventional crustal heating process [14] and slow neutrino cooling processes, including the modified Urca and bremsstrahlung (e.g., see [15]). Such cooling processes effectively increase $P_{\rm ign}$, because they decrease the overall temperature before the NS becomes very old ($\gtrsim 10^5$ yrs). Therefore, they change the burst behavior from Eq. (2). In the proceeding, however, we simply focus on the EOS dependence, which does not significantly change neutrino luminosity without any enhanced cooling processes.

To focus on the EOS uncertainties, we fix $\dot{M}_{-9} = 2.5$ and $Z_{\rm CNO} = 0.01$ in the proceeding, where $\dot{M}_{-9}$ is the accretion rate normalized by $10^{-9}\ M_\odot\ {\rm yr}^{-1}$ and $Z_{\rm CNO}$ is the initial metalicity. Note that the influence on light curves is crucially important, but the trend of EOS dependence does not change even if other parameters related to the companion are chosen. As the initial model for calculations, we use the steady-state model of accreting NSs (see [16]).

3 Results

In Figure 2, we show the dependence of Δt and $L_{\rm peak}$ on the EOS. As we see, if the radius is smaller, Δt and $L_{\rm peak}$ are lower because the amount of accreted matter is increased due to a stronger gravitational field. It finally shorten the ignition time, and therefore this trend is

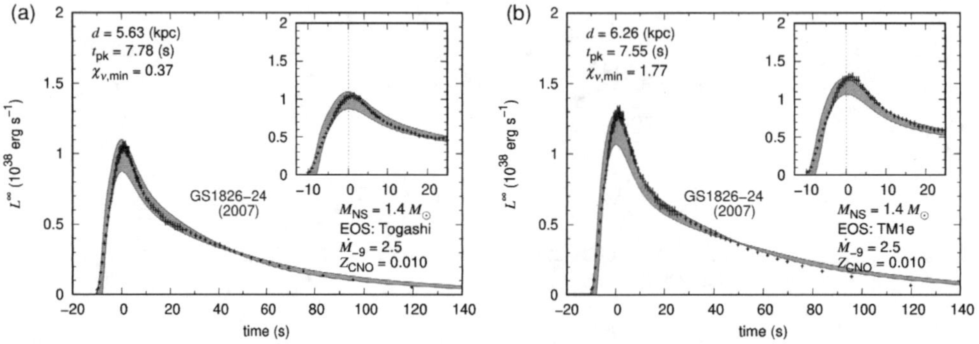

Figure 3. Comparison of calculated averaged burst light curves with observed ones of GS 1826–24 in 2007, assuming $\dot{M}_{-9} = 2.5, Z_{\rm CNO} = 0.01$ and $M_{\rm NS} = 1.4\ M_\odot$: (a) Togashi , (b) TM1e. We set $t = 0$ s at the peak time of light curves and plot within 1σ regions of many burst light curves. The distance including burst anisotropy d, the time after the peak $t_{\rm pk}$, and the minimum χ^2 value $\chi_{\nu,\rm min}$ are also shown.

consistent with the results of shell-flush model. For the Δt, we compare our models with the observation of GS 1826–24 in 2007 [17]. There is no model matching with observed Δt, but if $\dot{M}_{-9}, Z_{\rm CNO}$ and NS mass change, it could be reproduced (see the model of "L2n20Z1" in [7]; $\dot{M}_{-9} = 2.0, Z_{\rm CNO} = 0.01$, LS220 EOS, and $M_{\rm NS} = 1.58\ M_\odot$.).

We show that the comparison with the observed light curve of GS 1826–24 in 2007 in Figure 3. For the TM1e, the peak luminosity is higher than the Togashi due to the fuel amount difference. Thus, the Togashi agrees with the shape of observed light curves, while the TM1e disagrees. The luminosity is too high in large-radius NSs and sometimes reaches the Eddington luminosity. However, GS 1826–24 do not show photospheric radius expansion. Therefore, large-radius EOSs (e.g., TM1) tend to be rejected, although the model comparison highly depends on other input parameters, e.g., $\dot{M}_{-9}$ and $Z_{\rm CNO}$.

4 Conclusion

We study the dependence of the NS EOS on Type-I X-ray burst using multi-zone X-ray-burst models, with an emphasis on the observation of a clocked burster GS 1826–24. We find that Δt and $L_{\rm peak}$ are higher with larger-radius EOSs, whose trend are consistent with the shell-flush model. As a result of the comparison with the observation of GS 1826–24, we also find that larger-radius EOS is unpreferred due to higher luminosities. Since the EOS is related to internal heating and cooling processes, self-consistent burst calculations including the evolution of a NS are necessary. In the burst calculations presented here, we adopt the conventional heating and cooling processes, which should be updated. The more detailed analysis which are done with a series of X-ray burst calculations are shown in the paper [18].

References

[1] Z. Johnston, A. Heger, D.K. Galloway, Mon. Not. Roy. Astron. Soc. **494**, 4576 (2020), `arXiv:1909.07977`

[2] P.B. Demorest *et al.*, Nature **467**, 1081 (2010), `arXiv:1010.5788`

[3] J. Antoniadis *et al.*, Science **340**, 448 (2013), `arXiv:1304.6875`

[4] H.T. Cromartie *et al.*, Nat. Astron. **4**, 72 (2020), `arXiv:1904.06759`

[5] G.F. Burgio, H.J. Schulze, I. Vidaña, J.B. Wei, Prog. Part. Nucl. Phys. **120**, 103879 (2021), arXiv:2105.03747
[6] M.Y. Fujimoto, T. Hanawa, J. Iben, I., M.B. Richardson, Astrophys. J. **278**, 813 (1984)
[7] A. Dohi, M.a. Hashimoto, R. Yamada, Y. Matsuo, M.Y. Fujimoto, Prog. Theor. Exp. Phys. **2020**, 033E02 (2020), arXiv:2001.08367
[8] H. Togashi, K. Nakazato, Y. Takehara, S. Yamamuro, H. Suzuki, M. Takano, Nucl. Phys. A**961**, 78 (2017), arXiv:1702.05324
[9] J.M. Lattimer, D.F. Swesty, Nucl. Phys. A**535**, 331 (1991)
[10] H. Shen, F. Ji, J. Hu, K. Sumiyoshi, Astrophys. J. **891**, 148 (2020), arXiv:2001.10143
[11] H. Shen, H. Toki, K. Oyamatsu, K. Sumiyoshi, Astrophys. J. Suppl. **197**, 20 (2011), arXiv:1105.1666
[12] B.A. Li, B.J. Cai, W.J. Xie, and N.B. Zhang, Univ., **7**, 182 (2021)
[13] M.Y. Fujimoto, T. Hanawa, S. Miyaji, Astrophys. J. **247**, 267 (1981)
[14] P. Haensel, J.L. Zdunik, Astron. Astrophys. **227**, 431 (1990)
[15] A. Dohi, K. Nakazato, M.a. Hashimoto, M. Yasuhide, T. Noda, Prog. Theor. Exp. Phys. **2019**, 113E01 (2019), arXiv:1910.01431
[16] H. Liu, A. Dohi, M.a. Hashimoto, Y. Matsuo, G.L. Lü, T. Noda, Phys. Rev. D**103**, 063009 (2021), arXiv:2102.04012
[17] D.K. Galloway, A.J. Goodwin, L. Keek, Publ. Astron. Soc. Australia. **34**, e019 (2017), arXiv:1703.07485
[18] A. Dohi, N. Nishimura, M. Hashimoto, Y. Matsuo, T. Noda, S. Nagataki, Astrophys. J. **923**, 16 (2021),

EPJ Web of Conferences **260**, 06001 (2022) https://doi.org/10.1051/epjconf/202226006001
NIC-XVI

Type Ia SN progenitors: pre-explosion phase in nearly Chandrasekhar mass WDs

Inma Domínguez[1,*], *Luciano* Piersanti[2], *Eduardo* Bravo[3], *Oscar* Straniero[2], and *Sergio* Cristallo[2]

[1]Universidad de Granada 18071, Granada, Spain
[2]INAF-Osservatorio Astronomico d'Abruzzo, I-64100, Teramo, Italy
[3]E.T.S. Arquitectura del Vallés, Universitat Politècnica de Catalunya, 08173 Sant Cugat del Vallès, Spain

Abstract. Type Ia supernovae are used as distance indicators to measure the expansion rate of the Universe and to constrain the nature of dark energy. Current and upcoming surveys will allow to extend supernova Hubble diagrams to higher redshifts and to improve further their statistics.
It is accepted that Type Ia supernovae are thermonuclear explosions of carbon-oxygen white dwarfs in binary systems. However, the identification of their progenitors, the evolutionary path leading to the explosion and the explosion mechanism itself have not been identified yet. This is critical, as we need to understand the potential evolution of their luminosity with cosmic time and, thus, with their stellar progenitors.
We will review the current situation, considering observational hints. We will focus on our recent models, that follow the evolution of carbon-oxygen white dwarfs accreting mass up to thermonuclear runaway, and on their dependence with the initial metallicity of the white dwarf progenitors.

1 Introduction

Bright and homogeneous, Type Ia Supernovae (SNeIa) are our best extragalactic distance indicators. Since the 60s SNIa Hubble diagrams (HDs) have been used to measure the value of the Hubble-Lemaitre constant, which is still a subject of hot debate [e.g. 1]. SNIa HDs show the recent acceleration of the expansion rate of the universe [2, 3], and its previous deceleration [4]. The challenge today is to reach enough precision in SNIa HDs to reveal the nature of that enigmatic component of the Universe, which currently dominates its dynamics, the dark energy [see 5]. SNeIa are not equal, and the precision of SNeIa as distance indicators relies on calibration relations, such as the maximum-decline relation [6], that link maximum absolute magnitudes with the shape of the light curve peaks and colours. New calibration relations are proposed trying to increase the precision [7], which in current HDs is above 0.15 mag. Moreover, SNeIa play a central role in the origin and evolution of the nuclei, in particular, of the so called iron group nuclei, being responsible of about 60% of their universal abundances [e.g. 8, 9].

*e-mail: inma@ugr.es

In spite of their central role in cosmology, in nucleosynthesis, and in other fields, SNIa progenitors and explosion mechanisms are still unknown. The classical scenario consists of a thermonuclear explosion of a carbon-oxygen white dwarf (CO WD), which accreting mass from a companion reachs explosion conditions when its mass is very close to the Chandrasekhar mass limit [10], MCh scenario. Starting from C-ignition in the central inner zone, the whole WD is incinerated and its ashes contain intermediate-mass elements and a substantial amount of iron group nuclei. Among these nuclei, ^{56}Ni is mostly produced, and it is the radioactive energy from the cascade ^{56}Ni–^{56}Co–^{56}Fe that powers SNIa light curves (LCs). More ^{56}Ni implies more luminosity at maximum. This simple scenario perfectly agrees with SNIa homogenity and their tight calibration relation. It also implies that, to a great extent, the memory of the properties of the WD progenitor, such as the initial mass and chemical composition, is lost in the supernova outcomes, i.e., LCs and spectra.

At the other side, there is a growing amount of evidences pointing to a dependence of the LC properties with the local and/or general properties of the host galaxies. Brighter (post-maximum slow decline LCs) SNe Ia are associated with young stellar populations, in which the SNIa rate is higher. Dimmer (post-maximum fast decline LCs) also occur in galaxies in which star formation ceased several Gyr ago, and there, the SNIa rate is smaller. Moreover, after calibration, a dependence remains with the galaxy total stellar mass, and likely with specific star formation rate and metallicity, which may be interpreted as a dependence with the progenitor stellar population age [e.g. 11–19].

It may well be that two or more progenitor systems contribute to the normal (non peculiar) SNIa population. In fact, several progenitor systems have been proposed [see for detailed reviews, 20–23]. A critical property is the total mass of the WD at the explosion time, which is strictly connected to the explosion mechanism. Explosions of sub-Chandrasekhar mass WDs (sub-MCh) have been suggested to represent a significant part of SNe Ia. In these sub-MCh WDs, the explosion starts with a detonation at the external recently accreted He-rich layer. Then, an inward shock moves toward the center triggering a carbon detonation. Also, the nature of the companion object is still under debate: broadly speaking, it could be a normal star (Single Degenerate, or SD, scenario) or another WD (Double Degenerate, or DD, scenario). Other scenarios, that may explain some peculiar events, have been proposed, such as the merging of a CO WD with the CO core of an AGB companion star, or violent mergers of two white dwarfs.

2 Selected hints from observations and implications

Hints of a non-degenerate companion (MS, RG or AGB stars) have not been found yet. Hydrogen is not detected at maximum light epoch (otherwise it would be a Type II SNe) and, in most events, neither at late times, disfavouring H accretion. In the well studied nearby SNe, 2011fe and 2014J, a single degenerate companion has been excluded for 2011fe and a RG or an AGB companion for 2014J [24–27]. Late LCs are expected to show signatures of non-degenerate companions, but no detection has been claimed yet, as in the recent study [28] including above 100 SNe Ia. Although these observations are challenge and more high quality data are needed, they are easily explained within the DD scenario. Moreover, delayed time distributions and rates agree better with the DD scenario [19, 22, 29].

At the other side, SNIa nebular spectra [30, 31] and some Supernova Remnants (like Tycho, Kepler, 3C 397 and N103B) observations [e.g. 32–36] require burning at high densities and thus, better agree with a MCh explosion.

Taking together, observations suggest that the MCh DD scenario plays a major role. But then, Why the outcome of thermonuclear explosions of CO WDs with the same mass (MCh) should be related with the stellar population from which they come from ?

3 Models, results and conclusions

In this work we explore in detail the physical and chemical properties of the pre-explosive evolution of the MCh DD scenario. For all details we refer to our recent accepted article [37].

Based on the 1D hydrostatic code FUNS [38, 39], we follow the evolution of the WD progenitors, the accretion phase up to C-ignition and the subsequent convective C-burning phase up to the explosion, which is assumed to occur when the temperature reaches 800 MK. We include e-captures and all the relevant URCA reactions. The initial model is the one presented in [40], which consists of two CO WDs with initial masses 0.6 $M_\odot$ and 0.8 $M_\odot$ and Z=0.138. As the WD increases its mass due to accretion, it contracts and its density increases, reaching the threshold densities of different e-capture nuclear reactions. At the corresponding URCA shell, the rate of electron captures and beta decays are equal, while in the denser zones, bellow the URCA shell, e-captures dominate and in the less dense zones, above the URCA shell, the corresponding URCA pair nuclei decay. At the URCA shell, all terms in the energy budget cancel each other, except those corresponding to the neutrino and antineutrino emission, so URCA pairs produce cooling at the shell.

As the WD continues to accrete mass and contract, successive URCA shells form, while those previously formed move outwards.

After C-gnition, the situation becomes more complex, convection develops carrying URCA nuclei up and down. In fact, if convective URCA produced a net heating or a net cooling has been an open question for a long time. To calculate properly the URCA energy budget all terms should be consider. In particular, in the high density inner zones, heating is produced due to the thermal energy of the captured electrons, while cooling is produced due to the absorbed thermal energy needed to move down the electrons.

In order to include properly all physical and chemical processes we decide to adopt the following set-up for the simulations: (1) full coupling of physical and chemical evolution, including burning and mixing; (2) the Ledoux criterium to define the border of the convective zones, as the chemical gradients at the URCA shells are relevant and play a critical role; (3) full nuclear network, at least up to ^{32}P, including all relevant e-captures and URCA pairs.

In order to investigate the impact of different initial metallicities of the WD progenitors on the explosion ignition conditions, we adopt 5 different initial metallicities: Z=(0.0245, 0.6, 1.38, 2 and 4) $\times 10^{-2}$, and accrete CO rich matter according to the same procedure described in [40] at a rate of $=10^{-7} M_\odot$ /yr.

In agreement with previous works [40–44] we obtain that electron captures and URCA pairs imply a net cooling of the inner region, C-ignition occurs at higher densities and both, the explosion density and the neutronization levels, are higher as compared with models that do not include URCA reactions.

We identify, and this is a new result with respect to previous calculations, the relevant role played by the ^{21}(Ne,F) URCA pair [45]. The inclusion of this pair, along with the correct URCA energy budget and the Ledoux criterium, stops the growth of the convective core, and hence, stops the inward circulation of nuclei and electrons, and cooling dominates. Thus, the cooling of this isolated core implies that a larger amount of nuclear energy is needed to trigger the explosion, hence, the amount of C-consumption is also larger, as well as the explosion density and neutronization level. As the amount of ^{21}Ne depends on Z, the maximum mass attained by the convective core and the explosion density also depend on Z, except for the lower Z, in which the amount of initial ^{21}Ne is too low and does not play any significant role.

Our results show that:

(1) Convective core masses at the explosion depend on Z, increasing from 0.12 $M_\odot$ to 0.24 $M_\odot$ as Z increases, for the studied models. Note that different convective core masses imply different chemical gradients.

(2) Explosion densities increase with Z, from 5.02 to 6.34 10^9 g/cm^3. These high densities may be a hint of high accretion rates.

(3) Both, the central and average (over the convective core mass) neutronization levels at the explosion time also increase with Z with respect to the corresponding initial values. In principle, if the neutronization level of the inner zones of a SNIa could be deduced from the observed chemical pattern of neutronised elements, it would be possible to infer the initial metallicity of the WD progenitor:

$$Z_{pro} = 10.975 \times \eta_{exp} - 1.094 \times 10^{-3}$$

References

[1] E. Macaulay, R.C. Nichol, D. Bacon, D. Brout, T.M. Davis, B. Zhang, B.A. Bassett, D. Scolnic, A. Möller, C.B. D'Andrea et al., **486**, 2184 (2019), 1811.02376

[2] A.G. Riess, A.V. Filippenko, P. Challis, A. Clocchiatti, A. Diercks, P.M. Garnavich, R.L. Gilliland, C.J. Hogan, S. Jha, R.P. Kirshner et al., **116**, 1009 (1998), astro-ph/9805201

[3] S. Perlmutter, G. Aldering, G. Goldhaber, R.A. Knop, P. Nugent, P.G. Castro, S. Deustua, S. Fabbro, A. Goobar, D.E. Groom et al., **517**, 565 (1999), astro-ph/9812133

[4] A.G. Riess, L.G. Strolger, J. Tonry, S. Casertano, H.C. Ferguson, B. Mobasher, P. Challis, A.V. Filippenko, S. Jha, W. Li et al., **607**, 665 (2004), astro-ph/0402512

[5] M. Betoule, R. Kessler, J. Guy, J. Mosher, D. Hardin, R. Biswas, P. Astier, P. El-Hage, M. Konig, S. Kuhlmann et al., **568**, A22 (2014), 1401.4064

[6] M.M. Phillips, **413**, L105 (1993)

[7] C.R. Burns, M. Stritzinger, M.M. Phillips, E.Y. Hsiao, C. Contreras, S.E. Persson, G. Folatelli, L. Boldt, A. Campillay, S. Castellón et al., **789**, 32 (2014), 1405.3934

[8] F.X. Timmes, S.E. Woosley, T.A. Weaver, **98**, 617 (1995), astro-ph/9411003

[9] N. Prantzos, C. Abia, M. Limongi, A. Chieffi, S. Cristallo, **476**, 3432 (2018), 1802.02824

[10] F. Hoyle, W.A. Fowler, **132**, 565 (1960)

[11] M. Hamuy, M.M. Phillips, N.B. Suntzeff, R.A. Schommer, J. Maza, R. Aviles, **112**, 2398 (1996), astro-ph/9609062

[12] M. Hamuy, S.C. Trager, P.A. Pinto, M.M. Phillips, R.A. Schommer, V. Ivanov, N.B. Suntzeff, **120**, 1479 (2000), astro-ph/0005213

[13] F. Mannucci, M. Della Valle, N. Panagia, E. Cappellaro, G. Cresci, R. Maiolino, A. Petrosian, M. Turatto, **433**, 807 (2005), astro-ph/0411450

[14] M. Sullivan, D. Le Borgne, C.J. Pritchet, A. Hodsman, J.D. Neill, D.A. Howell, R.G. Carlberg, P. Astier, E. Aubourg, D. Balam et al., **648**, 868 (2006), astro-ph/0605455

[15] Y. Kang, Y.W. Lee, Y.L. Kim, C. Chung, C.H. Ree, **889**, 8 (2020), 1912.04903

[16] L. Kelsey, M. Sullivan, M. Smith, P. Wiseman, D. Brout, T.M. Davis, C. Frohmaier, L. Galbany, M. Grayling, C.P. Gutiérrez et al., **501**, 4861 (2021), 2008.12101

[17] M.E. Moreno-Raya, Á.R. López-Sánchez, M. Mollá, L. Galbany, J.M. Vílchez, A. Carnero, **462**, 1281 (2016), 1607.05526

[18] M. Rigault, Y. Copin, G. Aldering, P. Antilogus, C. Aragon, S. Bailey, C. Baltay, S. Bongard, C. Buton, A. Canto et al., **612**, C1 (2018)

[19] P. Wiseman, M. Sullivan, M. Smith, C. Frohmaier, M. Vincenzi, O. Graur, B. Popovic, P. Armstrong, D. Brout, T.M. Davis et al., arXiv e-prints arXiv:2105.11954 (2021), 2105.11954

[20] D. Branch, J.C. Wheeler, *Supernova Explosions* (2017)
[21] B. Wang, Z. Han, **56**, 122 (2012), 1204.1155
[22] D. Maoz, F. Mannucci, G. Nelemans, **52**, 107 (2014), 1312.0628
[23] M. Livio, P. Mazzali, **736**, 1 (2018), 1802.03125
[24] W. Li, J.S. Bloom, P. Podsiadlowski, A.A. Miller, S.B. Cenko, S.W. Jha, M. Sullivan, D.A. Howell, P.E. Nugent, N.R. Butler et al., **480**, 348 (2011), 1109.1593
[25] F. Patat, M.A. Cordiner, N.L.J. Cox, R.I. Anderson, A. Harutyunyan, R. Kotak, L. Palaversa, V. Stanishev, L. Tomasella, S. Benetti et al., **549**, A62 (2013), 1112.0247
[26] P.L. Kelly, O.D. Fox, A.V. Filippenko, S.B. Cenko, L. Prato, G. Schaefer, K.J. Shen, W. Zheng, M.L. Graham, B.E. Tucker, **790**, 3 (2014), 1403.4250
[27] R. Margutti, J. Parrent, A. Kamble, A.M. Soderberg, R.J. Foley, D. Milisavljevic, M.R. Drout, R. Kirshner, **790**, 52 (2014), 1405.1488
[28] M.A. Tucker, B.J. Shappee, P.J. Vallely, K.Z. Stanek, J.L. Prieto, J. Botyanszki, C.S. Kochanek, J.P. Anderson, J. Brown, L. Galbany et al., **493**, 1044 (2020), 1903.05115
[29] A. Castrillo, Y. Ascasibar, L. Galbany, S.F. Sánchez, C. Badenes, J.P. Anderson, H. Kuncarayakti, J.D. Lyman, A.I. Díaz, **501**, 3122 (2021), 2012.11958
[30] C.L. Gerardy, W.P.S. Meikle, R. Kotak, P. Höflich, D. Farrah, A.V. Filippenko, R.J. Foley, P. Lundqvist, S. Mattila, M. Pozzo et al., **661**, 995 (2007), astro-ph/0702117
[31] C.M. Telesco, P. Höflich, D. Li, C. Álvarez, C.M. Wright, P.J. Barnes, S. Fernández, J.H. Hough, N.A. Levenson, N. Mariñas et al., **798**, 93 (2015), 1409.2125
[32] C. Badenes, E. Bravo, J.P. Hughes, **680**, L33 (2008), 0805.3344
[33] S. Park, C. Badenes, K. Mori, R. Kaida, E. Bravo, A. Schenck, K.A. Eriksen, J.P. Hughes, P.O. Slane, D.N. Burrows et al., **767**, L10 (2013), 1302.5435
[34] H. Yamaguchi, C. Badenes, A.R. Foster, E. Bravo, B.J. Williams, K. Maeda, M. Nobukawa, K.A. Eriksen, N.S. Brickhouse, R. Petre et al., **801**, L31 (2015), 1502.04255
[35] H. Martínez-Rodríguez, C. Badenes, S.H. Lee, D.J. Patnaude, A.R. Foster, H. Yamaguchi, K. Auchettl, E. Bravo, P.O. Slane, A.L. Piro et al., **865**, 151 (2018), 1807.02990
[36] Y. Ohshiro, H. Yamaguchi, S.C. Leung, K. Nomoto, T. Sato, T. Tanaka, H. Okon, R. Fisher, R. Petre, B.J. Williams, **913**, L34 (2021), 2105.04101
[37] L. Piersanti, E. Bravo, O. Straniero, S. Cristallo, I. Dominguez, arXiv e-prints arXiv:2112.02943 (2021), 2112.02943
[38] O. Straniero, R. Gallino, S. Cristallo, Nuclear Physics A **777**, 311 (2006), astro-ph/0501405
[39] S. Cristallo, O. Straniero, R. Gallino, L. Piersanti, I. Domínguez, M.T. Lederer, **696**, 797 (2009), 0902.0243
[40] L. Piersanti, E. Bravo, S. Cristallo, I. Domínguez, O. Straniero, A. Tornambé, G. Martínez-Pinedo, **836**, L9 (2017), 1701.06453
[41] D.A. Chamulak, E.F. Brown, F.X. Timmes, K. Dupczak, **677**, 160-168 (2008), 0801.1643
[42] A.L. Piro, L. Bildsten, **673**, 1009-1013 (2008), 0710.1600
[43] H. Martínez-Rodríguez, A.L. Piro, J. Schwab, C. Badenes, **825**, 57 (2016), 1602.00673
[44] J. Schwab, H. Martínez-Rodríguez, A.L. Piro, C. Badenes, **851**, 105 (2017), 1711.04780
[45] J. Iben, I., **226**, 996 (1978)

Slow White Dwarf Mergers as a New Galactic Source of Trans-Iron Elements

Umberto Battino[1,2,*], *Claudia* Lederer-Woods[2], *Claudia* Travaglio[3], *Friedrich Konrad* Röpke[4], and *Brad* Gibson[1]

[1]E.A. Milne Centre for Astrophysics, University of Hull, Hull, HU6 7RX, UK
[2]School of Physics and Astronomy, University of Edinburgh, Edinburgh, EH9 3FD, UK
[3]INFN - Istituto Nazionale Fisica Nucleare, 10125 Turin, Italy
[4]Heidelberg Institute for Theoretical Studies, 69118 Heidelberg, Germany

Abstract. The astrophysical origins of the heaviest stable elements that we observe today in the Solar System are still not fully understood. Recent studies have demonstrated that H-accreting white dwarfs (WDs) in a binary system exploding as type Ia supernovae could be an efficient p-process source beyond iron. However, both observational evidence and stellar models challenge the required frequency of these events. In this work, we calculate the evolution and nucleosynthesis in slowly merging carbon-oxygen WDs. As our models approach the Chandrasekhar mass during the merger phase, the ^{22}Ne(α,n)^{25}Mg neutron source reaction is activated in the external layers of the primary WD, where the carbon-rich material accreted from the secondary WD is burned via the ^{12}C+^{12}C reaction, which provides the necessary α-particles via the ^{12}C(^{12}C,α)^{20}Ne channel. The resulting neutron capture abundance distribution closely resembles a weak s-process one and peaks at Zr, which is overproduced by a factor of 30 compared to solar. The mass of the most external layers enriched in first-peak s-process elements crucially depends on the ^{12}C+^{12}C reaction rate, ranging between 0.05 $M_\odot$ and ~0.1 $M_\odot$. These results indicate that slow white dwarf mergers can efficiently produce the lightest p-process isotopes (such as ^{74}Se, ^{78}Kr, ^{84}Sr, ^{92}Mo and ^{94}Mo) via γ-induced reactions if they explode via a delayed detonation mechanism, or eject the unburned external layers highly enriched in first peak s-process elements in the case of a pure deflagration. In both cases, we propose for the first time that slow WD mergers in binary systems may be a new relevant source for elements heavier than iron.

1 Introduction

Type Ia supernovae (SNIa), the violent death of white-dwarfs in a binary system, play a central role in the chemical evolution of galaxies. Despite their importance, the nature of SNIa progenitors is still a matter of debate, with two main scenarios proposed. In the single-degenerate scenario (see [1]), a WD accretes mass from the envelope of a less evolved stellar companion. When the object approaches the critical Chandrasekhar-mass (the maximum mass that can be sustained by electron degeneracy, around 1.39 $M_\odot$ for CO WD), C-burning ignites explosively causing a SNIa. An exception can occur when the

*e-mail: μU.Battino@hull.ac.uk

accreted matter is mainly He, which would trigger a detonation before the Chandrasekhar mass is reached (sub-Chandrasekhar SNIa, [2]). In the double-degenerate scenario (see [3]), the companion is another WD. Over time, the separation of the pair decreases until they merge, either via a violent merger event or the formation of an accretion disk around the primary WD if the mass ratio of the components is larger or lower than ~0.8, respectively.

The relevance of SNIa in galactic chemical evolution (GCE) depends on the fraction of SNIa made via each of the aforementioned channels. [4] showed how SNIa originating from the single-degenerate scenario can be a very important source of a class of trans-iron proton-rich isotopes known as *p*-nuclei. The main assumption of [4], recently confirmed by [5], was the presence of a trans-iron element distribution up to Pb (called "seed distribution") synthesized on the WD surface during the accretion of H-rich material from the companion. The formation of the surface seed distribution is crucial, as γ-induced reactions acting on it during the explosion can result in the efficient synthesis of the *p*-nuclei. Moreover, [4] found that if about 50–70% of all SNIa are made via the single-degenerate scenario, they can be responsible for at least 50% of the *p*-nuclei abundances in the Solar System.

On the other hand, this is not what was obtained by [6] who, based on the spectra from a large sample of early-type galaxies, found that the single-degenerate SNIa should be less than 6% of the total SNIa population. Furthermore, [7] computed the nuclesynthetic ejecta of sub-Chandrasekhar SNIa from the helium detonation scenario, re-determining the contributions of both near-Chandrasekhar and sub-Chandrasekhar events to the total SNIa rate by reproducing the observed [Ti/Fe], [Mn/Fe] and [V/Fe] in the Galaxy. They found that at least 30% of all SNIa should come from near-Chandrasekhar-mass explosions. This result has been recently qualitatively confirmed by [8], who showed how there may be a very small contribution needed from near-Chandrasekhar-mass explosions only to explain the solar [Mn/Fe], and at the same time how this conclusion, however, depends on the massive star yields adopted in the calculations. Therefore, if the relative contribution of the single-degenerate channel to the total SNIa rate is observationally constrained to be below ~6%, then ~24% of near Chandrasekhar-mass SNIa is coming from an alternative channel. Can near-Chandrasekhar-mass SNIa originate from double degenerate systems? This is still an open question, with two primary shortcomings identified. First, during the disk formation, carbon could be ignited near the accreting star's surface and the burning transferred very rapidly towards the inner regions, thus producing an ONeMg WD, and not a thermonuclear supernova. This has been excluded by [9], who found that C-burning indeed occurs, but it is soon extinguished, unless more than ~0.01 $M_\odot$ of He is present (see [10]). The second shortcoming consists of the fact that the mass transfer from the disc to the primary WD may occur at a high rate ($\sim 10^{-5}$ $M_\odot$ yr^{-1}), causing an off-centre explosive ignition of C-burning, well before the WD attains the Chandrasekhar mass. However, [11] showed that the effects of rotation, naturally arising in merging WDs, make the accretion rate $self\text{-}regulated$ and as low as a few 10^{-7} $M_\odot yr^1$.

In this work, we assumed that Chandrasekhar-mass WDs can actually be formed via the double-degenerate channel through a slow merger process via the accretion-disc, which we simulate using the stellar evolution code MESA (see [12]). We want to verify if a trans-iron seed distribution can be formed on the surface of the pre-supernova structure, which may result in a novel source of trans-iron elements in the Universe.

2 Computational methods and numerical results

The stellar models presented in this section are computed using the 1D stellar code MESA (revision 10108). The CO WD model 0.927-from-6.0-z2m2.mod, included in the MESA data folder, was cooled down for 10^5yr before being used as a starting model for our calculations.

The chemical composition in mass fraction of the accreted material, added at a constant accretion rate $\dot{M}$=2.4 10^{-6} $M_{\odot}yr^{-1}$, consisted of 0.48809 of ^{12}C, 0.48809 of ^{16}O, 0.002 of ^{20}Ne, 0.021 of ^{22}Ne, and the remaining 0.00082 in iron-group isotopes between ^{50}Cr and ^{64}Ni (with X(^{56}Fe)=0.0007). The nuclear reaction network used in our simulations is co-burn-extras.net, which mainly includes the isotopes needed for carbon burning up to ^{28}Si coupled by more than 50 reactions. The nuclear reaction rates recommended in the JINA REACLIB library were adopted (see [13]), except for the ^{12}C+^{12}C reaction rate from [14], the ^{12}C(α,γ)^{16}O from [15], and the ^{22}Ne(α,n)^{25}Mg and ^{22}Ne(α,γ)^{26}Mg from [16]. We extended the nuclear reaction network up to ^{141}Ba when the accreting WD approaches the Chandrasekhar mass and the temperature at the base of the accreted material exceeds 10^9K, in order to follow the neutron capture nucleosynthesis during the surface C-burning.

In Figure 1 the temperature and density profiles of our accreting WD are presented at different moments during the accretion towards the Chandrasekhar mass, closely resembling the ones obtained by [17]. When the WD mass reaches ~1.37 $M_{\odot}$, the temperature at the base of the accreted material exceeds 10^9K as a consequence of surface C-burning ignition. α particles are released via the ^{12}C(^{12}C, α)^{20}Ne reaction, which are then captured by ^{22}Ne and converted into neutrons via the ^{22}Ne(α,n)^{25}Mg reaction.

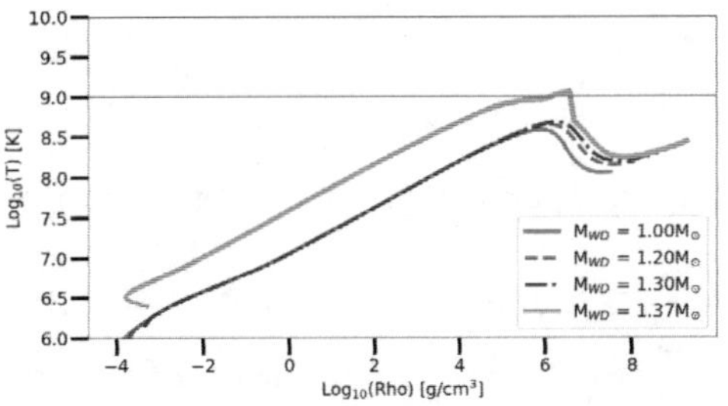

Figure 1. Temperature and density profiles of our accreting WD at different moments during the accretion towards the Chandrasekhar mass.

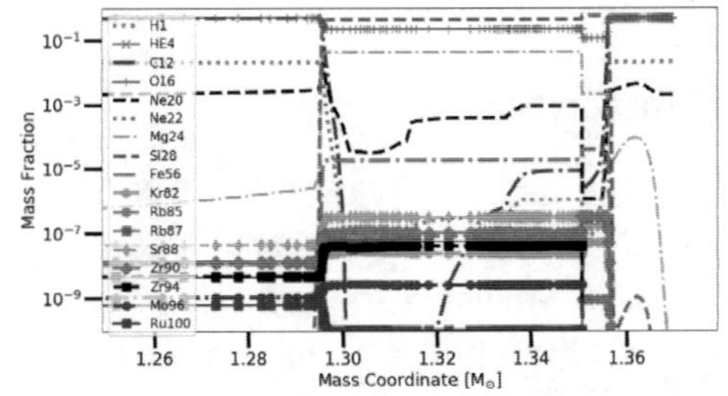

Figure 2. The neutron-capture isotopes produced on the top of the WD at the end of surface C-burning when adopting a ^{12}C+^{12}C reaction rate increased by a factor of two.

The neutron-capture seed products on the top of the WD at the end of surface C-burning are shown in Figure 2. Given the key role and still high uncertainty of the ^{12}C+^{12}C reaction rate, we performed two simulations, one adopting our standard rate and a second one with the reaction rate increased by a factor of two. In both cases, trans-iron elements are synthesized on the WD surface, forming essentially a weak *s*-process distribution peaked at Zr, which is overproduced by a factor of 30 compared to solar, and up to Mo. Moreover, depending on the ^{12}C+^{12}C reaction rate, the thickness in mass of the surface material enriched in heavy elements is 0.05 and 0.07 $M_{\odot}$ when adopting the standard and the increased ^{12}C+^{12}C reaction rate, respectively. This size is of the same order of what was found by [4] to be a prerequisite of an exploding near-Chandrasekhar WD in order to trigger an efficient *p*-process, and could include the production of the historically puzzling cases of Mo *p*-isotopes.

3 Conclusions

In this work, we calculated the evolution and nucleosynthesis in slowly merging CO WDs using the stellar code MESA. As our models approach the Chandrasekhar mass during the merger phase, the ^{22}Ne(,n)^{25}Mg neutron source reaction is activated in the external layers of the primary WD. The resulting neutron capture seed distribution closely resembles a weak s-process one and peaks at Zr, which is overproduced by a factor of 30 compared to solar. The mass of the most external layers enriched in first-peak s-process elements ranges between 0.05 $M_\odot$ and ~0.1 $M_\odot$, which would allow an efficient p-process during the subsequent explosion. Hence slow WD mergers could really be a novel trans-iron elements source, potentially being able to efficiently produce the lightest p-process isotopes (such as ^{74}Se, ^{78}Kr, ^{84}Sr, ^{92}Mo and ^{94}Mo) via γ-induced reactions if they explode via a delayed detonation mechanism, or eject the unburned external layers highly enriched in first peak s-process elements in the case of a pure deflagration. We then propose for the first time that slow WD mergers may be a new relevant astrophysics source for elements heavier than iron, pending a closer investigation by simulating the explosive nucleosynthesis originating from these progenitors. It must be stressed that only one solar-scaled metallicity composition of the accreted material has been studied so far. We will extend our investigation to different metallicities, as well as different C/O ratios and ^{22}Ne abundance in both the primary and secondary WD.

This work received support from the Science and Technology Facilities Council UK (ST/M006085/1 & ST/R000840/1), and the European Research Council ERC-2015-STG Nr. 677497. This article is based upon work from the ChETEC COST Action (CA16117). UB & BKG acknowledges support by ChETEC-INFRA (EU project no. 101008324).

References

[1] J. Whelan, I. Iben, Jr., ApJ **186**, 1007 (1973)
[2] L. Piersanti, A. Tornambé, L.R. Yungelson, MNRAS **445**, 3239 (2014), 1409.3589
[3] R.F. Webbink, ApJ **277**, 355 (1984)
[4] C. Travaglio, R. Gallino, T. Rauscher, F.K. Röpke, W. Hillebrandt, ApJ **799**, 54 (2015)
[5] U. Battino, M. Pignatari, C. Travaglio, C. Lederer-Woods, P. Denissenkov, F. Herwig, F. Thielemann, T. Rauscher, MNRAS **497**, 4981 (2020), 2007.13677
[6] J. Johansson, T.E. Woods, M. Gilfanov, M. Sarzi, Y.M. Chen, K. Oh, MNRAS **461**, 4505 (2016), 1607.02243
[7] F. Lach, F.K. Röpke, I.R. Seitenzahl, B. Coté, S. Gronow, A.J. Ruiter, A&A **644**, A118 (2020), 2010.14084
[8] S. Gronow, B. Cote, F. Lach, I.R. Seitenzahl, C.E. Collins, S.A. Sim, F.K. Roepke, arXiv e-prints arXiv:2103.14050 (2021), 2103.14050
[9] P. Lorén-Aguilar, J. Isern, E. García-Berro, A&A **500**, 1193 (2009)
[10] R. Pakmor, M. Kromer, S. Taubenberger, V. Springel, ApJ Lett. **770**, L8 (2013)
[11] L. Piersanti, S. Gagliardi, J. Iben, Icko, A. Tornambé, ApJ **598**, 1229 (2003)
[12] B. Paxton, L. Bildsten, A. Dotter, F. Herwig, P. Lesaffre, F. Timmes, ApJS **192**, 3 (2011), 1009.1622
[13] R.H. Cyburt, A.M. Amthor, R. Ferguson, Z. Meisel, K. Smith, S. Warren, A. Heger, R.D. Hoffman, T. Rauscher, A. Sakharuk et al., ApJS **189**, 240 (2010)
[14] L.R. Gasques, A.V. Afanasjev, E.F. Aguilera, M. Beard, L.C. Chamon, P. Ring, M. Wiescher, D.G. Yakovlev, Phys. Rev. C. **72**, 025806 (2005), astro-ph/0506386
[15] R. Kunz, M. Fey, M. Jaeger, A. Mayer, J.W. Hammer, G. Staudt, S. Harissopulos, T. Paradellis, ApJ **567**, 643 (2002)
[16] P. Adsley, U. Battino, A. Best, A. Caciolli, A. Guglielmetti, G. Imbriani, H. Jayatissa, M. La Cognata, L. Lamia, E. Masha et al., Phys. Rev. C. **103**, 015805 (2021)
[17] C. Wu, B. Wang, D. Liu, MNRAS **483**, 263 (2019), 1811.08638

Axion-like Particles from Nearby Type Ia Supernovae

Kanji Mori[1,*]

[1]Research Institute of Stellar Explosive Phenomena, Fukuoka University, 8-19-1 Nanakuma, Jonan-ku, Fukuoka-shi, Fukuoka 814-0180, Japan

Abstract. Axion-like particles (ALPs) are a class of hypothetical pseudo-scalar particles and can be created in hot astrophysical plasma through the interaction between photons. I calculate the production of ALPs in type Ia supernovae. It is found that many ALPs lighter than a few MeV can be produced in type Ia supernovae. Once produced, heavy ALPs decay into photons during propagation in the interstellar space. I calculate the flux of the decay photons and find that it may be detected by future MeV γ-ray telescopes if a type Ia supernova explodes near the Solar System.

1 Introduction

Axion-like particles (ALPs) are hypothetical pseudo Nambu-Goldstone bosons which can couple with photons. The ALP-photon interaction is described by the Lagrangian

$$\mathcal{L}_{a\gamma\gamma} = -\frac{1}{4} g_{a\gamma} a \tilde{F}^{\mu\nu} F_{\mu\nu}, \tag{1}$$

where $g_{a\gamma}$ is the coupling constant, a is the ALP field, $F_{\mu\nu}$ is the electromagnetic tensor, and $\tilde{F}^{\mu\nu}$ is its dual. Axions were first introduced to solve the strong CP problem in quantum chromodynamics (QCD) [1, 2]. Since then, the concept of QCD axions has been generalized as ALPs to describe similar exotic particles [e.g. 3].

Astrophysical objects such as stars and supernovae have been used to constrain heavy ALPs. Since ALPs can induce additional energy losses from stars and affect the stellar lifetime, observations of globular clusters have provided a constraint on ALPs [4]. Also, the additional energy losses may prevent neutrino bursts in core-collapse supernovae [5]. ALP parameters can be constrained by comparing the model calculation and neutrino observations of SN 1987A. Once produced in an astronomical object, heavy ALPs decay into photons during propagation in the interstellar space. The non-detection of decay photons from SN 1987A provided a stringent constraint [6].

Supernovae are classified into two classes: core-collapse and thermonuclear supernovae. Core-collapse supernovae are the death of massive stars. On the other hand, thermonuclear supernovae are thermonuclear explosions of white dwarfs and called type Ia supernovae (SNe Ia). Previous works on ALPs from supernovae have focused on core-collapse supernovae. In this study, we calculate the production of ALPs in a nearby SN Ia and the flux of decay photons [7].

*e-mail: kanji.mori@fukuoka-u.ac.jp

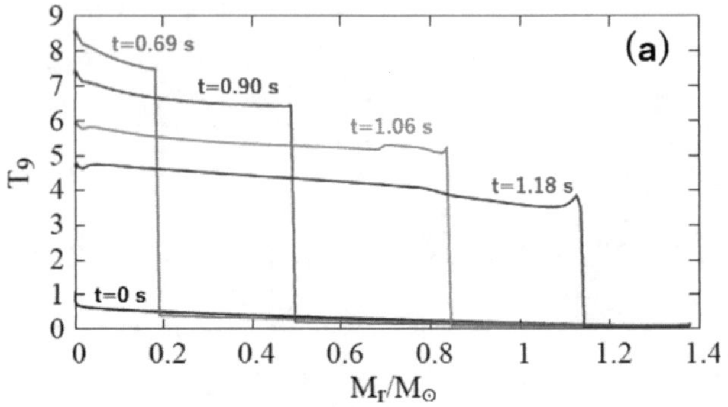

Figure 1. The temperature profile in the W7 model [7].

2 Method

We adopt the W7 model as a SN Ia model. W7 is a one-dimensional deflagration model with a near Chandrasekhar carbon-oxygen white dwarf [8]. Fig. 1 shows the temperature T_9 in unit of 10^9 K as a function of the mass coordinate M_r. The ignition starts at the center and the deflagration propagates toward the surface.

We consider the Primakoff process and the photon coalescence as ALP production processes. The rate of the Primakoff process is calculated as [7, 9]

$$\Gamma_{\gamma\to a} = g_{a\gamma}^2 \frac{T\kappa^2}{32\pi}\frac{p}{E}\left(\frac{((k+p)^2+\kappa^2)((k-p)^2+\kappa^2)}{4kp\kappa^2}\ln\left(\frac{(k+p)^2+\kappa^2}{(k-p)^2+\kappa^2}\right)\right.$$
$$\left. -\frac{(k^2-p^2)^2}{4kp\kappa^2}\ln\left(\frac{(k+p)^2}{(k-p)^2}\right)-1\right), \tag{2}$$

where T is the temperature, κ is the inverse Debye–Hückel length, p and E are the ALP momentum and energy, and k is the momentum of plasmons. The rate for the photon coalescence is calculated as

$$\frac{d^2 n_a}{dt dE} = g_{a\gamma}^2 \frac{m_a^4}{128\pi^3} p\left(1-\frac{4\omega_{\rm pl}^2}{m_a^2}\right)^{\frac{3}{2}} e^{-\frac{E}{T}}, \tag{3}$$

where m_a is the ALP mass and $\omega_{\rm pl}$ is the plasma frequency. The contribution of the photon coalescence to the total ALP production becomes significant when ALPs are heavier than ~ 1 MeV.

Heavy ALPs decay radiatively during the propagation. The mean free path is calculated as $\lambda = \beta\gamma/\Gamma_{a\to\gamma\gamma}$ [6, 7], where β is the ALP velocity, γ is the Lorentz factor, and $\Gamma_{a\to\gamma\gamma} = g_{a\gamma}^2 m_a^3/64\pi$.

3 Results

Fig. 2 shows the ALP luminosity defined as

$$L_a = 4\pi\int Q_a r^2 dr, \tag{4}$$

where r is the radius and Q_a is the ALP emissivity per unit volume and time. In the plot, the ALP-photon coupling is fixed to $g_{10} = 1$. It is seen that L_a reaches its peak at ~ 1 s

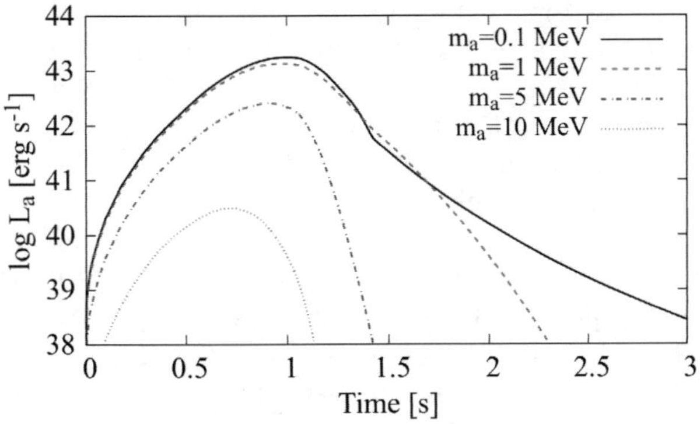

Figure 2. The ALP luminosity as a function of time after the ignition [7]. The ALP-photon coupling is fixed to $g_{10} = 1$.

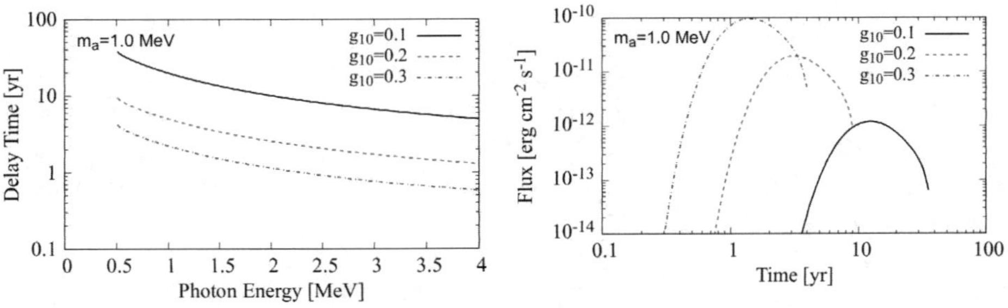

Figure 3. The delay time (left) and the energy flux (right) of decay photons from an SN Ia located 1 kpc away from the Solar System [7]. The ALP mass is fixed to $m_a = 1$ MeV.

after ignition. When ALPs are heavier than ~ 5 MeV, L_a is suppressed. This is because the temperature in SNe Ia is lower than 1 MeV and the plasma is not energetic enough to produce such heavy particles.

In this study, we assume that an SN Ia is located 1 kpc away from the Solar System. Since ALPs are slower than photons, decay photons would be observed later than the supernova event. The delay time is calculated as $t = L_1/\beta + L_2 - d_{\mathrm{SN}}$, where $L_1 \sim \lambda$ is the distance which ALPs travel, L_2 is the distance which decay photons travel, and d_{SN} is the distance between the SN Ia and the Solar System. Fig. 3 shows the delay time as a function of the energy of decay photons. One can see that more energetic photons arrive on Earth earlier. In this ALP parameter range, typical values of the delay time are $1 - 10$ years.

Fig. 3 also shows the energy flux $F(t) = E_\gamma(t) dn_\gamma(t)/dt$ as a function of time after the SN explosion. Here $E_\gamma(t)$ is the energy of decay photons and $n_\gamma(t)$ is the number flux of decay photons that arrive on Earth. The number flux is given as [7, 10]

$$\frac{dn_\gamma}{dt} = \frac{2}{4\pi d_{\mathrm{SN}}^2} \frac{dn_a}{dE_a} \left(\frac{dt}{dE_a} \right)^{-1} e^{-\frac{R_{\mathrm{SN}}}{\lambda(E_a)}} \left(1 - e^{-\frac{d_{\mathrm{SN}}}{\lambda(E_a)}} \right), \tag{5}$$

where the factor 2 represents two photons that are produced from an ALP, $d_{\mathrm{SN}} = 1$ kpc is the distance to the event, E_a is the ALP energy, n_a is the number of ALPs, and R_{SN} is the radius of the star. The formula takes into account the effect of the ALP decay in the star and in the interstellar space. From Fig. 3, we see that the flux increases as a function of g_{10}.

When $g_{10} \gtrsim 0.2$, the flux reaches $\sim 10^{-12}$ erg cm^{-2} s^{-1}. This flux may be observed by next-generation MeV γ-ray telescopes such as All-sky Medium Energy Gamma-ray Observatory [AMEGO; 11] and e-ASTROGAM [12]. (Non-)detection of decay photons would provide independent information on the ALP parameters.

In this study, we adopted a near-Chandrasekhar model of SNe Ia. However, the nature of SN Ia progenitors is still under debate. Recently, sub-Chandrasekhar models are being investigated as an alternative scenario [e.g. 13, 14]. Since the density in sub-Chandrasekhar white dwarfs is much lower, the ALP production there can be significantly different from the results for Chandrasekhar models. It is desirable to perform systematic studies for various models to understand the model dependence in future studies.

This work is supported by Research Institute of Stellar Explosive Phenomena at Fukuoka University and JSPS KAKENHI grant number JP21K20369.

References

[1] S. Weinberg, Phys. Rev. Lett. **40**, 223 (1978)
[2] F. Wilczek, Phys. Rev. Lett. **40**, 279 (1978)
[3] K. Choi, S.H. Im, C.S. Shin, arXiv e-prints arXiv:2012.05029 (2020), 2012.05029
[4] P. Carenza, O. Straniero, B. Döbrich, M. Giannotti, G. Lucente, A. Mirizzi, Physics Letters B **809**, 135709 (2020), 2004.08399
[5] G. Lucente, P. Carenza, T. Fischer, M. Giannotti, A. Mirizzi, J. Cosmol. Astropart. Phys. **2020**, 008 (2020), 2008.04918
[6] J. Jaeckel, P.C. Malta, J. Redondo, Phys. Rev. D **98**, 055032 (2018), 1702.02964
[7] K. Mori, Publ. Astron. Soc. Japan **73**, 1382–1388 (2021), 2107.09097
[8] K. Nomoto, F.K. Thielemann, K. Yokoi, Astrophys. J. **286**, 644 (1984)
[9] L. di Lella, A. Pilaftsis, G. Raffelt, K. Zioutas, Phys. Rev. D **62**, 125011 (2000), hep-ph/0006327
[10] A. Caputo, P. Carenza, G. Lucente, E. Vitagliano, M. Giannotti, K. Kotake, T. Kuroda, A. Mirizzi, Phys. Rev. Lett. **127**, 181102 (2021), 2104.05727
[11] J. McEnery, A. van der Horst, A. Dominguez, A. Moiseev, A. Marcowith, A. Harding, A. Lien, A. Giuliani, A. Inglis, S. Ansoldi et al., *All-sky Medium Energy Gamma-ray Observatory: Exploring the Extreme Multimessenger Universe*, in *Bulletin of the American Astronomical Society* (2019), Vol. 51, p. 245, 1907.07558
[12] A. de Angelis, V. Tatischeff, I.A. Grenier, J. McEnery, M. Mallamaci, M. Tavani, U. Oberlack, L. Hanlon, R. Walter, A. Argan et al., Journal of High Energy Astrophysics **19**, 1 (2018), 1711.01265
[13] K.J. Shen, D. Kasen, B.J. Miles, D.M. Townsley, Astrophys. J. **854**, 52 (2018), 1706.01898
[14] S.C. Leung, K. Nomoto, Astrophys. J. **888**, 80 (2020), 1901.10007

Screening Effects on Electron-Capture Rates and Type Ia Supernova Nucleosynthesis

Toshio Suzuki[1,8,*], *Kanji* Mori[2,8], *Michio* Honma[3], *Michael A.* Famiano[4,8], *Toshitaka* Kajino[5,6,8], *Motohiko* Kusakabe[5,8], and *A. Baha* Balantekin[7,8]

[1]Department of Physics, College of Humanities and Sciences, Nihon University, 3-25-40 Sakurajosui, Setagaya-ku, Tokyo 156-8550, Japan
[2]Research Institute of Stellar Explosive Phenomena, Fukuoka University, 8-19-1 Nanakuma, Jonan-ku, Fukuoka-shi, Fukuoka 814-0180, Japan
[3]Center for Mathematical Sciences, University of Aizu, Aizu-Wakamatsu, Fukushima 965-8580, Japan
[4]Department of Physics, Western Michigan University, Kalamazoo, MI 49008, USA
[5]School of Physics, Beihang University, Haidian-qu, Beijing 100083, People's Republic of China
[6]Graduate School of Science, The University of Tokyo, Hongo, Bunkyo-ku, Tokyo 113-0033, Japan
[7]Department of Physics, University of Wisconsin-Madison, Madison, WI 53706, USA
[8]National Astronomical Observatory of Japan, 2-21-2 Osawa, Mitaka, Tokyo 181-8588, Japan

Abstract. Nucleosynthesis of iron-group elements in Type Ia supernovae is studied for single-degenerate models with the use of electron-capture rates updated with the new shell-model Hamiltonian in pf-shell. An over-production problem of neutron-rich iron-group isotopes compared with the solar abundances is now found to be suppressed within a factor of about twice for the updated weak rates. Effects of screening on nucleosynthesis are investigated for explosion models of fast deflagration and slow deflagration with delayed detonation. The e-capture rates are reduced by the screening, especially by the screening effects on the ions. The production yields of most neutron-rich isotopes such as ^{50}Ti, ^{54}Cr and ^{58}Fe are found to be suppressed most by the screening. The inclusion of the screening is desirable for precise evaluation of abundances of neutron-rich nuclides.

1 Introduction

First, we show that a new shell-model Hamiltonian, GXPF1J [1], can describe Gamow-Teller (GT) strengths and e-capture rates in pf-shell nuclei quite well. Electron-capture rates at stellar environments are evaluated by using GT strength and phase space factor for the reaction. Electron chemical potential, thus the phase space factor also, increases as the density increases and the e-capture rates are enhanced at high densities. In the pioneering work of Fuller, Fowler and Newton (FFN) [2], a simple shell model was used to evaluate the GT strength. Now, large-scale shell-model calculations are possible for pf-shell nuclei. Typical Hamiltonians are KB3's and GXPF1's. While both can describe spin properties of pf-shell nuclei very well, experimental GT strength distributions in ^{58}Ni, for example, are better reproduced for GXPF1J than KB3G [3] in both n $\rightarrow$ p and p $\rightarrow$ n channels. The GXPF1J can also reproduce $B(M1)$ in ^{48}Ca and experimental e-capture rates in pf-shell nuclei best among various shell-model and RPA calculations [4].

*e-mail: suzuki.toshio@nihon-u.ac.jp

We here discuss ^{56}Ni in particular. The experimental GT strength obtained by (p, n) reaction has two-peak structure [5], and it can be well reproduced by GXPF1J [6] while KB3G and KBF [7] give only one peak. As the strength is more spread for GXPF1J, the e-capture rates of GXPF1J are smaller than those of KB3G [8, 9]. The weak rates conventionally used for stellar calculations is KBF [7] rather than KB3G. In KBF, experimental data of energies and GT strength available are taken into account, and thus the e-capture rates of KBF come close to those of GXPF1J.

In Sect. 2,the e-capture rates evaluated with the GXPF1J are applied to study nucleosynthesis of iron-group elements in Type Ia supernovae (SN). The screening effects on the rates and nucleosynthesis are discussed in Sect. 3.

2 Nucleosynthesis of iron-group nucleides in Type Ia SN

We discuss synthesis of iron-group nucleides in Type Ia SN. Here, we treat single-degenerate 'Chandra' model, where accretion of matter to a white dwarf (WD) from non-degenerate companion occurs. A SN explosion (SNe) occurs when WD mass approaches the Chandrasekhar limit, and much ^{56}Ni is produced in the explosion. Successive e-capture processes produce neutron-rich (n-rich) nuclei and the electron-fraction Y_e decreases. Decrease of e-capture rates leads to less production of n-rich nuclei and larger Y_e.

There was an over-production problem of neutron-excess iron-group isotopes compared with the solar abundances when e-capture rates of FFN were used. Neutron-rich nuclei such as ^{54}Cr and ^{58}Ni are produced several times more than the solar abundances when an explosion model W7 with fast deflagration with subsonic frame front propagation was used [10]. Another explosion model WDD2 [10] with slow deflagration and delayed detonation with supersonic shock wave will be used also. The production yields of elements are obtained with the e-capture rates of GXPF1J ($21 \leq Z \leq 32$) and KBF (other Z) for both the W7 and WDD2 explosion models [9]. The over-production of n-rich iron-group nuclides found for the FFN rates is now suppressed within a factor of about twice, and this improvement proves to be better for WDD2 [9] (see figures 1 and 2 in Sect. 3).

3 Screening effects on the weak rates and nucleosynthesis

3.1 Screening effects on the weak rates

There are two kinds of screening (SCR) effects. One is SCR effects of electrons, which is evaluated for relativistic degenerate electron liquid with the dielectric function obtained by RPA [11]. The ion-electron attraction is reduced by SCR, which leads to the reduction of e-capture rates and the enhancement of the β-decay rates. The other is the change of threshold energy due to the change of the chemical potential of the ion.

The correction of the chemical potential of the ion with charge Z is given by $\mu_C\ (Z) = kT f(\Gamma)$, where $\Gamma = Z^{5/3}(\frac{e^2}{kTa_e})$ with $a_e = (\frac{3}{4\pi n_e})^{1/3}$ and n_e the electron density is the Coulomb coupling parameter. For the strong-coupling regime ($\Gamma > 1$), results of Monte-Carlo calculations for one-component plasma [12] have been parametrized by a function of Γ [13]. For the weak-coupling regime ($\Gamma < 0.1$), $f(\Gamma)$ is obtained by cluster-expansion method [14].

$$
\begin{aligned}
f(\Gamma) &= -\frac{\Gamma^{3/2}}{\sqrt{3}} - \frac{\Gamma^3}{2}[\frac{3}{4}\ln(3\Gamma) + 0.57721 - \frac{11}{12}] \quad (\Gamma < 0.1) \\
f(\Gamma) &= -0.8980\Gamma + 3.8714\Gamma^{1/4} - 0.8828\Gamma^{-1/4} - 0.8610\ln\Gamma - 2.5269 \quad (\Gamma > 1) \quad (1)
\end{aligned}
$$

The first term for the weak-coupling regime is the Debye-Huckel limit, and the second term is the next-order correction. For the intermediate coupling ($0.1 < \Gamma < 1$), interpolation of weak and strong couplings is usually taken, but this leads to a rather complex formula [13]. Instead, a simple parametrization proposed in Ref. [15], $f(\Gamma) = -\Gamma^{3/2}/\sqrt{3} + \beta/\gamma\Gamma^{\gamma}$, is used here for $0 < \Gamma < 1$. For small Γ, it becomes the Debye-Huckel limit, and smoothly connected to the strong-coupling region at Γ=1 with β =0.2956 and γ =-1.9885. The correction of the ion chemical potential is negative and its magnitude increases as Z increases. This results in an enhancement of the Q-value, $\Delta Q = \mu_C(Z-1) - \mu_C(Z)$, which leads to the reduction (enhancement) of the e-capture (β-decay) rates [16]. Here, the SCR effects on thermonuclear reactions are also taken into account [17].

3.2 Screening effects on the synthesis of iron-group nucleides

Screening effects on the production yields of elements in Type Ia SNe compared with the solar abundances are investigated for the W7 and WDD2 explosion models [16]. Calculated results of the production yields obtained for the W7 and WDD2 models are shown in figure 1 and figure 2, respectively.

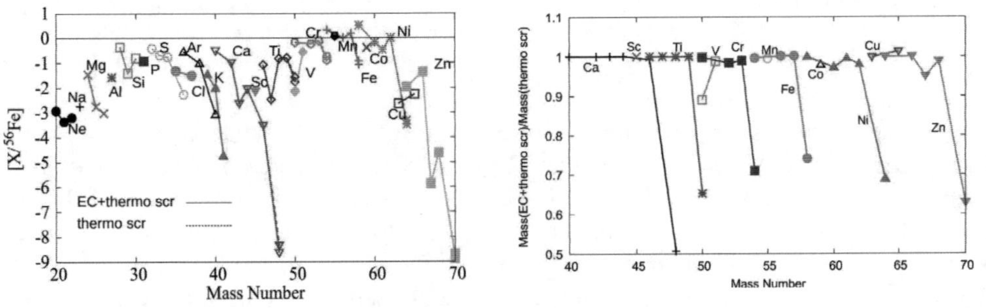

Figure 1. (Left) Abundances normalized by the solar and ^{56}Fe abundances obtained for the W7 model of SNe Ia. Solid and broken lines show results with and without the SCR effects, respectively. (Right) Abundance ratios between the cases with and without the SCR effects on e-capture rates obtained for the W7 model. Taken from [16].

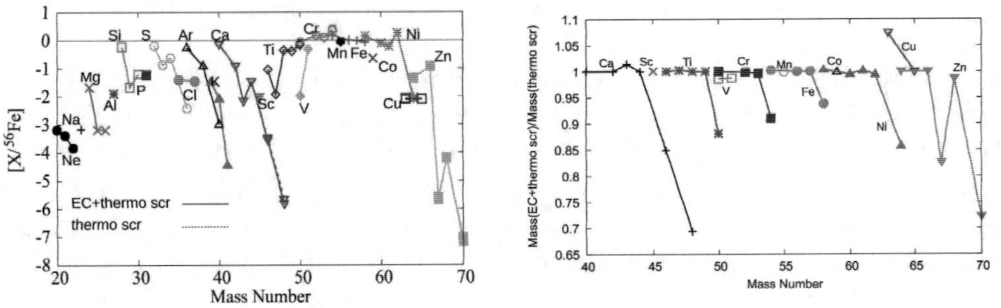

Figure 2. The same as in figure 1 for the WDD2 model of SNe Ia. Taken from [16].

The SCR effects are found to be large for the most n-rich isotopes such as ^{48}Ca, ^{50}Ti, ^{54}Cr, ^{58}Fe, ^{64}Ni and ^{70}Zn, and the reduction effects are smaller for WDD2 compared with W7. Most n-rich isotopes are produced more for WDD2. The yields of ^{54}Cr and ^{58}Fe are close to the solar abundances in case of WDD2, while they are one-order magnitude smaller in the W7 model. These features can be understood from the difference in Y_e, central density

ρ_c and central temperature T_c between the two models as shown in figure 3. Y_e is smaller for WDD2, which results in the production of more n-rich isotopes for the case of WDD2. ρ_c is higher for WDD2, which leads to less effects of SCR for WDD2.

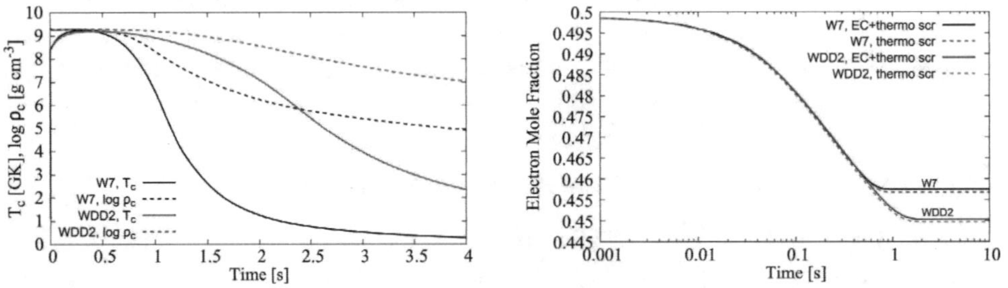

Figure 3. (Left) Time evolution of the central density ρ_c and temperature T_c. The solid lines show T_c and the broken lines show ρ_c. The upper (lower) lines are obtained by the WDD2 (W7) model of SNe Ia. (Right) Time evolution of the electron fraction Y_e at the center of the SN Ia models. The solid lines show the result with the SCR effects on both e-capture and thermonuclear reactions, while the broken lines show the result with the SCR effect only on thermonuclear reactions. Taken from [16].

Finally, we discuss the contribution of SNe Ia to the solar abundances of very n-rich nuclei such as ^{50}Ti, ^{54}Cr and ^{58}Fe, which are produced also in core-collapse SN. The contributions are 5-10% in W7 and 50-100% in WDD2. The suppressions due to the SCR effects are 30-50% in W7 and they are as small as 10-20% in WDD2. Besides the explosion models, SCR effects are important for precise evaluation of synthesis of very n-rich isotopes, ^{54}Cr and ^{58}Fe. ^{54}Cr is still over-produced in WDD2 model. Here, use is made of post-process calculations in 1D explosion models, where hydrodynamics and nucleosynthesis are decoupled. Improvement of this defect is left to future.

References

[1] M. Honma, T. Otsuka, T. Mizusaki et al., J. Phys. Conf. Ser. **20**, 7 (2005)
[2] G. M. Fuller, W. Fowler and M. J. Newman, Astrophys. J. **252**, 715 (1982); Astrophys. J. Suppl. **48**, 279 (1982)
[3] A. Poves, J. Sanchez-Solano, E. Caurier and F. Nowacki, Nucl. Phys. **A694**, 157 (2001)
[4] A. L. Cole, T. S. Anderson, R. G. T. Zegers et al., Phys. Rev. C **86**, 015809 (2012)
[5] M. Sasano, G. Perdikakis, R. G. T. Zegers et al., Phys. Rev. Lett. **107**, 202501 (2011)
[6] T. Suzuki, M. Honma, K. Higashiyama et al., Phys. Rev. C **79**, 061603(R) (2009)
[7] K. Langanke and G. Martinez-Pinedo, Atomic Data and Nucl. Data Tables **79**, 1 (2001)
[8] T. Suzuki, M. Honma, H. Mao et al., Phys. Rev. C **83**, 044619 (2011)
[9] K. Mori, M. A. Famiano, T. Kajino, T. Suzuki et al.,Astrophys. J. **833**, 179 (2016)
[10] K. Iwamoto, F. Brachwitz, K. Nomoto et al., Astrophys. J. Suppl. **125**, 439 (1999)
[11] N. Itoh, N. Tomizawa, M. Tamamura and S. Wanajo, Astrophys. J. **579**, 380 (2002)
[12] W, L. Slattery, G. D. Doolen and H. E. Dewitt, Phys. Rev. A **26**, 2255 (1982)
[13] S. Ichimaru, Rev. Mod. Phys. **65**, 255 (1993)
[14] R. Abe, Prog. Theor. Phys. **21**, 475 (1959)
[15] D. G. Yakovlev and D. A. Shalybkov, Astrophy. Space Phys. **7**, 311 (1989)
[16] K. Mori, T. Suzuki, M. Honma et al., Astrophy. J. **904**, 29 (2020)
[17] R. K. Wallace, S. E. Woosley and T. A. Weaver, Astrophys. J. **258**, 696 (1982)

EPJ Web of Conferences **260**, 07001 (2022) https://doi.org/10.1051/epjconf/202226007001
NIC-XVI

The Character of Three-Dimensional Core-Collapse Simulation Results

Adam Burrows[1,*] and *Matthew* Coleman[1,**]

[1]Department of Astrophysical Sciences, Princeton University, Princeton, NJ 08540

Abstract. We here in graphical form highlight some of the early phases in the explosion of core-collapse supernovae and the birth of neutron stars. The graphics presented summarizes the results of the almost fifty three-dimensional radiation/hydrodynamic simulations the Princeton group has simulated during the last three years using its state-of-the-art (though still evolving and improving) code FORNAX.

1 Introduction

The core-collapse supernova problem has been a multi-physics challenge in theoretical astrophysics for more than 60 years. Its theoretical study lies at the confluence of nuclear, particle, neutrino, statistical, and gravitational physics and has persistently necessitated, as a consequence of intrinsic instabilities and multi-dimensional effects, the use of state-of-the-art computational resources. In fact, theoretical progress has paralleled the evolution of high-performance computing, as well as the gradual maturation of our understanding of the neutrino and of neutrino-matter interactions [1]. Moreover, since the explosion is inaugurated in the dense core of a massive star, its initial development is shrouded in mystery by the profound opacity of the star to photons. Hence, the traditional synergy between theory and observation that characterizes much of astronomy is often denied us and we are forced to connect data and theory at least one step removed from direct (however remote) experience. Nevertheless, the theory that is emerging from the detailed simulations of the chaotic dynamics of collapse, bounce, delay to explosion, and explosion is beginning to be predictive of the systematics with progenitor properties of explosion energy, neutron star birth mass, nucleosynthesis, pulsar spins and kicks, and perhaps even birth magnetic fields [2, 3].

We will not, however, summarize in this brief offering the rich and complicated subject that is core-collapse supernova (CCSN) theory and its emerging explanatory power. Though these are heady times in the theory of one of the centermost problems in theoretical astrophysics, a vast and developing literature must substitute for such an ambitious endeavor. Rather, we here provide a storyboard in graphical form depicting some of the salient new aspects of CCSN explosions and neutron-star birth that we feel compelled to highlight, as revealed by our recent 3D simulation efforts [4–8]. Reviews of basic supernova physics can be found in [9–11]. Important 3D simulation work by other research groups (by no means exhaustive) can be found at [12–18].

*e-mail: burrows@astro.princeton.edu
**e-mail: msbc@astro.princeton.edu

2 Collapse, Bounce, Stall: A Turbulent Outer Convective Region Bounded by a Shock Wave

Figure 1 depicts a representative snapshot ~150 milliseconds after core bounce of the stalled shock wave (blue veil) bounding the neutrino-heating-driven convective shell that surrounds the inner proto-neutron star (PNS) (yellow ball) just prior to explosion. The tangle of trajectories in this turbulent region just interior to the stalled shock traces representative matter motions. The initial model is not rotating, so the collapsing matter comes in radially and this material is depicted on the outside of the plot with representative radial tracers. The hydrodynamic stress in the turbulent region interior to the shock, along with the neutrino energy deposition behind the shock, are together responsible for launching the explosion after a delay (depending upon the progenitor structure and mass) of generally 100–500 milliseconds. This figure summarizes the major regions in the engine of core-collapse supernova explosion. Note that the snapshot extends only ~300 kilometers (km) on a side, but resides in a star that might have a radius comparable to the radius of Jupiter's orbit. Hence, this small, dense, and hot core region is the highly energetic "mote in the eye of the storm." The explosion is launched into the rest of the star within seconds of bounce, but it is destined to take hours to a day to reach the progenitor star's periphery.

3 The Asymmetric Explosion Morphology: Bubbles and Fingers

Figure 2 depicts two different orientations of the same explosion structure after the ignition of the blast of an initially non-rotating 25-$M_\odot$ progenitor star near 550 ms after bounce. The pale blue veil is the shock wave and the surfaces are isoentropy surfaces colored by electron fraction (Y_e). The latter will translate roughly into element composition asymmetries. Clearly shown are the bubble-like elements that characterize explosions by the neutrino heating process that also generically drives convection. For models such as this 25-$M_\odot$, there is frequently a front-back dipolar structure and many finger structures. This is particularly in evidence for the more massive progenitors which might explode after a greater delay, allowing for more violent and vigorous turbulence. Also a feature of this figure and model, though not as obvious, is the fact that matter is still accreting in a ring structure, pinching to achieve a "wasp-waist" shape. Simultaneously accreting in one direction, while exploding in another, allows the maintenance at respectable levels of the accretion-powered component of the driving neutrino luminosity. This breaking of symmetry, not possible in 1D (spherical) explosions (if they do or could occur), is an important feature of the multi-dimensional CCSN explosion mechanism.

Figure 3 is a similar rendering of the explosion of a 9-$M_\odot$ model, with and without moderate rotation (initial core $\Omega_0 = 0.1$ rad s^{-1}) [19]. From the vantage of blast structure, only very rapid rotation is likely to have a noticeable effect. Figures 2 and 3 serve to indicate that all our models explode aspherically and that CCSN explosion debris distributions are inherently complicated. Figure 4 is the same 25-$M_\odot$ model portrayed in Figure 2, but at a later time and with representative debris matter trajectories. On the left side of the figure, the small bulbous structure is defined on its inner side by the pinching due to some residual accretion (via the "wasp waist").

4 The Birth of a Neutron Star and Inner Proto-neutron-star Convection

Most core-collapse supernova explosions leave behind a proto-neutron star residue. This object is hot and lepton-rich, but through neutrino losses over many seconds to a minute contracts, deleptonizes, and cools to leave a compact neutron star with a radius of ~12 km. Such an object might eventually be observed as either a radio pulsar or a magnetar. Figure 5 depicts the PNS and its surroundings created in the explosion of the 25-$M_\odot$ depicted earlier. The scale is about 300 km on a side. The outer tangle traces with representative mass trajectories the convective region just interior to the supernova blast wave (off the scene and not shown). The roughly spherical surface is an isodensity surface at $10^{9.5}$ g cm^{-3} painted by Y_e, interior to which is revealed after grahpical slicing the continuation of the outer convective tangle. Interior to that is a quiescent radiative zone and further in is a convective inner core. The latter might be the sight of magnetic dynamo action [3]. This PNS structure is a star in and of itself, whith convective and radiative zones, that our FORNAX code follows in 3D.

The same object a few hundred milliseconds later is portrayed at the same scale in Figure 6. We see that because of ongoing neutrino losses the PNS has shrunk and outer convection is beginning to subside. As noted, this process will continue for some time into the future, eventually leaving a cold, catalyzed, compact neutron star. One can estimate that currently there are ~10^8 such neutron stars in our galaxy.

5 The Possibility of Induced Pulsar Spin

Even for initially non-rotating models, the PNS core left behind can be left rotating. At late times after explosion, matter is still infalling. Due to the turbulence, such matter is not falling in spherically, but with non-zero impact parameters. Each accreted matter blob will impart angular momentum with a range of magnitudes and directions and it is the vector sum of these accretions that will determine the integrated resultant angular momentum given to the newly-born neutron star. The opposite signed vector angular momentum will be imparted to the ejected. Hence, through stochastic processes of random accretion, the PNS can obtain a spin when its initial spin, and that of the progenitor star, was zero [20, 21]. Figure 7 depicts the core near the end of our 25-$M_\odot$ simulation. The tracers enveloping it indicate clearly that many are spinning roughly coherently and that the outer PNS has a net spin. In fact, these and other CCSN calculations indicate that, given the stochastically accreted angular momentum of the core, the final period of the entire residual mass due to this induced process can be ~500 – 1000 ms. Though not very fast in the overall context of pulsars, this range is quite interesting, suggesting that induced rotation may have a role in a subset of core collapses.

6 Conclusion

In this short paper we have tried merely to indicate the complex character of the hydrodynamics that attends the dynamical evolution of the cores of massive stars in their terminal phases that are the energetic seat of core-collapse supernova explosions and neutron star birth. We have not attempted to provide detailed explanations of the radiation/hydrodynamics, the neutrino-matter couplings, nor the physics that underpins the phenomenon. It is merley our hope that the reader gleans through this graphical montage some sense of the flavor of the modern emerging theory of this central astrophysical problem.

References

[1] A. Burrows, arXiv e-prints arXiv:1812.05612 (2018), `1812.05612`

[2] Y. Masada, T. Takiwaki, K. Kotake, arXiv e-prints arXiv:2001.08452 (2020), `2001.08452`

[3] C.J. White, A. Burrows, M.S.B. Coleman, D. Vartanyan, arXiv e-prints arXiv:2111.01814 (2021), 2111.01814
[4] D. Radice, A. Burrows, D. Vartanyan, M.A. Skinner, J.C. Dolence, Ap. J. **850**, 43 (2017), 1702.03927
[5] D. Vartanyan, A. Burrows, D. Radice, M.A. Skinner, J. Dolence, M.N.R.A.S. **482**, 351 (2019), 1809.05106
[6] A. Burrows, D. Radice, D. Vartanyan, M.N.R.A.S. **485**, 3153 (2019), 1902.00547
[7] H. Nagakura, A. Burrows, D. Radice, D. Vartanyan, M.N.R.A.S. **492**, 5764 (2020), 1912.07615
[8] A. Burrows, D. Radice, D. Vartanyan, H. Nagakura, M.A. Skinner, J.C. Dolence, M.N.R.A.S. **491**, 2715 (2020), 1909.04152
[9] H.T. Janka, Annual Review of Nuclear and Particle Science **62**, 407 (2012), 1206.2503
[10] A. Burrows, Reviews of Modern Physics **85**, 245 (2013), 1210.4921
[11] A. Burrows, D. Vartanyan, Nature **589**, 29 (2021), 2009.14157
[12] E.J. Lentz, S.W. Bruenn, W.R. Hix, A. Mezzacappa, O.E.B. Messer, E. Endeve, J.M. Blondin, J.A. Harris, P. Marronetti, K.N. Yakunin, Ap. J. (Letters) **807**, L31 (2015), 1505.05110
[13] S.M. Couch, E.P. O'Connor, Ap. J. **785**, 123 (2014), 1310.5728
[14] T. Kuroda, T. Takiwaki, K. Kotake, Phys. Rev. D. **89**, 044011 (2014), 1304.4372
[15] T. Takiwaki, K. Kotake, Y. Suwa, M.N.R.A.S. **461**, L112 (2016), 1602.06759
[16] B. Müller, T.M. Tauris, A. Heger, P. Banerjee, Y.Z. Qian, J. Powell, C. Chan, D.W. Gay, N. Langer, M.N.R.A.S. **484**, 3307 (2019), 1811.05483
[17] R. Bollig, N. Yadav, D. Kresse, H.T. Janka, B. Müller, A. Heger, Ap. J. **915**, 28 (2021), 2010.10506
[18] M. Obergaulinger, M.Á. Aloy, M.N.R.A.S. **503**, 4942 (2021), 2008.07205
[19] M.S.B. Coleman, A. Burrows, C.J. White, arXiv e-prints arXiv:2111.00022 (2021), 2111.00022
[20] J.M. Blondin, A. Mezzacappa, Nature **445**, 58 (2007), astro-ph/0611680
[21] E. Rantsiou, A. Burrows, J. Nordhaus, A. Almgren, Ap. J. **732**, 57 (2011), 1010.5238

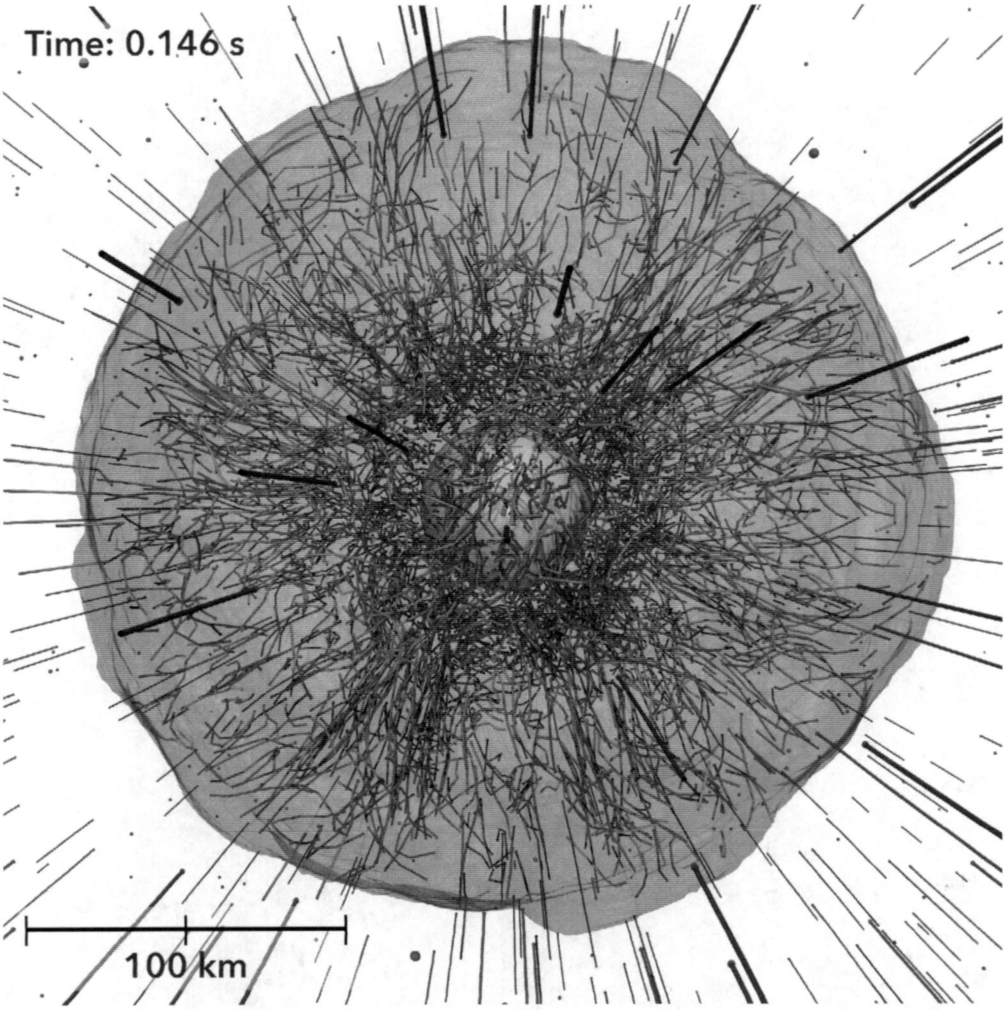

Figure 1. Ther inner structure of the proto-neutron star, bounded by the accretion shock wave (blue) just prior to explosion. The neutrino-driven turbulent region is just interior to the stalled shock is traced by the tangle of snapshot trajectories. See the text for more detail.

https://doi.org/10.1051/epjconf/202226007001

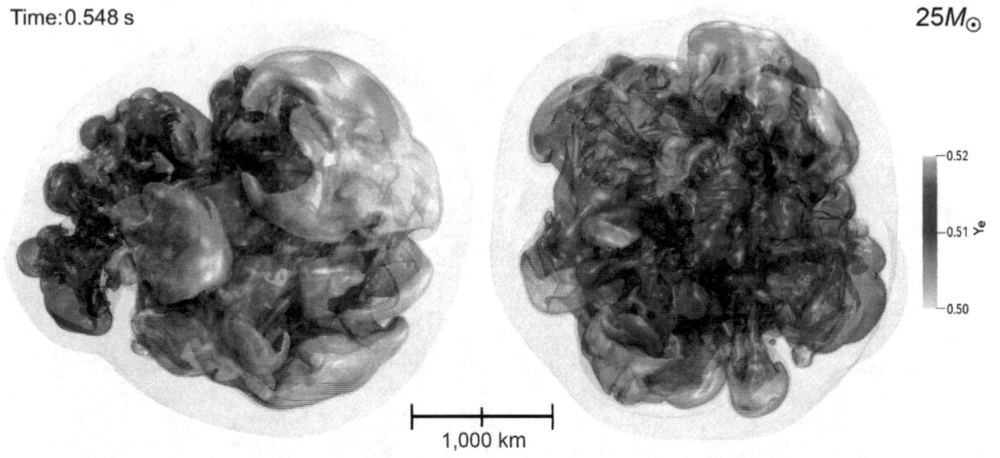

Figure 2. Two vantages of the exploding debris just 0.548 seconds after core bounce. The blue veil is the shock wave and the surfaces are isoentropy surfaces, colored by Y_e. The scale is ~3000 km. See the text for a discussion of this figure.

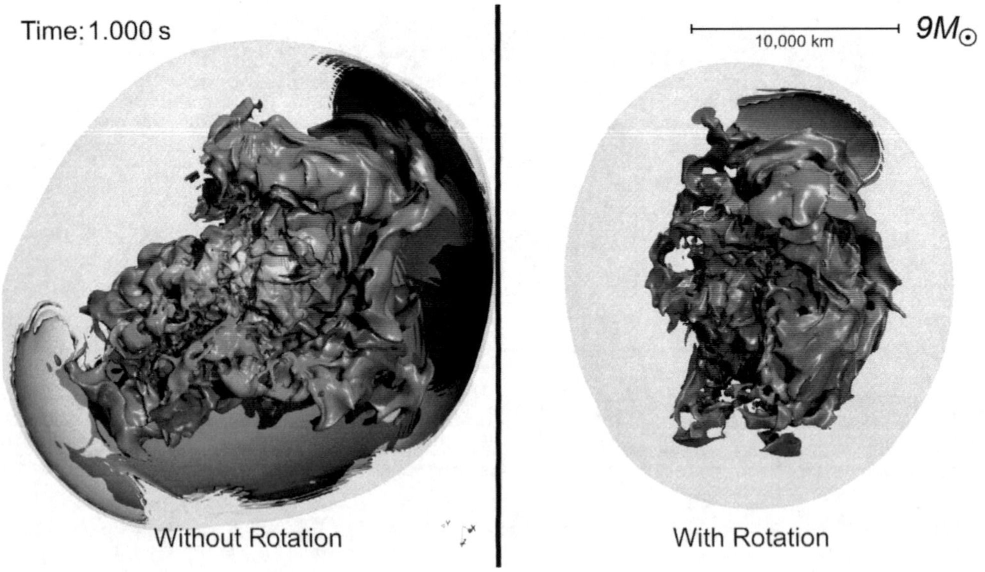

Figure 3. Similar to Figure 2, but comparing a modestly rotating model of a 9-$M_\odot$ explosion with the corresponding non-rotating model. See the text for more information.

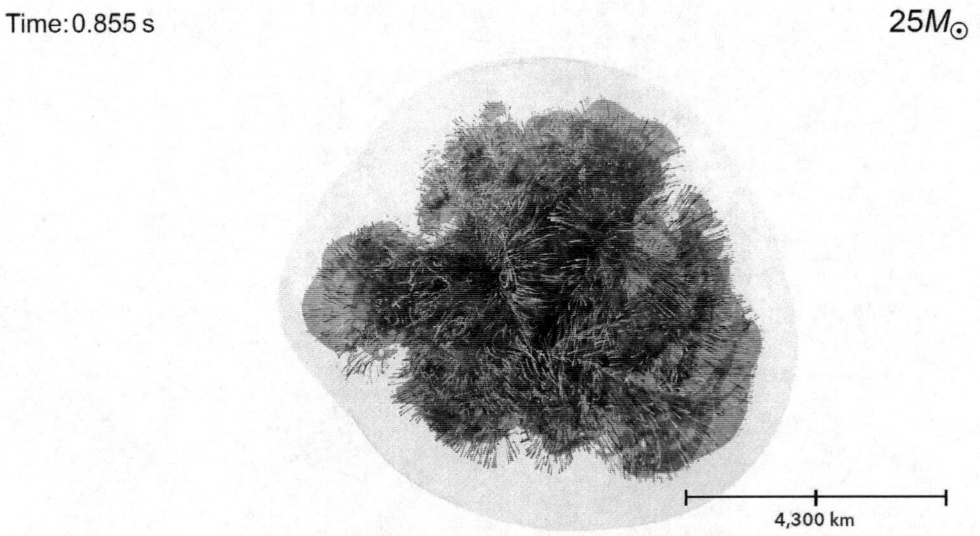

Figure 4. The blast wave with a traced debris cloud of the core of a 25-$M_\odot$ progenitor just one second into its explosion. The scale is near 10,000 km. See the text for a discussion.

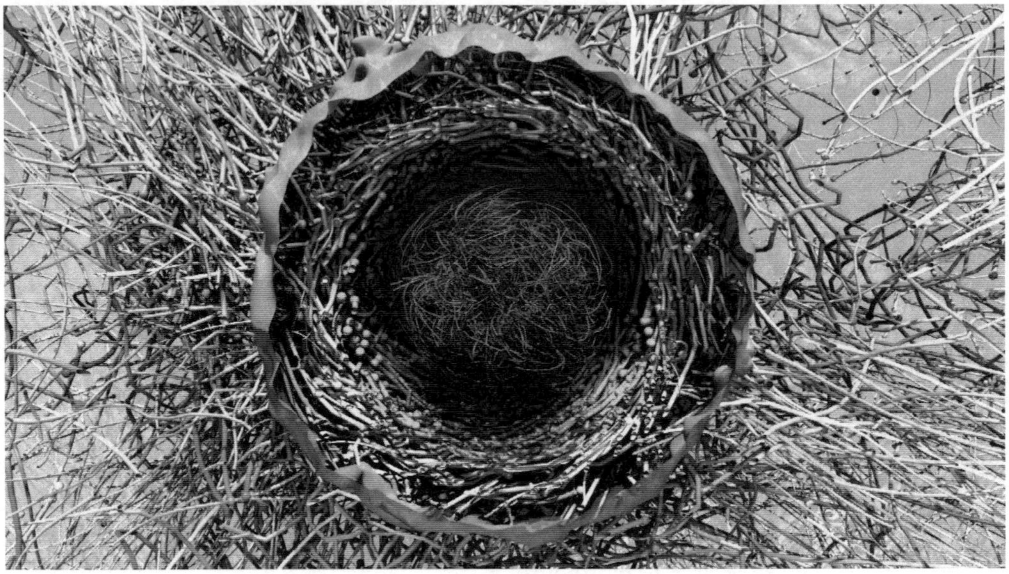

Figure 5. A snapshot from a simulation conducted by the Princeton supernova group [8] on Theta/ALCF of the supernova explosion a 25-$M_\odot$ star, depicting here the early evolution of the neutron star left behind. The scale is roughly 150 kilometers on a side and the time is roughly 500 milliseconds after core bounce and roughly 100 milliseconds after the inauguration of the explosion. The ragged pink shell is an isodensity surface at $10^{9.5}$ grams per cubic centimeter. The tangle in its exterior is of a representative collection of mass tracer particles in the outer convective zone interior to the blast wave (not shown, but beyond a thousand kilometers off the scene), and the inner "ball of yarn" traces the convective motions of the residual proto-neutron star. Convection in this inner region may be the site of a magnetic dynamo that generates magnetar and/or pulsar magnetic fields. Within one minute this core will have shrunk to a radius near 12 kilometers and cooled off to become one of the neutron stars that litter the galaxy. The colors depict the different electron fractions (Y_e) of the matter determined by the dynamical competition of electron-neutrino capture and electron capture on nucleons.

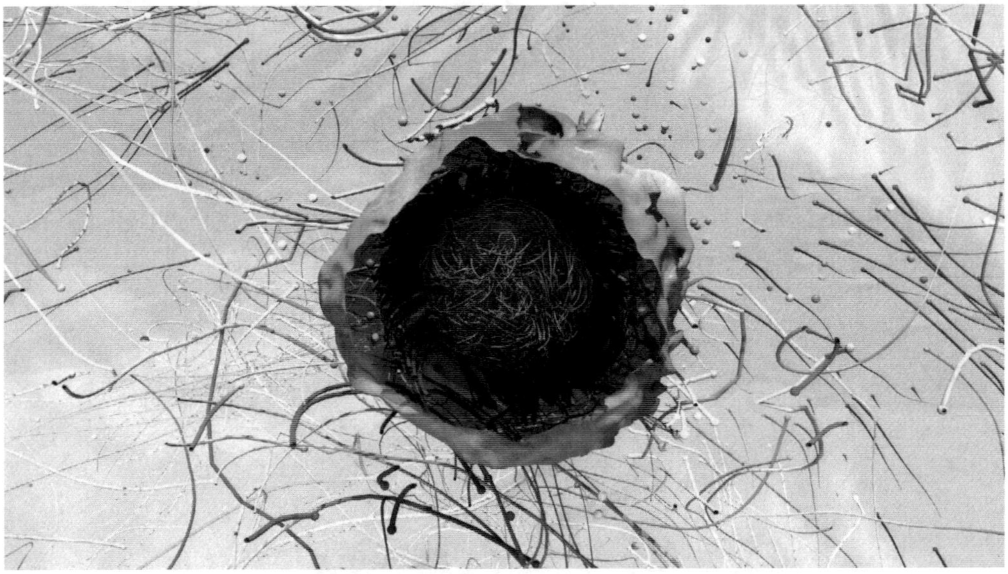

Figure 6. The same as Figure 5, but at a later time. See the text for a description and explanation.

Figure 7. An even later snapshot of the residual core, showing the degree to which the core has further shrunk and the rotating matter that now swirls at the periphery of the PNS. See the text for the context and an explanation.

Stellar elemental abundances constraining nucleosynthesis and chemical evolution of the universe

Wako Aoki[1,2,*]

[1]National Astronomical Observatory of Japan, 2-21-1, Osawa, Mitaka, Tokyo 181-8588, Japan
[2]Department of Astronomical Science, School of Physical Sciences, The Graduate University of Advanced Studies (SOKENDAI), 2-21-1, Osawa, Mitaka, Tokyo 181-8588, Japan

Abstract. Measurements of elemental abundances of stars in the Milky Way and surrounding dwarf galaxies in the past few decades have been providing useful constraints on modeling the nucleosynthesis and chemical evolution. Recent large spectroscopic surveys combined with studies on dynamical motions of stars in the Milky Way reveal connections between chemical enrichment and galaxy formation. This paper provides a brief overview of the recent update of the understanding of the distribution of elemental abundances and remaining issues for key elements, mostly focusing on the low metallicity range which should reflect nucleosynthesis by massive stars and early chemical evolution. This covers the abundance ratios of α-elements, iron-peak elements, and neutron-capture elements which have been measured for a large number of metal-poor stars.

1 Introduction

The cycle of formation, evolution, and death of stars, including interactions of binary stars, have been enriching heavier elements in the Universe. Low-mass stars formed in the early Universe could survive in the present Universe thanks to their long lifetime, and are observed as metal-poor stars. The elemental abundances of very metal-poor stars should reflect the results of nucleosynthesis by a small number of stars and phenomena in the early Universe. Hence, spectroscopic observations of metal-poor stars provide unique constraint on elemental abundances produced by individual nucleosynthesis events.

Since the evolution timescale of massive stars is short, the elemental abundances of very metal-poor stars are mostly determined by massive star evolution and supernova explosions. Taking account of the dependence of elemental abundance patterns produced by supernovae on their progenitor masses, abundance studies of very metal-poor stars are also able to estimate the masses of early generations of stars [1, 2].

Distributions of abundance ratios of individual elements for a large sample of metal-poor stars are useful constraints on chemical evolution models (Kobayashi, this volume). For instance, abundance ratios of α and iron-peak elements (e.g., [Mg/Fe])[1] as a function of metallicity are used as an indicator of contributions of core-collapse supernovae and type

*e-mail: aoki.wako@nao.ac.jp
[1] $[A/B] = \log(N_A/N_B) - \log(N_A/N_B)_\odot$, and $\log \epsilon_A = \log(N_A/N_H) + 12$ for elements A and B.

Ia supernovae, which reflect the timescale of chemical evolution of the substructures of the Milky Way Galaxy and dwarf galaxies.

Studies of elemental abundances for early generations of stars have been led by large surveys of metal-poor stars since 1980s. The HK survey [3] and the Hamburg/ESO survey [4] have applied objective prisms on Schmidt telescopes to select candidates for metal-poor stars from the absorption lines of ionized Ca (the so-called Ca H and K lines) and other features. High-resolution follow-up spectroscopy was conducted for the candidates in 1990s (e.g., [5, 6]), and later for larger samples with the 8-10 m-class telescopes [7–9]. Since 2000s, metal-poor stars have been also searched for using multi-object spectrometers, e.g., Sloan Digital Sky survey [10] and LAMOST [11]. Another approach is to conduct photometry using narrow-band filters designed to investigate absorption features that are sensitive to stellar metallicity (e.g., the Ca H-K lines). Skymapper survey [12] and Pristine survey [13] are examples of photomeric surveys that provide large samples of candidate very metal-poor stars.

We here discuss the recent results based on the follow-up high-resolution spectroscopy with the Subaru Telescope for candidates of metal-poor stars found by the LAMOST survey [14]. We note that the abundance measurements for a large sample of metal-poor stars found by the Skymapper survey recently reported similar results for many elements [15] .

2 Elemental abundances of metal-poor stars

A large sample ($\sim 10,000$ stars) of candidates for very metal-poor stars was obtained from the pilot and regular survey programs with LAMOST (e.g., [16]). High-resolution spectra of about 400 stars were obtained with the Subaru Telescope High Dispersion Spectrograph (HDS) in 2014-2017 [17]. Detailed elemental abundances are determined for these stars [14]. The sample mostly consists of main-sequence turn-off stars and red giants. The remaining stars are horizontal-branch stars that are on helium core burning phase in low-mass star evolution. The evolutionary status is well determined for most of the stars thanks to the accurate distances recently obtained by the trigonometric parallaxes provided by the Gaia mission [18].

The sample covers [Fe/H]< -2. In this metallicity range, stars in the halo component of the Milky Way is dominant.

Elemental abundances are determined using standard analysis of spectral lines using 1D LTE model atmospheres [19]. The analysis method and their behaviors have been well studied for metal-poor stars, and, hence, the results can be compared with previous studies. In this section, abundance trends are discussed for individual elements compared with chemical evolution models based on the recent results [14]. We here refer to three chemical evolution models [20–22].

2.1 α-elements (Mg, Si, Ca, Ti) and Sc

Abundance ratios between α-elements and Fe generally show over-abundances (i.e., [α/Fe]>0). This is a signature of dominant contributions of core-collapse supernovae in the early phase of chemical evolution, reflecting their short time-scale. Closer inspection of abundance ratios of individual elements provides useful constraints on, or suggestion for, the chemical evolution and the structure formation of the Milky Way halo.

The [Mg/Fe] and [Ca/Fe] values exhibit decreasing trends with increasing metallicity in [Fe/H]$\gtrsim -2.5$ (Fig. 1). A fraction of stars are known to have low [α/Fe] at higher metallicity ([Fe/H]$\gtrsim -1.5$)[23], and are interpreted as a result of accretion of stellar systems similar to

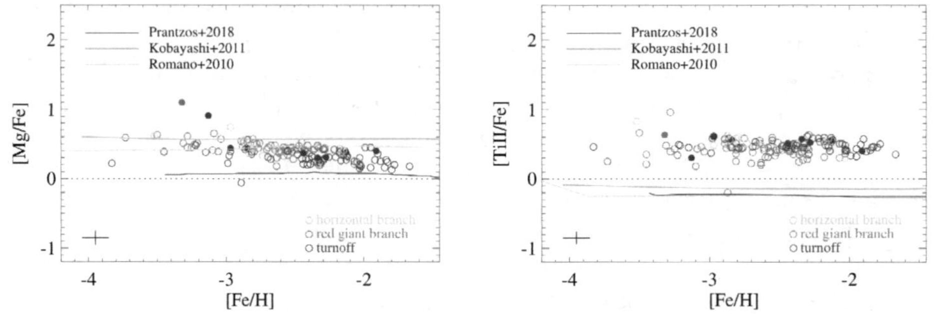

Figure 1. Abundance ratios ([Mg/Fe] and [Ti/Fe]) as a function of [Fe/H]. Red and blue circles indicate the abundance results for red giants and main-sequence turn-off stars, respectively [14]. Predictions of the three chemical evolution models [20–22] are shown by lines as presented in the panels.

current dwarf galaxies, in which the timescale of the chemical evolution is longer [24]. Such effects might already appear at lower metallicity. Si might show a similar trend, but it is not as clear as the above two elements due to the limited size of the sample and data quality. The situation will be improved by observations with new near-infrared spectrometers that covers the wavelength range in which many Si lines useful for accurate abundance measurements exist [25].

Ti also shows over-abundance as found for Mg, Si and Ca (Fig. 1). This result is obtained from analyses of neutral and ionized Ti lines, indicating that the result is robust. The [Ti/Fe] values are almost constant in this metallicity range. This overabundance is not explained by any of the three chemical evolution models. The discrepancy would be due to the limited understanding of the yields of core-collapse supernovae.

Sc, an element with odd atomic number, also exhibits over-abundances, which are not reproduced by chemical evolution models as in the case of Ti.

Abundance ratios of α-elements, with respect to Fe, show correlations each other. The correlations are particularly strong between [Mg/Fe] and [Si/Fe], and also between [Ca/Fe] and [Ti/Fe]. The former are lighter α-elements, representing the products during massive star evolution, whereas the later heavier ones are products of explosive nucleosynthesis. The correlations between the lighter and heavier α-elements are less significant. [Sc/Fe] shows clear correlations with [Ca/Fe] and [Ti/Fe], suggesting that the elements have similar origins.

2.2 Iron-peak elements

The [Cr/Fe] values determined for red giants show a clear decreasing trend with decreasing metallicity, whereas those for main-sequence stars are almost constant (Fig. 2). The trend found in red giants is considered to be caused by the non-LTE effect in the line formation of neutral Cr in very metal-poor stars. Taking account of this effect, Cr abundance ratios agree well with the solar value, as found for Ni. The [Mn/Fe] values are clearly lower than the solar value. These results are well reproduced by the chemical evolution models. On the other hand, Co shows slight over-abundance, which is not explained by models that predict the solar abundance ratio for very metal-poor stars. [Zn/Fe] shows increasing trend with decreasing metallicity (Fig. 2). The sample [14] does not well cover the lowest metallicity range [Fe/H]<-3.5 due to the limited data quality. This could lead selection bias in the abundance trend as discussed [15]. However, previous studies covering the low metallicity range (e.g., [7]) report that the trend continues down to [Fe/H]~ -4. The chemical evolution models with standard

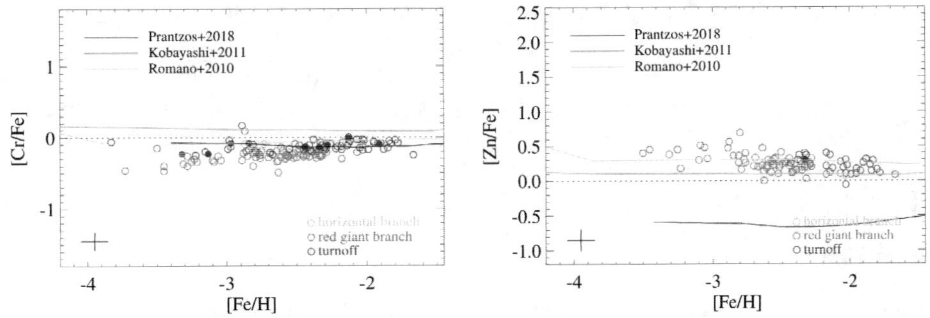

Figure 2. The same as Fig. 1, but for Cr and Zn.

yields predict lower abundance ratios of Zn. This suggests that contributions of energetic supernova explosions (e.g., hypernovae) are required to be included.

Zn abundance ratios do not show clear correlations with those of α-elements. The exception is the Ti abundance ratios, that exhibit a significant correlation with the Zn abundance ratios, suggesting a connection between the production site of these elements.

2.3 Neutron-capture elements

The heavy neutron-capture elements Sr and Ba are relatively abundant and have strong spectral lines in the optical range and, then, are useful for investigating the distributions of abundance ratios at very low metallicity. A large scatter in both [Sr/Fe] and [Ba/Fe] is a well-known property. However, a closer inspection reveals different features between them. In [Fe/H]> -3, most of the stars have [Sr/Fe] and [Ba/Fe] close to 0 (i.e., solar abundance ratio), but Sr shows stronger concentration. The large scatter of Sr and Ba abundance ratios ($-2 <\lesssim$[Sr/Fe],[Ba/Fe]$\lesssim +1$) is found in [Fe/H]$\lesssim -3$. While [Sr/Fe] values show a rather flat distribution in the above range, a large fraction of stars have [Ba/Fe]~ -1. The [Sr/Ba] values also show a wide distribution. This suggests the existence of different origins for Sr and Ba at very low metallicity. We note that the so-called carbon-enhanced mtal-poor stars with Ba excess (CEMP-s) are not included in this discussion.

A small fraction of stars with [Fe/H]< -3 have very low Sr and Ba abundance ratios ([Sr/Fe], [Ba/Fe]~ -2). There is no constraint on their origin from other elements (e.g., Eu). Further studies are required for these stars to understand the possible source of such small amount of neutron-capture elements. The origins of neutron-capture elements are also discussed in the conference (Tarumi et al., this volume).

3 Summary and prospects

Measurements of elemental abundances for a large sample of metal-poor stars have been revealing the origins and the evolution of elements in the Universe. The studies also identify problems to be solved in comparisons between observations and models for abundance distributions. For instance, the Ti and Sc over-abundances, and the slight over-abundance of Co and overabundance of Zn at extremely low metallicity are not well explained by the standard chemical evolution models. Correlations between elemental abundances (e.g. between Ti and Zn) could be useful hint to solve these problems.

Large scatter in abundance ratios of neutron-capture elements has been revealed by many studies. Investigation of the detailed distributions of Sr and Ba abundance ratios (e.g., existence of low-Sr and low-Ba stars) would contain useful information on the origins of these elements.

Some abundance trends (e.g., Mg/Fe ratios) could be affected by formation process of the Galactic halo, including accretions of dwarf galaxies. The recent rapid progress in the studies of the connection between abundance ratios and kinematics will shed new light also on understanding of the origins of elements in the Universe.

References

[1] Heger, A., & Woosley, S. E., Astrophysical Journal 724, 341. (2010)
[2] Ishigaki, M. N., Tominaga, N., Kobayashi, C., & Nomoto, K., Astrophysical Journal 857, 46 (2018)
[3] Beers, T. C., Preston, G. W., & Shectman, S. A., Astronomical Journal 103, 1987. (1992)
[4] Christlieb, N., Green, P. J., Wisotzki, L., & Reimers, D., Astronomy & Astrophysics, 375, 366. (2001)
[5] McWilliam, A., Preston, G. W., Sneden, C., & Searle, L., AJ, 109, 2757. (1995)
[6] Ryan, S. G., Norris, J. E., & Beers, T. C., ApJ, 471, 254. (1996)
[7] Cayrel, R., Depagne, E., Spite, M., Hill, V., Spite, F., François, P., Plez, B., Beers, T., Primas, F., Andersen, J., Barbuy, B., Bonifacio, P., Molaro, P., & Nordström, B., A&A, 416, 1117. (2004)
[8] Cohen, J. G., Christlieb, N., McWilliam, A., Shectman, S., Thompson, I., Wasserburg, G. J., Ivans, I., Dehn, M., Karlsson, T., & Melendez, J., ApJ, 612, 1107. (2004)
[9] Honda, S., Aoki, W., Kajino, T., Ando, H., Beers, T. C., Izumiura, H., Sadakane, K., & Takada-Hidai, M., ApJ, 607, 474. (2004)
[10] Yanny, B. et al., AJ, 137, 4377. (2009)
[11] Zhao, G., Zhao, Y.-H., Chu, Y.-Q., Jing, Y.-P., & Deng, L.-C., Research in Astronomy and Astrophysics, 12, 723. (2012)
[12] Keller, S. C., Schmidt, B. P., Bessell, M. S., Conroy, P. G., Francis, P., Granlund, A., Kowald, E., Oates, A. P., Martin-Jones, T., Preston, T., Tisserand, P., Vaccarella, A., & Waterson, M. F., PASP, 24, 1. (2007)
[13] Starkenburg, E., Martin, N., Youakim, K. et al., MNRAS, 471, 2587. (2017)
[14] Li, H., Aoki, W., Matsuno, T. et al., submitted to ApJ
[15] Yong, D., Da Costa, G. S., Bessell, M. S., et al., MNRAS, 507, 4102. (2021)
[16] Li, H., Tan, K., & Zhao, G., ApJS, 238, 16. (2018)
[17] Aoki, W., Li, H., Matsuno, T. et al., submitted to ApJ
[18] Gaia Collaboration et al., A&A, 616, A1. (2018)
[19] Castelli, F., & Kurucz, R. L., Modelling of Stellar Atmospheres, 210, A20. (2003)
[20] Romano, D., Karakas, A. I., Tosi, M., & Matteucci, F., A&A, 522, A32. (2010)
[21] Kobayashi, C., Karakas, A. I., & Umeda, H., MNRAS, 414, 3231. (2011)
[22] Prantzos, N., Abia, C., Limongi, M., Chieffi, A., & Cristallo, S., MNRAS, 476, 3432. (2018)
[23] Nissen, P. E., & Schuster, W. J., A&A, 511, L10. (2010)
[24] Helmi, A., ARA&A, 58, 205. (2020)
[25] Aoki, W. et al., submitted to PASJ

Exploring the uncertainties of (α, xn) reactions for the weak r-process

Athanasios Psaltis[1,*], *Almudena* Arcones[1,2], *Melina L.* Avila[3], *Maximillian* Jacobi[1], *Zach* Meisel[4], *Peter* Mohr[5], *Fernando* Montes[6,7], and *Wei Jia* Ong[8]

[1]Institut für Kernphysik, Technische Universität Darmstadt, Schlossgartenstr. 2, Darmstadt 64289, Germany
[2]GSI Helmholtzzentrum für Schwerionenforschung GmbH, Planckstr. 1, Darmstadt 64291, Germany
[3]Physics Division, Argonne National Laboratory, Argonne, IL 60439, USA
[4]Institute of Nuclear & Particle Physics, Department of Physics & Astronomy, Ohio University, Athens, OH, 45701 USA
[5]Institute for Nuclear Research (ATOMKI), H-4001 Debrecen, Bem tér 18/c, Hungary
[6]National Superconducting Cyclotron Laboratory, East Lansing, MI, 48824, USA
[7]Joint Institute for Nuclear Astrophysics – Center for the Evolution of the Elements, Michigan State University, East Lansing, Michigan 48824, USA
[8]Nuclear and Chemical Sciences Division, Lawrence Livermore National Laboratory, Livermore CA USA 94550

Abstract. "Light" heavy elements ($Z = 38 - 47$) can be synthesized in the neutrino–driven ejecta of core–collapse supernovae via the weak *r*–process. This nucleosynthesis scenario exhibits uncertainties from the absence of experimental data from (α, n) reactions on neutron–rich nuclei, and are mostly based on statistical model calculations. We present preliminary results from a recent sensitivity study, using the Atomki–V2 α-nucleus potential to identify the most important (α, n) reactions that can affect the production of "light" heavy elements between strontium and silver under different astrophysical conditions. We also discuss the planning of studies to experimentally determine (α, xn) reaction rates using the MUSIC detector at Argonne National Laboratory and the SECAR recoil separator at FRIB.

1 Introduction

The origin of the heavy elements in the cosmos is one of the most stimulating questions in modern physics. Half of them can be produced via the *r*–process in extreme astrophysical environments with a high neutron density, such as neutron star mergers (NSMs) or magneto-rotational supernovae explosions (MR-SNe), but there are still many open questions regarding which site has the dominant contribution in the Galactic *r*–process abundances [1]. Observations of metal–poor stars have shown a scatter in the "light" region between strontium and silver ($Z = 38 - 47$), which has sparked discussions for an additional nucleosynthesis process [2–4].

One promising nucleosynthesis scenario is the weak *r*–process (also known as α–process), which occurs in the neutron–rich, neutrino–driven wind ejecta of core–collapse

*e-mail: psaltis@theorie.ikp.physik.tu-darmstadt.de

supernovae [5, 6]. Nucleosynthesis in the weak r–process starts from Nuclear Statistical Equilibrium (NSE) with $Y_e = 0.40 - 0.49$[1]. As the ejecta expand and cool down, the plasma temperature decreases, and when it reaches $T \approx 5$ GK, an α–rich freeze–out occurs. α and proton captures – mainly (α, n), (p, n) and (α, γ) reactions – lead the reaction flow to heavier masses until the temperature falls to $T \approx 2$ GK. Refs. [7–9] have demonstrated that the main reaction channel which dominates this nucleosynthesis scenario is the (α, xn) reactions on unstable, neutron–rich nuclei. Nevertheless, the (α, xn) reaction rates that are used in nucleosynthesis studies rely on statistical model predictions, since few experimental cross sections are available. The most important input of such models is the α optical model potential (αOMP), which can produce differences in the calculated reaction rate of up to two orders of magnitude in the relevant temperature region [7, 8] (see Figure 1 for an example of the ^{93}Sr$(\alpha, xn)^{96}$Zr reaction).

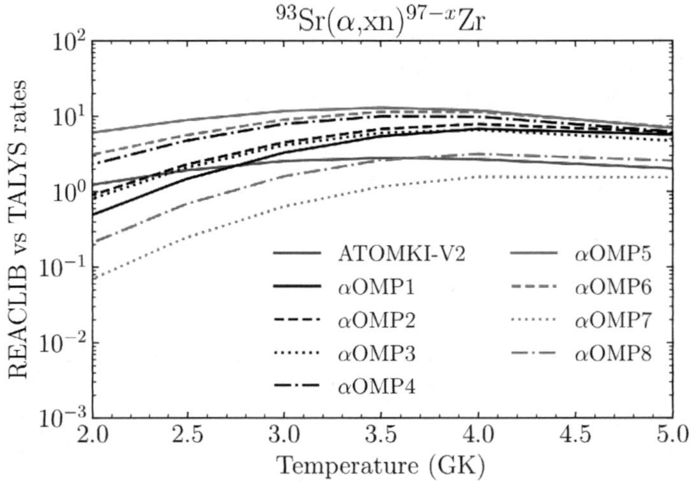

Figure 1. Comparison of the REACLIB ^{93}Sr$(\alpha, xn)^{97-x}$Zr reaction rate to the `TALYSv1.95` predictions using the 8 different α Optical Model potentials (αOMP) available in the code, including the Atomki–V2 potential [10] for the astrophysical window of the weak r–process.

Bliss *et al.* [11] recently studied the the effect of the (α, n) reaction rate uncertainty on the production of “light heavy” elements under different astrophysical conditions using Monte Carlo techniques. They used (α, xn) reaction rates calculated with the default αOMP of the reaction code `TALYS` and assumed uncertainties of a factor of 10. A list of (α, xn) reactions which produce the largest abundance differences was also identified.

2 The impact of new (α, n) reaction rates with an improved αOMP

Recently Mohr *et al.* [12] published a compilation of α–induced reaction rates for elements between iron and bismuth using the `TALYS` code and the Atomki–V2 αOMP [10, 13]. Recent experiments of (α, n) reactions on stable nuclei [14, 15] have shown agreement with the Atomki–V2 predictions to within a factor of two and thus provide reliability in the use of the αOMP when used on (α, n) reactions on unstable nuclei.

[1]The Y_e is defined as $Y_e = n_p(n_p + n_n)^{-1}$, where n_p and n_n are the number densities of protons and neutrons, respectively.

We have performed a Monte Carlo sensitivity study using the (α, xn) reaction rates from Ref. [12] following the methodology of Ref. [11]. The reaction rate uncertainties are lower compared to Ref. [11], since the rates based on the Atomki–V2 αOMP generally agree to within a factor of 2 to experimental data [14, 15]. Figure 2 shows some preliminary results of the predicted abundance ratios of Sr/Y and Y/Zr for a single astrophysical condition of the neutrino–driven wind. Production uncertainties have been greatly reduced and are now at the same order as the observational ones.

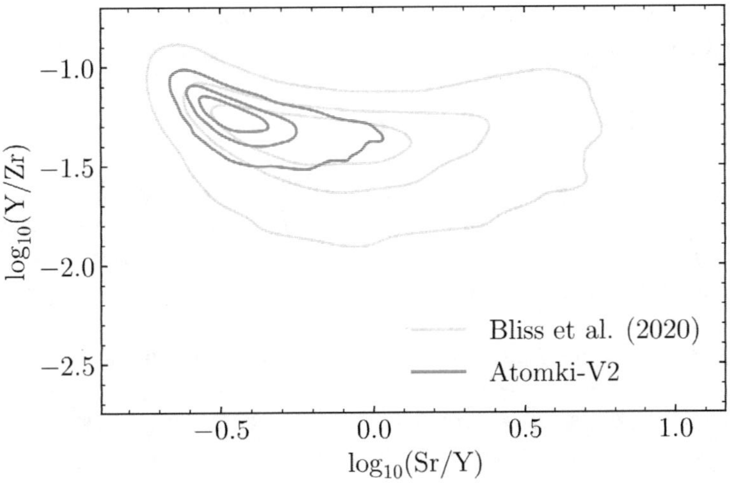

Figure 2. Kernel Density Estimate (KDE) of the Sr/Y and Y/Zr elemental ratios for the Monte Carlo study of Bliss *et al.* [11] using `TALYS v1.6` (α, n) reaction rates and the present work using rates based on the Atomki–V2 αOMP [12]. The contours show the 3σ uncertainties for each calculation. The conditions of the trajectory used are: $Y_e = 0.4$, entropy $s = 56\ k_B$/nucl. and expansion time $\tau = 63.8$ ms. Typical observational uncertainties are $\sim 0.1 - 0.2$ dex [2].

Using these new theoretical reaction rates, we have identified a list of (α, xn) reactions that affect the production of "light heavy" elements in different astrophysical conditions. For this, we used any (α, xn) reaction whose Pearson correlation coefficient between the rate variation and the abundance change of a particular nucleus is $r > |0.2|$ and the change is greater than a factor of 5 from the baseline case. Table 1 shows a summary of our preliminary results. Note that most of the reported target nuclei were also identified in the sensitivity study of Ref. [11].

Table 1. Target nuclei whose (α, xn) reactions affect the elemental abundances of $38 \leq Z \leq 47$.

Group #	Target nuclei	Notes
1	^{84}Se, ^{85}Se	Affect many elemental abundances under many astrophysical conditions
2	^{94}Sr, ^{97}Sr, ^{96}Zr	Affect few elemental abundances under many astrophysical conditions
3	^{100}Zr, ^{76}Zn	Affect many elemental abundances under few astrophysical conditions
4	^{60}Fe, ^{74}Zn, ^{104}Mo	Affect few elemental abundances under few astrophysical conditions

3 Experimental Plans

To better constrain the relevant astrophysical conditions in the weak *r*–process responsible for the production of elements between strontium and silver, experiments that can provide reaction rates with uncertainties lower than the current theoretical ones are needed. Some experiments, mainly in stable nuclei have recently been performed [14, 15], and few radioactive ion beam experiments are currently ongoing in various facilities worldwide, such as MUSIC at ANL/ATLAS, HABANERO and SECAR at NSCL/FRIB, EMMA at TRIUMF/ISAC–II. In the following we shall briefly discuss a limited set of future plans to experimentally determine (α, xn) reaction rates at temperatures relevant to the weak *r*–process.

3.1 ^{93}Sr$(\alpha, xn)^{96}$Zr at Argonne with MUSIC

The MUSIC detector is based at the Argonne National Laboratory[2] [16] and is an active–target system that is capable of measuring cross sections of (α, n) and (α, p) reactions relevant for astrophysics. It has a segmented anode which allows for highly efficient measurements of a large range of excitation functions of angle and energy integrated cross sections using single beam energy.

Recently, MUSIC successfully measured the ^{100}Mo(α, n) reaction and an experimental proposal to study the important ^{93}Sr(α, n) reaction rate was also approved by the ATLAS PAC (see also the rate discrepancies from theoretical predictions in Figure 1). The ^{93}Sr beam will be provided by the CARIBU spontaneous fission source [17].

3.2 ^{84,87}Se$(\alpha, n)^{87,90}$Kr at FRIB with SECAR

SECAR at FRIB[18] is a versatile recoil separator designed for radiative capture reactions, than can be used to measure (p, n) and (α, xn) reaction rates due to its relatively large recoil acceptance. The setup to study (α, xn) reaction rates at SECAR include helium from the JENSA gas target [19] and neutron–tagging from neutron detectors surrounding the SECAR target. Reaction products are detected at the final focal plane, where a pair of MCP position sensitive detectors provides a time–of–flight signal and an ionization chamber combined with a silicon detector provide time and energy recoil information. SECAR is currently under commissioning and the technique to measure (α, n) reactions is under development. Due to the unique capabilities of FRIB in terms of neutron–rich beams, the important ^{84,87}Se$(\alpha, n)^{87,90}$Kr reactions could be studied using SECAR in the near future.

4 Summary & Discussion

The origin of the "light heavy" elements is an exciting open question in nuclear astrophysics. The weak *r*–process is one of the plausible scenarios, however it suffers from uncertainties attributed both to the astrophysical conditions and the (α, xn) reaction rates.

We have performed a Monte Carlo (α, xn) reaction sensitivity study using reaction rates based on the Atomki–V2 αOMP. The reduced uncertainties of the (α, xn) reaction rates using the Atomki–V2 αOMP have reduced the production uncertainty of the "light" heavy elements via the weak *r*–process. Future experimental studies will help to further reduce the underlying nuclear physics uncertainties and shed light to their production in the universe.

[2]The detector can be transported to other facilities and recently a proposal to measure the ^{85}Se(α, xn) reaction was approved by the FRIB PAC.

References

[1] J.J. Cowan et al., Rev. of Mod. Phys. **93**, 015002 (2021)
[2] S. Honda et al., Astrophys. J **607**, 474 (2004)
[3] C. Sneden, J.J. Cowan, R. Gallino, Annu. Rev. Astron. Astrophys. **46**, 241 (2008)
[4] C. Hansen, F. Montes, A. Arcones, Astrophys. J **797**, 123 (2014)
[5] A. Arcones, F. Montes, Astrophys. J **731**, 5 (2011)
[6] J. Bliss et al., Astrophys. J **855**, 135 (2018)
[7] J. Pereira, F. Montes, Phys. Rev. C **93**, 034611 (2016)
[8] P. Mohr, Phys. Rev. C **94**, 035801 (2016)
[9] J. Bliss et al., J. Phys. G **44**, 054003 (2017)
[10] P. Mohr et al., Phys. Rev. Lett. **124**, 252701 (2020)
[11] J. Bliss et al., Physical Review C **101**, 055807 (2020)
[12] P. Mohr et al., At. Data Nucl. Data Tables **142**, 101453 (2021)
[13] P. Mohr, G. Gyürky, Z. Fülöp, Phys. Rev. C **95**, 015807 (2017)
[14] G.G. Kiss et al., Astrophys. J **908**, 202 (2021)
[15] T.N. Szegedi et al., Phys. Rev. C **104**, 035804 (2021)
[16] M. Avila et al., Nucl. Instrum. Methods. Phys. Res. A **859**, 63 (2017)
[17] G. Savard et al., Nucl. Instrum. Methods. Phys. Res. B **266**, 4086 (2008)
[18] G. Berg et al., Nucl. Instrum. Methods. Phys. Res. A **877**, 87 (2018)
[19] K.A. Chipps et al., Nucl. Instrum. Methods. Phys. Res. A **763**, 553 (2014)

Impacts of Strongly Magnetized Degenerate Plasma on the Electron-capture Rates

Yudong Luo[1,2,*], *Motohiko* Kusakabe[3,**], and *Toshitaka* Kajino[1,2,3,***]

[1]National Astronomical Observatory of Japan, 2-21-1 Osawa, Mitaka, Tokyo 181-8588, Japan
[2]Graduate School of Science, The University of Tokyo, 7-3-1 Hongo, Bunkyo-ku, Tokyo 113-0033, Japan
[3]School of Physics, International Research Center for Big-Bang Cosmology and Element Genesis, Beihang University, 37, Xueyuan Rd., Haidian-qu, Beijing 100083 China

Abstract. The strongly magnetized degenerate astrophysical plasma is investigated. A relativistic Hartree self-consistent field method is applied to calculate the screening potential. A profile from a 15 $M_\odot$ core collapsing supernova (SN) progenitor is applied to evaluate the electron-capture rates of ^{54}Fe. It is found that the screening potential at high field is enhanced compared with the previous study. If the field is high enough and only the lowest Landau level is allowed, two orders of magnitude reduction of the capture rates are found in the high-density region. Such deviations of the electron capture rates are essential since the rates determine the neutron richness of the progenitor model as well as the iron core mass, which are crucial for MHD-Jet SNe explosion calculation.

1 Introduction

The weak interactions determine the nucleosynthetic path in the supernovae (SNe) and also its explosion models [1–4]. For the MHD-Jet SNe, it is believed that the collapsing and bouncing phases have relativistic degenerate electron gas in a strong magnetic field of the order of 10^{14-16} G [5] in the inner region. The thermodynamics of the electrons (positrons) in such an environment is different from the field-free case and could further deviate the nuclear weak interactions significantly. Prompted by this fact, this work focuses on the microscopic effects of the relativistic, magnetized degenerate plasmas on the weak interactions. The general Hartree self-consistent field method of the relativistic degenerate electron gas [6, 7] is applied to calculate the strong screening potential in the background magnetic field. The impacts on the weak interaction rates from the highly magnetized degenerate plasma are also investigated. We apply a profile from the 15 $M_\odot$ progenitor model to calculate the electron-capture rates of iron group nuclei. The results are essential to determine the iron core mass before the explosion and the electron fraction Y_e during SN's collapsing phase.

*e-mail: yudong.luo@nao.ac.jp
**e-mail: kusakabe@buaa.edu.cn
***e-mail: kajino@buaa.edu.cn

2 Impacts of Strongly Magnetized Degenerate Plasma

The Electron capture rate of a nucleus (Z, A) is contributed by individual capture rates λ_{if} from one of the initial states i to all possible final states f [9–12]. Considering the background magnetic field, the integrand in phase space of the capture rate reads

$$f_{if} = \frac{eB}{4}\sum_{n=0}^{n_{max}}(2-\delta_{n0})\int_{\omega_l'}^{\infty} d\omega_B\, \frac{\omega_B}{p_B}(Q_{if}+\omega_B)^2 F(Z,\omega_B)S_e(\omega_B,\mu_B,T), \tag{1}$$

where the electron momentum transverse to the field direction is quantized into separate Landau levels, the maximum occupied level n_{max} is determined by the Fermi energy E_F of the electrons (positrons). The electron energy ω_B is given by $\omega_B^2 = p^2 + m_e^2 + 2eB$. μ_B denotes the chemical potential within the background magnetic field. The integration limits of Eq. 1 are set by the reaction threshold ω_l'. Q_{if} is the electron capture transition energy, determined from the nuclear masses. $F(Z,\omega)$ is the Fermi function which corrects the distortion of the electron wave function [13]. Inside the realistic astrophysical plasma, the interaction rates are varied from the vacuum case due to the many-body effect inside the plasma such as the Coulomb screening correction. The screening could mainly change: (1). the threshold energy for the capture is varied by the amount $\Delta Q_c = \mu_c(Z-1) - \mu_c(Z)$ [14], where μ_c is the Coulomb chemical potential [15]; (2). The energy of the captured electron ω is reduced by the screening potential $\Delta V(B)$, which is obtained under the relativistic degenerate condition [6, 7] (see Ref. [8] for more details).

In Fig. 1, we compare f_{ij} under different density (ρY_e) and field strength. For an extremely strong magnetic field, E_F decreases monotonically as a function of B, only the Lowest Landau Level (LLL) is occupied, and the value of f_{ij} shows the same trend since the Fermi energy determines the maximum energy of electrons. With decreasing magnetic field strength, the new Landau level pops up to cancel the extra energy carried by the magnetic field. Therefore, f_{ij} decreases firstly, and then increases since the cancelling effect decreases monotonically with decrease of the magnetic field. Until the next Landau level pops up, such a pattern will be repeated. The left panel of Fig. 1 shows the ratio between screened and unscreened f_{ij}: the screening could suppress f_{ij} for about 20% under high density and strong magnetic field case (the right bottom part), while for a weak magnetic field, screening could reduce f_{ij} only for few percent. This is consistent with the previous study [13] where only the field-free screening is taken into account. On the right panel, we compare the ratio of f_{ij} between the scenarios with and without magnetic field. For every $(\rho Y_e, B)$ combination that corresponds to a popped-up new Landau level, f_{ij} value could be enlarged for 1.5 times due to the larger value of E_F. Once the field strength becomes lower than 10^{14} G, it becomes the same value as the $B = 0$ scenario. Since $\lambda_{ij} \propto f_{ij}$, such impact on the phase space should leave the imprints in the final electron capture rates.

3 Electron Capture Rates of Iron Group Nuclei

The importance of the shell-model rates in the pre-SN stage has been investigated previously [2, 3]. It was shown that weak interaction rates significantly affect the central electron fraction Y_e at the onset of the core-collapse, which further affect the nucleosynthetic path. The detailed analysis of Y_e change after Oxygen burning also find that the critical flow during the final stages of stellar evolution is much closer to the valley of stability, and several essential electron capturing nuclei (e.g., ^{54,56,58}Fe, ^{55}Mn, ^{53}Cr) are identified [2, 3]. Inspired by the previous studies, we first focus on the rate of electron capture on ^{54}Fe. We calculate the reaction rates in different sets of stellar conditions (i.e., the combination of density, temperature

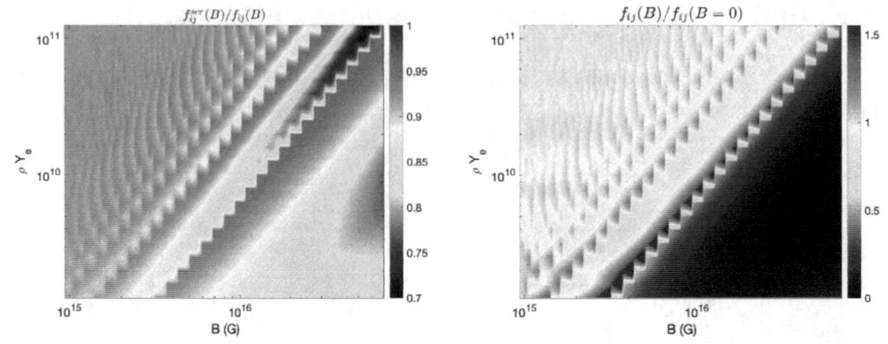

Figure 1. The left panel: The ratio between screened and unscreened f_{ij}. For the field-free situation, only a few percent deviations of f_{ij} could be found (i.e., left-upper region of this panel). For the high field and relatively low-density case ($B > 10^{16}$ G and $\rho Y_e \sim 10^9$), the screening effect could suppress f_{ij} about 20%. The right panel: The f_{ij} in a background magnetic field compared with a field-free case. f_{ij} could be enhanced about 1.5 times when the field strength just corresponds to a new popped-up Landau level. For the extremely high field case, only the LLL is allowed for the electrons to be occupied in, then f_{ij} is drastically reduced (i.e., left-upper region of this panel). The calculation here sets T = 10 GK. This figure is taken from [8].

and Y_e). Those combinations are appropriate for the inner core of a collapsing 15 $M_\odot$ star, the profile of which is taken from Ref. [13]. Fig. 2 shows the reaction rate of $^{54}Fe(e^-, \nu_e)^{54}Mn$

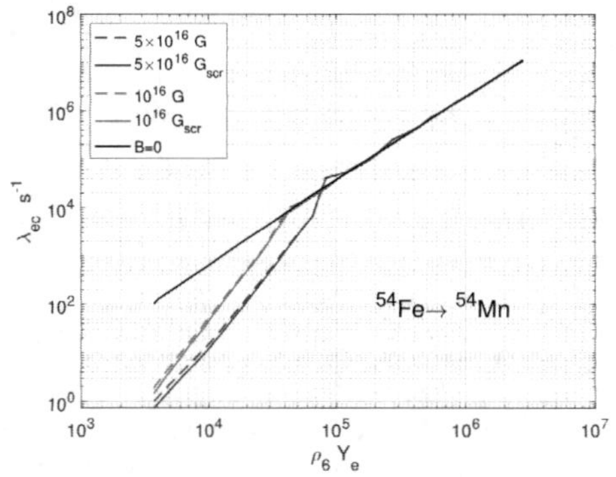

Figure 2. The reaction rate of $^{54}Fe(e^-, \nu_e)^{54}Mn$ as a function of ρY_e. Blue lines correspond to a 10^{15} G magnetic field. Red lines correspond to a 10^{16} G magnetic field, and black lines are the field-free case. The screening corrections are also included as solid lines. This figure is taken from [8].

as functions of ρY_e. All the rates are converged to the same value for electron chemical potentials μ_e larger than about 25 MeV, while at lower μ_e values, the capture rates are more sensitive to the magnetic field strengths. The screening could suppress the electron-capture rates as expected, and the suppression is stronger for the lower μ_e value. It is also worth mentioning that, due to the magnetic field, μ_e is also a function of B, as discussed. Therefore, for the same combination of (ρ, Y_e), μ_e decreased as the increasing field strength, the electron capture rate could be reduced by almost two orders of magnitude for the low density and high field region. In high-density region, the field strength that allowed only the LLL occupation is also high, therefore $B < B_{crit}$ makes the indistinguishable trend of the reaction rates in high-density region. The electron capture process could reduce the electron number density and the neutrino emission from electron capture could carry away the energy and entropy of the core. Although the dominant composition during the collapse phase of the core is protons and neutrons, it has been shown that the heavy nuclei play an equally important role for the Y_e evolution during the collapse [16]. Therefore, it is expected that such an influence of the magnetic field on the iron group nuclei could significantly impact the collapse and bouncing phase. It is worth extending to the sophisticated MHD-Jet SNe calculation.

4 Conclusion

In this work, the electron capture rates of iron group nuclei are re-evaluated for the strongly magnetized degenerate plasma. The degeneracy of the plasma makes the Thomas-Fermi approximation invalid. A relativistic generalization of the Hartree self-consistent field method is applied to calculate the strong screening potential. The screening potential could reduce the electron capture rate more significantly than the field-free situation. Moreover, the strong field strength only allows the lowest Landau level occupation. Therefore, the weak interaction rate is dramatically suppressed. A profile from a 15 $M_\odot$ core collapsing SN progenitor model is applied to calculate the electron-capture rates of ^{54}Fe. It is noticed that two orders of magnitude reduction is found between the rates for plasma with density $\rho \sim 10^9 \sim 10^{11}$ g cm^{-3} and magnetic field $B > 10^{14}$ G . These impacts have potential significance since electron capture on iron group nuclei is the leading reaction that reduces Y_e, which finally could affect the *r*-process path and also SNe explosion models. Then, such results should be applied to the collapse and bouncing phases of MHD-Jet SNe calculation.

Y. L. is supported by JSPS KAKENHI Grant No. 19J22167. M. K. is supported by NSFC Research Fund for International Young Scientists (11850410441). T. K. is supported in part by Grants-in-Aid for Scientific Research of JSPS (17K05457 and 20K03958).

References

[1] G. C. McLaughlin, G. M. Fuller and J. R. Wilson, Astrophys. J. **472**, 440 (1996).
[2] A. Heger, K. Langanke, G. Martínez-Pinedo and S. E. Woosley, Phys. Rev. Lett. **86**, 1678 (2001).
[3] K. Langanke and G. Martínez-Pinedo, Rev. Mod. Phys. **75**, 819 (2003).
[4] N. Nishimura, T. Takiwaki and F. -K. Thielemann, Astrophys. J. **810**, 109 (2015).
[5] V. M. Kaspi and A. M. Beloborodov, Annu. Rev. Astron. Astrophys. **55**, 261 (2017).
[6] A. E. Delsante and N. E. Frankel, Phys. Lett. A **67**, 279 (1978).
[7] A. E. Delsante and N. E. Frankel, Ann. Phys. **125**, 135 (1980).
[8] Y. Luo, Ph.D. thesis (The University of Tokyo, 2021); Y. Luo, M. Kusakabe, T. Kajino, M. A. Famiano and A. B. Balantekin, in preparation.
[9] G. M. Fuller, W. A. Fowler and M. J. Newman, Astrophys. J. Suppl. Ser. **42**, 447 (1980).
[10] G. M. Fuller, W. A. Fowler and M. J. Newman, Astrophys. J. **252**, 715 (1982).
[11] K. Langanke and G. Martínez-Pinedo, At. Data Nucl. Data Tables **79**, 1 (2001).
[12] J. Pruet and G. M. Fuller, Astrophys. J. Suppl. **149**, 189 (2003).
[13] A. Juodagalvis, K. Langanke, W. R. Hix, G. Martínez-Pinedo and J. M. Sampaio, Nucl. Phys. A **848**, 454 (2010).
[14] R. G. Couch and G. L. Loumos, Astrophys. J. **194**, 385 (1974).
[15] D. A. Shalybkov and D. G. Yakovlev, Astrophys. **27**, 562 (1987).
[16] J. M. Sampaio, K. Langanke, G. Martínez-Pinedo, E. Kolbe and D. J. Dean, Nucl. Phys. A **718**, 440 (2003).

EPJ Web of Conferences **260**, 08001 (2022) https://doi.org/10.1051/epjconf/202226008001
NIC-XVI

Commissioning of Underground Nuclear Astrophysics Experiment JUNA in China

Weiping Liu[1,*], *Zhihong* Li[1], *Jianjun* He[2], *Xiaodong* Tang[3], *Gang* Lian[1], *Jun* Su[2], *Yangping* Shen[1], *Zhu* An[4], *Fuqiang* Chao[1], *Jianjun* Chang[3], *Lihua* Chen[1], *Han* Chen[3], *Xiongjun* Chen[1], *Yunhua* Chen[5], *Zhijun* Chen[3], *Jianping* Cheng[2], *Baoqun* Cui[1], *Xiao* Fang[6], *Changbo* Fu[7], *Lin* Gan[8], *Bing* Guo[1], *Zhiyu* Han[1], *Xuyuan* Guo[5], *Guozhu* He[1], *Jinrong* He[5], *Alexander* Heger[9], *Suqing* Hou[3], *Hanxiong* Huang[3], *Ning* Huang[4], *Baolu* Jia[3], *Liyang* Jiang[1], *Shigeru* Kubono[10], *Jianmin* Li[11], *Mingchuan* Li[5], *Kuoang* Li[1], *Ertao* Li[8], *Tao* Li[3], *Yunju* Li[1], *Maria* Lugaro[12], *Xiaobing* Luo[4], *Hongyi* Ma[3], *Shaobo* Ma[3], *Dongming* Mei[13], *Wei* Nan[1], *Weike* Nan[1], *Ningchun* Qi[5], *Yongzhong* Qian[14], *Jiuchang* Qin[1], *Jie* Ren[1], *Changsong* Shang[5], *Liangting* Sun[3], *Wenliang* Sun[5], *Wanpeng* Tan[15], *Isao* Tanihata[16], *Shuo* Wang[17], *Peng* Wang[4], *Youbao* Wang[1], *Qi* Wu[3], *Shiwei* Xu[3], *Yao* Yang[3], *Xiangqing* Yu[3], *Qian* Yue[11], *Sheng* Zeng[1], *Long* Zhang[1], *Hao* Zhang[1], *Huanyu* Zhang[1], *Liyong* Zhang[2], *Ningtao* Zhang[3], *Peng* Zhang[5], *Qiwei* Zhang[1], *Tao* Zhang[7], *Xiaopeng* Zhang[18], *Xuezhen* Zhang[3], *Wei* Zhao[11], *Jifang* Zhou[5], and *Yong* Zhou[5]

[1]China Institute of Atomic Energy, Beijing, China
[2]Beijing Normal University, Beijing, China
[3]Institute of Modern Physics, Lanzhou, China
[4]Sichuan University, Chengdu, China
[5]Yalong River Hydro-Electronic Power Limited, Sichuan, China
[6]Zhongshan University, Guangzhou, China
[7]Shanghai Jiaotong University, Shanghai, China
[8]Shenzhen University, Shenzhen, China
[9]Monash University, Melbourne, Victoria, Australia
[10]RIKEN, Institute of Physical and Chemical Research, Wako, Japan
[11]Tsinghua University, Beijing, China
[12]Konkoly Observatory of the Hungarian Academy of sciences, Hungary
[13]South Dakota State University, Brookings, South Dakota, US
[14]Minnesota University, Minneapolis and Saint Paul, Minnesota, US
[15]University of Notre Dame, Notre Dame, Indiana, US
[16]Osaka University, Suita, Osaka, Japan
[17]Shandong University, Weihai campus, China
[18]Institute for High Energy Physics, Beijing, China

Abstract. Underground Nuclear Astrophysics Experiment in China (JUNA) has been commissioned by taking the advantage of the ultra-low background in Jinping underground lab. High current mA level 400 KV accelerator with an ECR source and BGO detectors were commissioned. JUNA studies directly a number of nuclear reactions important to hydrostatic stellar evolution at their relevant stellar energies. In the first quarter of 2021, JUNA performed the direct measurements of ^{25}Mg(p,γ)^{26}Al, ^{19}F(p,α)^{16}O, ^{13}C(α,n)^{16}O and ^{12}C(α,γ)^{16}O near the Gamow window. The experimental results reflect the potential of JUNA

*e-mail: wpliu@ciae.ac.cn

with higher statistics, precision and sensitivity of the data. The preliminary results of JUNA experiment and future plan are given.

1 Underground physics

Direct measurement of the cross sections for the key nuclear reactions crucial to hydrostatic stellar evolution within the Gamow window is important for obtaining benchmark data for stellar calculation, verifying extrapolation method, constraining theoretical calculations, explaining the abundance observations and solving key scientific questions in nuclear astrophysics[1].

The direct measurement of astrophysical reaction rates on stable nuclei that require high-intensity beams and extremely low background represents a major challenge at the frontiers of nuclear astrophysics due to the cosmic ray background and extremely low cross section, thus the combination of underground laboratory and high exposure accelerator/detector complex is the only solution. The first underground based low-energy accelerator facility, LUNA [2, 3] at Gran Sasso underground laboratory has successfully demonstrated the feasibility of meeting these challenges. Encouraged by the LUNA success, underground nuclear astrophysics has become one of the frontiers in the field of nuclear astrophysics. Relevant research programs are proposed in the long range plan in China, US and Europe, with high priorities, such as LUNA and CASPER.

China Jinping Underground Laboratory (CJPL) was established on the site of a hydropower plants in the Jinping mountain, Sichuan, China[4, 5]. The facility is located near the middle of a traffic tunnel. The facility is shielded by 2400 m of mainly marble overburden, with radioactively quiet rock. Its ultra-low cosmic ray background, which is about 2 orders of magnitude lower than that in Gran Sasso, makes it into an ideal environment for low background experiments. CJPL phase I (CJPL-I) is now housing CDEX [6] and PandaX dark matter experiments.

CJPL phase II[7] (CJPL-II) is the expansion followed by the success of CJPL-I, it currently available for temporary usage in year 2020-2021. It has much larger scale underground experiments space (300,000 m^3 volume), planned to house CDEX-II, PandaX-II, and JUNA[8]. The layout of JUNA in CJPL-II is shown in figure 1. The complete commissioning of CJPL-II is scheduled in March 2023, when all the three experiments can be restarted in a well equipped condition. In December of 2020, the JUNA collaboration installed the accelerator in CJPL before the long-period construction. Four reactions had been studied in the first quarter of 2021. Some preliminary results are presented in the following section.

The JUNA project was initiated in the year of 2015. The first ion source beam of 12 mA proton was delivered in 2016, and the first accelerated 260 keV 3 mA proton beam was delivered in 2017. All the beams reached their design goal in ground lab in 2020, with the intensity of 12 mA and 2.5 emA for proton and alpha beams in 350 keV. In the mean time, HPGe and BGO as well as neutron detectors and high power targets are all ready. With those ready parts we performed a couple test and calibrating experiment for the 4 reactions mentioned above. The test experiments results the energy and efficiency calibration, understand active background and deduce the high energy data for $^{25}Mg(p,\gamma)^{26}Al$ and $^{19}F(p,\alpha_\gamma)^{16}O$ reactions. The first underground JUNA beam in CJPL-II A1 hall was delivered in Dec. 2020, with proton beam of 2 mA and 300 keV from then one JUNA is ready for underground experiments.

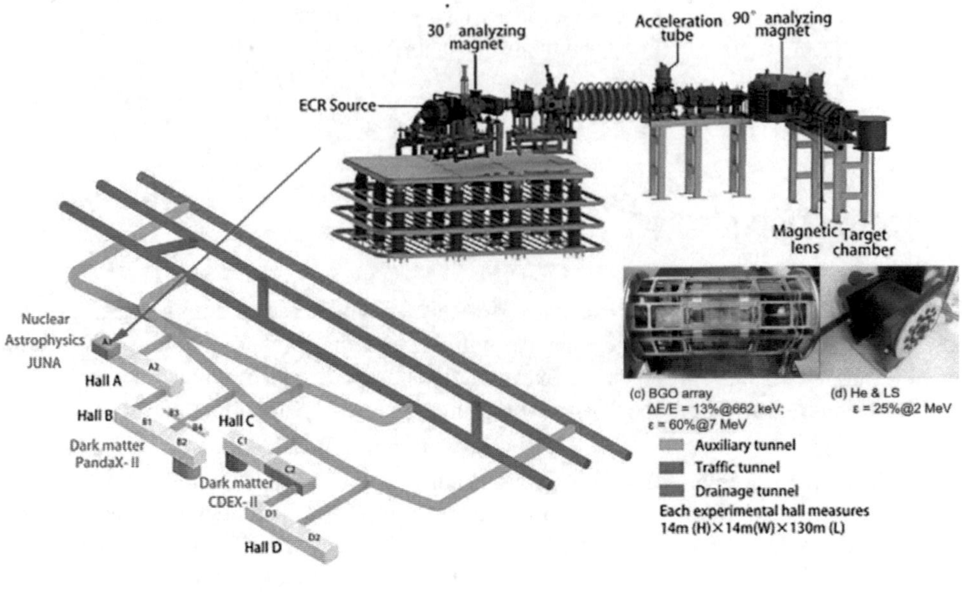

Figure 1. The layout of JUNA in CJPL-II

2 Nuclear astrophysics reactions measured

2.1 $^{12}C(\alpha,\gamma)^{16}O$ reaction

The $^{12}C(\alpha,\gamma)^{16}O$ reaction is quoted as the holy grail reaction in nuclear astrophysics. The uncertainty of this reaction affects not only the nucleosynthesis of elements up to iron, but also the evolution of the massive stars and their final fate (black hole, neutron star). The cross section of this reaction needs to be determined with an uncertainty of less than 10% at helium burning temperatures (T_9=0.2), corresponding to the Gamow window around $E_{c.m.}$=300 keV. It is extremely difficult to determine the reaction cross section (about 10^{-17} barn) at such low energy. A direct measurement near the Gamow window is performed in JUNA with high intensity $^4He^{2+}$ ion beam to provide better constraints for extrapolating models.

We developed the target with the condition of isotope pure ^{12}C implantation to reduce the disturbance of ^{13}C with the ratio of less than 10^{-5}. The BGO and LaBr detector array placed around the target can significantly increase the detection efficiency (absolute efficiency 30% at E_γ = 8 MeV for BGO with fired number selected) of γ-rays, with the flexibility of anti-coincidence to decrease the beam induced background. We used $^4He^{2+}$ beam with an intensity near 1 emA to achieve the most sensitive upper limit of 10^{-13} barn at $E_{c.m.}$ = 538 keV with potential signal higher than background. The data analysis is still in progress with expert evaluation to pin down the complicated background for either deducing upper limit or significant data points of this reaction near Gamow window, since the accelerator and target induced background is higher than we expected. In the near future we plan to measure this reaction with higher intensity beam, higher purity target and higher vacuum, planned in the coming second stage JUNA run from 2023.

2.2 $^{13}C(\alpha, n)^{16}O$

The $^{13}C(\alpha,n)^{16}O$ reaction is the main neutron source for the slow-neutron-capture (s-) and inter-mediate (i-) processes in AGB stars. Direct measurements at astrophysical energies in above-ground laboratories are hindered by the extremely small cross sections and vast cosmic-ray induced background. We performed a direct measurement at JUNA in CJPL using the most intense α beam available among the underground laboratories. The systematic uncertainties of target thickness and deterioration were eliminated using the thick target technique. Our measurement extended from the lowest energy ever measured $E_{c.m.}$ = 0.24 MeV up to 0.59 MeV, covering almost the entire i-process Gamow window with the highest precision. We also performed a consistent R-matrix analysis, which provides a reliable reaction rate for s-process nucleosynthesis with an improved uncertainty[9], see figure 2b.

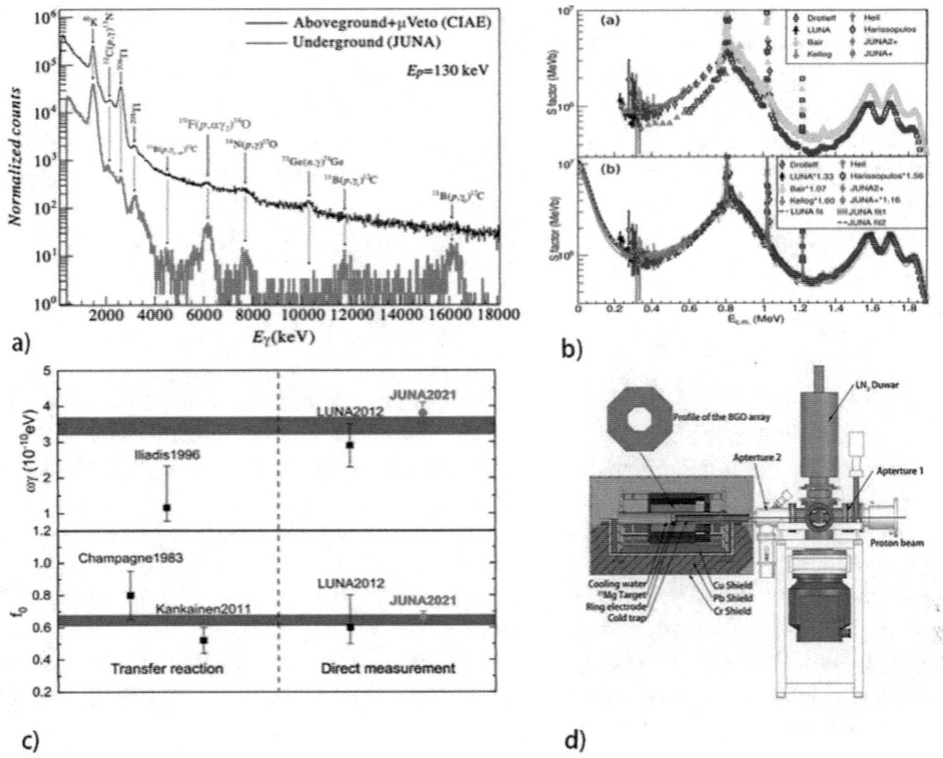

Figure 2. The experiment setup and data analysis

a) γ-ray spectra of the ^{19}F+p experiments measured by a 4π BGO array at a proton energy of E_p=130 keV.
b) The S factor of the $^{13}C(\alpha,n)^{16}O$ reaction (a) without and (b) with normalization factors.
c) Comparison of recent values of the resonance strength and ground state feeding factor of the 92 keV resonance.
d) Schematic diagram of detector setup.

2.3 ^{25}Mg(p, γ)^{26}Al

The ^{25}Mg(p,γ)^{26}Al reaction plays an important role in the study of cosmic 1.809 MeV γ-ray as a signature of ongoing nucleosynthesis in the Galaxy. At astrophysical temperature around 0.1 GK, the ^{25}Mg(p,γ)^{26}Al reaction rates are dominated by the 92 keV resonance capture process. We report a precise measurement of the 92 keV ^{25}Mg(p,γ)^{26}Al resonance in the dayone experiment at JUNA facility in CJPL. The resonance strength and ground state feeding factor are determined to be $(3.8 \pm 0.3) \times 10^{-10}$ eV and 0.66 ± 0.04, see figure 2c, respectively[10]. The results are in agreement with those reported in the previous direct underground measurement within uncertainty, but with significantly reduced uncertainties. Consequently, we recommend new ^{25}Mg(p,γ)^{26}Al reaction rates which are by a factor of 2.4 larger than those adopted in REACLIB database at the temperature around 0.1 GK. The new results indicate higher production rates of ^{26g}Al and the cosmic 1.809 MeV γ-rays. The implication of the new rates for the understanding of other astrophysical situations is very significant.

The thick-target yields of the 92 and 189 keV resonances of ^{25}Mg(p, γ)^{26}Al were measured with 2 pmA proton beams and 4π BGO γ-ray detector. The event rate of the ^{25}Mg(p, γ)^{26}Al reaction at $E_{c.m.}$ = 92 keV is about 200 events per day. The background counting rate is measured to be less than 5 events per day. Thus we accumulated around 1200 reaction events in two weeks, and the statistic uncertainty is about 3% with total exposure of 1225 C.

The resonance strength of 58 keV level is estimated by using the shell model calculation and the in-direct measurement by ^{25}Mg(^{7}Li,^{6}He)^{26}Al one nucleon transfer reaction[11]. The results show that the 58 keV resonance dominate the ^{25}Mg(p, γ)^{26}Al reaction rate at T < 0.06 GK[11]. Thanks for our indirect deduction of 58 keV, and our ground based experimental results for 304 keV, based on our underground data for 92 and 189 keV resonances, we are able to build up most precise reaction rate for all temperatures, together with that of LUNA data, the results will be published elsewhere.

2.4 ^{19}F(p,α)^{16}O reaction

Fluorine is one of the most interesting elements in nuclear astrophysics, where the ^{19}F(p,α_γ)^{16}O reaction is of crucial importance for Galactic ^{19}F abundances and CNO cycle loss in first generation Population III stars. As a dayone campaign at the JUNA, we finished direct measurements of the essential ^{19}F(p,α_γ)^{16}O reaction channel. The γ-ray yields were measured over $E_{c.m.}$=72.4-344 keV, see figure 2a [12], covering the Gamow window; our energy of 72.4 keV is unprecedentedly low, reported here for the first time. The experiment was performed under the extremely low cosmic-ray induced background environment of CJPL, one of the deepest underground laboratories in the world. The present low-energy S factors deviate significantly from previous theoretical predictions, and the uncertainties are significantly reduced. The thermonuclear ^{19}F(p,α_γ)^{16}O reaction rate has been determined directly at the relevant astrophysical energies.

Table 1. Basic parameters of four reactions studied

reaction	beam	intensity (emA)	$E_{c.m.}$ (keV)	cross section	efficiency %	CTS (/day)	BKD (/day)
^{12}C(α, γ)^{16}O	$^4He^{2+}$	1	538	10^{-13} barn	60	-	1.2
^{13}C(α, n)^{16}O	$^4He^{1+,2+}$	2	230-600	10^{-12} barn	26	97±24	113±5
^{25}Mg(p, γ)^{26}Al	$^1H^{1+}$	2	E_x=92	(3.8±0.3) $\times10^{-10}$ eV	38	200	10
^{19}F(p, α_γ)^{16}O	$^1H^{1+}$	0.5-1.0	72-344	10^{-12} barn	60	105	85

3 Summary

The parameters of four reactions are summarized in table 1. In summary, a new underground nuclear astrophysics experiment JUNA planned for the expanded space CJPL-II was performed. With a more powerful accelerator and a deeper location, JUNA has joined the research among underground nuclear astrophysics laboratories. The accelerator system and detector array was installed by the end of 2020, the experiments started in the beginning 2021 and the first four experimental data accumulated in the first quarter of 2021. The astrophysical implications base on those data is going on, with the published result hopefully available in the middle and end of 2021.

In the coming years before the re-opening in 2023, we will put more efforts to enhance beam intensity, target purity, as well as improving detectors. Future upgrading is planned (JUNA-II), by considering to build a MV level Van de Graaff for ^{12}C and ^{16}O beam in inverse kinematics and window-less targets as well as recoil spectrometers. We welcome the researchers world wide to join JUNA to perform more experiments by using the most favorable underground condition in Jinping.

References

[1] C. Iliadis, Nuclear Physics of Stars, Wiley-VCH Verlag GmbH (2007)

[2] A. Formicola, G. Imbriani, M. Junker, et al., Nuclear Instruments & Methods in Physics Research, 507, 609-616 (2003)

[3] H. Costantini, A. Formicola, G. Imbriani, et al., Reports on Progress in Physics, 972, 086301 (2009)

[4] H. S. Chen, Science(in Chinese), 62, 4 (2010)

[5] J. P. Cheng, S. Y. Wu, Q. Yue, et al., Physics (in Chinese), 03, 149-154 (2011)

[6] W. Zhao, Q. Yue, K. J. Kang, et al., Physical Review D, 88, 1201-1205 (2013)

[7] D. Normile, Science, 346, 1041-1041 (2014)Chicago Press (1988)

[8] W. P. Liu, et al., Science China Physics, Mechanics & Astronomy, 59, 642001-1-640001-7(2016)

[9] B. Gao,..., X. D. Tang,..., W. P. Liu, Direct measurement of the $^{13}C(\alpha,n)^{16}O$ reaction in the Gamow window of the i-process nucleosynthesis using intense beam in deep underground laboratory, submitted to Phys. Rev. Lett., 2021

[10] J. Su, H. Zhang, Z. H. Li,..., W. P. Liu, First result from the Jinping underground nuclear astrophysics experiment JUNA: precise measurement of the 92 keV $^{25}Mg(p,\gamma)^{26}Al$ resonance, Science Bulletin, 67, 2 (2022)

[11] Z. H. Li, J. Su, Y. J. Li, et al., Science China Physics, Mechanics & Astronomy, 58 (2015)

[12] L.Y. Zhang, J. Su, J.J. He,..., W. P. Liu, Direct Measurement of the Astrophysical $^{19}F(p,\alpha_\gamma)^{16}O$ Reaction in the Deepest Operational Underground Laboratory, Phys. Rev. Lett. 127, 152702 (2021).

Measurement of the low energy ^{25}Mg(p, γ)^{26}Al resonances

Jun Su[1,2*], *Zhihong* Li[2**], *Hao* Zhang[1,2], *Yunju* Li[2], *Ertao* Li[3], *Chen* Chen[2], *Yangping* Shen[2], *Jianjun* He[1], *Liyong* Zhang[1], *Xinyue* Li[1], *Gang* Lian[2], *Bing* Guo[2], *Luohuan* Wang[1], *Yaode* Sheng[1], *Yinji* Chen[1], *Long* Zhang[2], *Fuqiang* Cao[2], and *Weiping* Liu[2]

[1]Key Laboratory of Beam Technology of Ministry of Education, College of Nuclear Science and Technology, Beijing Normal University, Beijing 100875, China
[2]China Institute of Atomic Energy, P. O. Box 275(10), Beijing 102413, China
[3]College of Physics and Optoelectronic Engineering, Shenzhen University, Shenzhen 518060, China

Abstract. The cosmic 1.809 MeV γ-ray emitted by the radioactive nucleus ^{26}Al in the Galaxy is one of the key observation targets of the γ-ray astronomy. The ^{26}Al is mainly produced by the ^{25}Mg(p, γ)^{26}Al reaction in the stellar Mg-Al reaction cycle. At the astrophysical relevant temperatures, the reaction rates of ^{25}Mg(p, γ)^{26}Al are dominated by several narrow resonances at low energy. This work reports a measurement of the low energy ^{25}Mg(p, γ)^{26}Al resonances at Jinping Underground Nuclear Astrophysics experimental facility (JUNA) in the China Jinping Underground Laboratory (CJPL).

1 Introduction

The 1.809 MeV γ-ray emitted from the radioactive decay of ^{26}Al in the interstellar medium of the Galaxy has been observed by several satellites [1–3]. The half-life of ^{26}Al (7.17×10^5 y) is modest for the observation of the γ-ray astronomy. It is long enough that ^{26}Al synthesized in the interior of the star can survive and transfer to the interstellar medium. On the other hand, it is much shorter than the evolution time of the Galaxy so the observation of ^{26}Al decay is direct evidence of the ongoing nucleosynthesis in the Galaxy. Therefore, ^{26}Al is expected to be a tracer for stellar nucleosynthesis and a thermometer for its production site [4].

^{26}Al can be synthesized by Mg-Al reaction cycle in the H burning environments, e.g., asymptotic giant branch (AGB) stars, Wolf-Rayet stars, O-Ne-Mg novae and core collapse supernovae. It is believed that ^{26}Al in the interstellar medium of the Galaxy is mainly from massive stars and partly from AGB stars and novae [4]. In the Mg-Al reaction cycle, ^{26}Al is produced via the ^{25}Mg(p, γ)^{26}Al reaction and destroyed by β-decay or other reactions, e.g., ^{26}Al(p, γ)^{27}Si, ^{26}Al(n, p)^{26}Mg, ^{26}Al(n, α)^{23}Na, depending on the astrophysical sites. Hence, the ^{25}Mg(p, γ)^{26}Al reaction plays an important role in the study of the 1.809 MeV cosmic γ-ray and the origin of ^{26}Al in the Galaxy.

At astrophysical relevant temperatures, the reaction rates of ^{25}Mg(p, γ)^{26}Al are characterized by several narrow resonances at low energy, among which the main contributors are 58, 92, 189, and 304 keV resonances. Due to its extremely low resonance strength, the 58 keV

*e-mail: sujun@bnu.edu.cn
**e-mail: zhli@ciae.ac.cn

resonance has been only investigated by indirect method [5, 6]. The 304 keV resonance has been directly measured in several works [7–9] thus its resonance strength is well determined. Strieder et al. [10] performed the first direct underground measurement of the ^{25}Mg(p, γ)^{26}Al reaction and obtained the 92 and 189 keV resonance strengths. However, the 189 keV resonance strength result of the underground experiment [10] is larger than those obtained by ground measurement [11] and accelerator mass spectrometer(AMS) method [12].

In the present work, we performed a new direct measurement of the low energy resonances of ^{25}Mg(p, γ)^{26}Al at the 400 kV accelerator of Jinping Underground Nuclear Astrophysics experimental facility (JUNA) [13]. Since the result of the 92 keV resonance has already been published elsewhere [14], we present some supplementary results here.

2 Experiment

Figure 1 shows the experimental setup. The proton beam provided by the 400 kV accelerator was collimated by two apertures and then bombarded on the ^{25}Mg enriched isotopic target which was directly cooled by flowing water. In the upstream of the target, a cold trap cooled by liquid nitrogen was used to suppress the carbon deposition on the target. The target and the end of the vacuum tube composed a Faraday cup, which was used to measure the beam current. A ring electrode applied with a voltage of −300 V was installed between the target and cold trap to suppress secondary electrons.

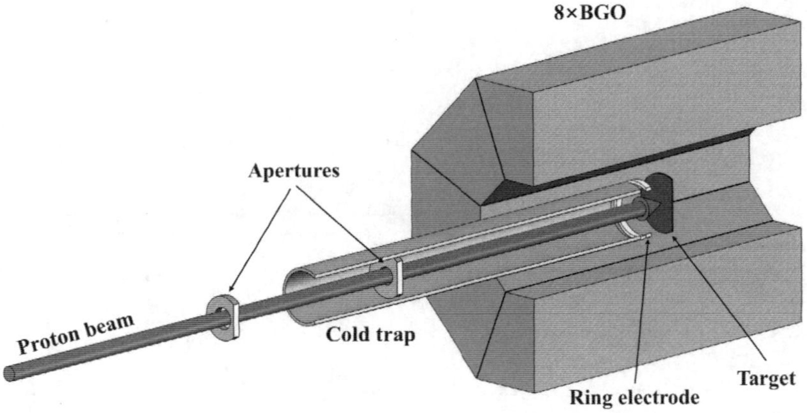

Figure 1: Schematic of the experimental setup.

^{25}Mg isotopic targets made by ^{25}Mg enriched magnesium metal (98.81 ± 0.02% abundance) were used. The thickness of targets was set to be ~60 μg/cm^2. A Cr layer with a thickness of ~40 μg/cm^2 was sputtered on the surface of each target to enhance the radiation resistance of the target. A near 4π BGO detector array was used to detect the γ-rays. The BGO detector array was calibrated by ^{137}Cs and ^{60}Co gamma sources. The single spectrum (sum of the spectra obtained by eight segments) and the sum spectrum (sum of the energies of all eight BGO segments) were obtained. The Monte Carlo simulation based on the GEANT4 software [15] was performed for the BGO detector array, which reproduced the calibration spectra well.

3 Data analysis

The resonance strength $\omega\gamma$ of a narrow resonance can be determined by measuring the maximum yield $Y_{\rm max}$ of a thick target

$$\omega\gamma = \frac{2\varepsilon_{\rm eff}(E_{\rm R})}{\lambda_{\rm R}^2} Y_{\rm max}, \tag{1}$$

where $\lambda_{\rm R}$ is the de Broglie wavelength and $\varepsilon_{\rm eff}(E_{\rm R})$ is the effective stopping power of proton in the target. The effective stopping power is related to the concentration of ^{25}Mg in the target, which changed with the bombardment of the proton beam on the target. Therefore, in the measurement with high-intensity beam, such as 92 keV resonance measurement, the effective stopping power was monitored by measuring the maximum yield of the 304 keV resonance regularly. The maximum yield was calculated by $Y_{\rm max} = A_{\rm sum}/\eta_{\rm sum}$, where the $A_{\rm sum}$ is the sum peak count and the $\eta_{\rm sum}$ is the sum peak efficiency. The sum peak efficiency is related to γ-ray decay scheme as well as the primary γ-ray branching ratios of the resonance.

Figure 2(a) shows the sum spectrum of the 304 keV resonance obtained by BGO detector array, in which the sum peak at $E_{\rm sum}$ = 6610 keV is clearly visible. Figure 2(b) shows the single spectrum obtained by putting a 6400-6800 keV gate on the sum peak shown in Fig. 2(a). In contrast to the sum spectrum, many individual γ peaks, which are closely related to the decay scheme of the resonance, appear in the single spectrum.

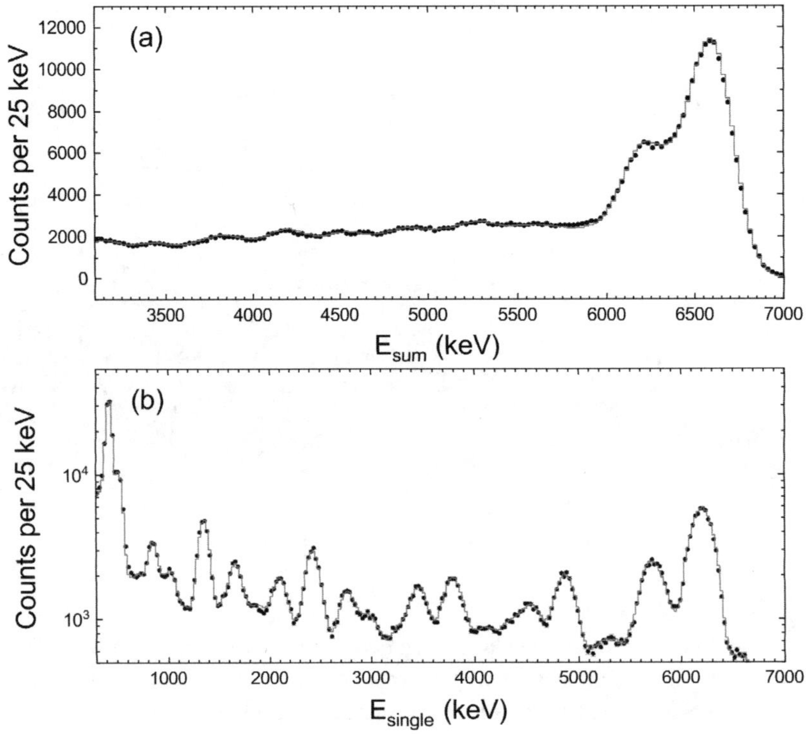

Figure 2: γ-ray spectra of the 304 keV ^{25}Mg(p, γ)^{26}Al resonance (black-filled circle). (a) Sum spectrum, the red line represents the GEANT4 simulation normalized to the experimental data. In the simulation the primary γ-ray branching ratios were extracted from the single spectrum. (b) Single spectrum, the red line represents the fitting results.

A fitting program based on the expectation-maximization method [16] was developed to extract the decay scheme of the resonance from the experimental single spectrum, in which the primary γ-ray branching ratios were set as adjustable input parameters. The fitting process is 1) Initial primary γ-ray branching ratios were fed in the GEANT4 simulation to produce a single spectrum, and 2) The simulation result was compared with the experimental data and the deviations were fed back to adjust the input parameters. After thousands of iterations, the primary γ-ray branching ratios will converge to a final result. By using this program, the primary γ-ray branching ratios of the 304 keV resonance were extracted by fitting the single spectrum shown in Fig. 2(b). The result is in good agreement with those reported by the previous measurements with HPGe detector [8, 9]. The GEANT4 simulation based on primary γ-ray branching ratios reproduced the sum spectrum obtained well, as shown in Fig. 2(a).

The experimental data of the 92 keV resonance were analyzed via the same process as the 304 keV resonance. Figure 3 shows the sum spectrum and single spectrum of the 92 keV resonance. The primary γ-ray branching ratios of the 92 keV resonance were extracted by fitting the single spectrum.

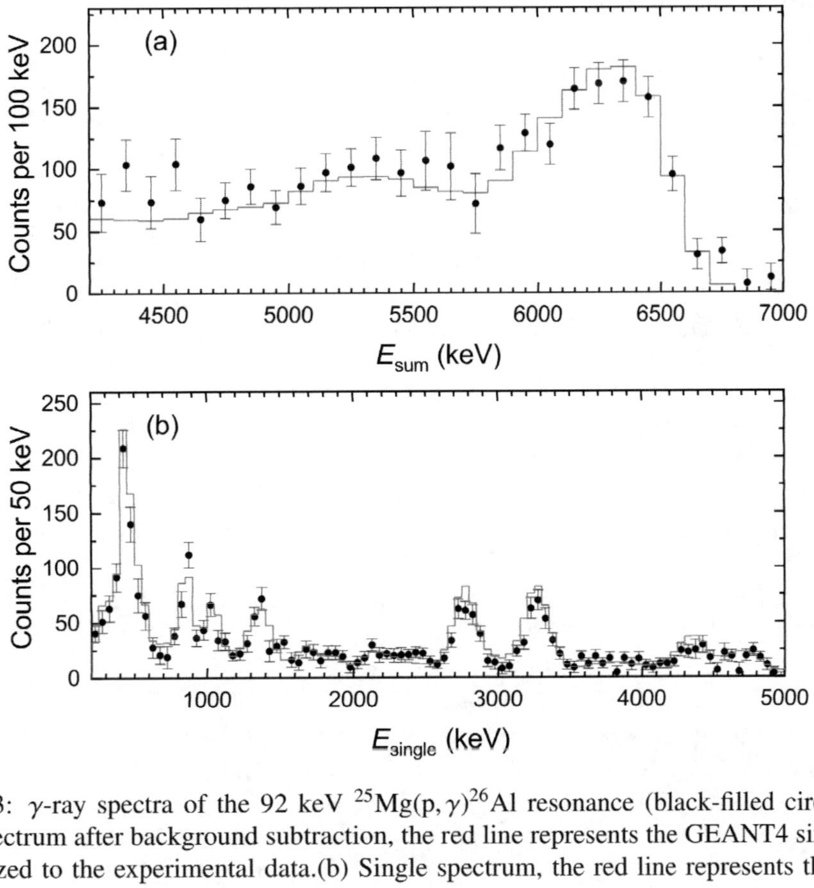

Figure 3: γ-ray spectra of the 92 keV $^{25}Mg(p,\gamma)^{26}Al$ resonance (black-filled circle). (a) Sum spectrum after background subtraction, the red line represents the GEANT4 simulation normalized to the experimental data.(b) Single spectrum, the red line represents the fitting results.

The 92 keV resonance strength was determined by a relative method as

$$\omega\gamma_{92} = \frac{\lambda_R^2(304)}{\lambda_R^2(92)} \frac{A_{92}/\eta_{92}}{A_{304}/\eta_{304}} \frac{N_{304}}{N_{92}} \frac{\varepsilon_{\rm eff}(92)}{\varepsilon_{\rm eff}(304)} \omega\gamma_{304}, \tag{2}$$

in which A_R, η_R, and N_R are the sum peak count, sum peak efficiency, and proton numbers of the corresponding resonance measurements, respectively. The sum peak efficiency ratio η_{304}/η_{92} is deduced to be 1.3 ± 0.04. The effective stopping power ratio $\varepsilon_{\rm eff}(92)/\varepsilon_{\rm eff}(304)$ is calculated to be 1.50 ± 0.01 over a large range of ^{25}Mg content in the target. With the $\omega\gamma_{304}$=(3.1±0.1)×10^{-2} eV value reported in Ref. [9], the resonance strength of 92 keV resonance is determined to be $\omega\gamma_{92}$=(3.8±0.3)×10^{-10} eV [14].

4 Summary

In summary, the ^{25}Mg(p, γ)^{26}Al reaction was directly measured at JUNA in CJPL. The primary γ-ray branching ratios of the 92 and 304 keV resonance were extracted from the single spectrum. The 92 keV resonance strength was determined with improved precision. The data analysis of other low energy resonances including the 189 keV resonance is in process and the results will soon be published.

5 Acknowledgements

The present work is supported by the National Natural Science Foundation of China (Grant Nos. 11490563, U1867211, 12125509, 11961141003, 11775133, and 12175152), and the Continuous Basic Scientific Research Project No. WDJC-2019-13.

References

[1] W.A. Mahoney, J.C. Ling, A.S. Jacobson, R.E. Lingenfelter, Astrophys. J **262**, 742 (1982)

[2] R. Diehl, C. Dupraz, K. Bennett, H. Bloemen, W. Hermsen, J. Knoedlseder, G. Lichti, D. Morris, J. Ryan, V. Schoenfelder et al., Astron. Astrophys. **298**, 445 (1995)

[3] R. Diehl, H. Halloin, K. Kretschmer, G.G. Lichti, V. Schönfelder, A.W. Strong, A. Von Kienlin, W. Wang, P. Jean, J. Knödlseder et al., Nature **439**, 45 (2006)

[4] R. Diehl, M.G. Krause, K. Kretschmer, M. Lang, M.M. Pleintinger, T. Siegert, W. Wang, L. Bouchet, P. Martin, New Astron. Rev. **92**, 101608 (2021)

[5] C. Iliadis, L. Buchmann, P.M. Endt, H. Herndl, M. Wiescher, Phys. Rev. C **53**, 475 (1996)

[6] Y.J. Li, Z.H. Li, E.T. Li, X.Y. Li, T.L. Ma, Y.P. Shen, J.C. Liu, L. Gan, Y. Su, L.H. Qiao et al., Phys. Rev. C **102**, 025804 (2020)

[7] P. Endt, P. De Wit, C. Alderliesten, Nucl. Phys. A **459**, 61 (1986)

[8] B. Limata, F. Strieder, A. Formicola, G. Imbriani, M. Junker, H.W. Becker, D. Bemmerer, A. Best, R. Bonetti, C. Broggini et al., Phys. Rev. C **82**, 015801 (2010)

[9] H. Zhang, Z.H. Li, J. Su, Y.J. Li, C. Chen, L. Zhang, F.Q. Cao, Y.P. Shen, W. Nan, W.K. Nan et al., Chin. Phys. C **45**, 084108 (2021)

[10] F. Strieder, B. Limata, A. Formicola, G. Imbriani, M. Junker, D. Bemmerer, A. Best, C. Broggini, A. Caciolli, P. Corvisiero et al., Phys. Lett. B **707**, 60 (2012)

[11] C. Iliadis, T. Schange, C. Rolfs, U. Schröder, E. Somorjai, H. Trautvetter, K. Wolke, P. Endt, S. Kikstra, A. Champagne et al., Nucl. Phys. A **512**, 509 (1990)

[12] A. Arazi, T. Faestermann, J.F. Niello, K. Knie, G. Korschinek, M. Poutivtsev, E. Richter, G. Rugel, A. Wallner, Phys. Rev. C **74**, 025802 (2006)

[13] W.P. Liu, Z.H. Li, J.J. He, X.D. Tang, G. Lian, Z. An, J.J. Chang, H. Chen, Q.H. Chen, X.J. Chen et al., Sci. China Phys. Mech. Astron. **59**, 642001 (2016)

[14] J. Su, H. Zhang, Z. Li, P. Ventura, Y. Li, E. Li, C. Chen, Y. Shen, G. Lian, B. Guo et al., Science Bulletin **67**, 125 (2021)

[15] S. Agostinelli, J. Allison, K. Amako, J. Apostolakis, H. Araujo, P. Arce, M. Asai, D. Axen, S. Banerjee, G.. Barrand et al., Nucl. Instr. Meth. A **506**, 250 (2003)

[16] J.L. Taín, D. Cano-Ott, Nucl. Instr. Meth. A **571**, 728 (2007)

Final results on the ^{13}C(α,n)^{16}O cross section at low energies at LUNA

Giovanni Francesco Ciani[1,*], *Laszlo* Csedreki[2,**], *David* Rapagnani[3], *Andreas* Best[3], *Alba* Formicola[4] for LUNA collaboration

[1]Università di Bari and INFN Division of Bari, Via E. Orabona 4, 40125 Bari, Italy
[2]Institute for Nuclear Research (ATOMKI), P.O. Box 51, HU-4001 Debrecen, Hungary
[3]Università degli Studi di Napoli and INFN Na,Strada Comunale Cintia, 80126 Napoli, Italy
[4]INFN Division of Roma La Sapienza, Piazzale Aldo Moro 2, 00185 Roma, italy

Abstract. It is well established that the $^{13}C(\alpha,n)^{16}O$ reaction (Q=2.215 MeV) is the major neutron source feeding the *s*-process in low mass ($1-3M_{\odot}$) Asymptotic Giant Branch (AGB) stars. In the last decades, several measurements have been performed. Nevertheless, no dataset reaches the Gamow window (140 keV $<E_{c.m.}<$250 keV). This is due to the exponential drop of the cross section σ(E) with decreasing energy. The consequence is that the reaction rate becomes so low that the cosmic background becomes predominant in surface laboratories. A recent measurement was carried out in deep underground laboratory of Laboratori Nazionali del Gran Sasso (LNGS) in the framework of the LUNA experiment.
To measure the $^{13}C(\alpha,n)^{16}O$ cross section at low energies, a multiple effort has been performed to suppress the background in the setup, to maximise the detector efficiency and to keep under control the target modification under an intense stable beam provided by the LUNA accelerator ($<I>= 200\,\mu$A). Thanks to these accuracies, the $^{13}C(\alpha,n)^{16}O$ cross section was measured in the center of mass energy range 230 keV $<E_{cm}<$305 keV with a maximum 20% overall uncertainty. This allowed to constrain the reaction rate at T=0.1 GK at 15% uncertainty and to lead the way for new possible astrophysical consequences.

1 Introduction

The $^{13}C(\alpha,n)^{16}O$ reaction is the main neutron source for the *s-process* in low mass AGB stars[1]. The reaction takes place subsequently complex convective motions in the so-called 13C pocket in a stellar environment of about $1-2\cdot10^8$ K, corresponding to a Gamow window between 140[1] and 250 keV, well below the Coulomb barrier of the reaction.

In the last 25 years, several direct cross section measurements of this reaction have been performed [2–4]. The lowest energy point, corresponding to E=265 keV, was measured by Drotleff et al. [3] with a 60% uncertainty. From the literature one can see that statistical error is mainly due to the low signal to noise ratio at low energies and a strong source of systematic uncertainty comes from the difficulty to keep under control target degradation.

*email: giovanni.ciani@ba.infn.it
**email: laszlo.csedreki@lngs.infn.it
[1]all the energies are in center of mass system, unless specified otherwise

In addition, the presence of a near threshold resonance at $E_R = -3 \pm 8$ keV, corresponding to $E_x = 6.356$ MeV state in ^{17}O, influences the cross section in the Gamow window, making extrapolations from higher energies complicated.
In 2008, Heil et al. [2] performed a direct measurement and a multichannel R-matrix analysis was carried out: the S(E) factor extrapolation uncertainty in the Gamow peak is at least of 20%. A recent work by deBoer et al.[5] states that one of the main source of uncertainty for R-matrix extrapolations is due to normalisation uncertainty of the $^{13}C(\alpha, n)^{16}O$ data, directly connected to systematic uncertainties on different datasets.
In this scenario the new LUNA (Laboratory for Underground Nuclear Astrophysics) measurement, whose goal is to reach the Gamow window with a direct measurement with an overall 20% uncertainty at maximum, finds a perfect collocation that allows to constrain the reaction rate for a better development of stellar evolution.

2 Experimental Setup

The LUNA experiment takes advantage from the 400 kV accelerator providing very stable proton and alpha beams in the energy range of 50 - 400 keV with a maximum on-target current up to 500 μA, with an energy resolution of 0.1 keV and a long term stability of 5 eV per hour [6].
Moreover, the deep-underground location guarantees a neutron environmental background reduction by three orders of magnitude lower than surface laboratories [7].
For the first time, the LUNA collaboration designed and installed a neutron detector array at the end of the solid target beamline.
The experimental setup used for this measurement is based on 18 ^3He counters with low intrinsic background arranged in two rings (6 in the inner ring, 12 in the outer ring) concentric with respect to the target chamber. The counters are embedded in a polyethylene moderator. The whole setup is surrounded by a 2 inches borated polyethylene absorber to further reduce the environmental background [8]. Counters are arranged in two different configurations (in a vertical and a horizontal orientation) to optimize neutron detection efficiency, target handling and target cooling. Figure 1a shows the horizontal arrangement mainly used for the measurement of the lowest energy points.
The alpha particle intrinsic background, coming from impurities of uranium and thorium in the counter cases, was reduced using stainless steel counters instead of standard aluminium ones.
Moreover, a wave functions pulse shape analysis (PSD) from the raw preamplifier from detectors allowed the rejection of remaining alpha signals [9] obtaining a final background in the whole setup of about 1 count/hours, more that two orders of magnitude lower than previous experiments in surface laboratories.
The moderator was designed with the aim to open it and insert a High-Purity Germanium (HPGe) detector in close geometry for the target monitoring as explained later.
The efficiency maximization for both experimental setups was achieved by Monte Carlo simulation based on a Geant4 code. Two different experimental campaigns were performed for a validation: at low neutron energies, below 1 MeV, the activation measurement of the ^{51}V(p,n)^{51}Cr reaction was performed at the Van De Graaff accelerator installed at the Institute for Nuclear Research ATOMKI (Debrecen, Hungary); at high energy a certificated AmBe radioactive source was used, whose average energy is at about 4 MeV. The interpolation of experimental data constrained the efficiency in the region of interest (E_n=2.5 MeV) to (34±3)% and (38±3)% for the vertical and horizontal detector arrangements,respectively [8].

Figure 1b shows the absolute neutron detection efficiency of the horizontal setup.

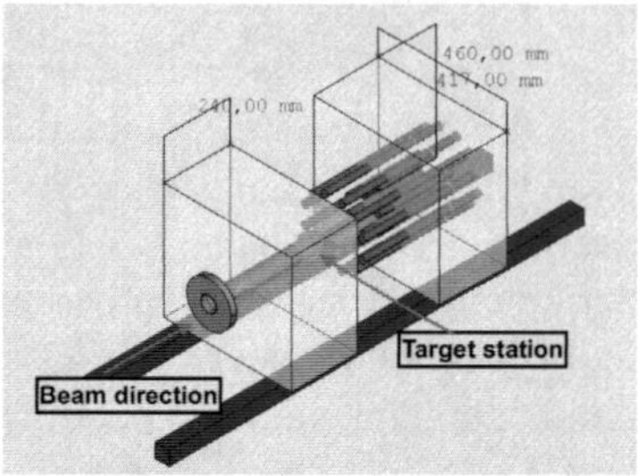

(a) Drawing of the horizontal neutron detector array used for the LUNA measurement[8].

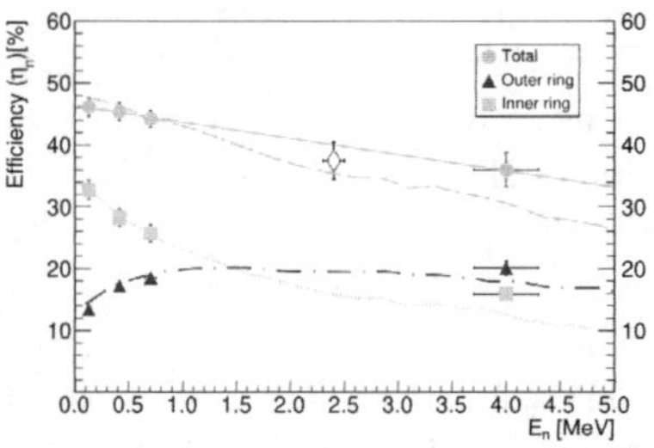

(b) Experimental (filled symbols) and simulated (dashed line) efficiency curves[8] of the horizontal setup. Efficiencies related to the inner (squares and dotted lines) and outer (triangles and dash-dotted lines) ring of the setups are also shown, respectively. The linear fit of the experimental data (solid lines) allows to interpolate the proposed efficiency value in the experimental region of interest (E_n=2.4 MeV) (empty diamond).

^{13}C targets used during the measurement at LUNA have been produced evaporating 99% ^{13}C isotopically enriched powder on tantalum backings.
To check target stochiometry, depth profile and uniformity immediately after the evaporation, an extensive target characterization was performed by means of Nuclear Resonant Reaction Analysis (NRRA) of the ^{13}C(p,γ)^{14}N reaction at 1.75 MeV at the Tandetron accelerator installed at ATOMKI [10].
The monitoring of the above mentioned quantities is crucial also during the cross section measurement performed at LUNA, where the NRRA technique is not applicable, due to the lack of resonances in the dynamic range of the accelerator.
For this reason, a new method of analysis was developed [11]. Data taking at LUNA consisted of long α-beam runs with accumulated charges of $\approx$ 1 C per run, interspersed by short proton-beam runs with open moderator and HPGe detector in close geometry, with typical accumulated charges of 0.2 C at most. During the last mentioned proton runs, the target degradation can be checked fitting the direct capture de-excitation to the ground state peak shape of ^{13}C(p,γ)^{14}N reaction with the HPGe detector. Proton beam runs were all performed at the same reference energy, E_p = 310 keV.

3 Results and future prospectives

Thanks to the impressive background suppression and the novel approach of target degradation monitoring, the LUNA collaboration measured the ^{13}C(α,n)^{16}O reaction cross section in an energy range 230 keV<$E_{c.m.}$<305 keV, apporaching the high energy edge of the Gamow window with an unprecedented uncertainty lower than 20% in the entire dataset[12]. Results are shown in Figure 2.

The new LUNA data were used together with data by Heil, Drotleff and Harissopulos for an R-matrix extrapolation using the Azure2 code. The latter dataset shows a clear discrepancy with respect to the other two. For this reason, in our analysis the Harissopulos dataset was scaled up by a factor of 1.37. Nevertheless, there are no strong motivations for doing so and in order to investigate the effect of the two different normalizations, we performed R-matrix calculation using data by Harissopulos et al. as a reference for the

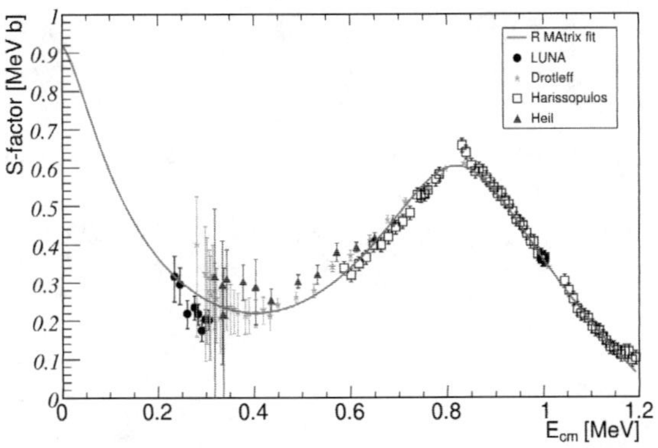

Figure 2: (Colour online) R-matrix extrapolation (red curve) of the astrophysical S(E) factor of $^{13}C(\alpha,n)^{16}O$ calculated using the new LUNA dataset together with the Heil, Drotleff and Harissopulos data. The figure is adapted from [12].

normalization of Heil and Drotleff. In the Gamow window we found comparable results in both scenarios.

Finally, the astrophysical reaction rate R=$N_A < \sigma v >$ as a function of stellar temperature was calculated by integration of the R-matrix cross section. Thanks to excellent quality of the data, the reaction rate At T= 0.1 GK is evaluated with an overall uncertainty of about 15%. In particular we took into consideration the reaction rate at -2σ uncertainty, where more ^{13}C survives and it is it burned at a higher temperature (~ 200 MK) into a convective shell powered by a subsequent thermal pulse. This generates a second neutron burst characterized by higher neutron density and lower exposure.

For stars of nearly solar composition (metallicity Z = 0.02, and Y = 0.27), this causes considerable variations of some isotopic abundances. In particular, the two radioactive nuclei ^{60}Fe and ^{205}Pb, as well as ^{152}Gd are influenced. This is due to the fact that the mentioned isotopes are close-by branching points that are sensitive to the neutron density. Even though the excellent results achieved, there are still open points due to different energy scales among datasets at higher energies (500 keV<$E_{c.m.}$<800 keV).
With the installation of the new LUNA MV facility, which will provide a maximum terminal voltage of 3.5 MV, the LUNA collaboration is planning to extend the measurement of the $^{13}C(\alpha,n)^{16}O$ at higher energies covering the energy range up to 1 MeV. This will give the unique possibility to provide a complete dataset over a wide energy range with well known uncertainties and avoiding normalizations [13].

References

[1] Cristallo S. *et al.*, The Astrophysical Journal, **859**, (2018)
[2] Heil M. *et al.*. Physical Review C **78**, 025803 (2008)
[3] Drotleff H.W. *et al.*, The Astrophysical Journal, **414**, 735 (1993)
[4] Harissopulos S. *et al.*, Physical Review C **72**, 062801(R) (2005)
[5] deBoer R.J. *et al.*, Physical Review C **101**, 045802 (2020)
[6] Formicola A. *et al.*, Nuclear Inst. and Methods in Physics Research A, **3**, 507 (2003)

[7] Best A. *et al.*, Nuclear Inst. and Methods in Physics Research A, **812**, 1-6 (2016)
[8] Csedreki L. *et al.*, Nuclear Inst. and Methods in Physics Research A, **994** 165081 (2021)
[9] Balibrea-Correa J. *et al.*, Nuclear Inst. and Methods in Physics Research A, **906** 103-109 (2018)
[10] Csedreki L. *et al.*, , Nuclear Inst. and Methods in Physics Research B, **478** 194-200 (2020)
[11] Ciani G.F. *et al.*, The European Physical Journal A **56** 75 (2020)
[12] Ciani G.F. *et al.*, Physical Review Letters **127**, 152701 (2021)
[13] Ferraro F. *et al.*, Front. Astron. Space Sci. **7**, 119 (2021)

EPJ Web of Conferences **260**, 08004 (2022) https://doi.org/10.1051/epjconf/202226008004
NIC-XVI

Direct measurement of the ^{19}F$(p, \alpha\gamma)^{16}$O reaction in JUNA

L.Y. Zhang[1,9], *J.* Su[1,9], *J.J.* He[1,9,*], *M.* Wiescher[2], *R.J.* deBoer[2], *D.* Kahl[3], *Y.J.* Chen[1], *X.Y.* Li[1], *J.G.* Wang[4], *L.* Zhang[5], *F.Q.* Cao[5], *H.* Zhang[5], *Z.C.* Zhang[6], *T.Y.* Jiao[4], *Y.D.* Sheng[1], *L.H.* Wang[1], *L.Y.* Song[1], *X.Z.* Jiang[1], *Z.M.* Li[1], *E.T.* Li[6], *S.* Wang[7], *G.* Lian[5], *Z.H.* Li[5], *X.D.* Tang[4], *H.W.* Zhao[4], *L.T.* Sun[4], *Q.* Wu[4], *J.Q.* Li[4], *B.Q.* Cui[5], *L.H.* Chen[5], *R.G.* Ma[5], *B.* Guo[5], *S.W.* Xu[4], *J.Y.* Li[4], *N.C.* Qi[8], *W.L.* Sun[8], *X.Y.* Guo[8], *P.* Zhang[8], *Y.H.* Chen[8], *Y.* Zhou[8], *J.F.* Zhou[8], *J.R.* He[8], *C.S.* Shang[8], *M.C.* Li[8], *X.H.* Zhou[4], *Y.H.* Zhang[4], *F.S.* Zhang[1,9], *Z.G.* Hu[4], *H.S.* Xu[4], *J.P.* Cheng[1], and *W.P.* Liu[5]

[1]Key Laboratory of Beam Technology and Material Modification of Ministry of Education, College of Nuclear Science and Technology, Beijing Normal University, Beijing 100875, China
[2]The Joint Institute for Nuclear Astrophysics, Department of Physics, University of Notre Dame, Notre Dame, Indiana 46556 USA
[3]Extreme Light Infrastructure – Nuclear Physics, Horia Hulubei National Institute for R&D in Physics and Nuclear Engineering (IFIN-HH), Bucharest-Măgurele 077125, Romania
[4]Institute of Modern Physics, Chinese Academy of Sciences, Lanzhou 730000, China
[5]China Institute of Atomic Energy, Beijing 102413, China
[6]College of Physics and Optoelectronic Engineering, Shenzhen University, Shenzhen 518060, China
[7]Shandong Provincial Key Laboratory of Optical Astronomy and Solar-Terrestrial Environment, Institute of Space Sciences, Shandong University, Weihai 264209, China
[8]Yalong River Hydropower Development Company, Chengdu 610051, China
[9]Beijing Radiation Center, Beijing 100875, China

Abstract. The ^{19}F$(p, \alpha\gamma)^{16}$O reaction is of crucial importance for Galactic ^{19}F abundances and CNO cycle loss in first generation Population III stars. Due to its extremely small cross sections, the ^{19}F$(p, \alpha\gamma)^{16}$O reaction has not been measured in the low energy part of the Gamow window(70-200 keV). As a day-one campaign, the experiment was performed under the extremely low cosmic-ray-induced background environment of the China JinPing Underground Laboratory(CJPL), one of the deepest underground laboratories in the world. The γ-ray yields were measured over $E_{c.m.}$=72.4–344 keV, covering the full Gamow window for the first time. The direct experimental data will help people to expound the fluorine over-abundances, energy generation, as well as heavy-element nuclosynthesis scenario in asymptotic giant branch (AGB) stars, with the astrophysical model on the firm ground.

1 Introduction

Fluorine is one of the most interesting elements in nuclear astrophysics, its astrophysical origin is puzzling. ^{19}F can be produced in the convective zone triggered by a thermal pulse in asymptotic giant branch (AGB) stars [1, 2]. So far, however, the astronomically observed fluorine over-abundances cannot be understood by using current AGB models [3–6]. In AGB

*e-mail: hejianjun@bnu.edu.cn

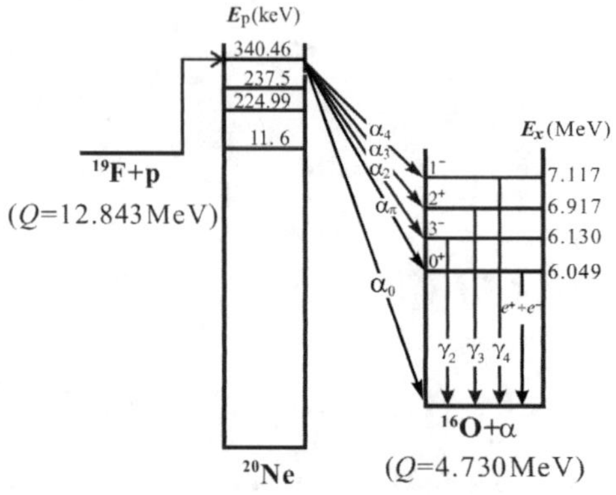

Figure 1. Level scheme of the ^{19}F$(p,\alpha)^{16}$O reaction. The total reaction rate is dominated by the (p,α_0) and $(p,\alpha\gamma)$ channel.

stars, ^{19}F is readily annihilated by hydrogen via the ^{19}F$(p,\alpha)^{16}$O reaction [3, 7, 8]. Given its importance, a precise and complete measurement of the total cross section appeared desirable. The ^{19}F$(p,\alpha)^{16}$O reaction occurs via three types of channels, i.e., (p,α_0), (p,α_π) and $(p,\alpha\gamma)$, as shown in Fig. 1. The total reaction rate is dominated by the (p,α_0) and $(p,\alpha\gamma)$ channel [3]. For the $(p,\alpha\gamma)$ channel, the energy range of $E_{c.m.} > 189$ keV has been studied which is much higher than the the low energy edge of the Gamow window(70-350 keV) [9, 10]. In this paper, we report on the progress of a direct measurement of the ^{19}F$(p,\alpha\gamma)^{16}$O reaction at Jinping Underground Nuclear Astrophysics experimental facility (JUNA). In the present work, the studies have been extended to $E_{c.m.}$=72.4-344 keV, the lowest to date. The results from present and previous work allow to calculate the reaction rates over a wide range of temperatures.

2 Experiment

The experiment was carried out on the JUNA accelerator at CJPL [11]. The experimental setup is similar to the one described in Ref. [2, 12]. The beam current was 1 mA for the low energy measurements. Two very strong and durable implanted ^{19}F targets were used [10, 12]. A 4π BGO detector array specially designed for the JUNA project [13] was equipped to detect the γ-rays, which was already used in previous work [12].

The γ-ray spectrum taken at a proton beam energy of E_p = 130 keV with the 4π BGO array was shown in Fig. 2. Two background peaks at 1460.8 keV (from ^{40}Ar) and 2614.5 keV (from ^{208}Tl) and the 6130 keV peak from the ^{19}F$(p,\alpha\gamma_2)^{16}$O reaction were used for energy calibration [2].

The γ-ray yield and hence the S Factors of the ^{19}F$(p,\alpha\gamma_2)^{16}$O reaction were determined by the integration of the 6130 keV peak of the spectrum(the red region in Fig. 2). The details can be found in Ref [2]. In conclusion, The present S Factors are much larger than the previous predictions. The thermonuclear ^{19}F$(p,\alpha\gamma_2)^{16}$O rate has been determined down to 0.05 GK

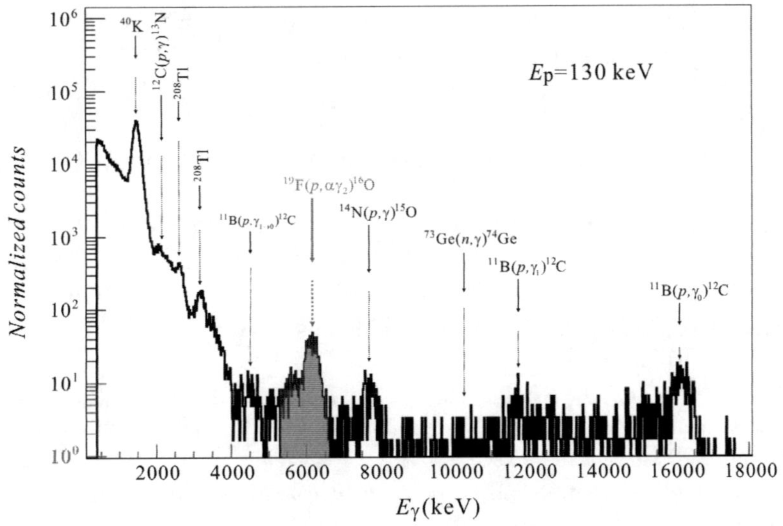

Figure 2. γ-ray spectrum of the ^{19}F+p experiments measured by a 4π BGO array at a proton energy of E_p = 130 keV [2].

and parameterized by the standard format of [14],

$$
\begin{aligned}
&N_A\langle\sigma v\rangle_{(p,\alpha\gamma)} \\
&= \exp(62.821 - \frac{0.022063}{T_9} - \frac{10.5347}{T_9^{1/3}} - 67.9612T_9^{1/3} + 50.592T_9 - 24.33T_9^{5/3} + 11.0325\ln T_9) \\
&+ \exp(30.5159 - \frac{0.097764}{T_9} + \frac{17.4599}{T_9^{1/3}} + 38.7519T_9^{1/3} - 134.383T_9 + 57.3453T_9^{5/3} + 37.5491\ln T_9) \\
&+ \exp(-18.6175 - \frac{0.349603}{T_9} + \frac{39.0245}{T_9^{1/3}} + 67.2527T_9^{1/3} - 116.029T_9 + 39.954T_9^{5/3} + 42.7072\ln T_9) \\
&+ \exp(-91.3551 - \frac{0.136527}{T_9} + \frac{0.16144}{T_9^{1/3}} + 21.6386T_9^{1/3} + 873.979T_9 - 1709.51T_9^{5/3} - 7.5102\ln T_9)
\end{aligned}
$$

with a fitting error of less than 1% over a temperature region of 0.01–1 GK [2].

We wish to thank the staff of the CJPL and Yalong River Hydropower Development Company for logistics support. This work was financially supported by the Fundamental Research Funds for the Central Universities and the National Natural Science Foundation of China (Nos. 11490562, 11825504, 12075027, 11961141004). RJD utilized resources from the Notre Dame Center for Research Computing and RJD and MW were supported by the National Science Foundation through Grant No. Phys-2011890, and the Joint Institute for Nuclear Astrophysics through Grant No. PHY-1430152 (JINA Center for the Evolution of the Elements). RJD acknowledges useful discussions with Daniel Odell regarding Bayesian uncertainty estimation applied to *R*-matrix theory. DK acknowledges the support of the Romanian Ministry of Research and Innovation under research contract 10N/PN 19 06 01 05.

References

[1] D. Clayton, *Handbook of Isotopes in the Cosmos: Hydrogen to Gallium*, 1st edn. (Cambridge University Press, 2003)

[2] L.Y. Zhang, J. Su, J.J. He, M. Wiescher, R. deBoer, D. Kahl, Y.J. Chen, X.Y. Li, J.G. Wang, L. Zhang et al., Physical Review Letters **127**, 152702 (2021)

[3] J.J. He, I. Lombardo, D. Dell'Aquila, Y. Xu, L.Y. Zhang, W.P. Liu, Chinese Physics C **42**, 015001 (2018)

[4] M. Lugaro, C. Ugalde, A.I. Karakas, J. Görres, M. Wiescher, J.C. Lattanzio, R.C. Cannon, The Astrophysical Journal **615**, 934 (2004), publisher: IOP Publishing

[5] A. Jorissen, V.V. Smith, D.L. Lambert, Astronomy and Astrophysics **261**, 164 (1992)

[6] S. Cristallo, O. Straniero, R. Gallino, L. Piersanti, I. Domínguez, M.T. Lederer, The Astrophysical Journal **696**, 797 (2009), publisher: American Astronomical Society

[7] S. Lucatello, T. Masseron, J.A. Johnson, M. Pignatari, F. Herwig, The Astrophysical Journal **729**, 40 (2011)

[8] C. Abia, K. Cunha, S. Cristallo, P.d. Laverny, I. Domínguez, A. Recio-Blanco, V.V. Smith, O. Straniero, The Astrophysical Journal **737**, L8 (2011), publisher: American Astronomical Society

[9] K. Spyrou, C. Chronidou, S. Harissopulos, S. Kossionides, T. Paradellis, C. Rolfs, W.H. Schulte, L. Borucki, The European Physical Journal A pp. 79–85 (2000)

[10] L.Y. Zhang, S.W. Xu, J.J. He, S. Wang, H. Chen, J. Hu, S.B. Ma, N.T. Zhang, S.Q. Hou, W.P. Liu, Nuclear Instruments and Methods in Physics Research Section B: Beam Interactions with Materials and Atoms **438**, 48 (2019)

[11] Q. Wu, L.T. Sun, B.Q. Cui, G. Lian, Y. Yang, H.Y. Ma, X.D. Tang, X.Z. Zhang, Z.M. Zhang, W.P. Liu, Nuclear Instruments and Methods in Physics Research Section A: Accelerators, Spectrometers, Detectors and Associated Equipment **830**, 214 (2016)

[12] L.Y. Zhang, Y.J. Chen, J.J. He, J. Su, X.Y. Li, H. Zhang, Y.J. Li, W.P. Liu, Nuclear Instruments and Methods in Physics Research Section B: Beam Interactions with Materials and Atoms **496**, 9 (2021)

[13] J. Su, Nuclear Instruments and Methods in Physics Research Section A: Accelerators, Spectrometers, Detectors and Associated Equipment, **to be submitted** (2022)

[14] T. Rauscher, F.K. Thielemann, Atomic Data and Nuclear Data Tables **75**, 1 (2000)

Probing the early Universe from deep underground

Francesca Cavanna[1,*] *for the LUNA collaboration*[2]

[1]INFN Sezione di Torino
[2]https://luna.lngs.infn.it

Abstract. Big Bang Nucleosynthesis (BBN) occurs during the first minutes of cosmological time in a rapidly expanding hot and dense Universe, where a fraction of protons and nearly all free neutrons end up bound in ^{4}He, while D, ^{3}H, ^{3}He, ^{6}Li, ^{7}Li and ^{7}Be nuclei form in trace quantities. Among these elements, deuterium is an excellent indicator of cosmological parameters because its abundance is highly sensitive to the primordial baryon density and to the number of relativistic species. Although astronomical observations of primordial deuterium abundance have reached percent accuracy, theoretical predictions based on BBN were hampered by large uncertainties on the cross-section of the deuterium burning D(p,γ)^{3}He reaction, before the LUNA measurement. In the following, I will report the results of the experimental campaign carried on at LUNA and its cosmological implications.

1 The primordial deuterium abundance: motivation for the measurement of the D(p,γ)^{3}He reaction

The BBN theory is one of the pillars of the picture we nowadays have of the Universe. In particular, for a given baryon density, it gives the primordial abundance of the light elements. Such predictions can then be compared with the results provided by spectroscopic observations of ancient astrophysical objects, like quasars or metal poor stars.

In the current precision era of cosmology, data from observations are characterized by higher uncertainties with respect to BBN predictions, except for deuterium. In particular, the D/H ratio is retrieved by astronomical observations with an excellent accuracy of about 1% [1], whereas the BBN prediction of D/H has an uncertainty of about 2% [2]. Actually, the relatively large error on the prediction of the D/H was dominated by the uncertainty on the cross sections of the nuclear reactions involved in the BBN network to first create and then destroy deuterium. In particular, before the LUNA measurement [3], the main uncertainty came from the D(p,γ)^{3}He reaction responsible for deuterium burning.

Several data sets on the D(p,γ)^{3}He reaction cross section, or equivalently its $S(E)$-factor were available in the literature before the LUNA measurement. The astrophysical $S(E)$ factor is defined as: $S(E) = E\sigma(E)\exp(2\pi\eta)$, where E is the energy of interaction in the centre of mass system, $\sigma(E)$ is the energy dependent cross-section, and η is the Sommerfeld parameter $\eta(E) = Z_1Z_2\alpha(\mu c^2/2E)^{1/2}$ (with Z_i atomic numbers of the interacting particles, α fine structure constant, μ reduced mass, and c speed of light). In the low-energy

*e-mail: francesca.cavanna@to.infn.it

range ($E_{C.M.} \simeq 2-20$ keV), mostly relevant to hydrogen burning in the Sun and in protostars, cross sections were obtained with a systematic error of at most 5.3% [4] using the 50 kV accelerator (now in disuse) of the Laboratory for Underground Nuclear Astrophysics (LUNA) at the Laboratori Nazionali del Gran Sasso (LNGS) in Italy [5]. At higher energies ($E_{C.M.} \simeq 30-700$ keV), available data sets were affected by systematic errors of 9% or higher [6–9]. The situation was further compounded by the fact that a recent *ab initio* calculation [10] disagrees at the 20-30% level with both the S-factor of Ma *et al.* [6] and a best fit [11] to selected data [4, 6, 12, 13], widely used in BBN calculations. These large uncertainties had significant impact on the comparison between predicted and observed primordial abundance, which is of crucial importance to access the baryon density of the Universe and to test more exotic scenarios beyond the Standard Model of fundamental interactions. As a consequence of the poor experimental data for the D(p,γ)^{3}He reaction, firm conclusions on the BBN stage and cosmological model couldn't be drawn with a satisfactory precision. For all these reasons, a new experimental campaign started at LUNA in 2016.

2 The D(p,γ)^{3}He cross section measurement: experimental setup and data analysis

The measurements were carried out using the LUNA facility located in the underground Gran Sasso National Laboratory (LNGS), Italy. The LUNA400 accelerator provided a high intensity proton beam (beam current 100-300 μA) in the 50-400 keV energy range. The beam entered in a cylindrical reaction chamber 33 cm long and with a 56 mm inner diameter, through a 40 mm long collimator having a inner diameter of 7 mm. High purity molecular deuterium at a 0.3 mbar pressure was present inside the reaction chamber. The beam was stopped by a copper calorimeter that measured its intensity. A large high-purity germanium detector (HPGe) faced to the reaction chamber at 90° to the beam axis was used to detect the photons emitted by the D(p,γ)^{3}He reaction. Possible gas impurities by in-leaking air were periodically checked by the strong $E_p = 278$ keV ^{14}N(p,γ)^{15}O resonance and were always below 0.5% [14].
The D(p,γ)^{3}He (Q = 5.493 MeV) reaction proceeds through direct capture to the ^{3}He ground state with the emission of a single gamma with an energy in the range 5.5 - 5.8 MeV. Since these energies are above the natural radioactivity endpoint, it is possible to fully exploit the suppression of cosmic ray induced background due to the Gran Sasso mountain [5, 15]. The incoming proton can be captured both in a s-wave or p-wave orbital angular momentum state and the emitted gamma-ray shows a not isotropic angular distribution dominated by the M1 and E1 components. To measure the cross section in the energy range of interest (30 keV < E_{cm} < 300 keV) for BBN we explored the full dynamic range of the LUNA-400 accelerator in 30-50 keV energy steps. The main source of beam induced background was due to the ^{19}F(p,$\alpha\gamma$)^{16}O reaction from the interaction of protons with fluorine contaminant usually present on collimators along the gas target and on the calorimeter. This beam-induced background was quantified in dedicated control runs in which inert ^{4}He gas was used instead of deuterium. A sample spectrum taken at 150 keV beam energy is shown in figure 1.

For an extended gas target setup, the cross section is given by the following expression:

$$\sigma(E) = \frac{N_\gamma}{N_p \int_0^L \rho(z) \cdot \epsilon(z) \cdot W(z) dz} \tag{1}$$

where N_γ is the number of counts after background subtraction. The integral extends over the target length, in which $\rho(z)$ represents the gas density profile, $\epsilon(z)$ is the gamma detection

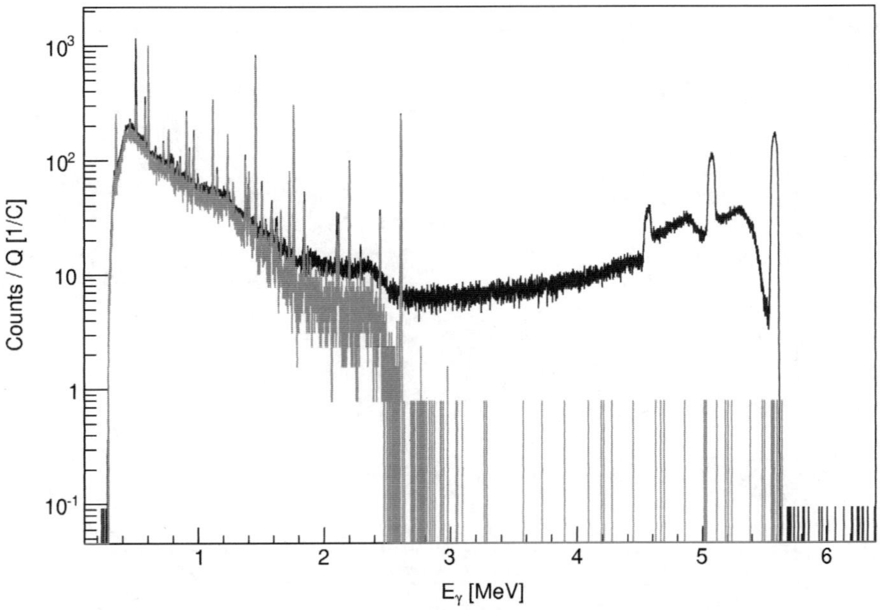

Figure 1: Typical γ-ray spectrum obtained with the HPGe detector at 150 keV beam energy: in blue the spectrum measured with D_2 gas and for comparison a beam induced background spectrum (grey) acquired with ^{4}He gas. The full-energy, single-escape and double-escape peaks from the D(p,γ)^{3}He reaction are clearly visible.

efficiency as a function of position along the beam axis, $W(z)$ accounts for the photon angular distribution and N_p is the number of projectiles impinging the target.

A major effort was devoted to prevent or minimize any source of systematic uncertainty in equation 1 to reach an accuracy of three percent on the S-factor. N_γ is essentially affected by the statistical error, that in our measurements was always lower than 1% for $E_{p>100}$ keV ($E_{C.M.} > 70$ keV), as a consequence of the suppression in the cosmic ray induced background with respect to surface experiments. The target density $\rho(z)$ was determined by measuring the pressure and temperature profiles along the target as well as the beam heating effect due to the power released by the incident beam in the gas target [16]. The integrated beam current N_p was measured with a consolidated technique in which the beam power was monitored during each run by measuring the power dissipated by the beam on the copper calorimeter where the beam was stopped [17]. In an extended gas target the interaction with proton beam can take place at different positions along the beam axis, resulting in different energies of the emitted photons and in different geonetrical angles subtended by the HPGe detector. Therefore, the γ-ray detection efficiency ϵ(z) must be carefully determined as a function of both position and energy. The γ-rays emitted by the D(p,γ)^{3}He reaction ($Q = 5.5$ MeV) have typical energies $E_\gamma = 5.5$-5.8 MeV, i.e. far away from the energy of the commonly used radioactive sources. The efficiency $\epsilon(z)$ of the HPGe detector was measured along the beam line, by filling the reaction chamber with N_2 and exploiting the resonance of the $^{14}N(p,\gamma)^{15}O$ reaction at E_r = 259 keV. This reaction produces cascade γ-rays in a large energy range (0.7-7 MeV), in such a way the efficiency were precisely determined [14]. Finally, the $W(z)$ function has

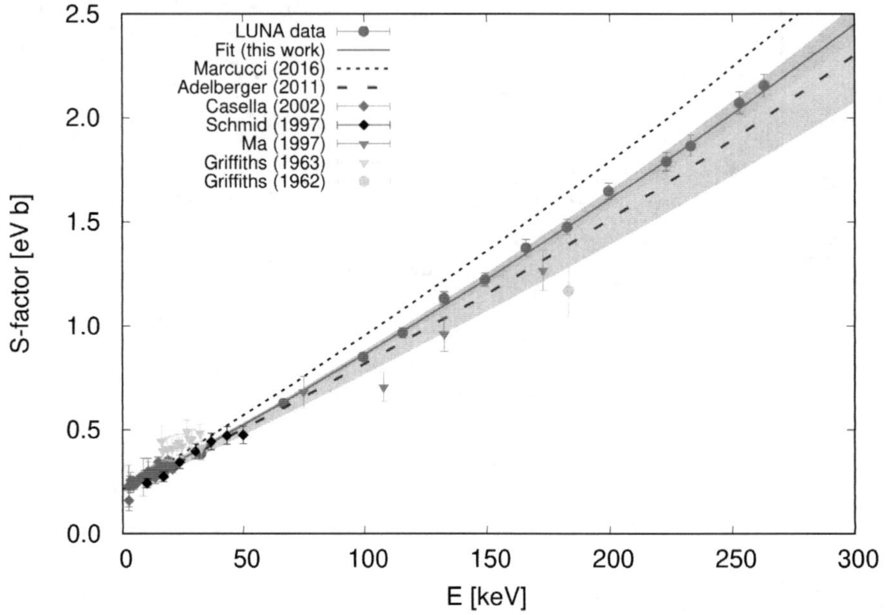

Figure 2: S-factor of the D(p,γ)^{3}He reaction; LUNA results are represented with filled red circles. Other experimental data are also shown. The best fit (red solid line) include all the reported experimental data. Band represents the 68% confidence level.

been unfolded from the shape of full energy peak of photons produced by the D(p,γ)^{3}He that depends on the angular distribution of emitted photons. The detailed description of the measurements performed to determine the above mentioned quantities and uncertainties can be found in [14].

After four years of data taking and a long data analysis campaign, it was possible to obtain the S-factor shown in Figure 2 where the LUNA data [3] are reported together with literature data. The new data show a significantly reduced uncertainty with respect to previous experimental dataset. Also shown is the polynomial best fit of S(E) for $E \sim 0 - 2$ MeV to previous dataset [11] (blue line) and the fit in which this new data set is considered (red line). The bands represent the quoted 1σ uncertainty of the best fit values.

3 Results and conclusions

To explore the impact of our D(p,γ)^{3}He S-factor on the predicted primordial deuterium abundance, we used the second release [18] of the numerical BBN code PArthENoPE. Under the assumption of the ΛCDM model, with $N_{\rm eff} = 3.045$, we performed a Bayesian likelihood analysis to derive $\Omega_{\rm b}h^2$ using the observed deuterium abundance, $({\rm D/H})_{\rm obs}$, and the theoretical behaviour of $({\rm D/H})_{\rm BBN}$ (now including the new LUNA data). We obtained $\Omega_{\rm b}h^2({\rm BBN}) = 0.02233 \pm 0.00036$, a value which is a factor of 2 more precise than that deduced using a previous S factor [11] and now in much better agreement with the $\Omega_b h^2$ based on CMB data [19].

If we use the baryon density provided by the PLANCK Collaboration [20], we derive a theoretical prediction on deuterium abundance $(D/H)_{BBN} = (2.52 \pm 0.03 \pm 0.06) \times 10^{-5}$, in excellent agreement with astronomical observations [1] $(D/H)_{obs} = (2.527 \pm 0.030) \times 10^{-5}$.

After the release of the LUNA results [3], several papers were published discussing the cosmological implications with the inclusion of the new $D(p,\gamma)^3He$ rate [2, 21–23]. All the authors agree on the need of new measurements of the $D(d,n)^3He$ and $D(d,p)^3H$ reactions which have became the most prominent source of uncertainty in the determination of the primordial deuterium abundance.

References

[1] R.J. Cooke, M. Pettini, C.C. Steidel, Astrophys. J. **855**, 102 (2018), 1710.11129

[2] O. Pisanti, G. Mangano, G. Miele, P. Mazzella, JCAP **04**, 020 (2021), 2011.11537

[3] V. Mossa, K. Stöckel, F. Cavanna, F. Ferraro, M. Aliotta, F. Barile, D. Bemmerer, A. Best, A. Boeltzig, C. Broggini et al., Nature **587**, 210 (2020)

[4] C. Casella, H. Costantini, A. Lemut, B. Limata, R. Bonetti, C. Broggini, L. Campajola, P. Corvisiero, J. Cruz, A. D'Onofrio et al., Nucl. Phys. A **706**, 203 (2002)

[5] F. Cavanna,, P. Prati, Int. J. Mod. Phys. A **33**, 1843010-346 (2018)

[6] L. Ma, H. Karwowski, C. Brune, Z. Ayer, T. Black, J. Blackmon, E. Ludwig, M. Viviani, A. Kievsky, R. Schiavilla, Phys. Rev. C **55**, 588 (1997)

[7] G. Schmid, B. Rice, R. Chasteler, M. Godwin, G. Kiang, L. Kiang, C. Laymon, R. Prior, D. Tilley, H. Weller, Phys. Rev. C **56**, 2565 (1997)

[8] G. Griffiths, M. Lal, C. Scarfe, Can. J. Phys. **41**, 724 (1963)

[9] I. Tišma, M. Lipoglavšek, M. Mihovilovič, S. Markelj, M. Vencelj, J. Vesić, Eur. Phys. J. A **55**, 137 (2019)

[10] L. Marcucci, G. Mangano, A. Kievsky, M. Viviani, Phys. Rev. Lett. **116**, 102501 (2016)

[11] E. Adelberger, A. García, R. Robertson, K. Snover, A. Balantekin, K. Heeger, M. Ramsey-Musolf, D. Bemmerer, A. Junghans, C. Bertulani et al., Rev. Mod. Phys. **83**, 195 (2011)

[12] G. Griffiths, E. Larson, L. Robertson, Can. J. Phys. **40**, 402 (1962)

[13] G. Schmid, M. Viviani, B. Rice, R. Chasteler, M. Godwin, K. G.C., L. Kiang, A. Kievsky, C. Laymon, R. Prior et al., Phys. Rev. Lett. **76**, 3088 (1996)

[14] V. Mossa, K. Stöckel, F. Cavanna, F. Ferraro, M. Aliotta, F. Barile, D. Bemmerer, A. Best, A. Boeltzig, C. Broggini et al., Eur. Phys. J. A **56**, 144 (2020), 2005.00002

[15] F. Ferraro, G.F. Ciani, A. Boeltzig, F. Cavanna, S. Zavatarelli, **7**, 119 (2021)

[16] F. Cavanna, R. Depalo, M.L. Menzel, M. Aliotta, M. Anders, D. Bemmerer, C. Broggini, C.G. Bruno, A. Caciolli, P. Corvisiero et al., **50**, 179 (2014), 1411.2888

[17] F. Ferraro, M. Takács, D. Piatti, V. Mossa, M. Aliotta, D. Bemmerer, A. Best, A. Boeltzig, C. Broggini, C. Bruno et al., Eur. Phys. J. A **54**, 44 (2018)

[18] R. Consiglio, P. de Salas, G. Mangano, G. Miele, S. Pastor, O. Pisanti, Comp. Phys. Comm. **233**, 237 (2018)

[19] B.D. Fields, K.A. Olive, T.H. Yeh, C. Young, Journ. Cosm. and Astropart. Phys. **2020**, 010 (2020)

[20] N. Aghanim, Y. Akrami, M. Ashdown, J. Aumont, C. Baccigalupi, M. Ballardini, A. Banday, R. Barreiro, N. Bartolo, S. Basak et al., arXiv e-prints arXiv:1807.06209 (2018)

[21] T.H. Yeh, K.A. Olive, B.D. Fields, JCAP **03**, 046 (2021), 2011.13874
[22] C. Pitrou, A. Coc, J.P. Uzan, E. Vangioni, Mon. Not. Roy. Astron. Soc. **502**, 2474 (2021), 2011.11320
[23] J. Moscoso, R.S. de Souza, A. Coc, C. Iliadis, Astrophys. J. **923**, 49 (2021)

Probing Massive Star Nucleosynthesis with Data on Metal-Poor Stars and the Solar System

Yong-Zhong Qian[1,*]

[1]School of Physics and Astronomy, University of Minnesota, Minneapolis, Minnesota 55455, USA

Abstract. Metal-poor stars were formed during the early epochs when only massive stars had time to evolve and contribute to the chemical enrichment. Low-mass metal-poor stars survive until the present and provide fossil records of the nucleosynthesis of early massive stars. On the other hand, short-lived radionuclides (SLRs) in the early solar system (ESS) reflect the nucleosynthesis of sources that occurred close to the proto-solar cloud in both space and time. Both the ubiquity of Sr and Ba and the diversity of heavy-element abundance patterns observed in single metal-poor stars suggest that some neutron-capture mechanisms other than the *r*-process might have operated in early massive stars. Three such mechanisms are discussed: the weak *s*-process in non-rotating models with initial carbon enhancement, a new *s*-process induced by rapid rotation in models with normal initial composition, and neutron-capture processes induced by proton ingestion in non-rotating models. In addition, meteoritic data are discussed to constrain the core-collapse supernova (CCSN) that might have triggered the formation of the solar system and provided some of the SLRs in the ESS. If there was a CCSN trigger, the data point to a low-mass CCSN as the most likely candidate. An $11.8\,M_\odot$ CCSN trigger is discussed. Its nucleosynthesis, the evolution of its remnant, and the interaction of the remnant with the proto-solar cloud appear to satisfy the meteoritic constraints and can account for the abundances of the SLRs ^{41}Ca, ^{53}Mn, and ^{60}Fe in the ESS.

1 Data on Metal-Poor Stars

Observations show that heavy elements such as Sr (with mass numbers $A \sim 86$–88) and Ba ($A \sim 134$–138) are prevalent in metal-poor stars [1]. At the early times when these stars were formed, low-mass stars would not have had time to evolve and produce such elements through the slow (s) neutron-capture process. Massive stars evolve quickly to end in core-collapse supernovae (CCSNe) and the neutrino-driven winds from the proto-neutron stars created in these events can produce Sr. Typical winds, however, do not have the appropriate conditions to make Ba (e.g., [2, 3]). While the rapid (r) neutron-capture process in rare CCSNe (e.g., [4]) or neutron star mergers (e.g., [5]) can produce Ba, it is difficult for such rare events to account for the wide-ranging enrichment at early times as indicated by the ubiquity of Ba in metal-poor stars (e.g., [6, 7]). Therefore, some other neutron-capture processes associated with massive stars are required.

Observations also show diverse abundance patterns of heavy elements in metal-poor stars. For example, there are substantial deviations of the abundance ratio La/Eu ($A = 139$ for La

*e-mail: qianx007@umn.edu

and A = 151, 153 for Eu) from the solar r-process value for stars with metallicities as low as [Fe/H] = log(Fe/H) − log(Fe/H)$_\odot$ ∼ −2.6 [8]. This diversity also calls for massive stellar sources for neutron-capture processes other than the r-process.

All of the above observations concern single metal-poor stars. So both the ubiquity of Ba and the diversity of La/Eu in these stars reflect enrichment of the general interstellar medium (ISM) by massive stellar sources, not some localized surface contamination through mass transfer of e.g., the s-process products from binary companions.

2 Neutron-Capture Processes in Early Massive Stars

We focus on three types of neutron-capture processes in early massive stars to address the above observations.

2.1 Enhanced Weak s-Process

Observations (e.g., [9]) show that there exist carbon-enhanced metal-poor (CEMP) stars with [C/Fe] ≥ 0.7, some of which do not have any enhancements of neutron-capture elements. These so-called CEMP-no stars constitute ∼ 20%, 40%, and 80% of all the metal-poor stars with [Fe/H] ≤ −2, −3, and −4, respectively. Such stars were most likely formed out of the general ISM that had been enriched by very early CCSNe with elevated abundances of C (and presumably also N and O) relative to Fe. While such stars have low masses in order to survive until the present, they must have been formed along with massive CEMP stars that already exploded as CCSNe.

During core H burning of a massive star, the initial CNO nuclei are converted into ^{14}N, which produces ^{22}Ne through $^{14}\mathrm{N}(\alpha,\gamma)^{18}\mathrm{F}(e^+\nu_e)^{18}\mathrm{O}(\alpha,\gamma)^{22}\mathrm{Ne}$ during core He burning. The activation of $^{22}\mathrm{Ne}(\alpha,n)^{25}\mathrm{Mg}$ during core He burning and subsequent evolution provides neutrons for the weak s-process that produces heavy elements typically up to Sr. For massive CEMP stars with initial abundances of [CNO/H] $\gtrsim -0.5$, even heavier elements up to Ba can be produced [10]. In general, the efficiency of the s-process in massive CEMP stars is sensitive to the initial enhancement of CNO and mass of the star while the yield increases approximately linearly with the initial Fe abundance. These results were obtained for non-rotating stars, and therefore, differ from similar results for fast-rotating massive metal-poor stars without C enhancement (e.g., [11, 12]) in that the former are independent of the uncertain rotation-induced mixing processes. The enhanced weak s-process in non-rotating CEMP stars of $\gtrsim 20\,M_\odot$ with [CNO/H] $\gtrsim -1.5$ can be an important source for heavy elements in the early Galaxy [10].

2.2 New s-Process Induced by Rapid Rotation

Massive metal-poor stars rotating above a critical speed can reach the so-called quasi-chemically homogeneous (QCH) state following core H burning (e.g., [13, 14]). Rotation-induced mixing results in primary production of ^{13}C. The subsequent occurrence of $^{13}\mathrm{C}(\alpha,n)^{16}\mathrm{O}$ during core He burning provides neutrons for a prolific s-process. Depending on the rotation speed and the mass loss rate, elements up to Bi (A = 209) can be produced for progenitors with initial metallicities of [Fe/H] $\lesssim -1.5$ [15]. This model of the s-process in rotating massive metal-poor stars differs from other studies of the s-process in such stars (e.g., [12]) because the QCH state was not reached for the latter.

The above model suggests that rapidly-rotating massive metal-poor stars are likely the first sites of the main s-process, which is usually associated with slowly-evolving low-mass stars. With the above massive stellar sources, the observed early onset of the *s*-process can be explained. In addition, the *s*-process contributions of these sources can account for at least some of the CEMP-*s* and CEMP-*r*/*s* stars with strong enrichments that are attributed to the *s*-process and a mixture of the *r*-process and the *s*-process, respectively [15].

2.3 Neutron-Capture Processes Induced by Proton Ingestion

During the last few years of the life of a metal-free or metal-poor star of $\sim$ 20–30 $M_\odot$, the He shell becomes convective following the depletion of C at the center. Some of the protons present at low levels in the outer He shell can be ingested into the inner He shell by convective boundary mixing. Subsequent transport to the hotter region initiates the reaction sequence $^{12}C(p,\gamma)^{13}N(e^+\nu_e)^{13}C(\alpha,n)^{16}O$, which provides neutron densities appropriate for driving a nuclear flow intermediate (*i*) between the *s*-process and the *r*-process, the so-called *i*-process [16]. As a result, elements up to Bi are produced [17]. Depending on the time available before core collapse, $^{17}O(\alpha,n)^{20}Ne$ may occur to provide much lower neutron densities typical of the *s*-process to facilitate further neutron capture on the products of the preceding *i*-process. Consequently, the final yield pattern can vary from *s*-like to *r*/*s*-like [17].

The above model of neutron-capture processes induced by proton ingestion in early non-rotating massive stars have important implications for the abundances of heavy elements in metal-poor stars. First of all, it provides a rapidly-evolving source that occurred frequently to enrich the early ISM with heavy neutron-capture elements. Therefore, it can account for the observed ubiquity of Ba in metal-poor stars (see Sect. 1). Further, because its yield pattern can vary from *s*-like to *r*/*s*-like, it can also explain the diversity of heavy-element abundance patterns in metal-poor stars (see Sect. 1). Finally, mass transfer of *s*-process and *i*-process products from an intermediate-mass binary companion is commonly invoked to account for CEMP-*s* and CEMP-*r*/*s* stars, respectively. For some CEMP-*s* and CEMP-*r*/*s* stars, however, there are no clear indications that they are in binaries. In fact, some CEMP-*s* stars are observed to be single [18], which is difficult for the binary mass transfer scenario to explain. Single CEMP-*s* and CEMP-*r*/*s* stars, however, can be accounted for by the above model because they could have been formed directly from the early ISM that had already been enriched by massive stars with neutron-capture processes induced by proton ingestion.

3 CCSN Contributions to the Solar System

In contrast to metal-poor stars formed at early times, both low-mass and massive stars contributed to the elemental abundances in the solar system, which was formed $\sim$ 9 Gyr after the big bang. The number of CCSNe among these contributing sources can be estimated as follows. A CCSN can enrich $\sim 3 \times 10^4\,M_\odot$ of ISM (e.g., [19]). For a Galactic CCSN rate of $\sim$ (30 yr)$^{-1}$ associated with $\sim 10^{10}\,M_\odot$ of gas, an average ISM would receive CCSN contributions once every $\sim$ 10 Myr. So the ISM out of which the solar system was formed, the proto-solar cloud, received contributions from $\sim$ 900 CCSNe over $\sim$ 9 Gyr.

Many short-lived radionuclides (SLRs) with lifetimes of $\sim$ 0.1–10 Myr are produced by CCSNe. Based on the above estimated interval of $\sim$ 10 Myr between successive CCSN contributions, the last few CCSNe contributing to the abundances in the solar system might have provided a significant amount of SLRs to the early solar system (ESS). Any SLRs from those CCSNe would have been incorporated into the meteorites formed in the ESS. Studies of such meteoritic samples indeed found a number of SLRs (e.g., [20]). These meteoritic data provide a strong constraint on the scenario that the last contributing CCSN also triggered the formation of the solar system because the SLRs contributed by that CCSN cannot exceed the corresponding abundances in the ESS.

Table 1. Example results on SLRs in the ESS.

R/I	τ_R (Myr)	Y_R ($M_\odot$)	$X_{I,\odot}$	$(N_R/N_I)_{\rm ESS}$	Data
^{41}Ca/^{40}Ca	0.147	7.11×10^{-6}	5.88×10^{-5}	4.31×10^{-9}	$(4.6\pm1.9)\times10^{-9}$
^{53}Mn/^{55}Mn	5.40	2.73×10^{-5}	1.29×10^{-5}	6.97×10^{-6}	$(7\pm1)\times10^{-6}$
^{60}Fe/^{56}Fe	3.78	4.17×10^{-6}	1.12×10^{-3}	1.04×10^{-8}	$(1.01\pm0.27)\times10^{-8}$

For a fraction f of the ejecta from the triggering CCSN injected into the proto-solar cloud and a time Δ between the CCSN and incorporation of SLRs into the ESS solids, the resulting number ratio of an SLR (R) to its stable isotope (I) would be

$$\left(\frac{N_R}{N_I}\right)_{\rm ESS} = \frac{fY_R/A_R}{X_{I,\odot}M_\odot/A_I}\exp\left(-\frac{\Delta}{\tau_R}\right), \tag{1}$$

where Y_R is the CCSN yield of R, τ_R and A_R are its lifetime and mass number, respectively, and $X_{I,\odot}$ and A_I are the solar abundance and mass number of I, respectively. The necessary condition for the triggering CCSN is that $(N_R/N_I)_{\rm ESS}$ match the data on some SLRs without exceeding the data on all the other SLRs.

An additional meteoritic constraint on the triggering CCSN comes from the lack of large deviations in the number ratios of stable isotopes of such major elements as Mg, Si, Ca, and Fe [21]. Both this constraint and the necessary condition regarding SLRs render it difficult for a high-mass CCSN to be the trigger because the large yields of such a source would have exceeded the data on some SLRs and caused large deviations in the number ratios of stable isotopes of some major elements [22]. Consequently, if a CCSN triggered the formation of the solar system, it was most likely at the lower end of the relevant mass range [22, 23].

4 Low-Mass CCSN Trigger for Solar System Formation

The nucleosynthesis of an 11.8 $M_\odot$ CCSN was studied in [22, 23]. Using a 3D model of the explosion, [23] found that the SLRs ^{41}Ca, ^{53}Mn, and ^{60}Fe in the ESS can be accounted for by such a CCSN (see Table 1) with $f = 3.6\times10^{-6}$ and $\Delta = 0.675$ Myr [see Eq. (1)]. Because only a very small fraction of the CCSN ejecta needs to have been injected into the proto-solar cloud, the contributions to the other SLRs and the stable isotopes were negligible.

The parameters f and Δ for the above low-mass CCSN trigger are consistent with the evolution of a CCSN remnant in an ISM and with the interaction between the associated shock and the proto-solar cloud. Simulations showed that a fraction $\epsilon_{\rm in}\sim0.05$ of the CCSN ejecta impacting the cloud would be injected into the cloud [24, 25]. The fraction f of the net CCSN ejecta injected into a cloud of radius $r_c\sim0.1$ pc [26] can be estimated as

$$f\sim\epsilon_{\rm in}\left(\frac{\delta\Omega_c}{4\pi}\right)\sim\frac{\epsilon_{\rm in}}{4}\left(\frac{r_c}{R_s}\right)^2\sim1.25\times10^{-4}\left(\frac{\epsilon_{\rm in}}{0.05}\right)\left(\frac{r_c}{0.1\ {\rm pc}}\right)^2\left(\frac{\rm pc}{R_s}\right)^2, \tag{2}$$

where $\delta\Omega_c$ is the solid angle subtended by the cloud from the center of the CCSN and R_s is the shock radius upon impact. To obtain $f\sim3.6\times10^{-6}$ requires $R_s\sim5.9$ pc. In addition, to trigger the collapse of the cloud requires a shock velocity $v_s\sim$ 20–40 km s^{-1} [24, 25].

The evolution of R_s and v_s for a CCSN remnant [27] can be estimated as

$$R_s = 14.0\left(\frac{E}{10^{51}\ {\rm erg}}\right)^{2/7}\left(\frac{\rm cm^{-3}}{n_0}\right)^{3/7}\left(\frac{4}{3}t_*-\frac{1}{3}\right)^{3/10}\ {\rm pc}, \tag{3}$$

$$v_s = 413\left(\frac{E}{10^{51}\ {\rm erg}}\right)^{1/14}\left(\frac{n_0}{\rm cm^{-3}}\right)^{1/7}\left(\frac{4}{3}t_*-\frac{1}{3}\right)^{-7/10}\ {\rm km\ s^{-1}}, \tag{4}$$

where E is the CCSN explosion energy, n_0 is the hydrogen number density of the ISM, and $t_* = t/t_{\rm unit}$ is the time t since the explosion in units of

$$t_{\rm unit} = 1.33 \times 10^4 \left(\frac{E}{10^{51}\ {\rm erg}}\right)^{3/14} \left(\frac{{\rm cm}^{-3}}{n_0}\right)^{4/7} \ {\rm yr}. \quad (5)$$

For $E = 2 \times 10^{50}$ erg [28] and $n_0 \sim 50$ cm^{-3} [26], $R_s \sim 5.9$ pc is reached for $t \sim 5.3 \times 10^4$ yr with an impact velocity $v_s \sim 33$ km s^{-1}. Because the above evolution time is far less than Δ, the latter would be essentially the time needed for condensation of solids in the ESS.

5 Discussion and Conclusions

Observations of single metal-poor stars reveal both the ubiquity of Sr and Ba and the diversity of heavy-element abundance patterns, which appear to require some neutron-capture mechanisms other than the r-process to have operated in early massive stars. Among various possibilities, perhaps the weak s-process in non-rotating massive CEMP stars is the best understood. Its contributions might have been significant at low metallicities if a substantial fraction of the stars of $\gtrsim 20\,M_\odot$ had been formed with [CNO/H] $\gtrsim -1.5$ [10]. The new s-process induced by rapid rotation can potentially account for the observed early onset of the s-process and at least some of the CEMP-s and CEMP-r/s stars [15]. The modeling of this process depends on our understanding of the evolution of rapidly-rotating metal-poor stars and the associated mixing. Likewise, the neutron-capture processes induced by proton ingestion in early non-rotating massive stars depend on the treatment of mixing during the evolution of such stars. These processes can potentially account for both the observed ubiquity of Ba and the diversity of heavy-element abundance patterns in metal-poor stars [17]. In addition, their relatively frequent contributions to the ISM may have resulted in the formation of a significant fraction of the CEMP-s and CEMP-r/s stars by birth, which could have been the dominant channel to form such stars in single configuration [17].

Meteoric data demonstrate the presence of SLRs in the ESS. They also show a lack of large deviations in the number ratios of stable isotopes of such major elements as Mg, Si, Ca, and Fe. These data render it difficult for a high-mass CCSN to be the trigger for the formation of the solar system because the large yields of such a source would have exceeded the data on some SLRs and caused large deviations in the number ratios of stable isotopes of some major elements [22]. Consequently, if there was a CCSN trigger, a low-mass CCSN would be the most likely candidate [22, 23]. The nucleosynthesis of an $11.8\,M_\odot$ CCSN, the evolution of its remnant, and the interaction of the remnant with the proto-solar cloud appear to satisfy the meteoritic constraints and can account for the abundances of the SLRs ^{41}Ca, ^{53}Mn, and ^{60}Fe in the ESS.

Acknowledgments

I thank Projjwal Banerjee, Wick Haxton, Alexander Heger, Bernhard Müller, and Andre Sieverding for collaboration. This work was supported in part by the U.S. Department of Energy under Grant No. DE-FG02-87ER40328.

References

[1] I. U. Roederer, Astron. J. **145**, 26 (2013).
[2] Y.-Z. Qian and S. E. Woosley, Astrophys. J. **471**, 331 (1996).

[3] R. D. Hoffman, S. E. Woosley, and Y.-Z. Qian, Astrophys. J. **482**, 951 (1997).
[4] T. Fischer, M.-R. Wu, B. Wehmeyer, N.-U. F. Bastian, G. Martínez-Pinedo, and F.-K. Thielemann, Astrophys. J. **894**, 9 (2020).
[5] F.-K. Thielemann, M. Eichler, I. V. Panov, and B. Wehmeyer, Annu. Rev. Nucl. Part. Sci. **67**, 253 (2017).
[6] Y.-Z. Qian, Astrophys. J. **534**, L67 (2000).
[7] D. Argast, M. Samland, F.-K. Thielemann, and Y.-Z. Qian, Astron. Astrophys. **416**, 997 (2004).
[8] J. Simmerer, C. Sneden, J. J. Cowan, J. Collier, V. M. Woolf, and J. E. Lawler, Astrophys. J. **617**, 1091 (2004).
[9] W. Aoki, T. C. Beers, N. Christlieb, J. E. Norris, S. G. Ryan, and S. Tsangarides, Astrophys. J. **655**, 492 (2007).
[10] P. Banerjee, Y.-Z. Qian, and A. Heger, Mon. Not. R. Astron. Soc. **480**, 4963 (2018).
[11] M. Pignatari, R. Gallino, G. Meynet, R. Hirschi, F. Herwig, and M. Wiescher, Astrophys. J. **687**, L95 (2008).
[12] U. Frischknecht *et al.*, Mon. Not. R. Astron. Soc. **456**, 1803 (2016).
[13] S.-C. Yoon and N. Langer, Astron. Astrophys. **443**, 643 (2005).
[14] S. E. Woosley and A. Heger, Astrophys. J. **637**, 914 (2006).
[15] P. Banerjee, A. Heger, and Y.-Z. Qian, Astrophys. J. **887**, 187 (2019).
[16] J. J. Cowan and W. K. Rose, Astrophys. J. **212**, 149 (1977).
[17] P. Banerjee, Y.-Z. Qian, and A. Heger, Astrophys. J. **865**, 120 (2018).
[18] T. T. Hansen *et al.*, Astron. Astrophys. **588**, A3 (2016).
[19] K. Thornton, M. Gaudlitz, H. -Th. Janka, and M. Steinmetz, Astrophys. J. **500**, 95 (1998).
[20] A. M. Davis and K. D. Mckeegan, *Meteorites and Cosmochemical Processes, Treatise on Geochemistry* vol. 1 (Elsevier, 2014) 361–395.
[21] G. J. Wasserburg, M. Busso, R. Gallino, and K. M. Nollett, Nucl. Phys. **A777**, 5 (2006).
[22] P. Banerjee, Y.-Z. Qian, A. Heger, and W. C. Haxton, Nat. Commun. **7**, 13639 (2016).
[23] A. Sieverding, B. Müller, and Y.-Z. Qian, Astrophys. J. **904**, 163 (2020).
[24] A. P. Boss and S. A. Keiser, Astrophys. J. **788**, 20 (2014).
[25] A. P. Boss and S. A. Keiser, Astrophys. J. **809**, 103 (2015).
[26] E. A. Bergin and M. Tafalla, Annu. Rev. Astron. Astrophys. **45**, 339 (2007).
[27] D. F. Cioffi, C. F. McKee, and E. Bertschinger, Astrophys. J. **334**, 252 (1988).
[28] B. Müller *et al.*, Mon. Not. R. Astron. Soc. **484**, 3307 (2019).

r-Process Contributions to Low-Metallicity Stars

Friedrich-K. Thielemann[1,2,*], *Khalil* Farouqi[3], *Stephan* Rosswog[4], and *Karl-Ludwig* Kratz[5,6]

[1]Dept. of Physics, University of Basel, Klingelbergstrasse 82,CH-4056 Basel, Switzerland
[2]GSI Helmholtz Center for Heavy Ion Research, Planckstrasse 1, D-64291 Darmstadt, Germany
[3]ZAH, Landessternwarte, University of Heidelberg, Königstuhl 12, D69117 Heidelberg, Germany
[4]The Oskar Klein Centre, Department of Astronomy, Stockholm University, Stockholm, Sweden
[5]Dept. of Chemistry, Pharmacy & Geosciences, University of Mainz, D-55126 Mainz, Germany
[6]Max Planck Institute for Chemistry (Otto Hahn Institute), D-55128 Mainz, Germany

Abstract. Various nucleosynthesis studies have pointed out that the rapid neutron capture r-process elements in very metal-poor (VMP) halo stars might have different origins. It has been known that an r-process can either be obtained in neutron-rich low Y_e conditions or in high entropy environments [see e.g. 1–5], an overview over many investigations has appeared recently [6]. In the present article we analyze with statistical methods the observational abundance patterns from trans-Fe elements up to the actinides and come to the conclusion that four to five categories of astrophysical events must have contributed. These include the ejection of Fe and trans-Fe elements Sr, Y, Zr (continuing possibly beyond to slightly higher mass numbers) in category 0 events (hereafter "C0"), Fe and weak r-process contributions (including Eu in moderate to slightly larger but varying amounts) in CI and CII events, strong r-process abundance patterns with no or negligible (in comparison to solar) Fe production in CIIIa and CIIIb events, where category CIIIb shows a tendency for an actinide boost behavior. When comparing these categories with presently existing nucleosynthesis predictions, we suggest to identify them (despite remaining uncertainties) with regular core-collapse supernovae, quark deconfinement supernovae, magneto-rotational supernovae, neutron star mergers, and outflows from black hole accretion tori.

1 Introduction

Stellar (surface) abundances are inherited from the interstellar medium in which the stars were born, i.e. one can look back in time. The early phases of galactic evolution are characterized by a large degree of abundance scatter and spatial inhomogeneities, which cannot be explained by traditional chemical evolution studies [7]. To account for this, inhomogeneous “chemical evolution“ models do not assume that ejecta mix completely and instantaneously with the whole galactic interstellar medium. Instead, they consider only the pollution of about 5×10^4 M$_\odot$ (via a Sedov-Taylor blast wave) of the interstellar medium [see e.g. 8, 9]. This is based on an analysis that the ejection of 0.1 M$_\odot$ of Fe leads in a CCSN remnant to [Fe/H]≈-2.7 [10]. If only 10% or 1% of supernova ejecta contribute to the formation of the next star, one expects, therefore, [Fe/H] values as small as -4 or -5. More extended recent chemical

*e-mail: f-k.thielemann@unibas.ch

evolution studies take this into account in a more general approach [see e.g. 11, 12]. We will use as a starting hypothesis that stars with metallicities below -2.5 might be interpreted as the result from the contribution of only a single nucleosynthesis event. Can we therefore see in such low-metallicity stars the results of individual but different astrophysical events and which of these events are responsible for the heavy element r-process production?
One of the major questions about r-process contributions to solar abundances is whether they are universal and stem from a unique astrophysical site. Early signs from low-metallicity stars seemed to show such a unique r-process pattern, therefore pointing to a universal site [13]. This interpretation started to become questionable when actinide-boost stars were observed [14] and the first limited-r stars were discovered [15]. As an extension of explaining these different patterns we suggest, despite substantial uncertainties, an identification of the observational categories listed in the abstract with regular core-collapse supernovae (C0), quark deconfinement supernovae (CI), magneto-rotational supernovae (CII), neutron star mergers with a long-lived central remnant (CIIIa) and black hole-accretion torus systems (CIIIb) formed either by collapsars, neutron star black hole binaries or massive neutron star mergers that rapidly form a central black hole. But before doing so and checking these suggestions via utilizing ejecta composition predictions for comparisons with abundance patterns of low-metallicity stars, we want to briefly summarize the statistical methods which led to these identifications.

2 Statistical Analysis of r-Process Element Abundances in Low-Metallicity Stars

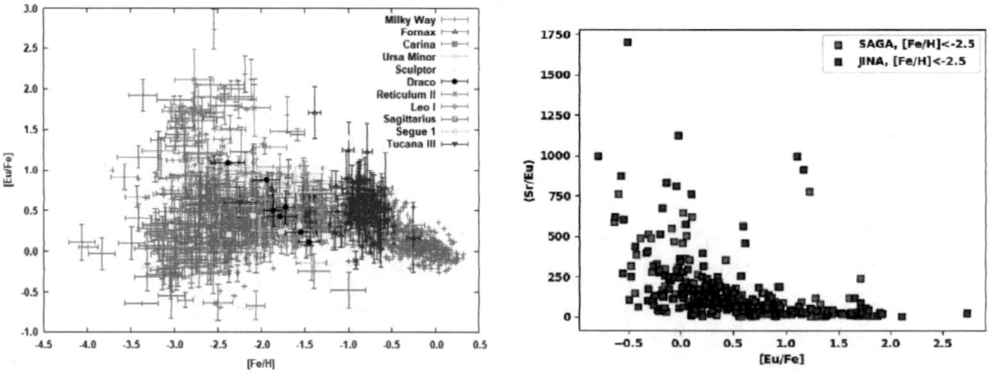

Figure 1. Left: [Eu/Fe] ratios of 1572 stars with Eu detections from the SAGA database [16]. A hugh scatter can be observed before at about [Fe/H]=-2 an averaging (smaller spread) sets in, which continues with a different gradient when SNe Ia start to contribute at -1. Right: Sr/Eu ratios for all stars with [Fe/H <-2.5, which show a drastic change at about [Eu/Fe]=0-0.3, i.e. the division between limited-r stars and r-enriched stars [for these observational classifications and the definition of actinide boost stars see e.g. 17–26]. (Notice that some points are identical for SAGA and JINA and only the entry plotted for JINA - in blue – is visible, especially the value of about 1100 for [Eu/Fe]$\approx$0.)

If abundance patterns of low-metallicity stars result from the pollution by only a single type of event, this would cause correlations between those elements resulting from the originating event [for a recent example see 27]. Such correlations (i.e. linear relations, respective constant ratios, between their abundances) underline the co-production of these elements. If (at low metallicities) elemental abundances are interpreted as imprints of individual explosions,

a large scatter as a function of [Fe/H] points to either rare events with different amounts of admixtures and/or to different types of events with strong variations in their ejected composition. The variation in [Eu/Fe] vs. [Fe/H] could be interpreted by such explanations, the large Sr/Eu in limited-r stars vs. the low values in r-enriched stars hints at different types of events (see Fig.1). There exist of course uncertainties in existing data, especially if an easier to detect element like Sr is compared to a harder to detect element like Eu, that might be close to detection limits. One of the possible options to verify the size of these uncertainties is to compare two databases which contain collected but differently processed information. While the SAGA database is known to enter data from different authors and observational facilities without much censorship, the JINAbase library has been compiled, addressing also systematic errors. When utilizing the SAGA or JINA databases for Fig,1 (right panel) we get slightly different but very similar results with the same trend. Therefore, we think that the conclusions pointing to different origins, due to this trend, are sound.
In late phases of galactic evolution, a large number of different events have taken place, leading, due to superpositions and admixtures of the ejecta of many different types of events, also to close to constant element ratios. These represent, opposite to the behavior at lowest metallicities, "spurious" correlations. However, they are just the result of averaging over many events and do not support the interpretation as co-production in an individual event. In order to analyze the observed abundance patterns in low-metallicity stars in a safe way among the above listed interpretations, we utilize mainly three methods to gain insight into the possibly different origins of the r-process elements: cluster analysis, rank tests, and correlations.

2.1 Clustering

Cluster analysis is a statistical technique to identify how data can be grouped because of characteristics they have in common [28]. It sorts data into groups which have a maximal number of characteristics in common. One method is the K-means Clustering: it distributes data into K mutually exclusive clusters so that with a given measure data points are as close to each other as possible within each cluster, while at the same time they as far from other clusters as possible. Each cluster is characterized by a central point.

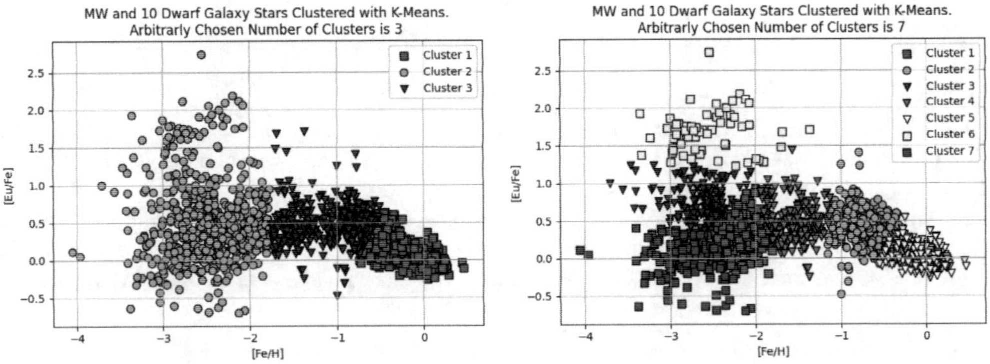

Figure 2. 3-cluster and 7-cluster analysis of [Eu/Fe] ratios as a function of [Fe/H], see text.

In the 3-cluster analysis of the Eu vs. Fe abundances of Fig.2 one can recognize in cluster 1 the low-metallicity stars where the large scatter indicates contributions from individual rare and/or different events, in cluster 2 for [Fe/H]>-2 that already an averaging/mixing of event products sets in and the scatter becomes small, and in cluster 3 that Fe is contributed also from

SNe Ia and changes the general Eu/Fe trend. In the 7-cluster analysis one recognizes different stages of the approach to averaging/mixing (clusters 1, 2, 4) and for low-metallicity stars with [Fe/H]<-2.5 three clusters (3, 6, 7) with properties close to the observational limited-r, r-I, and r-II classes of [Eu/Fe]<0-0.3, < 1, and >1.

2.2 Rank Test

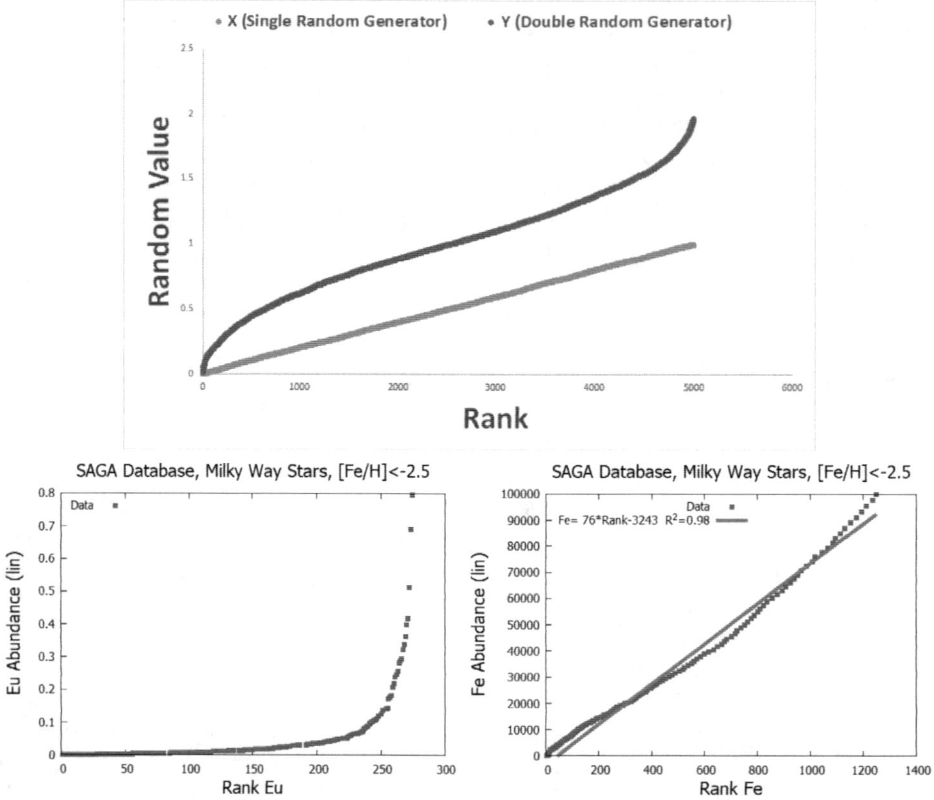

Figure 3. Top: data ordered along their ranks from a single number generator X, showing (a) a linear relation between X-values and their rank, and (b) the sum $X_1 + X_2$ of two single number generators vs. its rank, deviating from a linear relationship. Bottom left: Eu abundances vs. their rank for low-metallicity stars with [Fe/H]<-2.5, at high ranks (i.e. high Eu abundances) one recognizes an additional/different strong contribution. Bottom right: The same for Fe abundances, displaying almost a linear relationship, pointing to one identical source, but at low as well as high Fe abundances slight deviations are seen.

We show examples for utilizing a single random number generator X and ordering the values according to their size, giving them integer entries (ranks). When plotting them as a function of their ranks, one can see a linear relation between the random numbers and their ranks (see the orange line in Fig.3, top panel). The blue line is obtained when utilizing two random number generators, adding their values $Y = X_1 + X_2$, and plotting Y as a function of its ranks. It deviates from a linear relation between Y and its ranks. The deviation at low ranks depends on the fact whether the second contribution includes many low-value entries. The deviation at high ranks depends on the number of high-value entries. If one applies this method to abundance observations of individual elements as a function of rank, it can indicate whether only one type of nucleosynthesis source or two or several contributed to this element.

We utilized this method also to display Eu abundances versus their corresponding ranks in Fig.3 (bottom left). The integer rank passes, with increasing Eu abundances, through all observational points from the smallest to the highest abundance. The fit deviates strongly from a linear behavior at high Eu ranks. This argues for a superposition with an additional Eu source, especially responsible for the high Eu abundances. A missing deviation from the linear slope at small ranks underlines that this additional Eu source contains essentially no contributions with negligible Eu production in Fig.3 (bottom right). A different behavior can be noticed for Fe abundances versus their corresponding ranks. The figure shows a close to linear relation. This would argue for a single or dominating production site (explosive Si-burning in CCSNe?). The slight deviation at high ranks could be interpreted as an additional source with high Fe production (hypernovae?), the slight deviation at low ranks as an additional source with small Fe production (magneto-rotational supernovae?)

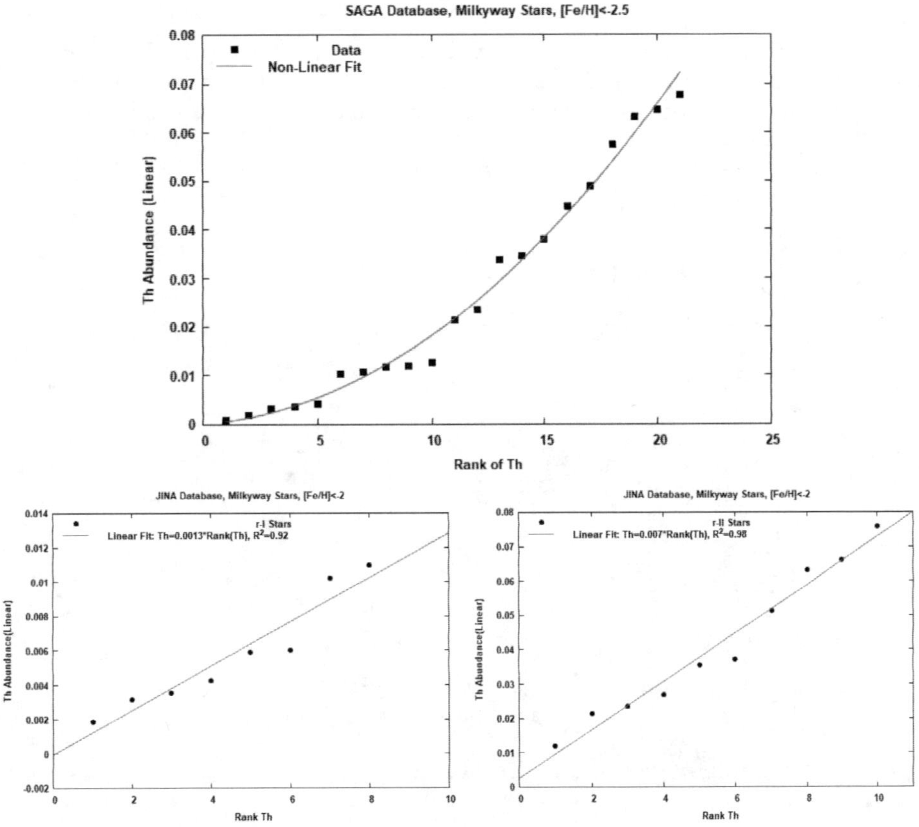

Figure 4. Top: Th abundances vs. their rank for low-metallicity stars, i.e. r-enriched r-I and r-II stars. Bottom left and right: The same but considering r-I and r-II stars separately. The linear relationship points to a single nucleosynthesis source for each of these two sets.

If one performs a similar rank test for Th (Fig,4) in r-enriched stars (r-I and r-II), it can be noticed that a superposition of two types of events is required. But one finds a linear rank relation when plotting Th abundances as a function of their ranks separately for r-I and r-II stars (with a slight scatter). Thus, r-I and r-II stars seem each to be dominated by contributions from one specific type of event, each of them being responsible for their Th production independently. This is not the case when performing a rank tests for Eu in r-I and r-I stars (Fig.5). Opposite to Th, Eu does not show a linear rank relation, when considering r-I and r-II stars separately. This suggests that also among r-enriched (strong r-process r-I and r-II) stars weak r-process events «spill in» an additional Eu contribution.

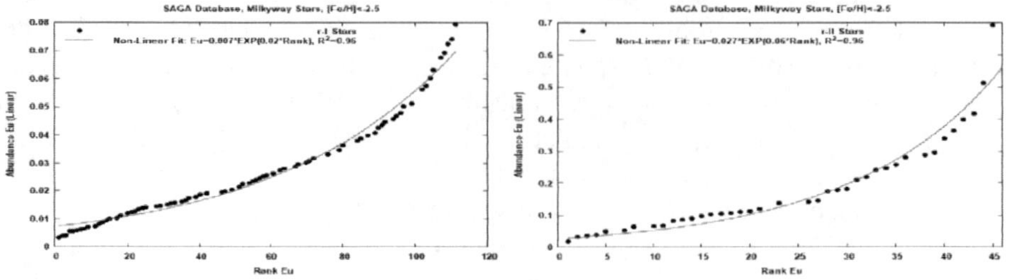

Figure 5. Rank tests for Eu abundances in r-I and r-II stars separately. For Eu a further nucleosynthesis source is required in addition to the two different strong sources dominating Th in r-I and r-II stars. This indicates that Eu from a weak r-process event (dominating limited-r stars) contributes as well.

2.3 Correlations

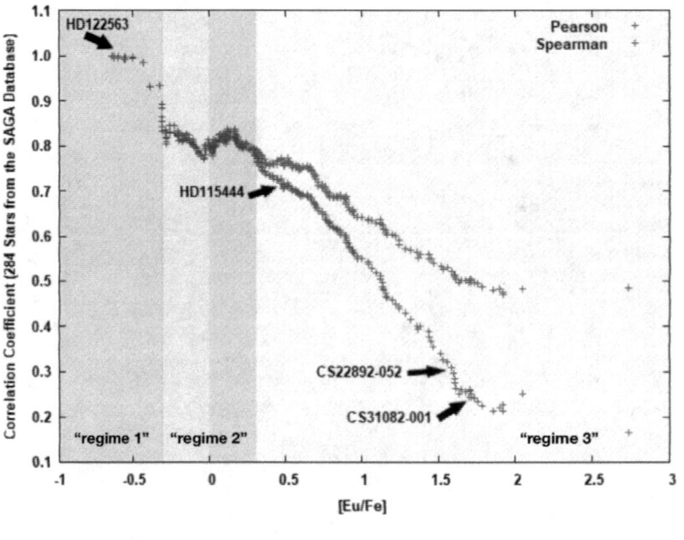

Figure 6. Shown are the Pearson and Spearman correlation coefficients for low-metallicity stars ([Fe/H <-2.5) from the JINAbase database [29], considering stars in an [Eu/Fe] interval with its upper limit displayed on the x-axis. We see for limited-r stars of regime 1 and 2 a high correlation, indicating Fe and Eu co-production, while for r-enriched stars of regime 3 the correlations vanish, arguing for separate nucleosynthesis sources for the Fe and Eu found in these stars. Four typical stars (HD122563, limited-r), HD115444 (r-enriched, r-I), CS22892-052 (r-enriched, r-II), and CS31082-001 (r-enriched, r-II with actinide boost) are displayed as well.

When analyzing data sets of (x,y) points, the Pearson correlation coefficient measures a possible linear relationship (constant ratio) between their values. In such a case the value of the correlation is either +1 or -1, depending on whether a linear fit to this relation has a positive or negative slope. The correlation measures also a possible noisiness, the extreme case of a linear fit with a huge χ^2, i.e. which does not fit the data at all, stands for a correlation value of 0. The Spearman correlation coefficient measures correlations in a slightly «milder» way, based on ranks rather than a linear relation. When both methods deviate, this indicates the start of a different correlation pattern [for more information see 30, 31].

When applying this method to correlations between Eu and Fe we find the following results (Fig.6) for the Pearson and Spearman correlation coefficients. Among the limited-r stars two types of contributions seem to dominate, we introduced them as categories CI and CII. For both related regimes (1 and 2) we find a high correlation between Fe and Eu, indicating the co-production of both elements in the responsible events. In regime 3 (r-enriched stars) a vanishing correlation is strongly supporting sites which produce heavy r-process elements like Eu uncorrelated (i.e. not co-produced) with Fe. We identify these sites with category CIII events of two types (CIIIa and CIIIb), as indicated from the different strength of r-process production without or with an actinide boost (see Fig.7).

3 Identifying Astrophysical Sites, their Ejecta Predictions, and Imprint in Low-Metallicity Stars

If concluding from our statistical analysis that C0 stars correspond to regular CCSNe, CI and CII stars to quark deconfinement supernovae and magneto-rotational supernovae, that are mainly responsible for the weak r-process in limited-r stars of regime 1 and 2, this suggestion should be tested by comparing the theoretical element predictions for these events with observations of regime 1 and 2 stars. In Table 1 we show this comparison by utilizing the predictions for quark deconfinement (QD) SNe [32] as well as for two sets of magneto-rotational (MR) SNe 1 and 2 [33, 34]. QD and MR SNe can reproduce the regime 1 and 2 Sr/Eu observations well, but in order to reproduce also the observed Sr/Fe and Eu/Fe values, an additional contribution of n x 0.1 $M_\odot$ of Fe with about n=200-300 regular CCSNe is required per QD SN while n=10 regular CCSNe per MR SN are needed (comparable with the magnetar fraction of neutron stars). We do not show results from electron-capture supernovae [35], because their Sr/Eu ratios are extremely high and their actual existence has been questioned recently [36].

Table 1. Candidates for category CI and CII events

Element ratio	Regime 1	Regime 2	MRS1	MRS2	QD
Sr/Eu	200-750	200-800	438-1300	340	142-468
Sr/Fe	$< 2 \times 10^{-6}$	$< 2.5 \times 10^{-5}$	$0.7 - 5 \times 10^{-2}$	2.6×10^{-2}	$(5-6) \times 10^{-3}$
Eu/Fe	2×10^{-8}	10^{-7}	$1.5 - 3.6 \times 10^{-5}$	7.5×10^{-5}	$(1-4) \times 10^{-5}$
Fe [$M_\odot$]			3×10^{-2}	2×10^{-2}	5.5×10^{-2}

The rank tests in the analysis section have shown that the strong r-process events of category CIII fall into two subcategories without and with an actinide boost. The question is which physical conditions lead to such an actinide boost.

There have been many investigations relating this to fission. Fission cycling affects the Th/Eu ratio and one can find different productions of Eu, U, and Th for different Y_e conditions in the r-process. Fig.7 shows the results of r-process abundance predictions [38, 39] as a function

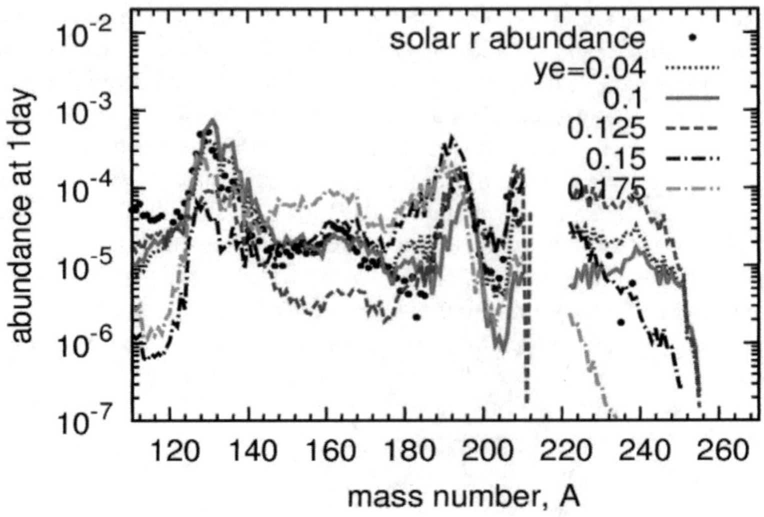

Figure 7. Utilizing the DZ (Duflo-Zucker) mass model [37] one sees large variations of actinide production as a function of Y_e in the ejecta. The highest actinide production takes place for a range of 0.125 to 0.15, decreasing strongly for higher as well as lower $Y_e's$ [from 38, 39].

of Y_e in the ejecta. In order to explain actinide boosts it requires Y_e-superpositions in an event which include a sufficient amount of Y_e=0.15 to 0.125 matter. Results similar to Fig.7 with high Th/Eu values have been obtained in other simulations [40, 41], consistent with Y_e-values for the ejecta of BH accretion tori [42, 43].
Actinide-boost stars with Th/Eu supersolar (>0.42, [44]) exist mostly among r-II stars. This answers the question how we can relate the observed behavior to category CIIIa and CIIIb events by identifying them with NS-mergers and BH accretion tori outflows, e.g. collapsars. This is consistent with the separate rank tests for Th in r-I and r-II stars of Fig.4. Independent of the actinide boost question we test in Table 2 the predicted Eu production for these events (about $10^{-6}M_\odot$ vs. $10^{-4}M_\odot$, when utilizing the results from [43, 45]), combined with about $0.5M_\odot$ of Fe for the latter collapsar case. We find that in order to reproduce the observed Eu/Fe values in r-I stars (by NS mergers) and r-II stars (by collapsars) requires in both cases the addition of about n x $0.1M_\odot$ of Fe with n≈120, i.e. category CIII events are in the 8 per mil range in relation to CCSNe [consistent with 46].

Table 2. Candidates for category CIIIa and CIIIb events for complete r-process stars of regime 3

Element ratio	average r-I	average r-II	NS mergers (with Fe from 1 SN)	collapsars
Eu/Fe	3×10^{-7}	3×10^{-6}	3.7×10^{-5}	7.4×10^{-5}

This section followed the initial statistical analysis of the abundance pattern of low-metallicity stars, that came to the conclusion that there exist at least three types of sites for the limited-r, r-I, and r-II stars. We actually found a possible division among limited-r stars and suggested four categories: I for QD supernovae (a hint from Fig.6 in regime 1, although not clearly proven), II for magneto-rotational supernovae for explaining regime 2, IIIa for neutron star mergers, and IIIb for collapsars/hypernovae for explaining regime 3. For those four categories there exist abundance patterns (and ejecta mass) predictions from theoretical models. Utilizing these predictions we asked the question how the observed Sr/Eu, Sr/Fe, and Eu/Fe could be reproduced among limited-r, r-I, and r-II stars. The result was that (with a slightly unclear option for QD-supernovae) limited-r stars could be reproduced when magneto-rotational supernovae would occur with a frequency of 1/10 (n=10 in Table 1) of regular supernovae (which is consistent with the fraction of magnetars among neutron stars), and r-I and r-II stars could be reproduced with neutron star merger and collapsars if the occur with a frequence of a few permille of core-collapse supernovae (n≈120 in Table 2). For neutron star mergers this is consistent with the frequency based on LIGO and kilonova observations. For collapsars this makes sense but the models are not sufficiently advanced, yet, in order for a full proof.

4 Summary

Is the large scatter in r-process abundances with respect to those of other elements (e.g. Eu vs. Fe) at low metallicities only the result of inhomogeneous admixture from rare events, or do the also observed variations in r-process abundance patterns point to different origins? Such questions have been addressed in many galactic chemical evolution studies [e.g. 8, 9, 11, 12]. There exist attempts to explain the variation in abundance patterns all by neutron star merger events with varying neutron star masses and also mass symmetries/asymmetries in the binary systems [47], including systems with fast BH formation which are part of our category CIIIb. With the correlation / no correlation tests of e.g. Eu with Fe, we suggest different sources resulting either in a weak r-process with Fe co-production or a strong r-process with no or negligible Fe co-production. The effect that the r-process (or Eu) production in these events (magneto-rotational supernovae, NS mergers, collapsars) also increases among these categories by about a factor of 10 for each, permits an explanation of the limited-r, r-I, and r-II patterns, combined with the inhomogeneous evolution in the early Galaxy.

We can formulate a consistent picture, utilizing existing abundance pattern predictions for these events, related to r-process ejecta masses and event frequencies, which are in accordance with category CI and CII events being less frequent than regular core-collapse supernovae by about a factor of 10, and category CIIIa and CIIIb events being less frequent by a factor of 100 to 1000. This seems consistent with the first Fe from core-collapse supernovae showing up at metallicities as low as [Fe/H]$\lesssim$-5, the first input from category CI and CII events at about [Fe/H]≈-4, and the first input from category CIIIa and CIIIb events at about [Fe/H]≈-3. This implies that stars with metallicities less than [Fe/H]≈-2.5 do not necessarily show the imprint of only one nucleosynthesis event, but possibly a superposition. This indicates that most of the Fe is going back to frequent core-collapse supernovae, opposite to the majority of r-process elements, originating from a variety of rarer events. These conclusions are still somewhat speculative, they should be tested against further observational and modeling progress. More details of our present study can be found in ref. [48].

Acknowledgement. The investigations of this paper would not have been possible without the publicly available observational SAGA and JINA data bases [16, 29], and we want to express our thanks to the authors of these tremendous research resources. Furthermore, the COST actions ChETEC (Chemical Elements as Tracers of the Evolution of the Cosmos, CA16117), GWvese (Gravitational waves, black holes and fundamental physics, CA16104), and Pharos (The multi-messenger physics and astrophysics of neutron stars, CA16214) provided an inspiring atmosphere for thoughts along the lines discussed here. SR has been supported by the Swedish Research Council (VR) under grant numbers 2016-03657_3 and 2020-05044, by the Swedish National Space Board under grant number Dnr. 107/16, the research environment grant "Gravitational Radiation and Electromagnetic Astrophysical Transients (GREAT)" funded by the Swedish Research council (VR) under Dnr 2016-06012 and by the Knut and Alice Wallenberg Foundation (KAW 2019.0112).

References

[1] J.M. Lattimer, D.N. Schramm, Astrophys. J. **192**, L145 (1974)
[2] R.D. Hoffman, S.E. Woosley, Y.Z. Qian, Astrophys. J. **482**, 951 (1997)
[3] C. Freiburghaus, J.F. Rembges, T. Rauscher, E. Kolbe, F.K. Thielemann, K.L. Kratz, B. Pfeiffer, J.J. Cowan, Astrophys. J. **516**, 381 (1999)
[4] C. Freiburghaus, S. Rosswog, F.K. Thielemann, Astrophys. J. **525**, L121 (1999)
[5] K. Farouqi, K. Kratz, B. Pfeiffer, T. Rauscher, F. Thielemann, J.W. Truran, Astrophys. J. **712**, 1359 (2010)
[6] J.J. Cowan, C. Sneden, J.E. Lawler, A. Aprahamian, M. Wiescher, K. Langanke, G. Martinez-Pinedo, F.K. Thielemann, Rev. Mod. Phys. **93**, 015002 (2021)
[7] F. Matteucci, D. Romano, A. Arcones, O. Korobkin, S. Rosswog, Mon. Not. Roy. Astron. Soc. **438**, 2177 (2014)
[8] G. Cescutti, D. Romano, F. Matteucci, C. Chiappini, R. Hirschi, Astron. & Astrophys. **577**, A139 (2015)
[9] B. Wehmeyer, M. Pignatari, F.K. Thielemann, Mon. Not. Roy. Astron. Soc. **452**, 1970 (2015)
[10] S.G. Ryan, J.E. Norris, T.C. Beers, Astrophys. J. **471**, 254 (1996)
[11] C. Kobayashi, A.I. Karakas, M. Lugaro, Astrophys. J. **900**, 179 (2020)
[12] F. van de Voort, R. Pakmor, R.J.J. Grand, V. Springel, F.A. Gómez, F. Marinacci, Mon. Not. Roy. Astron. Soc. **494**, 4867 (2020)
[13] D.L. Burris, C.A. Pilachowski, T.E. Armandroff, C. Sneden, J.J. Cowan, H. Roe, Astrophys. J. **544**, 302 (2000)
[14] R. Cayrel, V. Hill, T.C. Beers, B. Barbuy, M. Spite, F. Spite, B. Plez, J. Andersen, P. Bonifacio, P. François et al., Nature **409**, 691 (2001)
[15] S. Honda, W. Aoki, Y. Ishimaru, S. Wanajo, S.G. Ryan, Astrophys. J. **643**, 1180 (2006)
[16] T. Suda, Y. Katsuta, S. Yamada, T. Suwa, C. Ishizuka, Y. Komiya, K. Sorai, M. Aikawa, M.Y. Fujimoto, Publ. Astron. Soc. Japan **60**, 1159 (2008)
[17] H. Schatz, R. Toenjes, B. Pfeiffer, T.C. Beers, J.J. Cowan, V. Hill, K.L. Kratz, Astrophys. J. **579**, 626 (2002)
[18] I.U. Roederer, K.L. Kratz, A. Frebel, N. Christlieb, B. Pfeiffer, J.J. Cowan, C. Sneden, Astrophys. J. **698**, 1963 (2009)
[19] I.U. Roederer, C. Sneden, J.E. Lawler, J.J. Cowan, Astrophys. J. **714**, L123 (2010)
[20] L. Mashonkina, N. Christlieb, K. Eriksson, Astron. & Astrophys. **569**, A43 (2014)
[21] I.U. Roederer, Astrophys. J. **835**, 23 (2017)
[22] E.M. Holmbeck, T.C. Beers, I.U. Roederer, V.M. Placco, T.T. Hansen, C.M. Sakari, C. Sneden, C. Liu, Y.S. Lee, J.J. Cowan et al., Astrophys. J. **859**, L24 (2018)
[23] T.T. Hansen, E.M. Holmbeck, T.C. Beers, V.M. Placco, I.U. Roederer, A. Frebel, C.M. Sakari, J.D. Simon, I.B. Thompson, Astrophys. J. **858**, 92 (2018)
[24] C.M. Sakari et al., Astrophys. J. **868**, 110 (2018)
[25] E.M. Holmbeck, T.T. Hansen, T.C. Beers, V.M. Placco, D.D. Whitten, K.C. Rasmussen, I.U. Roederer, R. Ezzeddine, C.M. Sakari, A. Frebel et al., Astrophys. J. Suppl. **249**, 30 (2020)
[26] R. Ezzeddine, K. Rasmussen, A. Frebel, A. Chiti, K. Hinojisa, V.M. Placco, A.P. Ji, T.C. Beers, T.T. Hansen, I.U. Roederer et al., Astrophys. J. **898**, 150 (2020)
[27] D. Yong, C. Kobayashi, G.S. Da Costa, M.S. Bessell, A. Chiti, A. Frebel, K. Lind, A.D. Mackey, T. Nordlander, M. Asplund et al., Nature **595**, 223 (2021), 2107.03010

[28] B.S. Everitt, S. Landau, M. Leese, D. Stahl, *Cluster Analysis, 5th Edition*, Wiley Series in Probability and Statistics (Wiley, New York, 2011), ISBN 978-0-470-97844-3
[29] A. Abohalima, A. Frebel, Astrophys. J. Suppl. **238**, 36 (2018)
[30] A. Tamhane, D. Dunlop, *Statistics and Data Analysis* (Prentice-Hall, Upper Saddle River, 2000)
[31] D. Spiegelhalter, *The art of statistics* (Penguin Random House, Milton Keynes, 2019)
[32] T. Fischer, M.R. Wu, B. Wehmeyer, N.U.F. Bastian, G. Martínez-Pinedo, F.K. Thielemann, Astrophys. J. **894**, 9 (2020)
[33] N. Nishimura, H. Sawai, T. Takiwaki, S. Yamada, F.K. Thielemann, Astrophys. J. **836**, L21 (2017)
[34] M. Reichert, M. Obergaulinger, M. Eichler, M.Á. Aloy, A. Arcones, Mon. Not. Roy. Astron. Soc. **501**, 5733 (2021)
[35] S. Wanajo, H.T. Janka, B. Müller, Astrophys. J. **726**, L15 (2011)
[36] O.S. Kirsebom et al., Phys. Rev. Lett. **123**, 262701 (2019)
[37] J. Duflo, A.P. Zuker, Phys. Rev. C **52**, R23 (1995)
[38] F.K. Thielemann, B. Wehmeyer, M.R. Wu, *r-Process Sites, their Ejecta Composition, and their Imprint in Galactic Chemical Evolution*, in *Journal of Physics Conference Series* (2020), Vol. 1668 of *Journal of Physics Conference Series*, p. 012044
[39] M.R. Wu et al. (2017), unpublished
[40] E.M. Holmbeck, T.M. Sprouse, M.R. Mumpower, N. Vassh, R. Surman, T.C. Beers, T. Kawano, Astrophys. J. **870**, 23 (2019)
[41] M. Eichler, W. Sayar, A. Arcones, T. Rauscher, Astrophys. J. **879**, 47 (2019)
[42] R. Fernández, A. Tchekhovskoy, E. Quataert, F. Foucart, D. Kasen, Mon. Not. Roy. Astron. Soc. **482**, 3373 (2019)
[43] D.M. Siegel, J. Barnes, B.D. Metzger, Nature **569**, 241 (2019)
[44] K. Lodders, Space Sci. Rev. **217**, 44 (2021)
[45] B. Côté, K. Belczynski, C.L. Fryer, C. Ritter, A. Paul, B. Wehmeyer, B.W. O'Shea, Astrophys. J. **836**, 230 (2017)
[46] S. Rosswog, U. Feindt, O. Korobkin, M.R. Wu, J. Sollerman, A. Goobar, G. Martinez-Pinedo, Class. Quantum Gravity **34**, 104001 (2017)
[47] E.M. Holmbeck, A. Frebel, G.C. McLaughlin, R. Surman, R. Fernández, B.D. Metzger, M.R. Mumpower, T.M. Sprouse, Astrophys. J. **909**, 21 (2021)
[48] K. Farouqi, F.K. Thielemann, S. Rosswog, K.L. Kratz, arXiv e-prints arXiv:2107.03486 (2021)

EPJ Web of Conferences **260**, 09003 (2022) https://doi.org/10.1051/epjconf/202226009003
NIC-XVI

Light Element Abundances Constrain Primordial Black Holes in the Critical Collapse Model

Chao Chen[1,*], *Yudong* Luo[2,3,**], *Motohiko* Kusakabe[4,***], and *Toshitaka* Kajino[2,3,4,****]

[1]Jockey Club Institute for Advanced Study, The Hong Kong University of Science and Technology, Hong Kong S.A.R., China
[2]Graduate School of Science, The University of Tokyo, 7-3-1 Hongo, Bunkyo-ku, Tokyo 113-0033, Japan
[3]National Astronomical Observatory of Japan, 2-21-1 Osawa, Mitaka, Tokyo 181-8588, Japan
[4]School of Physics, International Research Center for Big-Bang Cosmology and Element Genesis, Beihang University, 37, Xueyuan Rd., Haidian-qu, Beijing 100083 China

Abstract. We study the photodisintegration process triggered by the nonthermal electromagnetic Hawking radiation from primordial black holes (PBHs) in the critical collapse model. The presence of a low-mass tail of critical collapse mass function could enhance energetic photon emissions from Hawking radiation of PBHs. Nuclear photodisintegration rates are calculated with a nonthermal photon spectrum derived by solving the Boltzmann equation iteratively. With the newest observational limit on the ^{3}He abundance in Galactic H II regions, the updated ^{3}He constraints on PBH mass spectrum in the horizon mass range $10^{12} - 10^{13}$ g are derived. Our results show that ^{3}He constraints on the critical mass function are about one order of magnitude severer than the monochromatic one, although the fraction of PBHs in the low-mass tail region is relatively small.

1 Introduction

Primordial black holes (PBHs) may be formed from density fluctuations in the very early Universe and have been studied over more than 50 years [1, 2]. PBHs could be small enough for Hawking radiation to be observationally significant [3]. It has been well understood that the mass distribution of PBHs is extended in many scenarios [4]. Even for PBHs forming at a single epoch, they would have an extended mass distribution when critical phenomena of gravitational collapse are taken into account [5]. We show in this paper that the small amounts of PBHs within the low-mass tail in the mass function of the critical collapse model would affect the elemental abundances. More stringent BBN constraints are then acquired than those on the monochromatic one due to the simple fact that the Hawking radiation in the high energy region is stronger for low-mass PBHs. This leads to a caution that other constraints for the critical collapse model may also be altered compared to the monochromatic case.

*e-mail: iascchao@ust.hk
**e-mail: yudong.luo@nao.ac.jp
***e-mail: kusakabe@buaa.edu.cn
****e-mail: kajino@buaa.edu.cn

2 Critical Collapse Mass Function and Nonthermal Nucleosynthesis via Hawking Radiation

It is well known that PBH formation is associated with the critical phenomena of gravitational collapse, and the early studies showed that the PBH mass at the formation epoch has a universal scaling property [6]

$$M = KM_H(\delta - \delta_c)^{\gamma}, \tag{1}$$

near the threshold $\delta \simeq \delta_c$ for PBH formation. The rescaling relation between PBH masses M and density contrast δ implies that the PBH masses are naturally extended due to the Gaussian distribution of δ, the differential comoving number density is given by

$$n(M) = \frac{1}{\beta} n_{\rm PBH}(t_f) P[\delta(M)] \frac{d\delta(M)}{dM},$$

for each horizon mass M_H, β (or the redefined β' in the following part) is the initial PBH abundances. Considering PBHs with an extended mass function, the total Hawking emission should be contributed by all individual mass scales. Figure 1 plots the instantaneous emission

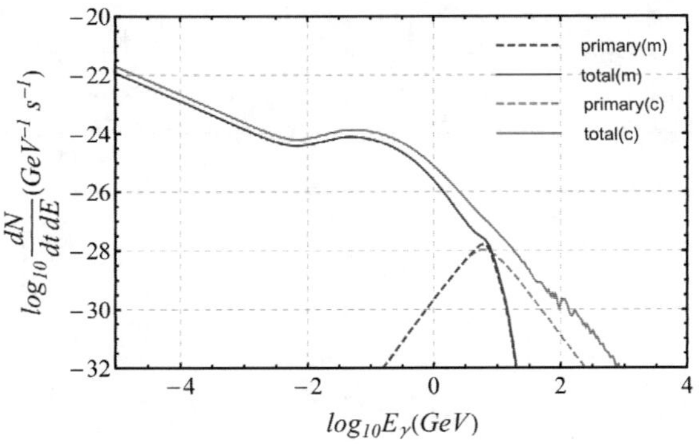

Figure 1. Instantaneous emission rate of photons for PBHs with horizon mass 10^{13} g, for the same energy fraction of the monochromatic (denoted by the script "m", the blue curve) and critical collapse mass functions (denoted by the script "c", the red curve).

rate of photons for PBHs with the horizon mass 10^{13} g for monochromatic (blue) and critical collapse mass functions (red) at the formation epoch, in which we set the total number density of PBHs with these two types of mass functions to be equal.

In this study, we mainly focus on PBHs within horizon mass $10^{12}\text{g} \leq M_{\rm H} \leq 10^{13}\text{g}$ and their lifetime $10^{8}\text{s} \leq \tau \leq 10^{13}\text{s}$. For PBHs within this range, proton and neutron emissions contribute to < 20% of the final production [12] so that the hadrodissociation process could not play a dominant role, the most important process is the photodisintegration triggered by non-thermal photons. The injections of energetic photons via Hawking radiation would initiate the electromagnetic (EM) cascade showers which are mainly made of energetic photons and pairs of electron and positron. At different cosmic temperature T, the photons would form a quasi-static equilibrium spectrum $f_\gamma(E_\gamma; T)$ very quickly, and $f_\gamma(E_\gamma; T)$ is given by the solution of the steady Boltzmann equation:

$$\begin{aligned}\frac{\partial f_\gamma(E_\gamma;T)}{\partial t} =& \frac{\partial}{\partial E}\left[b_{exp}(E;T) f_\gamma(E_\gamma;T)\right] \\ &- \Gamma_\gamma(E_\gamma;T) f_\gamma(E_\gamma;T) + S(E_\gamma;T) \equiv 0. \end{aligned} \tag{2}$$

Here, $S(E_\gamma; T)$ is the source term that depends on the energies of the Hawking-radiated photons at a cosmic temperature T, i.e., the time evolution of Hawking spectrum in terms of T. $b_{exp}(E; T)$ is the photon energy loss rate via the cosmic expansion, and $\Gamma_\gamma(E_\gamma; T)$ describes the degradation rate of zeroth generation nonthermal photons, respectively. The photodisintegration of nuclei can be triggered by photons with energies larger than the photodisintegration energy threshold. The time evolution of the nuclear abundances is governed by

$$\frac{dY_A}{dt} = \sum_T N_{AC}[T\gamma]_A Y_T(T) - \sum_P [A\gamma]_P Y_A(T), \quad (3)$$

where $[T\gamma]_A$ and $[A\gamma]_P$ are producing and destroying reaction rates for nuclei A of reactions $\gamma + T \to A$ and $\gamma + A \to P$, respectively. N_{AC} is the number of the identical nuclear species in the final state. Secondary nonthermal reactions also occur if products of the primary photodisintegration reactions are energetic enough.

3 Light Element Abundance Constraints on PBHs

We use a nonthermal BBN nuclear reaction network code and have updated the reaction rates of nuclei with mass numbers $A \leq 10$ using the JINA REACLIB Database [8]. For the nonthermal photodisintegration reaction rates, we take into account both the primary and the secondary photodisintegration reactions. We run the nonthermal BBN code to put constraints

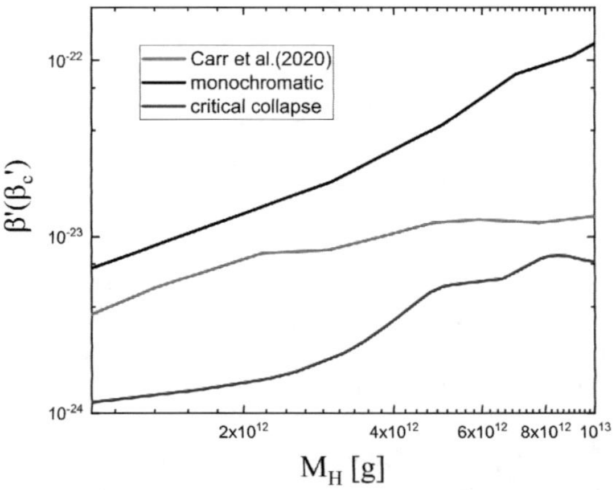

Figure 2. The upper bounds on $\beta'(M_H)$ (the black curve) and $\beta'_c(M_H)$ (the blue curve) in the horizon mass range $M_H = 10^{12} - 10^{13}$ g for the monochromatic and critical collapse mass functions, respectively, from the ^{3}He/H ratio in the present Galaxy. The red curve refers to the constraint from Carr et al. [9].

on the initial mass spectrum for PBHs within the horizon mass range $10^{12} - 10^{13}$ g. By using the ^{3}He abundance, we plot the upper limits of $\beta'(M_H)$ and $\beta'_c(M_H)$ for the monochromatic and critical collapse mass functions in figure 2, respectively. The black curve shows our updated constraints on β' for the monochromatic mass function of PBHs within mass range $10^{12} - 10^{13}$ g by using the observational limit ^{3}He/H $< 1.5 \times 10^{-5}$. Compared with the constraint in Ref. [9] (the red curve), there exits one order of magnitude difference between our constraints and theirs. Considering that they used the more conservative constraint ^{3}He/D<1.37, the constraint derived in this study is much weaker. The hadrodissociation is ignored in this work, however, for the realistic PBH evaporation for this mass range, the branching ratio of proton energy normalized to the total available mass of PBH is less than 20%. Therefore, even considering the hadrodissociation process in our calculation, our result could only move down utmost by half an order of magnitude, which still cannot explain the present inconsistency. The analytical expressions of the ^{3}He constraints on β'including both 100% photodisintegration and 100% hadrodissociation, are performed in Ref. [13], the estimated bounds should be the conceivably strongest, our result is consistent with the analytical estimations. We also see that the constraint on $\beta'_c(M_H)$ (blue curve) is lower than that of the monochromatic one due to the enhancement of high-energy photons from the low-mass tail.

4 Conclusion

This work studies the nuclear photodisintegration reactions triggered by nonthermal electromagnetic Hawking radiation from PBHs in the critical collapse model, with the initial horizon masses of $10^{12} - 10^{13}$ g. As a consequence of Hawking radiation, the low-mass tail of the critical collapse mass function would enhance high-energy photon emission compared with the monochromatic mass function. We find out in this work that the high-energy photons radiated from the low-mass tail of the critical collapse mass function could significantly affect the primordial ^{3}He abundances via nonthermal BBN processes. We update the constraint on the initial mass spectrum β' of PBHs in the monochromatic and critical collapse model, respectively. Our result indicates that the tail contribution of the critical collapse model should be referred to with caution when considering the constraints on PBH evaporation.

Y.L. is supported by JSPS KAKENHI Grant No. 19J22167. M. K. is supported by NSFC Research Fund for International Young Scientists (11850410441). T. K. is supported in part by Grants-in-Aid for Scientific Research of JSPS (17K05457 and 20K03958).

References

[1] Y. B. Zel'dovich and I. D. Novikov, Sov. Astron. **10**, 602 (1967).
[2] S. Hawking, Mon. Not. Roy. Astron. Soc. **152**, 75 (1971).
[3] S. W. Hawking, Nature **248**, 30 (1974).
[4] B. Carr, F. Kuhnel and M. Sandstad, Phys. Rev. D **94**, no. 8, 083504 (2016).
[5] M. W. Choptuik, Phys. Rev. Lett. **70**, 9-12 (1993).
[6] J. C. Niemeyer and K. Jedamzik, Phys. Rev. Lett. **80**, 5481-5484 (1998).
[7] M. Kawasaki, K. Kohri and T. Moroi, Phys. Rev. D **63**, 103502 (2001).
[8] P. Descouvemont, A. Adahchour, C. Angulo, A. Coc and E. Vangioni-Flam, Atom. Data Nucl. Data Tabl. **88**, 203-236 (2004).
[9] B. Carr, K. Kohri, Y. Sendouda and J. Yokoyama, [arXiv:2002.12778 [astro-ph.CO]].
[10] M. Kusakabe, A. B. Balantekin, T. Kajino and Y. Pehlivan, Phys. Rev. D **87**, no.8, 085045 (2013).
[11] H. Ishida, M. Kusakabe and H. Okada, Phys. Rev. D **90**, no.8, 083519 (2014).
[12] J. H. MacGibbon, Phys. Rev. D **44**, 376-392 (1991).
[13] Y. Luo, C. Chen, M. Kusakabe and T. Kajino, J. Cosmol. Astropart. Phys. **05**, 042 (2021).

Gamma-ray observations of cosmic nuclei

Roland Diehl[1,2,*]

[1]Max Planck Institut für extraterrestrische Physik, D-85748 Garching, Germany
[2]Excellence Cluster 'Origins' and Technical University of Munich, D-85748 Garching, Germany

Abstract. Gamma rays from nuclear processes such as radioactive decay and de-excitations are among the most-direct tools to witness the production and existence of specific nuclei and isotopes in and near cosmic nucleosynthesis sites. With space-borne instrumentation such as NuSTAR and SPI/INTEGRAL, and experimental techniques to handle a substantial instrumental background from cosmic-ray activations of the spacecraft and instrument, unique results have been obtained, from diffuse emissions of nuclei and positrons in interstellar surroundings of sources, as well as from observations of cosmic explosions and their radioactive afterglows. These witness non-sphericity in supernova explosions and a flow of nucleosynthesis ejecta through superbubbles as common source environments. Next-generation experiments that are awaiting space missions promise a next level of observational nuclear astrophysics.

1 Introduction

Observations of gamma rays from nuclear processes are one of the most-direct tools to witness the production and existence of specific nuclei and isotopes in and near cosmic nucleosynthesis sites. Processes that cause characteristic gamma-ray lines from specific nuclides are radioactive decay and energetic collisions, both resulting in de-excitations of specific (daughter-) isotopes. Observations in the gamma-ray band require space-borne instrumentation, because the atmosphere of the Earth presents a thick target to gamma rays and their interactions with nuclei and electrons[1] In this paper, we will briefly address the required instrumentation and analysis methods, and then review what has been learned from diffuse emissions of nuclei in interstellar surroundings of sources, as well as from observations of cosmic explosions and their radioactive afterglows.

2 Telescopes for nuclear line emissions

Space-borne instrumentation are a technical challenge, as they must be built to survive their launch into space, and operations in a harsh environment characterised by high irradiation from cosmic rays and extreme temperature gradients between sun-exposed components and

*e-mail: rod@mpe.mpg.de

[1]This is exploited for ground based gamma-ray astronomy at TeV energies and above, using the atmosphere as a detector, though the electromagnetic shower of particles and photons that is created by high-energy gamma-rays interacting in the atmosphere.

those in the shadow and cold space. Experimental techniques are required to handle a substantial instrumental background in the measured detector events, caused from cosmic-ray activations of the spacecraft and instrument (see [1] for a detailed example).

Successful pioneering sky imaging and spectroscopy surveys with Compton-CGRO [2] and INTEGRAL [3], respectively, have shown the potential of this astronomical window. Currently, only the INTEGRAL and NuSTAR[4] missions are in operation for measurements of nuclear lines from cosmic sources. Both these missions are in their late, extended, phases, and may be terminated soon. No future mission for nuclear astronomy is currently planned by any of the international space agencies [see however 5–7, for recent project proposals and prototypes].

3 Supernova explosions

3.1 Thermonuclear supernovae

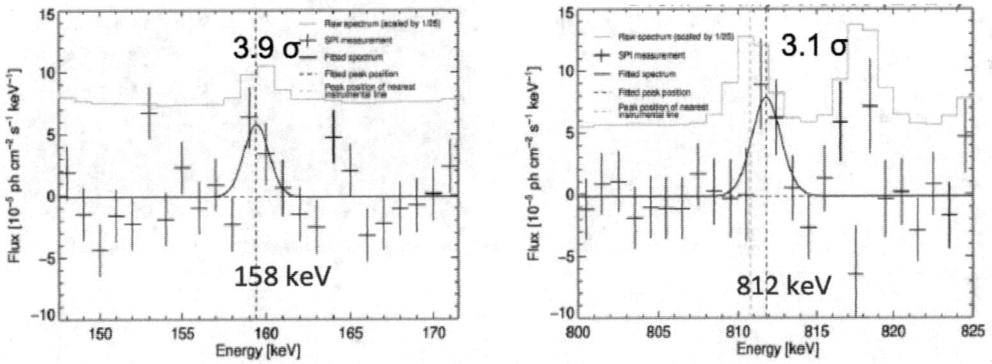

Figure 1. The early detection of decay lines from ^{56}Ni after 20 days past explosion signify appearance of ^{56}Ni at surprisingly-low optical depth [14]. Although the significance is at the 4σ level only, this signifies a deviation from the standard assumption of ^{56}Ni being at the center of a nearly-spherical explosion. The instrumental background spectrum is indicated as grey histogram, scaled down to fit; the nearest instrumental line positions are indicated by dashed lines (green), while the fitted line centers of the SN2014J signal are also indicated (dashed black lines).

Fourty years after their prediction from supernova models [e.g. 8], finally the characteristic ^{56}Co gamma-ray lines from the decay chain of ^{56}Ni could be measured with the INTEGRAL observatory [9] and in particular its gamma-ray spectrometer SPI [10, 11]. Supernova SN2014J was close enough to provide sufficiently-intense gamma-ray flux for this observational proof of a key ingredient of modeling supernova light from thermonuclear supernovae, from ^{56}Co radioactivity with its characteristic 11 day decay time. The amount of ^{56}Ni inferred from the gamma-ray flux of 0.49 ± 0.09 $M_\odot$ [10] is in agreement with the amount inferred from the optical brightness of the supernova, as based on the empirical peak-brightness/^{56}Ni heating-rate relation called *Arnett's law* [12]. This is reassuring.

On the other hand, there were indications in the time series of gamma-ray spectra from the ^{56}Co decay lines that individual clumps of ^{56}Co decay at different bulk velocities appeared at different times, signifying substantial deviations from spherical symmetry of explosion or ^{56}Co distribution [see 10, for more detail on the ^{56}Co gamma-ray signal]. Even more clear and spectacular was the detection of ^{56}Ni decay lines early-on (Fig. 1): With a radioactive lifetime of about 9 days, it was believed that ^{56}Ni would always be embedded so deeply within the supernova's core that even gamma-rays could not leak out before ^{56}Ni was converted to ^{56}Co already. Only some *He cap* models included the possibility of early gamma-ray emission from ^{56}Ni decay [13], as helium deposition on the surface of the white dwarf could cause a helium surface explosion triggering the thermonuclear supernova. Thus, this discovery of early ^{56}Ni gamma-ray lines was discussed as a support for such a double detonation [see 14, for more detail], and double-degenerate progenitor models in general.

3.2 Core-collapse supernovae

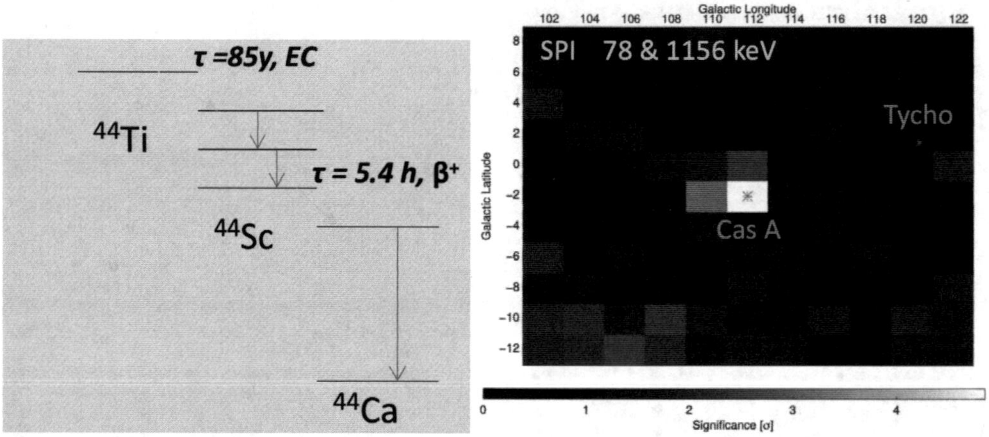

Figure 2. *Left:* The signal of ^{44}Ti decay results in lines at 68, 78, and 1156 keV. *Right:* The SPI ^{44}Ti line signal is clearly attributed to the Cas A source (right). [From 15].

SN1987A still is the core-collapse supernova with best observational details of the aftermath of such an explosion of a massive star after its gravitational collapse. However, the Cas A supernova remnant comes close in that impressive observational detail of the supernova remnant at its young age of about 350 years allows for interpretations and tests of our supernova explosion models in great detail. At this age, the energy input to the supernova remnant from radioactive decay of ^{44}Ti is most important. With its decay time of 89 year, ^{44}Ti decay (see Figure 2 left) provides lines at 68 and 78 keV from the first decay stage, i.e. de-excitations of ^{44}Sc, and another line at 1156 keV from the ^{44}Ca daughter isotope, which had led to its discovery with the COMPTEL instrument [16]. NuSTAR with its imaging hard X-ray telescope can observe the two low-energy lines from ^{44}Ti decay, and the image derived from these observations has been an illuminating lesson for us [17, 18]. The image shows several clumps of ^{44}Ti emission in the central region of the remnant, and imaging-resolved spectroscopy demonstrates that among these clumps different bulk velocities can be discriminated. But in spite of its technological advantages with hard-X-ray optics, also NuSTAR detector signals suffer from a large background, from a diffuse cosmic X-ray background and from cosmic-rays that activate spacecraft materials and lead to a large instrumental background. Independent measurements with other instruments are important to address potential systematics in such data analyses, as they are dependent on models for instrumental backgrounds and response behaviour. INTEGRAL/SPI conducted a deep observational campaign to provide high-quality data for Cas A, accumulated now to 11.2 Ms of exposure.

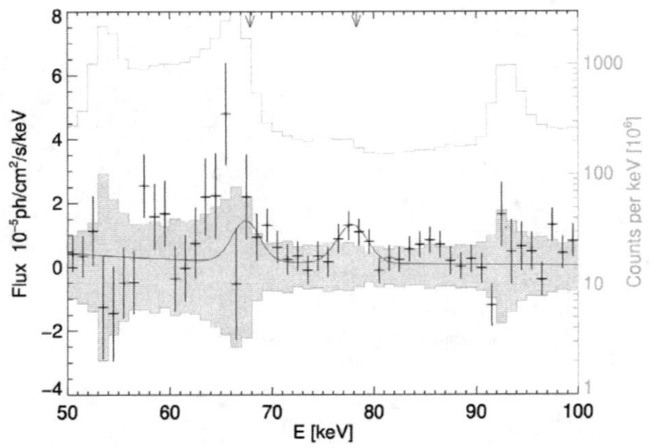

Figure 3. The signal of lines from ^{44}Sc de-excitation in Cas A is shown here, together with uncertainties remaining from the instrumental background model (redish hatched area). The instrumental background shape is shown as a histogram in grey. [From 15].

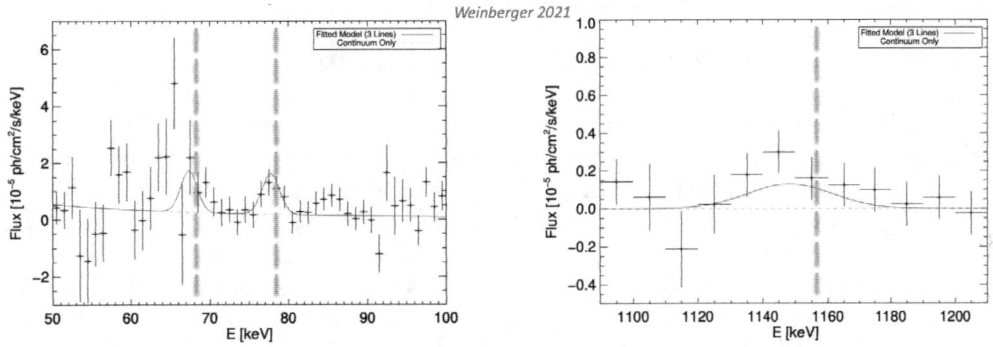

Figure 4. The signal in all lines from ^{44}Ti in Cas A show a bulk redshift (dashed lines mark the laboratory values of the respective line energies. [From 15].

Figure 2 (right) shows from this analysis [19] that the ^{44}Ti signal in SPI spectra can safely be attributed to the Cas A source, rather than diffuse emission or another nearby source. Figure 3 shows the spectrum around the low-energy lines measured also by NuSTAR, indicating the uncertainties incurred by the instrumental background. In Figure 4, we show the latest spectra for the two lines with high-quality signals, i.e., the 78 keV and 1156 keV lines. In both spectra, it is evident that the fitted line energy is offset towards lower energies, which is interpreted as Doppler shift from net bulk motion. The effect is more obvious in teh 1156 keV line, as Doppler shifts scale with photon energy. These spectra independently thus confirm the NuSTAR finding of predominant motion of ^{44}Ti away from us, resulting in a net redshift for all of the observed ^{44}Ti. Note that for SPI on INTEGRAL, Cas A appears as an unresolved point-like source, as imaging resultion with the SPI coded-mask telescope is about 3 degrees only.

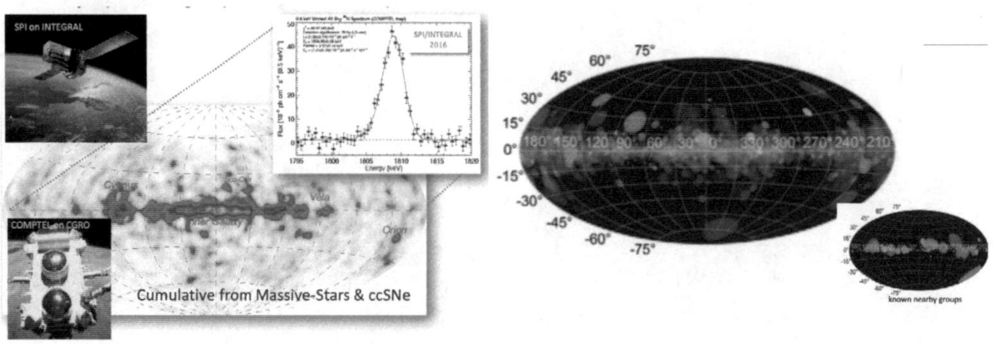

Figure 5. *Left:* The ^{26}Al decay in the Galaxy has been imaged across the sky with data from COMPTEL; high-resolution spectroscopy was added with measurements from INTEGRAL . *Right:* A bottom-up model, based on known massive-star cluster locations nearby and MC-sampled clusters at large distances, and using stellar and supernova yields and line-of-sight integrated intensities, predicts an ^{26}Al appearance of the sky that is close to observations. The inset at the bottom right shows the contributions from known nearby clusters alone. [From 30]

4 Source environments and ejecta flow

More long-live radioactivities, with decay times that exceed the occurrence frequency of nucleosynthesis events that may eject those, will accumulate in the interstellar medium around the sources. These will appear as diffuse emission, from the neighbourhood of sources. For lifetimes longer than the mixing timescale of the interstellar medium, which is estimated to be around 100 Myrs, only diffuse extended or large-scale emission will signify such nucleosynthesis, and no astronomical information is retained. But for lifetimes between 0.1 and several Myrs, one may expect to be able to trace the flow of nucleosynthesis ejecta on their early path from the sources towards spreading in the general galactic interstellar medium, before cooling down to be able to form clouds and stars of a next generation herein. ^{26}Al and ^{60}Fe , with radioactive lifetimes of 1.04 and 3.8 Myrs, repsectively, are suited for such tracing of ejecta flows, once their radioactivity gamma-ray lines can be measured with adequate significance.

4.1 Diffuse ^{26}Al emission of the Galaxy

The gamma-ray line from ^{26}Al decay has first been measured in 1978 [20], and has been imaged across the sky from data of the NASA Compton Observatory and its COMPTEL instrument (Figure 5 left, [21, 22]. From the image, it had been clear that massive-star groups and their supernovae were the most plausible sources [23]. INTEGRAL's gamma-ray spectrometer had adequate energy resolution to add spectroscopic information on ^{26}Al (Figure 5 left, [24–27]) (SPI resolution 2-3 keV, versus 200 keV for COMPTEL).

One of the key results derived from these observations has been the rate of core-collapse supernovae within our Galaxy [25]. This is based on the measured ^{26}Al line flux, and its interpretations along the line of sight in terms of an adopted source distribution model for our Galaxy, to infer the absolute amount of ^{26}Al within our Galaxy. Using yields from supernova and massive-star models, plus an adopted distribution of stars across their entire mass range (i.e., from 8-10 to about 120 $M_\odot$), one can then determine a supernova rate, that is based on a Galaxy-wide measurement (as opposed to measurements of massive-star census in the solar vicinity with extrapolation, or measurements of galaxies other than our Milky Way Galaxy; [see 25, for a discussion].

Originally, we adopted a first-order Galactic-structure model in the form of a double-exponential disk, with scale parameters galactocentric radius R_{disk} = 4 kpc and scale height z_{disk} = 0.18 kpc, as found to be an acceptable fit for the inner Galaxy region $-30^0 \leq l \leq 30^0$ degrees and $-10^0 \leq b \leq 10^0$ degrees. We found that other spatial models such as spiral arm models and models with plausibly well-fitting scale heights in the range 100-400 pc did alter the result by less than 4%.

We have recently deepened our investigations of the first step in this analysis, i.e., the determination of the ^{26}Al flux and its allocation to sources along the line of sight. Thus, we expanded the region used for this analysis from 'inner-Galaxy only' to 'all sky'. Then, we used our results from fitting a variety of massive-star tracers in the Galaxy [see 26, 28, 29, for details on tracer fitting] to better constrain sources along lines of sight by knowledge about our Galaxy and in particular the vicinity of the solar system. The ^{26}Al flux measurement can be extended to the entire sky and to the use of both single and multiple events in SPI's germanium detectors [30]. The latter requires careful modeling of instrumental backgrounds, following the directives of our studies of instrumental background and response behaviour [1, 31]. For a reliable flux determination using the entire sky, imaging deconvolution and the integration of deconvolved flux in the image may appear as a straightforward method. However, as the instrumental response is complex for gamma-ray instruments and far from merely a somewhat blurred point spread function, and the Poissonian noise from a high intensity of background counts is severe, such deconvolutions require iterative revisions of forward-convolved model using maximum-likelihood criteria for determination of their parameters. The resulting images therefore are biased from the way models have been constructed [see 32, for a detailed discussion of these issues for the case of the COMPTEL telescope]. In the end, for best flux determination one must find a spatial model that is satisfactory in terms of goodness of fit and in terms of what we know about ^{26}Al sources and about their candidate tracers in the neighbourhood of the solar system and out to large lines of sight through our Galaxy. This is based on catalogues of massive stars and massive-star clusters for the distances out to which those are reliable [30]. Then, these were combined with a population synthesis approach [following 33], and adopting some spatial blurring from ^{26}Al propagation within a cluster's age, to derive a *bottom-up* expectation for ^{26}Al emission from the massive-star groups in the solar neighbourhood; these are most important for the flux determination as flux scales with $\frac{1}{distance^2}$. Figure 5, right, shows the contribution from these nearby clusters in the lower-right part of the image. Then, Monte-Carlo sampling of candidate massive star clusters was used for the more-distant part of the Galaxy [30] , drawing from a Galaxy model as before. Now, this part of spatial modeling was then used only for the part of the Galaxy where other information is not available. Comparing results for a variety of tracers commonly used for massive stars, starting from exponential disks of different sizes, then different spiral-arm based models were used [30], all with scale heights in the potential range of interest as found for ^{26}Al data scale height fitting [26]. Finally, this population synthesis modeling determines expected ^{26}Al fluxes using yields from models of massive stars and core-collapse supernovae, and an appropriate mass range and mass weighting. The latter includes different initial-mass functions for massive stars, and different supernova explodability models [see 30, for details]. Globally, yields then are scaled with a star formation rate factor for the entire Galaxy. Figure 5, right, displays the morphology of the ^{26}Al sky as composed from such a *bottom-up* modeling effort, for one set of plausible parameters.

One result from these studies is that the total mass of ^{26}Al in our Galaxy is reduced by as much as 25 to 30%; this is due to a better account of nearby source regions, such as the Scorpius-Centaurus stars or other groups out kpc distances such as in the Cygnus region.

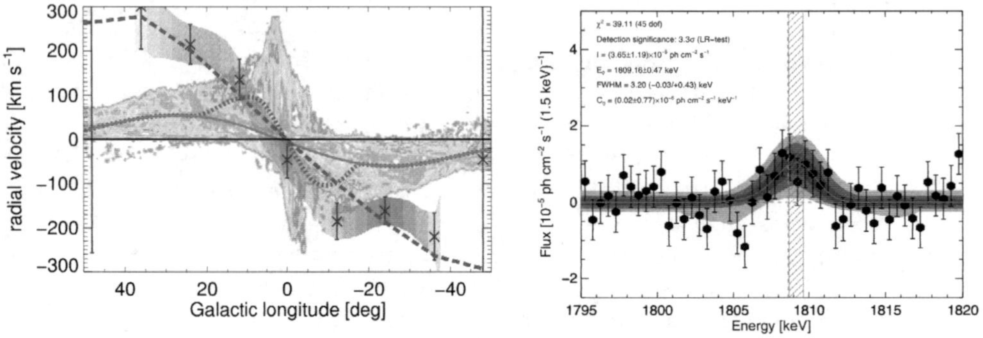

Figure 6. *Left:* The ^{26}Al decay line shows systematic shifts along Galactic longitudes, which are attributed to large scale Galactic rotation [34]. Excess velocities of about 200 km s^{-1} are observed with respect to other tracers of Galactic rotation (colour scale: CO clouds). This is attributed to on-average larger streaming velocities in cavities elongated into the forward direction of galactic rotation [35]. *Right:* ^{26}Al line spectrum from the Orion region, indicating blue shifted emission from bulk motion towards the observer [27].

Thus, plausible Galactic ^{26}Al masses do not exceed 2 $M_\odot$, for plausible values of the star formation rate not exceeding 5 $M_\odot$ y^{-1}. Also, these studies show a remarkable sensitivity to largely-extended ^{26}Al flux from also high-latitude regions, such as could occur if the nearby massive-star groups had produced superbubbles that extend over large regions of the sky; the 'Local Bubble' is one such candidate. All these results [see 30, and Diehl *et al., in preparation*] point into the direction that the rate of core-collapse supernovae derived from ^{26}Al measurements and their modeling will fall into a range of 1.0 to 1.3 century^{-1}, equivalent to a recurrence rate of around 75 years.

4.2 ^{26}Al kinematics and superbubbles

The analysis of the ^{26}Al gamma-ray line and its dependency with viewing directions had led to the finding that systematic line shifts reminiscent of large-scale galactic rotation was observed [25, 34]. Comparing the ^{26}Al data with other tracers of such large-scale Galactic rotation (Figure 6, left), it was evident that ^{26}Al data showed a systematic velocity excess with respect to all other sources. This was understood and modelled as being the result of ^{26}Al propagation within large cavities (superbubbles) that extended asymmetrically around the massive-star clusters at their origins [35] (see green dashed model line in Figure 6, left). Thus, ^{26}Al kinematics results are supportive of our suggestion that massive star nucleosynthesis is preferentially injected into circumstellar surroundings in the form of large superbubbles, that have been blown by winds of the most-massive, hence rapidly-evolving, stars of the clusters, and later increased further by successive supernova explosions within. The density gradient away from spiral arms as the birth sites of these stellar clusters leads to a large extent of these superbubbles towards lower densities in inter-arm regions and the Galactic halo. This allows ^{26}Al that is ejected into directions following the large-scale rotation can propagate for a longer time freely until decay, while ^{26}Al ejcted into the opposite direction will be decelerated by higher density environments such as the back end of the superbubble wall before its radioactive decay, so that decay occurs at a lower velocity; the net effect of this kinematic situation will be a systematic velocity bias in the direction of galactic rotation, as observed.

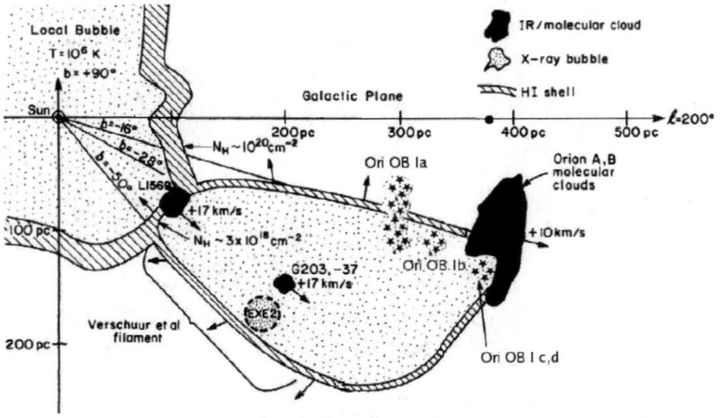

Figure 7. The Orion region is characterised by two massive molecular clouds at about 450 pc distance, with recent massive-star formation preferentially on its near side (star symbols), and a large cavity bounded by denser HI emission, which extends from the clouds towards the direction of the Sun. ([Updated from 37] [by 36].

The Orion region with the Orion A and B molecular clouds at about 450 pc distance and its OB1 association subgroups presents a situation that again supoorts this picture. The Eridanus cavity has been mapped in HI and X rays to extend from the Orion molecular clouds far and asymmetrically towards the Sun, as shown in the sketch of Figure 7 [36]. From the geometry shown, it appears plausible that ^{26}Al emission also should have a velocity bias from such propagation within the Eridanus cavity. INTEGRAL/SPI observations show a tantalizing hint for such a blue shift of the line [27], although not significant in current results (Figure 6, right). More observational data are being accumulated, towards this test of our superbubble picture.

4.3 Diffuse ^{60}Fe emission of the Galaxy

^{60}Fe dffuse gamma-ray line emission had been discovered with RHESSI [38], later also measured with INTEGRAL/SPI [39, 40]. From models of massive-star nucleosynthesis, these sources of ^{26}Al were also expected to eject ^{60}Fe radioactivity at the same time, and with similar amounts [41]. But these observations quickly made clear that ^{60}Fe gamma-ray intensities are well below those of ^{26}Al gamma rays, by as much as a factor of ten. This presents a challenge to astrophysical models of massive stars and their supernovae [see 42, 43, for a thorough discussion].

In an effort to better determine the intensity ratio between ^{26}Al and ^{60}Fe , we analysed both these isotopes from a dataset that included identical observations and data selections, in order to minimize any systematic effects, such as could occur from using slightly different sky exposures or data selections [44]. We found spectral intensities is shown in Figure 8, left, for one plausible sky intensity distribution that may represent both emissions. Their analysis then used a variety of candidate tracers for ^{26}Al and for ^{60}Fe , comparing qualities of the fit to SPI observations. It turned out that morphologies of ^{26}Al and of ^{60}Fe emissions are not necessarily identical, however with clear similarities; ^{60}Fe emission also is plausibly diffuse and extended, rather than originating from very few sources only. From astrophysical models, indeed some differences could be expected. Simulations of simple galaxy models illustrate this [45, 46]. Also, in our population synthesis models [30], differences are found,

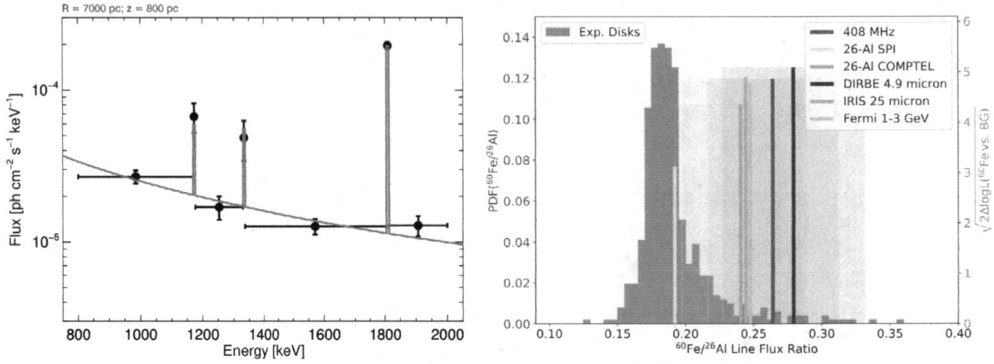

Figure 8. *Left:* The ^{60}Fe signal and ^{26}Al decay line intensities can be compared in INTEGRAL analyses from a consistent set of observations [44]. *Right:* From modeling comparisons using a wide set of different spatial distribution and tracer models, the gamma-ray intensity ratio ^{60}Fe /^{26}Al is constraints to 0.4 as an upper bound . [44].

upon a closer look, although here we adopted the massive-star/core-collapse supernova origin for ^{60}Fe . From this analysis, the total Galactic amount of ^{60}Fe was found to be around 4-6 $M_\odot$ [30]. We studied systematic effects from different modelings of spatial distributions (see Figure 8, right) [44], and concluded that the ^{60}Fe /^{26}Al intensity ratio is constrained to be below a value of 0.4, as an upper limit. As stated above, this is a challenge for current models [see also 30].

5 Conclusions

Gamma-ray lines from cosmic sources of nucleosynthesis have been accessible to observations since more than four decades. Imaging information is limited, as gamma-ray telescopes have fundamental limits in the degree range.

Spectral lines, however, reveal unique details about sources of nucleosynthesis, in particular about supernova explosions. Lines from the ^{56}Ni and ^{44}Ti decay chains have confirmed the fundamental understanding that nucleosynthesis ejecta with radioactive isotopes power the afterglow emission of supernova explosions through their decay gamma rays and leptons. Gamma-ray spectroscopy has also shown that supernova explosions of both types, thermonuclear, and core collapse, are likely not spherically symmetric; data from SN2014J, and from Cas A, respectively, show clear signs of bulk motion deviant from spherical symmetry.

Long-lived radioactive admixtures of nucleosynthesis ejecta in the forms of ^{26}Al and ^{60}Fe have shown that massive-star clusters are prominent sources of these trace isotopes. Analysing the ^{26}Al emission morphology of the Galaxy, and kinematic Doppler information from imaging spectroscopy, it is found that superbubbles of large extents are characteristic environments wherein massive-star ejecta are launched, towards eventually cooling and mixing with ambient gas towards later formation of next-generation stars. ^{60}Fe emission is found considerably fainter than ^{26}Al emission, limited to about 40% of ^{26}Al . It is important to combine these gamma-ray results with other messengers of interstellar gas, to learn how nucleosynthesis ejecta propagate to lead to galactic and cosmic chemical evolution.

Acknowledgements. This work is supported through the Max Planck Society/MPE, through the European COST action number 16117 'ChETEC', the Excellence Clusters 'Universe' and 'Origins', and DLR as well as ESA, supporting the INTEGRAL mission.

References

[1] R. Diehl, T. Siegert, J. Greiner, M. Krause, K. Kretschmer, M. Lang, M. Pleintinger, A.W. Strong, C. Weinberger, X. Zhang, Astron. Astrophys. **611**, A12 (2018)

[2] N. Gehrels, E. Chipman, D.A. Kniffen, Astron. Astrophys. Suppl. **97**, 5 (1993)

[3] C. Winkler, T.J.L. Courvoisier, G. Di Cocco, N. Gehrels, A. Giménez, S. Grebenev, W. Hermsen, J.M. Mas-Hesse, F. Lebrun, N. Lund et al., Astron. Astrophys. **411**, L1 (2003)

[4] F.A. Harrison, W.W. Craig, F.E. Christensen, C.J. Hailey, W.W. Zhang, S.E. Boggs, D. Stern, W.R. Cook, K. Forster, P. Giommi et al., Astrophys. J. **770**, 103 (2013), 1301.7307

[5] A. de Angelis, V. Tatischeff, I.A. Grenier, J. McEnery, M. Mallamaci, M. Tavani, U. Oberlack, L. Hanlon, R. Walter, A. Argan et al., Journal of High Energy Astrophysics **19**, 1 (2018), 1711.01265

[6] T. Tanimori, arXiv e-prints arXiv:2002.05362 (2020), 2002.05362

[7] J.A. Tomsick, S.E. Boggs, A. Zoglauer, E. Wulf, L. Mitchell, B. Phlips, C. Sleator, T. Brandt, A. Shih, J. Roberts et al., arXiv e-prints arXiv:2109.10403 (2021), 2109.10403

[8] D.D. Clayton, S.E. Woosley, Reviews of Modern Physics **46**, 755 (1974)

[9] E. Churazov, R. Sunyaev, J. Isern, J. Knödlseder, P. Jean, F. Lebrun, N. Chugai, S. Grebenev, E. Bravo, S. Sazonov et al., Nature **512**, 406 (2014), 1405.3332

[10] R. Diehl, T. Siegert, W. Hillebrandt, M. Krause, J. Greiner, K. Maeda, F.K. Röpke, S.A. Sim, W. Wang, X. Zhang, Astron. Astrophys. **574**, A72 (2015), 1409.5477

[11] J. Isern, P. Jean, E. Bravo, J. Knödlseder, F. Lebrun, E. Churazov, R. Sunyaev, A. Domingo, C. Badenes, D.H. Hartmann et al., Astron. Astrophys. **588**, A67 (2016), 1602.02918

[12] W.D. Arnett, Astrophys. J. **253**, 785 (1982)

[13] L.S. The, A. Burrows, Astrophys. J. **786**, 141 (2014), 1402.4806

[14] R. Diehl, T. Siegert, W. Hillebrandt, S.A. Grebenev, J. Greiner, M. Krause, M. Kromer, K. Maeda, F. Röpke, S. Taubenberger, Science **345**, 1162 (2014), 1407.3061

[15] C. Weinberger, Ph.D. thesis, TU Munich (2021)

[16] A.F. Iyudin, R. Diehl, H. Bloemen, W. Hermsen, G.G. Lichti, D. Morris, J. Ryan, V. Schoenfelder, H. Steinle, M. Varendorff et al., Astron. Astrophys. **284**, L1 (1994)

[17] B.W. Grefenstette, F.A. Harrison, S.E. Boggs, S.P. Reynolds, C.L. Fryer, K.K. Madsen, D.R. Wik, A. Zoglauer, C.I. Ellinger, D.M. Alexander et al., Nature **506**, 339 (2014), 1403.4978

[18] B.W. Grefenstette, C.L. Fryer, F.A. Harrison, S.E. Boggs, T. DeLaney, J.M. Laming, S.P. Reynolds, D.M. Alexander, D. Barret, F.E. Christensen et al., Astrophys. J. **834**, 19 (2017), 1612.02774

[19] C. Weinberger, R. Diehl, M.M.M. Pleintinger, T. Siegert, J. Greiner, Astron. Astrophys. **638**, A83 (2020), 2004.12688

[20] W.A. Mahoney, J.C. Ling, A.S. Jacobson, R.E. Lingenfelter, Astrophys. J. **262**, 742 (1982)

[21] R. Diehl, Experimental Astronomy **6**, 103 (1995)

[22] S. Plüschke, R. Diehl, V. Schönfelder, H. Bloemen, W. Hermsen, K. Bennett, C. Winkler, M. McConnell, J. Ryan, U. Oberlack et al., *The COMPTEL 1.809 MeV survey*, in *Exploring the Gamma-Ray Universe*, edited by A. Gimenez, V. Reglero, C. Winkler (2001), Vol. 459 of *ESA Special Publication*, pp. 55–58

[23] N. Prantzos, Astron. Astrophys. Suppl. **120**, C303+ (1996)

[24] R. Diehl, K. Kretschmer, S. Plüschke, V. Schönfelder, A.W. Strong, M. Cerviño, D. Hartmann, Astronomische Nachrichten Supplement **324**, 18 (2003)

[25] R. Diehl, N. Prantzos, P. von Ballmoos, Nuclear Physics A **777**, 70 (2006), `arXiv:astro-ph/0502324`

[26] W. Wang, M.G. Lang, R. Diehl, H. Halloin, P. Jean, J. Knödlseder, K. Kretschmer, P. Martin, J.P. Roques, A.W. Strong et al., Astron. Astrophys. **496**, 713 (2009), `0902.0211`

[27] T. Siegert, R. Diehl, *The* ^{26}*Al Gamma-ray Line from Massive-Star Regions*, in *14th International Symposium on Nuclei in the Cosmos (NIC2016)*, edited by S. Kubono, T. Kajino, S. Nishimura, T. Isobe, S. Nagataki, T. Shima, Y. Takeda (2017), p. 020305, `1609.08817`

[28] R. Diehl, K. Bennett, C. Dupraz, J. Knoedlseder, G. Lichti, D. Morris, U. Oberlack, J. Ryan, V. Schoenfelder, A. Strong et al., Astron. Astrophys. Suppl. **120**, C321+ (1996)

[29] J. Knödlseder, K. Bennett, H. Bloemen, R. Diehl, W. Hermsen, U. Oberlack, J. Ryan, V. Schönfelder, P. von Ballmoos, Astron. Astrophys. **344**, 68 (1999)

[30] M.M.M. Pleintinger, Ph.D. thesis, Technische Universität München (2020)

[31] T. Siegert, R. Diehl, C. Weinberger, M.M.M. Pleintinger, J. Greiner, X. Zhang, Astron. Astrophys. **626**, A73 (2019), `1903.01096`

[32] J. Knödlseder, D. Dixon, K. Bennett, H. Bloemen, R. Diehl, W. Hermsen, U. Oberlack, J. Ryan, V. Schönfelder, P. von Ballmoos, Astron. Astrophys. **345**, 813 (1999), `arXiv:astro-ph/9903172`

[33] R. Voss, R. Diehl, D.H. Hartmann, M. Cerviño, J.S. Vink, G. Meynet, M. Limongi, A. Chieffi, Astron. Astrophys. **504**, 531 (2009), `0907.5209`

[34] K. Kretschmer, R. Diehl, M. Krause, A. Burkert, K. Fierlinger, O. Gerhard, J. Greiner, W. Wang, Astron. Astrophys. **559**, A99 (2013), `1309.4980`

[35] M.G.H. Krause, R. Diehl, Y. Bagetakos, E. Brinks, A. Burkert, O. Gerhard, J. Greiner, K. Kretschmer, T. Siegert, Astron. Astrophys. **578**, A113 (2015), `1504.03120`

[36] K.M. Fierlinger, A. Burkert, E. Ntormousi, P. Fierlinger, M. Schartmann, A. Ballone, M.G.H. Krause, R. Diehl, Mon. Notices Royal Astron. Soc. **456**, 710 (2016), `1511.05151`

[37] D.N. Burrows, K.P. Singh, J.A. Nousek, G.P. Garmire, J. Good, Astrophys. J. **406**, 97 (1993)

[38] D.M. Smith, *Gamma-Ray Line Observations with RHESSI*, in *5th INTEGRAL Workshop on the INTEGRAL Universe*, edited by V. Schoenfelder, G. Lichti, C. Winkler (2004), Vol. 552 of *ESA Special Publication*, pp. 45–+

[39] M.J. Harris, J. Knödlseder, P. Jean, E. Cisana, R. Diehl, G.G. Lichti, J.P. Roques, S. Schanne, G. Weidenspointner, Astron. Astrophys. **433**, L49 (2005), `arXiv:astro-ph/0502219`

[40] W. Wang, M.J. Harris, R. Diehl, H. Halloin, B. Cordier, A.W. Strong, K. Kretschmer, J. Knödlseder, P. Jean, G.G. Lichti et al., Astron. Astrophys. **469**, 1005 (2007), `0704.3895`

[41] F.X. Timmes, S.E. Woosley, D.H. Hartmann, R.D. Hoffman, T.A. Weaver, F. Matteucci, Astrophys. J. **449**, 204 (1995), `arXiv:astro-ph/9503120`

[42] R. Diehl, M.G.H. Krause, K. Kretschmer, M. Lang, M.M.M. Pleintinger, T. Siegert, W. Wang, L. Bouchet, P. Martin, New Astron. Rev. **92**, 101608 (2021), `2011.06369`

[43] R. Diehl, M. Lugaro, A. Heger, A. Sieverding, X. Tang, K. Li, E. Li, C.L. Doherty, M.G.H. Krause, A. Wallner et al., arXiv e-prints arXiv:2109.08558 (2021),

2109.08558

[44] W. Wang, T. Siegert, Z.G. Dai, R. Diehl, J. Greiner, A. Heger, M. Krause, M. Lang, M.M.M. Pleintinger, X.L. Zhang, Astrophys. J. **889**, 169 (2020), 1912.07874

[45] Y. Fujimoto, M.R. Krumholz, S. Tachibana, Mon. Notices Royal Astron. Soc. **480**, 4025 (2018), 1802.08695

[46] D. Rodgers-Lee, M.G.H. Krause, J. Dale, R. Diehl, Mon. Notices Royal Astron. Soc. **490**, 1894 (2019), 1909.10978

Compton Imaging and Machine-Learning techniques for an enhanced sensitivity in key stellar (n,γ) measurements

J. Lerendegui-Marco[1,⋆], V. Babiano-Suárez[1], J. Balibrea-Correa[1], V. Babiano-Suárez[1], L. Caballero[1], D. Calvo[1], C. Domingo-Pardo[1], I. Ladarescu[1], D. Real[1], F. Calviño[2], A. Casanovas[2], A. Tarifeño-Saldivia[2], V. Alcayne[3], C. Guerrero[4,5], M.A. Millán-Callado[4,5], T. Rodríguez-González[4,5], M. Barbagallo[6,7], N.M Chiera[8], R. Dressler[8], S. Heinitz[8], E.A Maugeri[8], D. Schumann[8], and U. Köster[9]

[1] *Instituto de Física Corpuscular, CSIC-University of Valencia, Spain*
[2] *Universitat Politecnica de Catalunya (UPC), Barcelona, Spain*
[3] *Centro de Investigaciones Energéticas, Medioambientales y Tecnológicas (CIEMAT), Madrid, Spain.*
[4] *Dpto. Física Atómica, Molecular y Nuclear, Universidad de Sevilla, Sevilla, Spain*
[5] *Centro Nacional de Aceleradores(CNA), Sevilla, Spain*
[6] *European Organization for Nuclear Research (CERN), Meyrin, Switzerland*
[7] *Instituto Nazionale di Fisica Nucleare, Sezione di Bari, Italy.*
[8] *Paul Scherrer Institut (PSI), Villigen, Switzerland*
[9] *Institut Laue-Langevin ILL, Grenoble, France*

Abstract. Neutron capture cross-section measurements are fundamental in the study of astrophysical phenomena, such as the slow neutron capture (s-) process of nucleosynthesis operating in red-giant stars. To enhance the sensitivity of such measurements we have developed the i-TED detector. i-TED is an innovative detection system which exploits the Compton imaging technique with the aim of obtaining information about the incoming direction of the detected γ-rays. The imaging capability allows one to reject a large fraction of the dominant γ-ray background, hence enhancing the (n,γ) detection sensitivity.

This work summarizes the main results of the first experimental proof-of-concept of the background rejection with i-TED carried out at CERN n_TOF using an early i-TED demonstrator. Two state-of-the-art C_6D_6 detectors were also used to benchmark the performance of i-TED. The i-TED prototype built for this study shows a factor of ~3 higher detection sensitivity than C_6D_6 detectors in the ~10 keV neutron-energy range of astrophysical interest. This works also introduces the perspectives of further enhancement in performance attainable with the final i-TED array and new analysis methodologies based on Machine-Learning techniques. The latter provide higher (n,γ) detection efficiency and similar enhancement in the sensitivity than the analytical method based on the Compton scattering law. Finally, we present our proposal to use this detection system for the first time on key astrophysical (n,γ) measurements, in particular on the s-process branching-point ^{79}Se, which is especially well suited to constrain the thermal conditions of Red Giant and Massive Stars.

⋆e-mail: jorge.lerendegui@ific.uv.es

1 Introduction

Neutron capture cross-section measurements are fundamental in the study of astrophysical phenomena, such as the slow neutron capture (s-) process of nucleosynthesis operating in red-giant stars [1]. The best suited method to measure neutron-capture cross sections over the full stellar range of interest is the time-of-flight (TOF) technique, such as the CERN n_TOF Facility [2]. For such experiments, liquid scintillators, such as C_6D_6 are particularly convenient because of their fast time-response and low intrinsic sensitivity to scattered neutrons [3]. However, an important constrain in many TOF capture experiments in the 1 keV to 100 keV neutron-energy interval of relevance for astrophysics arises from the background induced by neutrons that are scattered in the sample and get subsequently captured around in the experimental hall or in the detectors (see e.g. Refs. [4, 5]). Indeed, according to recent MC studies [6], in the 1 keV to 100 keV neutron-energy interval of relevance for astrophysics, this type of background represents one of the dominant contributions in many neutron capture experiments.

In order to reduce this dominant source of background a total energy detector with γ-ray imaging capability, so-called i-TED, has been recently proposed [7]. i-TED is an array of four Compton cameras, each of them consisting of 5 position-sensitive detectors (PSDs) made of $LaCl_3$(Ce) monolytic crystals and distributed in two detection planes, hereafter called scatter and absorber planes. This system exploits the Compton imaging technique with the aim of obtaining information about the incoming direction of the detected γ-rays and, hence, selecting only the true capture events originated in the sample under investigation, as it is illustrated in Fig. 1.

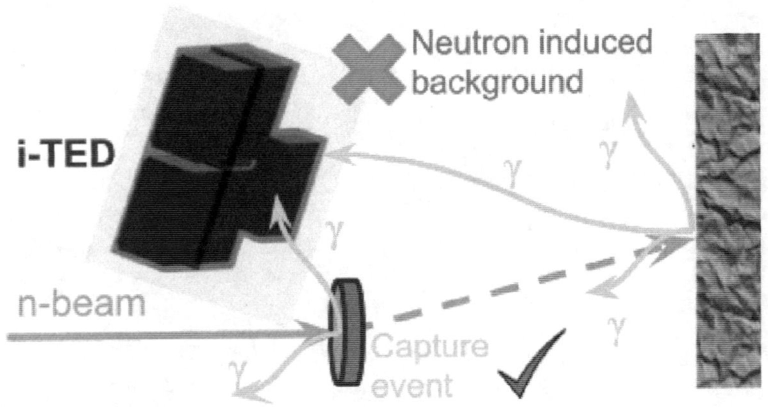

Figure 1. Scheme of a neutron-capture TOF experiment illustrating the concept of the i-TED Compton imager applied to the suppression the dominant neutron induced background.

In the first part of this work we present the proof-of-concept (PoC) experiments carried out with an i-TED demonstrator at CERN n_TOF intended to validate the background rejection concept. Secondly, we investigate the prospects for the final i-TED array and new analysis methodologies for the background rejection based on Machine-Learning techniques, which allow to remarkably improve its performance when compared to the analytical method used so far. The last part of this manuscript summarizes the plans to measure the key stellar ^{79}Se(n,γ) reaction using the i-TED detector.

2 Compton imaging for background rejection: Proof-of-concept

Aiming at the experimental validation of the background reduction in a capture measurement with i-TED, the ^{56}Fe(n,γ) reaction was studied at CERN n_TOF [2] using an i-TED demonstrator based on 3 position-sensitive detectors (PSDs). The objective of this PoC experiment was to quantify the attainable enhancement in terms of signal-to-background with respect to state-of-the-art C_6D_6 detectors [3]. The details on the apparatus and the experiment can be found in [8]. The reconstruction of the γ-ray energy depositions and interaction points in i-TED were based on our previous works [9, 10].

Fig. 2 shows the measured counts as a function of the neutron energy obtained with C_6D_6 detectors and i-TED. The isolated resonance at a neutron energy of 1.15 keV is well suited to evaluate the signal-to-background ratio (SBR) in the neutron-energy range of interest for astrophysics. To this aim, all spectra have been normalized to the peak of this resonance. The results indicate that a comparable SBR is obtained for both C_6D_6-detectors and the S-detector of i-TED. Moreover, the SBR is enhanced in a factor 2.7 when the A- and S-planes are operated in time coincidence, shown in Fig. 2 as a solid red line. Finally, an additional suppression of the background is obtained by means of the Compton imaging capability of i-TED. A maximum SBR of 3.5 (red dashed line) was achieved by applying cuts in the imaging parameter λ, defined in Eq. (2) of Ref. [8].

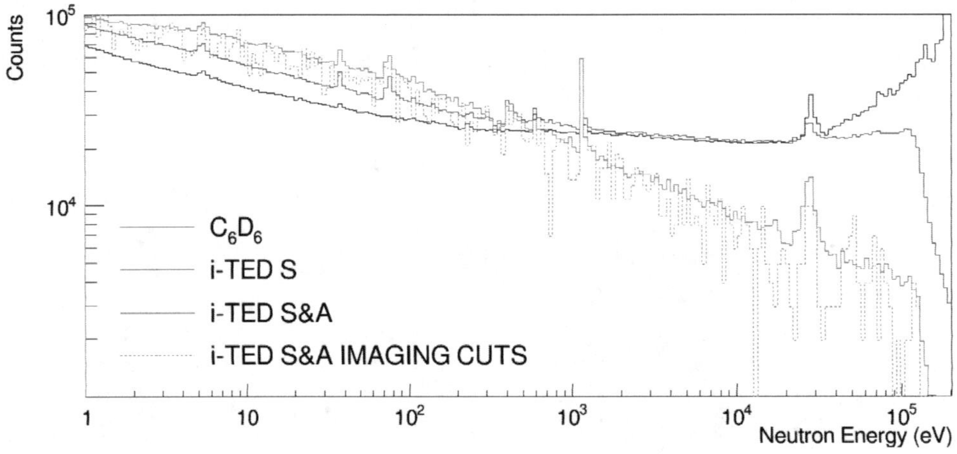

Figure 2. Neutron-energy spectra measured with the ^{56}Fe sample using the i-TED S-detector in singles-mode (blue line) and i-TED with S- and A-detectors in time-coincidence mode (solid red line). The spectrum measured with the C_6D_6 detectors is shown in black. The dashed-red spectrum shows the best result obtained after a Compton imaging selection.

3 Prospects based on Machine Learning techniques

The main drawback of the analytical method used in the analysis of the PoC experiment is the sharp drop in (n,γ) efficiency associated to the imaging selections. For this reason, alternative analysis techniques, based on Machine Learning (ML) algorithms have been explored. The complexity of the data acquired with the i-TED Compton camera make these algorithms a powerful tool.

Accurate MC simulations of the response of the final i-TED array to neutron capture and background events were carried out in this work [8] to train several ML-classification algorithms in the

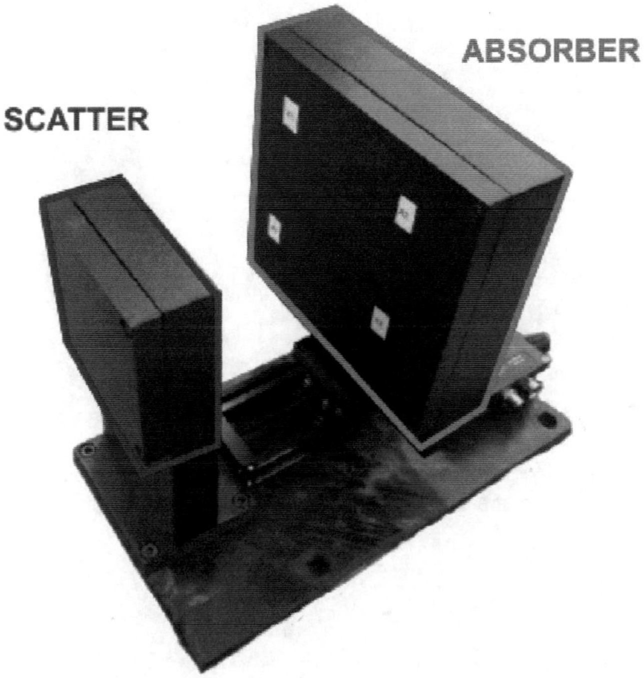

Figure 3. i-TED Compton module composed of one Scatter (50 × 50 × 10 *mm* $LaCl_3$) and four Absorber (50 × 50 × 25 *mm* $LaCl_3$) detectors.

discrimination between the two kind of events. The details on the technical implementation of the i-TED detector, displayed in Figure 3, are given in Refs. [8, 9].

Among the trained ML classifiers, one of the best performing ones, XGBoost, has been compared with the analytical imaging method used in the ^{56}Fe(n,γ) PoC experiment. Two figures of merit have been analyzed, which are shown in Figure 4:

- **Relative (n,γ) efficiency**: fraction of capture events which are correctly identified.
- **Relative SBR gain factor**: fraction of correct (n,γ) events over the fraction of background events wrongly predicted as capture.

The advantages of the ML-based method compared to the analytical imaging approach are evident from Fig. 4, as it provides relatively high efficiency (60-70%) and SBR (1.7-3) simultaneously, being this situation unreachable with the analytical imaging method.

In the meantime, the development of the final i-TED detector has been completed, and after its recent commissioning at CERN n_TOF is ready to tackle the first measurement of the key stellar ^{79}Se(n,γ) branching point reaction, as it is presented in the following Section.

4 Future plans for key astrophysical measurements: ^{79}Se(n,γ)

The unstable ^{79}Se (terrestrial half-life $t_{1/2}$ = 3.27(8)×10^5 years) represents one of the most relevant and debated s-process branching nuclei [11]. The ^{79}Se branching is particularly interesting because it

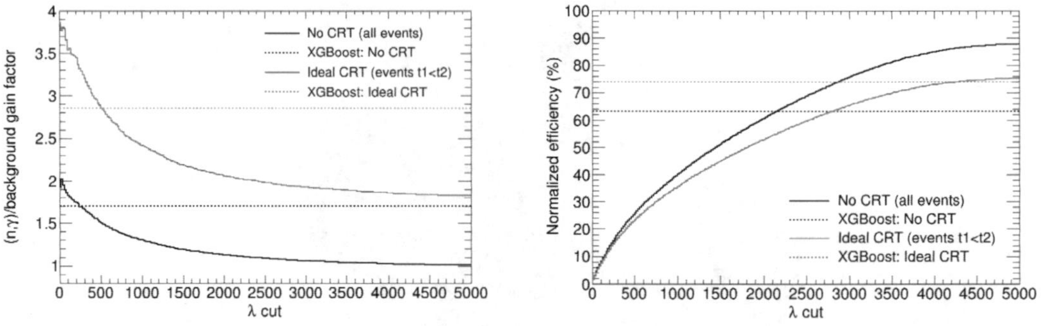

Figure 4. Relative SBR gain factor (left) and relative (n,γ) (right) as a function of λ_{max}. The reference SBR = 1 and efficiency = 100% correspond to i-TED (S&A) with no imaging selection. The black and red solid lines correspond to the analytical imaging case with no time resolution and ideal coincidence-resolving time (CRT), respectively. The dashed lines show the results obtained with XGBoost (constant values since the λ cut does not apply).

is located in the transition region between weak (MSs) and the main (AGBs) s-process. The s-process path bifurcates at ^{79}Se due to the competing action of neutron capture and beta decay of this isotope (see Fig. 5). Thus, a detailed knowledge of the ^{79}Se capture cross section is fundamental to fix the branching ratio to ^{82}Kr (neutron capture) or to ^{80}Kr (beta decay). This Kr isotopic ratio has been measured in bulk SiC acid residues [12], which give the most precise data currently available on s-process nucleosynthesis. The branching at ^{79}Se is particularly well suited for determining the thermal conditions of the stellar environment thanks to the strong thermal dependence of the beta decay rate of this isotope [13]. In this context, the first neutron capture measurement on this isotope was proposed to the Isolde and n_TOF Committee at CERN (INTC) [14] and will be carried out in the first half of 2022.

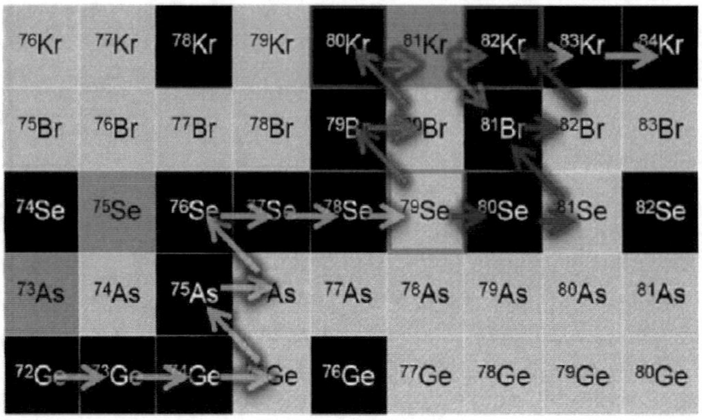

Figure 5. S-process path branching at ^{79}Se, which determines the ratio between the s-only ^{80}Kr and ^{82}Kr (Figure from Ref. [11]).

The ^{79}Se sample for this experiment was produced by means of neutron irradiation of an enriched ^{208}Pb^{78}Se alloy-sample in the high-flux reactor at ILL. For the final sample, 3 g of metallic Se powder enriched to 99.34% in ^{78}Se were mixed with highly enriched lead (99% ^{208}Pb) to produce a pellet of 14×5 mm, a mass of 3.9028 g and encapsulated in a 0.5 mm thick aluminum casing. The expected amount of ^{79}Se is of about 3 mg. More details of the sample preparation can be found in Ref. [15].

In the preparation of the proposal, we have carefully studied the experiment feasibility and the expected results. Fig. 6 shows an example of the expected count rates for the two Se isotopes in the sample, which resembles the actual experimental capture yield.

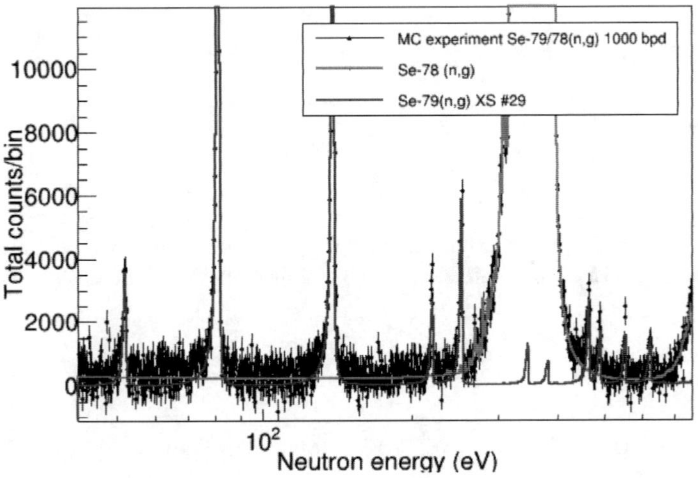

Figure 6. Total expected counts of ^{78}Se+^{79}Se(n,γ) for the experiment at CERN n_TOF illustrating the feasibility to identify ^{79}Se resonances.

The R-Matrix analysis of the extracted yield, followed by a statistical analysis of the individual parameters, will lead to a set of average resonance parameters. These parameters can be then used to perform a Hauser-Feshbach calculation of the semi-empirical cross section up to 300 keV, from which the MACS at different k_BT can be determined (see for instance Ref. [16]). The final uncertainty in the MACS, associated to the envisaged number of observed resonances, is expected to strongly constrain the value compared to the actual spread of theoretical calculations compiled in KaDoNiS (see Fig. 7).

Acknowledgements

This work has been carried out in the framework of a project funded by the European Research Council (ERC) under the European Union's Horizon 2020 research and innovation programme (ERC Consolidator Grant project HYMNS, with grant agreement No. 681740). The authors acknowledge support from the Spanish Ministerio de Ciencia e Innovación under grants PID2019-104714GB-C21, FPA2017-83946-C2-1-P, FIS2015-71688-ERC, CSIC for funding PIE-201750I26.

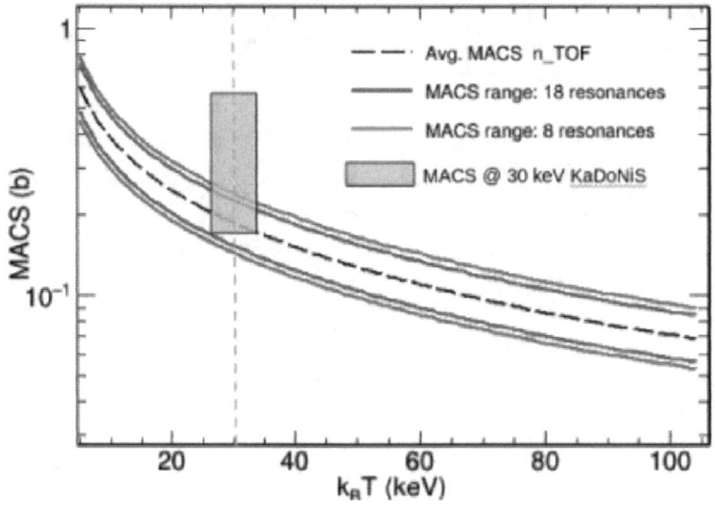

Figure 7. Expected MACS at different stellar temperatures (black dashed) and uncertainty range related to the limited number of observed resonances compared to the range of theoretical calculations compiled in KaDoNiS.

References

[1] F. Käppeler, R. Gallino, S. Bisterzo, W. Aoki, Reviews of Modern Physics **83**, 157 (2011), 1012.5218

[2] C. Guerrero, A. Tsinganis, E. Berthoumieux, M. Barbagallo, F. Belloni, F. Gunsing, C. Weiß, E. Chiaveri, M. Calviani, V. Vlachoudis et al., European Physical Journal A **49**, 27 (2013)

[3] R. Plag, M. Heil, F. Käppeler, P. Pavlopoulos, R. Reifarth, K. Wisshak, n TOF Collaboration, Nuclear Instruments and Methods in Physics Research A **496**, 425 (2003)

[4] C. Domingo-Pardo, U. Abbondanno, G. Aerts, H. Álvarez-Pol, F. Alvarez-Velarde, S. Andriamonje, J. Andrzejewski, P. Assimakopoulos, L. Audouin, G. Badurek et al., Physical Review C **74**, 025807 (2006)

[5] G. Tagliente, et al., Phys. Rev. C **87**, 014622 (2013)

[6] P. Žugec, N. Colonna, D. Bosnar, S. Altstadt, J. Andrzejewski, L. Audouin, M. Barbagallo, V. Bécares, F. Bečvář, F. Belloni et al., Nuclear Instruments and Methods in Physics Research Section A: Accelerators, Spectrometers, Detectors and Associated Equipment **760**, 57 (2014)

[7] C. Domingo-Pardo, Nuclear Instruments and Methods in Physics Research A **825**, 78 (2016)

[8] V. Babiano, J. Lerendegui-Marco, et al., The European Physical Journal A (accepted) **57** (2021)

[9] V. Babiano, J. Balibrea, L. Caballero, D. Calvo, I. Ladarescu, J. Lerendegui, S. Mira Prats, C. Domingo-Pardo, Nuclear Instruments and Methods in Physics Research A **953**, 163228 (2020), 1908.08533

[10] J. Balibrea-Correa, J. Lerendegui-Marco, D. Calvo, L. Caballero, V. Babiano, I. Ladarescu, M.L. Redondo, J.L. Tain, A. Tolosa, C. Domingo-Pardo et al., Nuclear Instruments and Methods in Physics Research A **985**, 164709 (2021)

[11] F. Kappeler, H. Beer, K. Wisshak, Reports on Progress in Physics **52**, 945 (1989)

[12] R.S. Lewis, S. Amari, E. Anders, Nature **348**, 293 (1990)

[13] N. Klay, F. Käppeler, Phys. Rev. C **38**, 295 (1988)
[14] J. Lerendegui-Marco, C. Domingo-Pardo, et al., CERN-INTC-2020-065;INTC-P-580 (2020)
[15] N.M. Chiera, et al., Nuclear Instruments and Methods in Physics Research A **submitted** (2021)
[16] C. Guerrero, J. Lerendegui-Marco, M. Paul, M. Tessler, S. Heinitz, C. Domingo-Pardo, S. Cristallo, R. Dressler, S. Halfon, N. Kivel et al., Phys. Rev. Lett. **125**, 142701 (2020)

Cross-section measurements relevant for the astrophysical p process at the University of Cologne

Felix Heim[1,*], *Martin* Müller[1], *Philipp* Scholz[2], *Svenja* Wilden[1], and *Andreas* Zilges[1]

[1]University of Cologne, Institute for Nuclear Physics, Germany
[2]Department of Physics, University of Notre Dame, Indiana 46556-5670, USA

Abstract. The astrophysical p process unites all processes that have been introduced to explain the abundance of a group of 30 to 35 neutron-deficient nuclei which are referred to as p nuclei. In general, these p processes include large networks of nuclear reactions and a complete understanding of the individual reaction rates is required to describe the abundance of the p nuclei qualitatively and quantitatively. In many cases the involved nuclear reactions are not accessible in the laboratory, either due to their low cross sections or because they involve unstable or exotic isotopes. For those purposes, the motivation of cross-section measurements performed at the University of Cologne is twofold: First, experimentally constrained reaction rates are of direct relevance for nucleosynthesis network calculations. And second, experimental cross-section values are required to test existing theoretical descriptions and to improve their predictive power. In this work, we present the experimental setups and methods that are used to measure nuclear cross-sections at very low sensitivities and we show a detailed overview of proton- and α-induced reactions that have been measured in Cologne in the last decade.

1 Introduction

The majority of stable isotopes is created by neutron-capture processes within various astrophysical scenarios [1, 2]. However, a small group of nuclei on the neutron-deficient side of the valley of stability - the p nuclei - demands for a different kind of nucleosynthesis mechanism [3, 4]. The γ process has been introduced for that purpose which represents a huge network of photodisintegration reactions which act on pre-produced nuclei near the valley of stability. Since the γ process is a huge network of thousands of nuclear reactions a large database of experimentally constrained cross sections is necessary to model the abundance flow within the respective astrophysical scenario. Unfortunately, in many cases one has to rely on theoretical predictions and extrapolations because no experimental data are available within the astrophysical relevant energy region. For those reasons, the improvement of the sensitivity limit of experimental setups used for astrophysical cross-section measurements is of high importance. In addition, testing the predictive power of theoretical descriptions which are used to predict cross-section data in regions that are not accessible in the laboratory is one of the key motivations for cross-section experiments relevant for the p-process nucleosynthesis. In this contribution, different experimental methods and setups are briefly introduced that

*e-mail: fheim@ikp.uni-koeln.de

https://doi.org/10.1051/epjconf/202226011001

Figure 1. Left side: The γ-ray spectrometer HORUS and the target chamber dedicated for nuclear astrophysics [5]. The setup is used for in-beam cross-section measurements or for the irradiation of targets which can be transported to the dedicated offline counting setup (right side) for a subsequent analysis. The counting setup consists of two clover-type detectors which are shielded by lead and copper.

are used at the University of Cologne for cross-section measurements. In Sec. 3, an overview is given over proton- and α-induced reactions that have been measured in Cologne over the last decade.

2 Experimental setups and methods for cross-section measurements in Cologne

In general, there are two main techniques to measure the cross-section of a nuclear reaction in the laboratory. Both methods are based on the idea to count the number of nuclear reactions and normalize them to the number of incoming projectiles and target nuclei (see. Eq. 1).

$$\sigma = \frac{N_{\text{Reactions}}}{N_P \cdot N_T} \tag{1}$$

The idea of the first method is to measure the *prompt* emission of particles emitted by the reaction product. In the case of emitted γ-radiation, this technique is called in-beam γ-ray spectroscopy. The method and its applicability is presented in detail in Ref. [5]. The second type of methods can be applied if the produced nuclei are radioactive and decay via the emission of any kind of radiation, e.g., via γ-emission or α-emission. Via measuring the produced activity of the reaction products the number of produced nuclei during irradiation and hence, the nuclear cross section σ can be derived. This method is therefore called activation method [6].

2.1 In-beam γ-ray spectroscopy

At the University of Cologne a dedicated target chamber is available that has been explicitly tailored to the needs of astrophysical cross-section measurements. The chamber is mounted within the γ-ray spectrometer HORUS and is very compact in its dimension (about 7 cm in length). That allows to move all detectors very close to the target position and yields a full-energy peak efficiency of about 3 % at E_γ = 1.3 MeV. The target chamber, the γ-ray spectrometer and the experimental methods are explained in detail in Ref. [5] and the chamber is shown on the left side in Fig. 1. During the last decade, a multitude of proton- and

α-induced reactions has been measured in Cologne by means of in-beam γ-ray spectroscopy, see Fig. 2 or Tab. 1. The obtained cross sections have not only been used to be compared to existing theoretical models but the detection of the prompt γ-rays also allowed to study more fundamental nuclear properties such as the γ-ray strength function or the nuclear level density, as described in Refs. [7, 8].

2.2 Activation experiments

If the reaction product is unstable and decays via the emission of particles or γ rays, the activated material can be transported to a dedicated counting setup after the end of irradiation. At the University of Cologne a low-background counting setup is available which consists of two HPGe clover detectors which are shielded by lead and copper. The setup has a full-energy peak efficiency of about 5.3 % at a γ-ray energy of E_γ = 1.3 MeV [9] and is shown on the right side of Fig. 1. Using the γ-ray counting setup, various proton- and α-induced reactions have been investigated in Cologne within the last years. In addition to single-target irradiation, we performed stacked-target experiments for the ^{170,172}Yb(α,n)^{173,175}Hf reactions using water-cooled stacks which consist of four targets each [10].

Apart from γ-ray counting experiments, we have been able to measure the astrophysical relevant ^{144}Sm(α, γ)^{148}Gd reaction at energies close to the Gamow window. As ^{148}Gd does not emit any γ rays within its α-decay, we performed an α-counting experiment [11].

3 Overview of investigated reactions

Over the last decade, a multitude of nuclear reactions involving heavy nuclei has been investigated by means of different techniques at the University of Cologne. The derived cross-section data are inevitable for nucleosynthesis network calculations, in particular they are relevant for p-process nucleosynthesis. Additionally, the experimental data are required to benchmark theoretical models that are used to predict reaction rate data in nuclear regions that are not (yet) accessible via experiments and to put the models on a firm basis. The explicit nuclear reactions investigated in Cologne are listed in Tab. 1 and depicted in Fig. 2. In the future, detailed investigations of nuclear reactions relevant for nuclear astrophysics will be continued in Cologne, aiming at extending the available database of measured cross sections.

Table 1. List of proton- and α-induced reaction cross sections relevant for nuclear astrophysics measured at the University of Cologne over the last decade.

(p,γ)	(p,n)	(α,γ)	(α,n)
^{85}Rb$(p,\gamma)^{86}$Sr [12]	^{96}Mo$(p,n)^{96m,g}$Tc [9]	^{107}Ag$(\alpha,\gamma)^{111}$In [20]	^{107}Ag$(\alpha,n)^{110}$In [20]
^{87}Rb$(p,\gamma)^{88}$Sr [12]		^{108}Cd$(\alpha,\gamma)^{112}$Sn [21]	^{108}Cd$(\alpha,n)^{111}$Sn [21]
^{89}Y$(p,\gamma)^{90}$Zr [5, 8, 13]		^{112}Sn$(\alpha,\gamma)^{116}$Te [24]	^{141}Pr$(\alpha,n)^{144}$Pm [22]
^{92}Mo$(p,\gamma)^{93}$Tc [14]		^{144}Sm$(\alpha,\gamma)^{148}$Gd [11]	^{168}Yb$(\alpha,n)^{171}$Hf [23]
^{93}Nb$(p,\gamma)^{94}$Mo [7, 15]		^{168}Yb$(\alpha,\gamma)^{172}$Hf [23]	^{170}Yb$(\alpha,n)^{173}$Hf [10]
^{96}Mo$(p,\gamma)^{97}$Tc [16]			^{172}Yb$(\alpha,n)^{175}$Hf [10]
^{107}Ag$(p,\gamma)^{108}$Cd [17]			
^{109}Ag$(p,\gamma)^{110}$Cd [18]			
^{130}Ba$(p,\gamma)^{131}$La [19]			

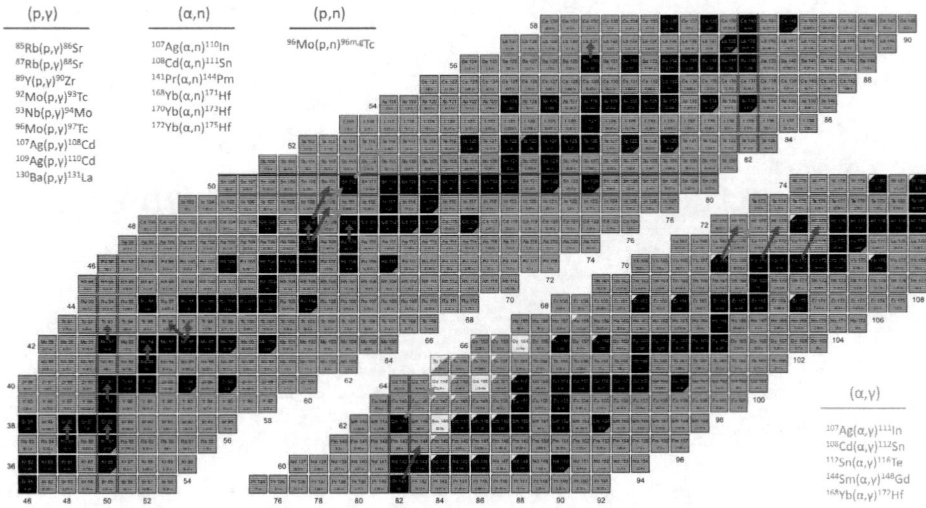

Figure 2. Illustration of proton- and α-induced reaction cross sections relevant for nuclear astrophysics measured at the University of Cologne over the last decade.

References

[1] F. Käppeler *et al.*, Rev. Mod. Phys. **83** 157 (2011)
[2] J. J. Cowan *et al.*, Rev. Mod. Phys. **93** 015002(2021)
[3] M. Arnould and S. Goriely, Phys. Rep. **384** 1 (2003)
[4] T. Rauscher *et al.*, Rep. Prog. Phys. **76** 76 (2013)
[5] F. Heim *et al.*, Nucl. Instr. Meth. Phys. Res. A **966** 163854 (2020)
[6] Gy. Gyürky *et al.*, Eur. Phys. J. A **55** 41 (2019)
[7] F. Heim *et al.*, Phys. Rev. C **103** 025805 (2021)
[8] L. Netterdon *et al.*, Phys. Lett. B **744** 358 (2015)
[9] F. Heim *et al.*, Phys. Rev. C **103** 054613 (2021)
[10] M. Müller *et al.*, to be published
[11] P. Scholz *et al.*, Phys. Rev. C **102** 045811 (2020)
[12] S. Wilden *et al.*, to be published
[13] L. Netterdon *et al.*, Nucl. Instr. Meth. Phys. Res. A **754** 94 (2014)
[14] J. Mayer *et al.*, Phys. Rev. C **93** 045809 (2016)
[15] F. Heim *et al.*, Phys. Rev. C **101** 035807 (2020)
[16] F. Heim *et al.*, to be published
[17] F. Heim *et al.*, Phys. Rev. C **101** 035805 (2020)
[18] F. Heim *et al.*, Phys. Rev. C **103** 055803 (2021)
[19] L. Netterdon *et al.*, Phys. Rev. C **90** 035806 (2014)
[20] Y. Wang *et al.*, to be published
[21] P. Scholz *et al.*, Phys. Lett. B **761** 247 (2016)
[22] A. Sauerwein *et al.*, Phys. Rev. C **84** 045808 (2011)
[23] L. Netterdon *et al.*, Nucl. Phys. A **916** 149 (2013)
[24] L. Netterdon *et al.*, Phys. Rev. C **91** 035801 (2015)

^{3}He(α,γ)^{7}Be cross section measurement around ^{7}Be known energy levels

Ákos Tóth[1,2] and Tamás Szücs[1]
[1]Institute for Nuclear Research (ATOMKI), Debrecen, Hungary
[2]University of Debrecen, Doctoral School of Physics, Debrecen, Hungary
E-mail: toth.akos@atomki.hu

Abstract. The ^{3}He(α,γ)^{7}Be reaction plays an important role in two astrophysical scenarios. It is a key reaction in lithium production during the Big Bang Nucleosynthesis and one of the central reaction in the p-p chain in stars. In the case of the former event, the Gamow energy of the reaction is around 0.2 MeV, while in the case of the p-p chain in the Sun, an order of magnitude less, around 0.023 MeV. Experimental investigation at such low energies is very difficult, if possible at all, thus low energy extrapolation inevitable to predict the reaction rate at these energies. The extrapolation and its uncertainty are influenced by the precision and covered energy range of the data used. There are many precision datasets between $E_{c.m.}$ = 0.3–3.1 MeV, but only one below and one above. At higher energies known levels of ^{7}Be exist, which motivates the study of that energy range. Therefore, we performed investigations in the energy range of $E_{c.m.}$ = 4.3–8.3 MeV, where the radiative cross section has not been studied so far. For the cross section determination, the activation technique was used utilising a thin-windowed gas cell and the MGC-20 cyclotron of ATOMKI.

1 Introduction

The study of nuclear reaction cross sections is of interest in various aspects of astrophysics, such as the formation of the elements, stellar evolution, the formation of the solar system and the evolution of the early universe. The ^{3}He(α,γ)^{7}Be nuclear reaction is important in two different astrophysical scenarios. On the one hand, it plays an important role in the production of elements in the Big Bang Nucleosynthesis (BBN), through the production of ^{7}Li. On the other hand, it is one of the branching reactions of the proton-proton chain therefore significantly affects the flux of neutrinos from the Sun's core. It is very important to know the cross section of the reaction with as little uncertainty as possible, because this reduces the uncertainty of the model calculations, which are currently less precise than the experimental solar neutrino flux determinations.

In the relevant energy range in stars, the so-called Gamow window, experimental data are difficult to provide because of the extremely low cross sections caused by the Coulomb barrier between the interacting charged particles.

A part of the energy range for the BBN has been experimentally determined by the LUNA Collaboration in the range of $E_{c.m.} = 0.09 - 0.17$ MeV [1, 2, 3, 4]. Currently, this is the only measurement that has been performed within the Gamow window of this reaction.

However, if we have experimental data at higher energies, it is possible to extrapolate to the low energy range and predict the reaction rate. There are many experimental data available in the $E_{c.m.} = 0.3 - 3.1$ MeV range [5, 6, 7, 8, 9, 10]. In 2019, the reaction cross section was measured around the proton separation energy of the compound ^{7}Be nucleus in $E_{c.m.} = 4.0-4.4$ MeV [11].

To ensure the accuracy of the extrapolation, experimental data are needed not only around the relevant energy range, but also at higher energies. In our experiment, we investigated the cross section of the reaction around the known energy levels of ^{7}Be, covering the energy range of $E_{c.m.} = 4.3 - 8.3$ MeV.

2 Experimental details

The cross section of the ^{3}He(α,γ)^{7}Be reaction was determined using the activation technique [12]. Usually the first step is to create radioactive reaction products by irradiation and then detect the yield of radiation emitted after the decay of these nuclei. The experiments were carried out with a thin-windowed gas cell. High putiry Al foils served as entrance and exit windows for the cell of 4.19-cm active lengths. The thickness of the entrance foils used was 10 μm, while that of the exit foils was 10, 20 and 25 μm, depending on the energy of the ^{7}Be produced during irradiation. Despite the foil high purity, its contains some impurities (Cu, Fe, Mg, Si...). On these impurities can undergo (α,n) reactions, which can produce nuclei with short half-lives ($<$ 1 day). The decay of these nuclei increases the undesirable background in the spectrum.

High purity (99.999 %) ^{3}He gas was used as the target, with an initial pressure of about 100 mbar during the irradiations. In the cell, an appropriately sized O-ring was placed under the foil, clamped down with a metal plate and secured with screws, giving the foil a diameter of 12 mm.

The irradiations were performed by the ATOMKI MGC-20 cyclotron accelerator. To create adequate activity, the length of the irradiations was nearly 20 hours. The activation chamber was isolated from the rest of the beamline. A voltage of -300 V was applied to apertures right between the entrance of the chamber and the beam defining aperture to eliminate the disturbing effect of the generated secondary electrons during irradiation. The function of the collimator is to form the diverging beams. The gas cell is part of the activation chamber, which works as a Faraday cup, so it is possible to determine the number of bombarding particles by charge measurement.

The pressure in the cell has been traced continuously with a pressure gauge. Most of the alpha beam unreacted passed through the cell, those were stopped in a water cooled tantalum sheet. The energy loss of the beam in the entrance foil was calculated with the SRIM program [13]. Depending on the initial beam energy the energy loss varied between 0.65 MeV and 1.0 MeV. The exit foil was used as a catcher foil. The produced ^{7}Be nuclei are implanted in the aluminium foil. Following irradiation, the catcher aluminium foil with the implanted ^{7}Be nuclei was placed in front of a high-purity germanium detector at a distance of 1 cm. To reduce background radiation, the detector was surrounded by 8 cm thick lead. The γ-ray detection of the 477.6 keV γ-transition was started after a cooling time of at least one day, to let the disturbing parasitic activities created on the foil impurities decay out. Each sample was usually measured for 7-12 days, in several periods to achieve good statistics.

The detector efficiency-energy function was determined using calibrated sources of known activity (^{133}Ba, ^{152}Eu and ^{60}Co). To avoid the so called true-coincidence-summing, the calibration measurements were carried out in far geometry. Subsequently, a high activity ^{7}Be source was used to determine the efficiency ratio for the close-far geometry. This ratio was used to determine the efficiency in the 1 cm geometry for the γ-transition under investigation. In this way, the uncertainty of the efficiency could be given with an accuracy of 2%.

The beam loses energy when it passes through the entrance foil, so this must be taken into account when determining the centre-of-mass energy. It is therefore necessary to determine the exact thickness of the entrance foil. This was done using an Ortec Soloist type alpha spectrometer, which includes a silicon ion implanted detector and various collimators. The detector and the radiation source are located in a common vacuum area. The first step of the thickness measurement is to perform the calibration with a mixed alpha source (containing ^{239}Pu, ^{241}Am and ^{244}Cm). Then the arrangement is modified by placing the entrance foil between the alpha source and the detector, and a collimator is placed along the source-foil-

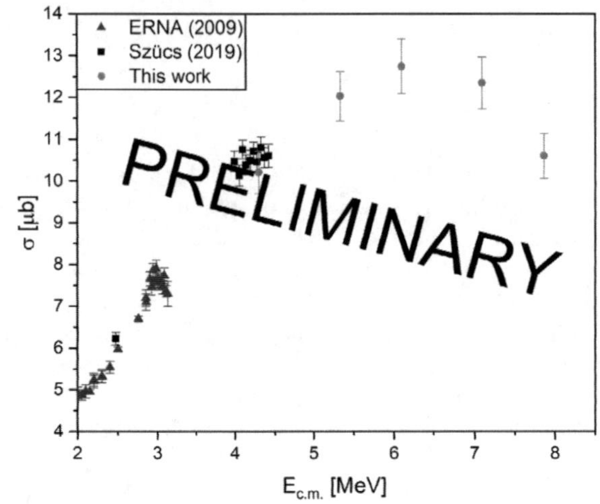

Figure 1. The new preliminary data compared to literature values [7, 11]. The total uncertainty of the data is shown.

detector axis. In this way, different points on the foil (4-5 locations) were tested, and the thickness of the entrance foils were determined with a few percent accuracy.

3 Preliminary results

The cross section of the ^{3}He(α,γ)^{7}Be reaction has been determined over an energy range never investigated experimentally before, around the known energy levels of the ^{7}Be nucleus.

In **Figure 1.**, preliminary cross sections and their associated statistical uncertainties as a function of centre-of-mass energy are provided. The first point was a benchmark test, which shows good agreement with the results of previous measurements. The energy range $E_a = 11 - 20$ MeV is studied in detail. The ATOMKI MGC-20 cyclotron is capable of accelerating twice positively charged alpha particles in the $E_a = 3.5 - 10.4$ and $10.4 - 20$ MeV energy range [14]. Accordingly, irradiation at energies higher than 20 MeV is not possible with this device. The cross section reaches its maximum at 15 MeV and then shows a decreasing trend at higher energies.

References

[1] D. Bemmerer et al. (LUNA Collaboration), Phys. Rev. Lett. **97**, 122502, (2006).
[2] F. Confortola et al. (LUNA Collaboration), Phys. Rev. C **75**, 065803, (2007).
[3] Gy. Gyürky et al. (LUNA Collaboration), Phys. Rev. C **75**, 035805, (2007).
[4] H. Costantini et al. (LUNA Collaboration), Nucl. Phys. A **814**, 144, (2008).
[5] B. S. Nara Singh, M. Hass, Y. Nir-El, and G. Haquin, Phys. Rev. Lett. **93**, 262503, (2004).
[6] T. A. D. Brown, C. Bordeanu, K. A. Snover, D. W. Storm, D. Melconian, A. L. Sallaska, S. K. L. Sjue, S. Triambak, Phys. Rev. C **76**, 055801 (2007).
[7] A. Di Leva et al., Phys. Rev. Lett. **102**, 232502, (2009).
[8] M. Carmona-Gallardo et al., Phys. Rev. C **86**, 032801(R) (2012).
[9] C. Bordeanu, Gy. Gyürky, Z. Halász, T. Szücs, G. Kiss, Z. Elekes, J. Farkas, Z. Fülöp, and E. Somorjai, Nucl. Phys. A **908**, 1 (2013).
[10] A. Kontos, E. Uberseder, R. deBoer, J. Görres, C. Akers, A. Best, M. Couder, and M. Wiescher, Phys. Rev. C **87**, 065804, (2013).
[11] T. Szücs, G. G. Kiss, Gy. Gyürky, Z. Halász, T. N. Szegedi, and Zs. Fülöp, Phys. Rev. C **99**, 055804, (2019).
[12] Gy. Gyürky, Z. Fülöp, F. Käppeler, G. G. Kiss, and A. Wallner, Eur. Phys. J. A **55**, 41 (2019).
[13] J. F. Ziegler, M. Ziegler, and J. Biersack, Nucl. Instrum. Methods Phys. Res., Sect. B **268**, 1818 (2010).
[14] S. Biri, I. K. Vajda, P. Hajdu, R. Rácz, A. Csík, Z. Kormány, Z. Perduk, F. Kocsis, and I. Rajta, Eur. Phys. J. Plus **136**, 247 (2021).

EPJ Web of Conferences **260**, 11003 (2022) https://doi.org/10.1051/epjconf/202226011003
NIC-XVI

The challenging direct measurement of the 65 keV resonance strength of the ^{17}O(p,γ)^{18}F reaction at LUNA

Giovanni Francesco Ciani[1,*] *Denise* Piatti[2,**] *Riccardo Maria* Gesuè[3] for the LUNA collaboration

[1]Università degli Studi di Bari and INFN Division of Bari, Via E. Orabona 4, 40125 Bari (Italy)
[2]Università degli Studi di Padova and INFN Division of Padova, Via Marzolo 8, 35136 Padova (Italy)
[3]Università degli Studi di Napoli and INFN Division of Napoli,Via Cintia, 80126 Napoli (Italy)

Abstract. A precise determination of proton capture rates on oxygen is mandatory to predict the abundance ratios of the oxygen isotopes in a stellar environment where hydrogen burning is active. The ^{17}O(p,γ)^{18}F reaction, specifically, plays a crucial role in AGB nucleosynthesis as well as in explosive hydrogen burning occurring in type Ia novae. At temperatures of interest for the former scenario (20 MK $\leq$ T $\leq$ 80 MK) the main contribution to the astrophysical reaction rate comes from the $E_{\mathrm{c.m.}}$ = 65 keV resonance. The strength of this resonance is presently determined only through indirect measurements, with an adopted value of $\omega\gamma = (1.6 \pm 0.3) \times 10^{-11}$ eV. Thanks to the low background environment of the Laboratori Nazionali del Gran Sasso, the intense and stable beam provided by the LUNA 400 kV accelerator and the experience in oxygen target production, the LUNA collaboration is aiming the first direct measurement of the above mentioned resonance strength. In the present work details of challenging direct measurement planned at LUNA will be described.

1 Introduction

The ^{17}O(p,γ)^{18}F reaction (Q = 5607.1(5)[1]) participates in the NO section of the CNO cycle, which is the main H-burning cycle in H-shell of giant stars or in explosive scenarios as type Ia novae.

Oxygen isotopic ratios in H-burning shell of Red Giant Branch (RGB) and Asymptotic Giant Branch (AGB) stars only depend on the rate of proton capture reaction on oxygen. Variations of the isotopic ratio occur at the surface as result of convective instabilities in the stellar envelope. This isotopic ratio can be observed in stellar IR spectra [2] and it probes the depth attained by the mixing episodes as well as the additional processes that may contribute to the chemical surface variation [3] Moreover, the oxygen isotopic ratio is measured in pre-solar meteoritic grains [4] providing high-precision snapshots of the progenitor isotopic composition as well as of nuclear and stellar parameters.

In both cases the reliability of probing the deep mixing and of the pre-solar grain origin depend on our knowledge of the rates of the nuclear reactions involved.

The ^{17,18}O(p,γ)^{18,19}F and ^{17,18}O(p,α)^{14,15}N reaction cross sections have been measured at the Laboratory for Underground Nuclear Astrophysics (LUNA), located deep underground at

*e mail: giovanni.ciani@ba.infn.it
**e mail: denise.piatti@pd.infn.it

Laboratori Nazionali del Gran Sasso (LNGS) where muonic background is reduced by 6 orders of magnitude with respect to the surface experiments [5]. Results obtained by the LUNA collaboration [6–11] had paramount impact on the interpretation of the oxygen isotopic ratio observed in giant stars and in stardust grains [12, 13]. Moreover, a study of the $^{16}O(p,\gamma)^{17}F$ reaction is also planned at LUNA 400 kV.

The $^{17}O(p,\gamma)^{18}F$ reaction study completed at LUNA was mainly focused at Gamow energies for classical novae, 160 keV $\leq E_{c.m.} \leq$ 370 keV. At temperature of interest for the AGB star H-burning shell (20 MK $\leq$ T $\leq$ 80 MK), however, the main contribution to the astrophysical reaction rate comes from the $E_{c.m.}$ = 65 keV resonance. An accurate measurement of the resonance strength can improve the reaction rate determination and will help constraining the RGB and AGB models.

The strength of the $E_{c.m.}$ = 65 keV resonance is presently determined only through indirect measurements [14–16]. In a recent reevaluation by Fox et al. [17], the $\omega\gamma$ was calculated from the reported resonance strengths of $^{17}O(p,\alpha)^{14}N$, $^{14}N(\alpha,\gamma)^{18}F$ and from the Γ_α partial width obtained by the $^{14}N(\alpha,\alpha')$ scattering data [14]. The value reported by Fox et al. [17] for the resonance strength $\omega\gamma = (1.6 \pm 0.3) \times 10^{-11}$ eV. Recent measurements of the $^{17}O(p,\gamma)^{18}F$ reaction performed at LUNA using both the prompt γ-ray detection and the activation method led to a precise determination of the $E_{c.m.}$ = 183 keV resonance strength [7], while a direct measurement of the $E_{c.m.}$ = 65 keV resonance strength remained prohibitive because of background and efficiency limitations.

2 Experimental Setup

Starting from the resonance strength reported in [17], the expected yield is less then 1 count/C. For this reason the setup must be optimized in terms of background reduction to overcome a disfavoured signal-to-noise ratio.

In order to minimize the γ-ray absorption the scattering chamber and target holder of the setup are made of aluminum. To maximize the detection efficiency a 4π BGO detector is installed all around the target. Solid targets are produced by anodization of tantalum backings in enriched, 90% ^{17}O, water with well-known stochiometry, Ta_2O_5 [18]. In order to characterize and monitor target thickness they will be doped in ^{18}O (at the level of 5%). This will allow to periodically scan the $E_{c.m.}$ = 143 keV resonance of the $^{18}O(p,\gamma)^{19}F$ reaction as monitor for target degradation. Finally in order to further reduce the environmental background the whole setup is surrounded by a shielding made of borated polyethylene and lead, as neutron and gamma absorber, respectively [19]. Figure 1 shows the upgraded setup.

3 Preliminary Results

Tests have been performed to fine tune and characterize the present setup.
Preliminary GEANT4 simulations show a 30% improvement in the detection efficiency at the E_γ = 662 keV with respect to the previous setup configuration, with a steel chamber and a brass target holder. The comparison is shown in Figure 2.
The 30 day long observed background count rate in the region of interest is (0.08 ± 0.01) counts per hour, reduced by a factor of (4.2 ± 0.8) with respect to using the BGO unshielded [20]. The average current for proton beam energy of about 75 keV is $I_b = 100\,\mu A$. According to the above preliminary evaluations the detection limit was calculated following [21] and a resonance strength of the order of 10^{-11} eV can be detected.
LUNA now has the good opportunity of directly measuring this resonance in spite of its low

 https://doi.org/10.1051/epjconf/202226011003

strength, with expected large impact on the $^{17}O(p,\gamma)^{18}F$ reaction rate and our understanding of AGB nucleosynthesis processes.

Figure 1: Present setup for the $^{17}O(p,\gamma)^{18}F$ cross section measurement at LUNA. Left: the external borated Polyethylene shielding. Right: the three-layer shielding opened to show the aluminum chamber and target holder to which the water cooling is connected. Part of the BGO detector is also visible inside the shielding.

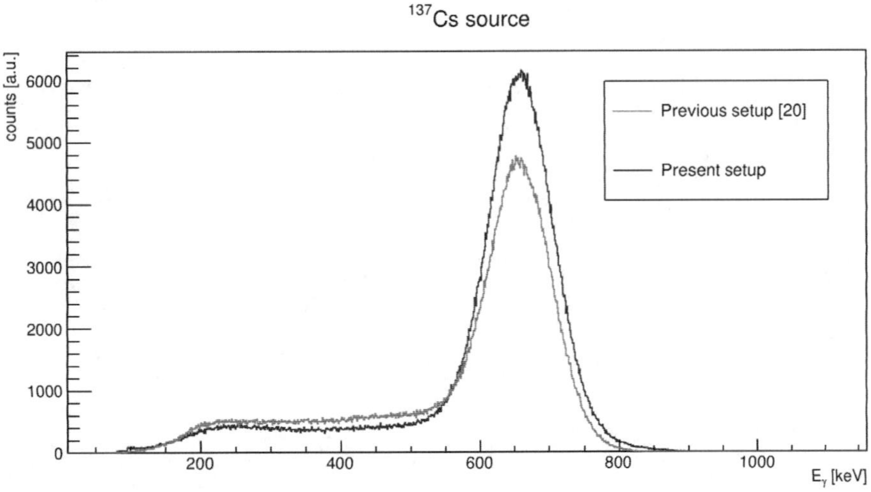

Figure 2: Comparison between ^{137}Cs spectrum simulated with present setup, blue line, and with previous setup, red line.

References

[1] M.Wang *et al.*, Chinese Physics C **41**, 030003 (2017)
[2] T. Lebzelter *et al.*, Astronomy and Astrophysics **578**, A33 (2015)

[3] C. Charbonnel *et al.*, Astrophysical Journal Letters **453**, L41 (1995)
[4] E. Zinner *et al.*, Presolar Grains. In Davis, A. M. (ed.) Microphysics of Cosmic Plasmas, 181–213 (2014).
[5] C. Broggini *et al.*, Progress in Particle and Nuclear Physics **98**, 55 (2018).
[6] D. A. Scott *et al.*, Physical Review Letters **109**, 202501 (2012)
[7] A. Di Leva *et al.*, Physical Review C **89**, 015803 (2014)
[8] C. G. Bruno *et al.*, European Physical Journal A **51**, 94 (2015)
[9] C. G. Bruno *et al.*, Physical Review Letters **117**, 142502 (2016)
[10] A. Best *et al.*, Physics Letters B **797**, 134900 (2019)
[11] F.R. Pantaleo *et al.*, Physical Reviw C **104**, 025802 (2021)
[12] O. Straniero *et al.*, Astronomy and Astrophysics **598**, A128 (2017)
[13] M. Lugaro *et al.*, Nature Astronomy **1**, 0027 (2017)
[14] H.-B. Mak *et al.*, Nuclear Physics A **343**, 79 (1980)
[15] V. Landre *et al.*, Physical Review C **40**, 1972 (1989)
[16] J.C. Blackmon *et al.*, Physical Review Letter **74**, 2642 (1995)
[17] C. Fox *et al.*, Physical Review C **71**, 055801 (2005)
[18] A. Caciolli *et al.*, EPJ A **48**, 144 (2012)
[19] A. Boeltzig *et al.*, Journal of Physics G **45**, 025203 (2018)
[20] A. Boeltzig *et al.*, Physics Letters B **795**, 122-128 (2019)
[21] G. Gilmore, *Practical γ-ray spectrometry, 2nd edition* (John Wiley and Sons, New York 2008)

Binary neutron star mergers of quark matter based nuclear equations of state

Atul Kedia[1,*], *Grant* Mathews[1,**], *Hee Il* Kim[2], and *In-Saeng* Suh[3,1]

[1]Center for Astrophysics, Department of Physics, University of Notre Dame, Indiana, USA
[2]Center for Quantum Spacetime, Sogang University, Seoul 04107, Korea
[3]Center for Research Computing, University of Notre Dame, Indiana, USA

Abstract. With observations of gravitational wave signals from binary neutron star mergers (BNSM) by LIGO-Virgo-KAGRA (LVK) Collaboration and NICER, the nuclear equation of state (EOS) is becoming increasingly testable by complementary numerical simulations. Numerous simulations currently explore the EOS at different density regimes for the constituent neutron stars specifically narrowing the uncertainty in the sub-nuclear densities. In this paper we summarize the three-dimensional general relativistic-hydrodynamics based simulations of BNSMs for EOSs with a specific emphasis on the quark matter EOS at the highest densities.

1 Introduction

Neutron stars are ideal locations to probe the properties of matter at the highest density. The microphysics of nuclear interactions in a neutron star are reflected in its large structural features like its mass-radius relation. These interactions also modulate the evolution of neutron-star binary systems. This amplification of subatomic physics makes probing the physics at scales 10^{-14} m plausible through the observation of neutron star typically of sizes $\sim 10^4$ m. In particular, neutron star binary systems provide a means to analyze the pressure of nuclear matter at all domains of nuclear densities. The first detection of gravitational waves from the binary neutron star merger GW170817 by the LIGO-Virgo Collaboration and the pulsar PSR J0740 by NICER have provided new insights into the form and structure of dense neutron-star matter. [1, 2] The gravitational wave signal detected thus is dependent upon the tidal distortion of the neutron stars as they approach merger. The LIGO analysis [1, 3] of the tidal polarizability [4, 5] deduced from post-Newtonian dynamics implies that the radius of the stars of 1.4 $M_\odot$ is in the range 10.5 km $\leq R \leq$ 13.3 km, which has placed tight constrains on the equation of state (EOS) for nuclear matter as the stars approach merger.

Here we summarize the literature in the regime of quantum chromodynamics (QCD) attained in the core of the short-lived massive neutron star formed after the merger of a neutron star binary, and the potential locations to look at for signs from the QCD regime. The reason to be excited about this, particularly now, is that the detection of the gravitational radiation during the *postmerger* could be used as a sensitive probe of both the order of the quark-hadron phase transition and the properties of matter in the non-perturbative regime of QCD.

*e-mail: atulkedia93@gmail.com
**e-mail: gmathews@nd.edu

The prospect of the postmerger evolution being used to explore EOS issues has been proposed for some time (e.g. [6]). There have also been several investigations into the effects of the formation of quark matter in the BNSM. [7–10]. For the most part these studies have considered effects of a first-order phase transition and the formation of a mixed quark-hadron phase. However, a first-order transition can soften the EOS and hasten the formation of the imminent black hole.

The postmerger gravitational wave strain lies in a frequency range of 2 kHz to 5 kHz and as a result the strength for a binary at 50 Mpc is not easily accessible to aLIGO/aVirgo/KAGRA. However, the third generation GW observatories, such as the Einstein Telescope and the Cosmic Explorer, will have enhanced sensitivities in this frequency range and will be capable of observing the postmerger evolution of a BNSM. Furthermore, there is a suggestion in the literature (see Refs. [11, 12]) of post merger energy output in gravitational radiation from the GW170817 event that appears to be an extended ringdown (see however Ref. [13].) In Ref. [12] it was hypothesized that such extended emission might result from spin down of a magnetar.

2 Dense matter Equation of State

To describe the evolution of matter in a closed form, the hydrodynamics equations require an additional constraint that relates the various state variables of the matter, i.e. pressure, density, electron fraction, chemical potentials, etc. in a neutron star. [14, 15] Constraints on the low density (sub-nuclear) equation of state (EOS) have been placed by aLIGO based upon the tidal polarizability deduced from the chirp associated with event GW170817. [1, 3] However the EOS at higher densities attained during a binary neutron star merger must also include the consequences of a transition between hadronic matter and quark matter. As the merged system collapses to a black hole it unavoidably encounters certain high dense phases of matter, particularly the transition to quark matter.

It is worth noting that as the baryon density and chemical potential increase the QCD strong coupling constant α_s approaches unity and a nonperturbative approach to QCD is imperative. In particular, there is rich physics in this region of the quark-matter phase diagram including the generation of constituent quark masses, due to chiral symmetry breaking [16], and quark pairing leading to color superconductivity [17]. The evolution of these effects until the asymptotic regime must be described during the collapse.

The transition from hadronic matter to quark matter is not yet fully understood. [18] However, people have proposed many interesting models to describe the transition from hadronic matter to quark matter at densities between 2-5 ρ_{nuc}. [19, 20] Considering strongly interacting quark states for the crossover density region, people could have made the EOSs satisfying 2 times solar mass observational bound. [21] Also, many models having the first order phase transition features have been proposed to describe the transition.

3 Gravitational Waves and Power Spectral Density

In addition to f_{max} which could be obtained from the instantaneous change of the phase of h at the chirp, the power spectral density (PSD) of the strain, poses to reveal further features about the star's EOS. Refs. [10, 22] shows the presence of high frequency spectral features in the f_1, f_2 and f_3 (as defined in Ref. [22]) modes for binaries with different EOS description. The positioning and strength of the frequencies depends on the nature of the EOS. One of the important aspects about the power spectra of the mergers is that the strength of the PSD at 2 kHz to 5 kHz frequencies is weak enough that the current LVK detectors are unable to discern them from the noise. It is also pointed out that the third generation gravitational wave interferometers, the Einstein Telescope and the Cosmic Explorer, would have higher sensitivities at these frequencies making them more likely to resolve the f-modes. Further, Ref. [22] notes the occurrence of 'quasiuniversal' relations relating f-modes with compactness and the maximum frequency of chirp with tidal deformability of a single stable NS for the given EOS. These relations can be explained partially by the stiffness of the equation of state, softer EOSs leading to higher f-modes. Recent studies of the BNSM with the phase transition have examined whether the features of the phase transition could be imprinted in the f-mode frequencies, for example, the shift of the peak frequencies. [8, 23] We are currently investigating the spectral features of the BNSM with a specific crossover EOS and have found some interesting postmerger behaviour showing elongated duration owning to the stiffened EOS. [18, 19, 24]

4 Conclusion

We briefly summarize recent studies of BNSM with quark matter EOSs. An observation of the postmerger and the collapse in a BNSM could be used to determine the nature of the phase transition and the physics at the crossover densities. The next generation GW detectors will probe and unveil the physics of this dense matter physics with their highly improved power of detectabilities at higher GW frequencies.

Finally, noting a caveat that there have been no studies fully considering aspects such as magnetohydrodynamics and neutrino transport for the simulations, however, we expect that essential features of the non-perturbative characters of the QCD would be revealed and confirmed without needing to take these considerations into account.

References

[1] B.P. Abbott et al. (LIGO Scientific Collaboration and Virgo Collaboration), Phys. Rev. Lett. 119, 161101 (2017).

[2] M. C. Miller et al 2021, *The Astrophysical Journal Letters*, 918 L28

[3] B. P. Abbott, et al. "Tests of general relativity with GW170817." Phys. Rev. Lett. 123.1 (2019): 011102.

[4] T. Damour, M. Soffel, and C. Xu, Phys. Rev. D 45, 1017 (1992).

[5] E. E. Flanagan and T. Hinderer, Phys. Rev. D 77, 021502 (2008).

[6] A. Bauswein, H.-T. Janka, K. Hebeler, and A. Schwenk, Phys. Rev. D 86, 063001 (2012).

[7] E. R. Most, L. Jens Papenfort, V. Dexheimer, M. Hanauske, H. Stoecker, and L. Rezzolla, On the de-confinement phase transition in neutron-star mergers, Eur. Phys. J. A 56, 59 (2020)

[8] L. R. Weih, M. Hanauske, and L. Rezzolla, Postmerger Gravitational-Wave Signatures of Phase Transitions in Binary Mergers, Phys. Rev. Lett. 124, 171103 (2020).

[9] S. L. Liebling, C. Palenzuela, and L. Lehner, Effects of High Density Phase Transitions on Neutron Star Dynamics, Class. Quant. Grav. 38, 115007 (2021)

[10] A. Prakash, D. Radice, D. Logoteta, et al., arXiv:2106.07885 (2021)

[11] H. M. P van Putten and D. V. Massimo, MNRAS, 482, L46 (2019).

[12] H. M. P van Putten, D. V. Massimo, and A. Levinson, Astrophys. J. Lettt, 876, L2 (2019).

[13] M. Oliver, D. Keitel, A. Miller, et al. MNRAS, 485, 843 (2019).

[14] D. Radice, A. Perego, F. Zappa, and S. Bernuzzi Ap. J. Lett. 852.2 (2018): L29.

[15] J. Olson, *et al.*, arXiv:1612.08992 (2016).
[16] Hatsuda, T., and Kunihiro, T. 1994, PhR, 247, 221
[17] M. G. Alford, A. Schmitt, K. Rajagopal, and T. Schäfer, 2008, RvMP, 80, 1455.
[18] Baym, G., Hatsuda, T., Kojo, T., et al. 2018, RPPh, 81, 056902
[19] G. Baym, S. Furusawa, T. Hatsuda, T. Kojo, H. Togashi, ApJ, 885, 42 (2019)
[20] E. Annala, E., T. Gorda, A. Kurkela, et al., Nature Physics, (2020), 16(9), 907-910.
[21] P. B. Demorest, T. Pennucci, S. M. Ransom, et al. , Nature (London) 467, 1081 (2010).
[22] K. Takami, L. Rezzolla, and L. Baiotti, Physical Review D 91.6 (2015): 064001
[23] S. Blacker, NUF Bastian, A. Bauswein, et al., Physical Review D 102.12 (2020): 123023.
[24] A. Kedia, H.I. Kim, I.S. Suh, G. Mathews, (2022) in preparation.

EPJ Web of Conferences **260**, 11005 (2022) https://doi.org/10.1051/epjconf/202226011005
NIC-XVI

Investigation of the ^{7}Li(p,n) neutron fields at high energies

Benjamin Brückner[1,*], *Philipp* Erbacher[1], *Kathrin* Göbel[1], *Tanja* Heftrich[1], *Kafa* Khasawneh[1], *Deniz* Kurtulgil[1], *Christoph* Langer[3], *Ralf* Nolte[2], *Markus* Reich[1], *René* Reifarth[1], *Benedikt* Thomas[1], *Mario* Weigand[1], *Michael* Wiescher[4], and *Meiko* Volknandt[1]

[1]Goethe-Universität Frankfurt, Max-von-Laue-Str. 1, 60438 Frankfurt, Germany
[2]Physikalisch-Technische Bundesanstalt, Bundesallee 100, 38116 Braunschweig, Germany
[3]FH Aachen, Heinrich-Mußmann-Straße 1, 52428 Jülich, Germany
[4]University of Notre Dame, 46556 Indiana, USA

Abstract. The neutron activation method is well-suited to investigate neutron-capture cross sections relevant for the main s-process component. Neutrons can be produced via the ^{7}Li(p,n) reaction with proton energies of 1912 keV at e.g. Van de Graaff accelerators, which results in a quasi-Maxwellian spectrum of neutrons corresponding to a temperature of $k_B T = 25$ keV. However, the weak s-process takes place in massive stars at temperatures between 25 and 90 keV. Simulations using the PINO code [2] suggest that a Maxwellian spectrum for higher energies, e.g. $k_B T = 90$ keV, can be approximated by a linear combination of different neutron spectra. To validate the PINO code at proton energies $E_p \neq 1912$ keV, neutron time-of-flight measurements were carried out at the PTB Ion Accelerator Facility (PIAF) at the Physikalisch-Technische Bundesanstalt in Braunschweig, Germany.

1 Introduction

The s-process is divided into the main and weak component. Differences lie in the neutron densities, neutron-to-seed ratios and the temperatures [1]. The main s-process component, which contributes mostly to nuclei with mass numbers above $A \approx 90$, takes place at $k_B T = 25$ keV in thermal pulsing low mass asymptotic giant branch (TP-AGB) stars. In contrast, the weak s-process takes place in massive stars with more than 8 solar masses at temperatures between $k_B T = 25$ keV and $k_B T = 90$ keV. Nuclei in the mass number range between 60 and 90 are mostly created in this environment. Since the neutron fluence in the weak s-process is too low to achieve reaction flow equilibrium, a particular neutron-capture cross section not only influences the abundance of the respective isotope, but affects the abundances of all heavier isotopes as well [3].

1.1 Activation technique

The activation technique has been proven to be a well-suited tool to investigate neutron-capture cross sections relevant for the main s-process component: A quasi-stellar neutron

*e-mail: brueckner@iap.uni-frankfurt.de

spectrum corresponding to a temperature of $k_B T = 25\,\text{keV}$ can be obtained by bombarding a several μm thick metallic lithium target with protons at an energy of 1912 keV. The neutrons are emitted in a forward cone of 120° opening angle, which has to be covered by the sample [5]. The neutron capture cross section is determined as an integral value over the neutron energy distribution. A quasi-stellar neutron spectrum similar to a $k_B T = 90\,\text{keV}$ Maxwellian neutron energy spectrum cannot be reproduced by only one proton beam energy. However, simulations using PINO [2] suggest a linear combination of neutron distributions to approximate a quasi-stellar distribution with $k_B T = 90\,\text{keV}$ or other temperatures using the ^{7}Li(p,n)^{7}Be reaction. The results of the simulations are shown in Fig. 1. Hence, samples have to be activated with neutron distributions generated with various proton energies. The resulting spectrum-averaged cross sections (SACS) for each activation energy can be linearly combined to calculate the Maxwellian-averaged cross section (MACS).

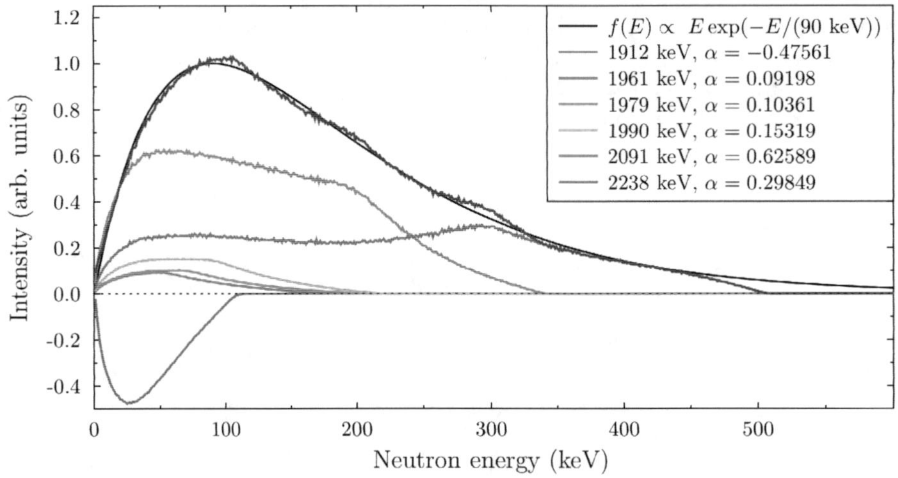

Figure 1. Simulated neutron energy distributions for the synthesis of an energy distribution with $k_B T = 90\,\text{keV}$.

2 Experiment

We want to verify and improve the accuracy of the PINO simulations for energies other than $E_p = 1912\,\text{keV}$. Therefore, we investigated the angular energy distribution of the neutrons from the ^{7}Li(p,n)^{7}Be reaction for different proton energies at the PTB Ion Accelerator Facility (PIAF) at the Physikalisch-Technische Bundesanstalt in Braunschweig. The unique experimental setup of the low-scatter facility allows the production of neutron reference fields in open geometry [4]. The neutron energy was determined by the time-of-flight method, which relies on the periodical production of neutrons at a well-defined frequency [3], given by the fast pulsed proton beam which was delivered by a 2 MV Tandem accelerator with a maximum repetition rate of 2.5 MHz. The proton beam had a nominal pulse width of about 1.5 – 2 ns and a frequency of 1.25 MHz. The beam impinged on a metallic lithium target on tantalum backing with a thickness of 5 μm, assuming a nominal density of 0.534 g/cm^3. By measuring the neutron time-of-flight t for a certain length of flight path L the neutron energy E_n can be determined by the classic approximation:

$$E_n = \frac{1}{2} m_n \left(\frac{L}{t}\right)^2 \tag{1}$$

Since the neutron-energy resolution can be enhanced by an increased flight path, but at the expense of neutron flux at the detector position, a flight path of $L = 0.7$ m was selected as a compromise. Ten different proton energies were measured: 1.887 MeV, 1.897 MeV, 1.907 MeV, 1.912 MeV, 2 MeV, 2.1 MeV, 2.2 MeV, 2.3 MeV, 2.5 MeV and 2.8 MeV. Measurements below 1.912 MeV were conducted for the energy calibration of the accelerator and to investigate the neutron production threshold. Neutron spectra were recorded using three ^{6}Li-Glass scintillation detectors mounted on movable arms. The neutron flux was monitored with two long counters positioned at 16° and 98° relative to the ion beam.

3 Results and Outlook

For each proton beam energy time-of-flight spectra were acquired at angles between 0° and 95° in 5° steps. The measurement durations varied between 15 and 60 min depending on the proton current and neutron yield of the reaction. During the measurements the proton current on target was 0.7-1.5 μA. The time-of-flight spectra were background corrected and the neutron yield was normalized using the recorded number of protons on the lithium target.

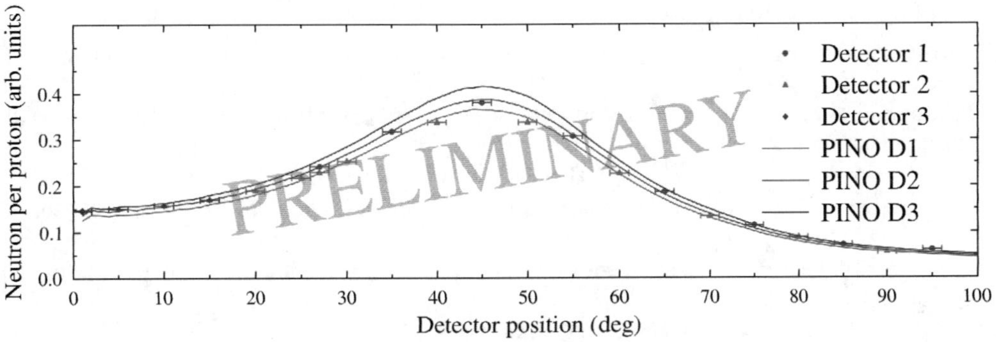

Figure 2. Neutron-per-Proton distribution over the measured angles. Each detector was simulated individually due to unique specifications and minor differences in target-detector distance.

For proton energies smaller than 2.3 MeV the resulting spectra as shown in Fig. 2 are in very good agreement with the PINO simulation. Above this energy, however, larger deviations are observable. Their origin is probably the uncertain description of the second neutron production channel ^{7}Li(p,n)^{7}Be^{m}. As soon as the reason for the deviations becomes clear the PINO code will be adjusted.

References

[1] P. A. Seeger, W. A. Fowler and D. D. Clayton, The Astrophysical Journal **11** 121 (1965)
[2] R. Reifarth, M. Heil, F. Käppeler and R. Plag, Nuclear Instruments and Methods in Physics Research A **608** 139 - 143 (2009)
[3] R. Reifarth, C. Lederer and F. Käppeler, Journal of Physics G **41** 053101 (2014)
[4] S. Roettger, R. Boettger, F. D. Brooks, A. Buffler, J. P. Meulders, R. Nolte, F. D. Smit and F. Wissmann, AIP Conference Proceedings **1175** 375 - 381 (2009)
[5] W. Ratynski and F. Käppeler, Physical Review C **37** 595 (1988)

Measurement of the ^{7}Be(p, γ)^{8}B reaction cross section with the recoil mass separator ERNA

Raffaele Buompane[1,2,*], *Antonino* Di Leva[2,3], *Lucio* Gialanella[1,2], *Antonio* D'Onofrio[1,2], *Mario* De Cesare[4], *Jeremias* G. Duarte[1,2], *Zsolt* Fülöp[5], *Leandro* R. Gasques[6], *György* Gyürky[5], *Lizeth* Morales-Gallegos[2], *Fabio* Marzaioli[1,2], *Giancarlo* Palumbo[2,7], *Giuseppe* Porzio[1,2], *David* Rapagnani[2,3], *Vincenzo* Roca[1,2], *Detlef* Rogalla[8], *Mauro* Romoli[2], *Claudio* Santonastaso[1,2], and *Daniel* Schürmann[1,2].

[1]Dipartimento di Matematica e Fisica, Università della Campania "L. Vanvitelli", Viale Lincoln, 5, Caserta, Italy, EU

[2]Istituto Nazionale di Fisica Nucleare, Sezione di Napoli, Via Cinthia snc, Napoli, Italy, EU

[3]Dipartimento di Fisica "E. Pancini", Università di Napoli "Federico II", Via Cinthia snc, Napoli, Italy, EU

[4]Dipartimento di Metodologie e Tecnologie per le Osservazioni e Misure, Centro Italiano Ricerche Aerospaziali, Capua, Italy, EU

[5]Institute for Nuclear Research (Atomki), H-4001 Debrecen, Hungary, EU

[6]Departamento de Física Nuclear, Instituto de Física da Universidade de São Paulo, São Paulo, SP, Brazil

[7]Dipartimento di Economia, Management Istituzioni - Laboratorio Chimico-Merceologico, Università degli Studi di Napoli "Federico II", Napoli, Italy, EU

[8]RUBION, Ruhr-Universität Bochum, Bochum, Germany, EU

Abstract. The cross section of ^{7}Be(p, γ)^{8}B represents one of the most important nuclear inputs for the prediction of the high energy component of solar neutrinos and it has also a direct impact on the ^{7}Li abundance after the Big Bang Nucleosynthesis. The importance of this reaction triggered an intense experimental work over the last decades, where discrepancies were observed between the results of different measurements. In addition, a question remains about possible common systematic effects, considering that all measurements share the same experimental approach, i.e. an intense proton beam impinging on a ^{7}Be radioactive target. Inverse kinematics, i.e. a ^{7}Be ion beam and a hydrogen target, with the direct measurement of the total reaction cross section by means of the detection of the ^{8}B recoils, can shed light on such systematic effects. Efforts attempted so far were limited by the low ^{7}Be beam intensity. We present here the results of a new measurement at E_{cm}= 376 to 819 keV using a high intensity ^{7}Be beam in combination with a windowless gas target and the recoil mass separator ERNA (European Recoil mass separator for Nuclear Astrophysics) at CIRCE (Center for Isotopic Research on Cultural and Environmental heritage), Caserta, Italy. Our results, including the systematic error, are compatible with previous measurements that yields lower value of $S_{17}(0)$ and are compatible with the currently accepted value from [1] only at a 2-σ level.

*e-mail: raffaele.buompane@unicampania.it

1 Introduction

The solar neutrinos of ^{7}Be to ^{8}B have a great interest, since they represent a large fraction of the high energy solar neutrino flux. The rate of the reaction ^{7}Be(p, γ)^{8}B relative to the ^{7}Be electron capture decay rate in the Sun determines the ratio of these neutrino. The importance of ^{7}Be to ^{8}B neutrinos ratio triggered several experiments to determine cross section of ^{7}Be(p, γ)^{8}B at the relevant astrophysical energies [2–12] and the ^{7}Be half-life in different environments [13–16].

All direct ^{7}Be(p, γ)^{8}B cross section measurements producing results with sufficient precision to constrain astrophysical models are measured in direct kinematics i.e. using an intense proton beams on radioactive targets. The complicated target stoichiometry and its modification under beam bombardment, as well as the difficulties of the detection of the reaction products are possibly the origin of the discrepancy between different data sets, that persists even if a selection based on a careful review of the experiments is done as proposed in [1]. An alternative approach using a radioactive ion beam and a hydrogen gas target was attempted in [12], and later in [17]. In both cases a recoil mass separator was employed to detect the ^{8}B recoils. However, those experiments could not achieve a sufficient precision because of the low ion beam intensity ($\approx 10^7$ pps). We present here a new experiment exploiting the same method with the much higher ion beam intensity available at the Tandem Accelerator Laboratory at CIRCE (Center for Isotopic Reasearch on Cultural and Environmental heritage), Dept. of Mathematics and Physics, University of Campania "L. Vanvitelli", Italy, where a ^{7}Be ion beam is routinely produced [18] with an intensity up to 10^9 pps.

2 Experimental setup

Until 2009 the recoil mass separator ERNA was installed at the Dynamitron Tandem Laboratorium of the Ruhr Universität Bochum, Germany. In 2009 ERNA was moved to the Tandem accelerator laboratory at CIRCE, University of Campania "L. Vanvitelli", Caserta, Italy. The new installation gave the opportunity to upgrade the experimental setup. In particular, a new windowless hydrogen gas target was built [19] and an additional magnet (Charge State Selection Magnet, CSSM in the following) was inserted between the gas target and the first focusing element of the separator. The experimental apparatus has been described in details in [19–22]. The measurements are performed in inverse kinematic, the ^{8}B nuclei formed in the reaction emerge from the target at forward angles together with the beam ions in an intensity ratio of 10^{-11} to 10^{-13}, depending on the cross section. The mass separator ERNA filter the recoils that are then directly detected in the ΔE – E telescope detector [22], example of matrices obtained is shown in figure 1. The number of ^{8}B recoils N_r observed in the final detector having an efficiency ϵ is given by:

$$N_r = N_b \cdot T_{RMS} \cdot \Phi_{q_r} \cdot \epsilon \cdot \int_{E_b-\Delta E_b}^{E_b} \frac{\sigma(E)}{|dE/dN_t|} dE \quad (1)$$

where N_b is the number of ^{7}Be projectiles measured by the elastic scattering of beam ions in the target region, Φ_{q_r} is the probability of the charge state q_r selected for the separation, T_{RMS} is the transmission of the recoils from the target to the final detector. The interaction cross section σ(E) is integrated over the beam energy loss in the target ΔE_b at the beam energy E_b, that is determined by the beam ions stopping power in the target $\frac{dE}{dN_t}$, where N_t is the number of nuclei per unit target area. All the relevant quantities have been determined experimentally, the systematic uncertainty in the determination of cross section is about 4%, stemming from the systematic uncertainties in the determination of the target thickness (2%),

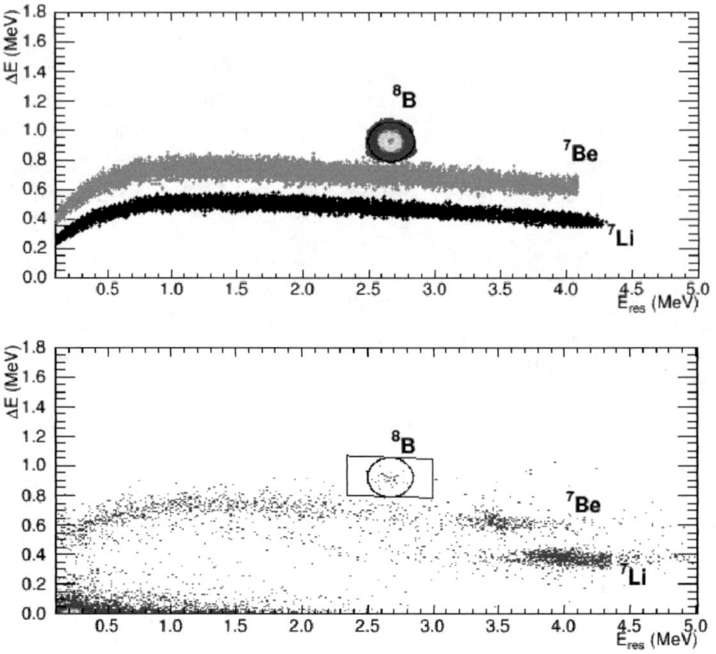

Figure 1. $\Delta E - E_{res}$ matrices of the end detector of ERNA for $^7Be(p,\gamma)^8B$ at E_b=675.5 keV. The upper matrix represent the predictions of a Monte Carlo simulation for the different ion species. The lower matrix the experimental results. Both ^{7}Li and ^{7}Be leaky beam ions are visible with long low energy tails. The rectangular region around the elliptical area where the ^{8}B recoils are expected is used to estimate the background.

number of projectiles (2%), transmission of the recoils (2%), recoil detection efficiency (1%), and charge state probability (2%). A detailed discussion of measurements and of the uncertainties affecting the relevant quantities is given in [20]. The beam energy has a systematic uncertainty of 0.06% [26].

3 Results

The resulting cross sections are shown in figure 2. The data are compared with a fit function obtained summing the nonresonant component S_{MN} from Table V of [23] and a resonant component S_{res}:

$$S_{factor}(E) = A \cdot S_{MN}(E) + \pi \lambda\!\!\!^{-2} \cdot \frac{2J+1}{(2J_p+1)\cdot(2J_t+1)} \cdot \frac{\Gamma_p(E)\cdot\Gamma_\gamma(E)}{(E-E_{res})^2 + (\Gamma_{tot}(E)/2)^2} \cdot E \cdot e^{2\pi\eta(E)} \quad (2)$$

where A is a scaling factor for the nonresonant component. $J_p = 1/2$, $J_t = 3/2$, and $J = 1$ are the spin of the ^{1}H, ^{7}Be, and ^{8}B relevant states for the resonant capture, respectively. The energy dependent widths $\Gamma_p(E)$ and $\Gamma_\gamma(E)$ are calculated using the approximations given in [24]:

$$\Gamma_p(E) = \Gamma_p(E_{res}) \frac{e^{2\pi\eta(E)}}{e^{2\pi\eta(E_{res})}} \quad (3)$$

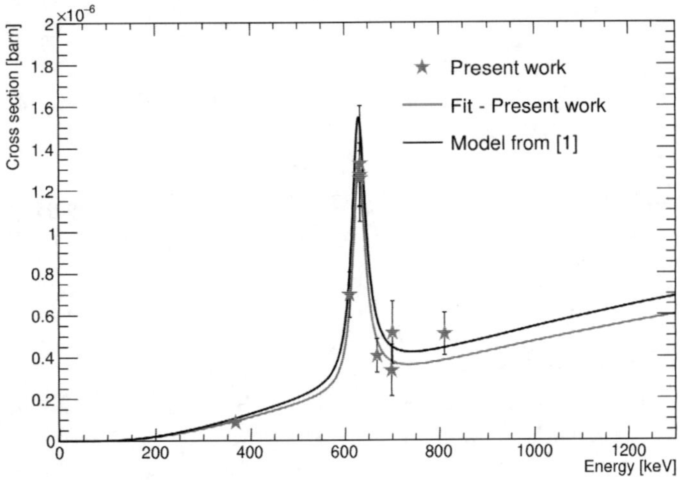

Figure 2. The red stars represent the cross section values obtained in the present work. The dotted black line represents the fit of [1]. The solid red line represents the model from equation 2 fitted to present work data.

Table 1. Fit results to our data using the function in eq. 2. ν indicates the number of degrees of freedom.

Data set	ν	χ^2	A	E_{res}(keV)	Γ_p(keV)	Γ_γ(meV)	$S_{17}(0)(eV \cdot b)$
Present work	6	3.8	0.67±0.09	630[c]	33±6	20±4	16.6± 2.1

[c] Fixed from [25].

$$\Gamma_\gamma(E) = \Gamma_\gamma(E_{res})\left(\frac{E_\gamma}{E_{\gamma res}}\right)^3 \tag{4}$$

The fit function included one parameter (A) for the nonresonant component and three parameters (E_{res}, Γ_p and Γ_γ) for the resonant component. The data on the resonance were not sufficient to fix the E_{res} that is fixed to the values from [25]. The free parameters were obtained using Maximum Likelyhood minimizing the least square function through the routine MINUIT (Cernlib). Best fit values based on our experimental data are reported in table 1.

4 Conclusion

The total cross section of $^7Be(p,\gamma)^8B$ was measured in the energy range E_{eff}=367.2 keV to 812.2 keV using a radioactive ^7Be beam and a recoil mass separator. For the first time, this approach provides results with adequate precision to determine the cross section of $^7Be(p,\gamma)^8B$ at astrophysical energy. Our result, including the 4% systematic error combined in quadrature, yields $S_{17}(0)$=16.6±2.2 eV·b compatible only with a part of previous measurements that yields lower value of $S_{17}(0)$. Our $S_{17}(0)$ is compatible with the currently accepted range $S_{17}(0)$=20.8±0.7 eV·b [1] only at a 2-σ level, thus strengthening the discrepancy between existing data sets. A detailed analysis including all datasets [2–12] is be presented in [26].

References

[1] E.G. Adelberger, A. García, R.G.H. Robertson, K.A. Snover, A.B. Balantekin, K. Heeger, M.J. Ramsey-Musolf, D. Bemmerer, A. Junghans, C.A. Bertulani et al., Rev. Mod. Phys. **83**, 195 (2011)

[2] B.W. Filippone, A.J. Elwyn, C.N. Davids, D.D. Koetke, Phys. Rev. C **28**, 2222 (1983)

[3] L.T. Baby, C. Bordeanu, G. Goldring, M. Hass, L. Weissman, V.N. Fedoseyev, U. Köster, Y. Nir-El, G. Haquin, H.W. Gäggeler et al. (ISOLDE Collaboration), Phys. Rev. C **67**, 065805 (2003)

[4] L.T. Baby, C. Bordeanu, G. Goldring, M. Hass, L. Weissman, V.N. Fedoseyev, U. Köster, Y. Nir-El, G. Haquin, H.W. Gäggeler et al. (ISOLDE Collaboration), Phys. Rev. Lett. **90**, 022501 (2003)

[5] F. Hammache, G. Bogaert, P. Aguer, C. Angulo, S. Barhoumi, L. Brillard, J.F. Chemin, G. Claverie, A. Coc, M. Hussonnois et al., Phys. Rev. Lett. **80**, 928 (1998)

[6] F. Hammache, G. Bogaert, P. Aguer, C. Angulo, S. Barhoumi, L. Brillard, J.F. Chemin, G. Claverie, A. Coc, M. Hussonnois et al., Phys. Rev. Lett. **86**, 3985 (2001)

[7] M. Hass, C. Broude, V. Fedoseev, G. Goldring, G. Huber, J. Lettry, V. Mishin, H. Ravn, V. Sebastian, L. Weissman et al., Physics Letters B **462**, 237 (1999)

[8] F. Strieder, L. Gialanella, G. Gyürky, F. Schümann, R. Bonetti, C. Broggini, L. Campajola, P. Corvisiero, H. Costantini, A. D'Onofrio et al., Nuclear Physics A **696**, 219 (2001)

[9] A.R. Junghans, E.C. Mohrmann, K.A. Snover, T.D. Steiger, E.G. Adelberger, J.M. Casandjian, H.E. Swanson, L. Buchmann, S.H. Park, A. Zyuzin, Phys. Rev. Lett. **88**, 041101 (2002)

[10] A.R. Junghans, E.C. Mohrmann, K.A. Snover, T.D. Steiger, E.G. Adelberger, J.M. Casandjian, H.E. Swanson, L. Buchmann, S.H. Park, A. Zyuzin et al., Phys. Rev. C **68**, 065803 (2003)

[11] A.R. Junghans, K.A. Snover, E.C. Mohrmann, E.G. Adelberger, L. Buchmann, Phys. Rev. C **81**, 012801 (2010)

[12] L. Gialanella, F. Strieder, L. Campajola, A. D'Onofrio, U. Greife, G. Gyurky, G. Imbriani, G. Oliviero, A. Ordine, V. Roca et al., Eur. Phys. J. A **7**, 303 (2000)

[13] C. Santonastaso, R. Buompane, A.D. Leva, L. Morales-Gallegos, N. Itaco, G. Landi, H.C. Neitzert, D. Rapagnani, L. Gialanella, Il Nuovo Cimento **44**, 1 (2021)

[14] E. Segre, C.E. Wiegand, Physical review **75**, 40 (1949)

[15] L. gun Liu, C.A. Huh, Earth and Planetary Science Letters **180**, 163 (2000)

[16] T. Ohtsuki, H. Yuki, M. Muto, J. Kasagi, K. Ohno, Phys. Rev. Lett. **93**, 112501 (2004)

[17] D.W. Bardayan, K.A. Chipps, R.P. Fitzgerald, J.C. Blackmon, K.Y. Chae, A.E. Champagne, U. Greife, R. Hatarik, R.L. Kozub, C. Matei et al., The European Physical Journal A **42**, 457 (2009)

[18] B.N. Limata, L. Gialanella, A.D. Leva, N.D. Cesare, A. D'Onofrio, G. Gyurky, C. Rolfs, M. Romano, D. Rogalla, C. Rossi et al., Nuclear Instruments and Methods in Physics Research Section B: Beam Interactions with Materials and Atoms **266**, 2117 (2008), accelerators in Applied Research and Technology

[19] Schürmann, D., Di Leva, A., Gialanella, L., De Cesare, M., De Cesare, N., Imbriani, G., D´Onofrio, A, Romano, M., Romoli, M., Terrasi, F., Eur. Phys. J. A **49**, 80 (2013)

[20] Buompane, R., De Cesare, N., Di Leva, A., D´Onofrio, A., Gialanella, L., Romano, M., De Cesare, M., Duarte, J. G., Fülöp, Zs., Morales-Gallegos, L. et al., Eur. Phys. J. A **54**, 92 (2018)

[21] A. Di Leva, G. Imbriani, R. Buompane, L. Gialanella, A. Best, S. Cristallo, M. De Cesare, A. D'Onofrio, J. Duarte, L. Gasques et al., Physical Review C **95** (2017)

[22] D. Rogalla, S. Theis, L. Campajola, A. D'Onofrio, L. Gialanella, U. Greife, G. Imbriani, A. Ordine, V. Roca, C. Rolfs et al., Nuclear Instruments and Methods in Physics Research Section A: Accelerators, Spectrometers, Detectors and Associated Equipment **437**, 266 (1999)

[23] P. Descouvemont, Phys. Rev. C **70**, 065802 (2004)

[24] C.E. Rolfs, W.S. Rodney, Physical review (1988)

[25] D. Tilley, J. Kelley, J. Godwin, D. Millener, J. Purcell, C. Sheu, H. Weller, Nuclear Physics A **745**, 155 (2004)

[26] R.Buompane, A.D. Leva, L. Gialanella, A. D'Onofrio, M.D. Cesare, J.G. Duarte, Z. Fülöp, L.R. Gasques, G. Gyürky, L. Morales-Gallegos et al., Physics Letters B (2021), manuscript accepted for publication

Study of the ^{20}Ne(p,γ)^{21}Na reaction at LUNA

Antonio Caciolli[1,2,*] and *Sandra* Zavatarelli[3,**] for the LUNA collaboration

[1]Physics and Astronomy Department, University of Padova, Via Marzolo 8, 35131, Padova (Italy)
[2]INFN Section of Padova, via Marzolo 8, 35131, Padova (Italy)
[3]INFN Section of Genova, Via Dodecaneso, 33, 16146 Genova (Italy)

Abstract. The ^{20}Ne(p,γ)^{21}Na is the first reaction of the NeNa cycle and having the slowest reaction rate, it controls the speed at which the entire cycle proceeds. This reaction is important in different stellar scenarios such as red giants stars (during H shell-burning), asymptotic giant branch stars, novae, and massive stars. Existing uncertainties on the ^{20}Ne(p,γ)^{21}Na are now the ones that severely affect the estimated elements production in the NeNa cycle. In particular, such uncertainties affect the amount of produced ^{22}Na, a stellar γ-ray signature, and ^{22}Ne, an important neutron source in the s-process, via the ^{22}Ne(α,n)^{25}Mg. Here, we propose to study the resonance at E_R = 366 keV and to measure the ^{20}Ne(p,γ)^{21}Na reaction cross section at proton energies below 400 keV, in particular the contribution from the sub-threshold state, in order to better constrain the overall astrophysical rate of this important reaction. In this contribution, the experimental setup will be described and the expected results discussed.

1 Introduction

The NeNa-MgAl cycles are involved in the synthesis of Ne, Na, and Mg isotopes. Proton capture on ^{20}Ne nuclei may occur in different stellar scenarios such as red giants stars (during H shell-burning), asymptotic giant branch stars, novae, and massive stars. These reactions can proceed in these environments if the temperature reaches $T \sim 0.05$ GK [1].

Until recently, the major uncertainty on the NeNa cycle was given by the ^{22}Ne(p,γ)^{23}Na reaction. Now, thanks to the latest LUNA measurements [2–6], it has been reduced to negligible values [7]. As a result, existing uncertainties on the ^{20}Ne(p,γ)^{21}Na are now the ones that severely affect the estimated elements production in the NeNa cycle.

The ^{20}Ne(p,γ)^{21}Na reaction has already been studied experimentally by several experiments with results published in referred journals [8–13] with both direct and indirect approaches. In the recent years, two PhD theses have been reported on this reaction from the TUNL and DRAGON nuclear astrophysics groups [14, 15]. Quoted data show some discrepancies on the resonance strength of the 366 keV resonance and only the data in the PhD thesis [15] report cross section data below 400 keV, the range of the LUNA accelerator [16]. For this reason, the LUNA collaboration has decided to approach the study of this reaction by using a gas target [17] with natural neon gas inside the target chamber. In this contribution we describe the setup used at LUNA for the ^{20}Ne(p,γ)^{21}Na reaction study.

*e-mail: antonio.caciolli@unipd.it
**e-mail: sandra.zavatarelli@ge.infn.it

2 Experimental setup

The setup has been installed at the LUNA400 accelerator facility under the Gran Sasso Laboratories where the shielding of the 1.4 km of rock greatly reduces the cosmic rays flux [18]. The experimental approach followed in this experimental study is equal to the one already adopted in [6]. The proton beam enters the scattering chamber by passing through three apertures that regulates also the differential pumping system allowing to work without any window before the scattering chamber: a possible layer before entering the chamber could deteriorate the beam characteristics introducing problematic uncertainty in the analysis. The beam current is measured by using a calorimeter and the gamma rays emitted by the reaction are detected by two different germanium detectors fully shielded. The Q-value of the ^{20}Ne(p,γ)^{21}Na reaction is about 2.4 MeV and the emitted gamma radiation falls in an energy range dominated by the radioactive decays of the experimental room materials [19]. All systematics related to the gas characterisation and response against beam heating are already well known, since also in the previous experiment a neon gas has been used (with a different isotopical enrichment). The efficiency has been tested again using radioactive calibrated sources and by scanning the 278 keV resonance in the ^{14}N(p,γ)^{15}O reaction. With respect to the setup used in [6], in the actual experiment we have decided to remove internal collimators and to use the scattering chamber as an extended gas target. Recently, the LUNA collaboration have worked with extended gas target to constraint the d+p reaction [20, 21] with optimal results and we have observed by an extensively study with simulations that this configuration can solve the problem related with resonance strength evaluation with important straggling effect [22]. During the data taking in 2021 the resonance has been measured and the correction for beam straggling well evaluated. In this study the branching to different excited states of ^{21}Na have been also detected by combining also the results of the two detectors. A preliminary investigation of the direct capture component has been done demonstrating the feasibility of our study down to 330 keV and even below. A new experimental campaign is foreseen for February 2022 with the aim of investigating the direct capture component. During 2022, the ^{21}Ne(p,γ)^{22}Na reaction will be also studied with the same setup.

References

[1] J. Marion, W. Fowler, Astrophys. J. **125**, 221 (1957)

[2] F. Cavanna, R. Depalo, M. Aliotta, M. Anders, D. Bemmerer, A. Best, A. Boeltzig, C. Broggini, C.G. Bruno, A. Caciolli et al., Phys. Rev. Lett. **115**, 252501 (2015), 1511.05329

[3] R. Depalo, F. Cavanna, M. Aliotta, M. Anders, D. Bemmerer, A. Best, A. Boeltzig, C. Broggini, C.G. Bruno, A. Caciolli et al. (LUNA Collaboration), Phys. Rev. C **94**, 055804 (2016)

[4] F. Ferraro, M.P. Takács, D. Piatti, F. Cavanna, R. Depalo, M. Aliotta, D. Bemmerer, A. Best, A. Boeltzig, C. Broggini et al. (LUNA Collaboration), Phys. Rev. Lett. **121**, 172701 (2018)

[5] F. Cavanna, R. Depalo, M. Aliotta, M. Anders, D. Bemmerer, A. Best, A. Boeltzig, C. Broggini, C.G. Bruno, A. Caciolli et al. (The LUNA Collaboration), Phys. Rev. Lett. **120**, 239901 (2018)

[6] F. Cavanna, R. Depalo, M.L. Menzel, M. Aliotta, M. Anders, D. Bemmerer, C. Broggini, C.G. Bruno, A. Caciolli, P. Corvisiero et al., Eur. Phys. J. A **50**, 179 (2014), 1411.2888

[7] A. Slemer, P. Marigo, D. Piatti, M. Aliotta, D. Bemmerer, A. Best, A. Boeltzig, A. Bressan, C. Broggini, C.G. Bruno et al., Monthly Notices of the Royal Astronomical Society **465**, 4817 (2017)

[8] N. Tanner, Phys. Rev. **114**, 1060 (1959)

[9] C. Van der Leun, W. Mouton, Physica **30**, 333 (1964)

[10] J. Keinonen, M. Riihonen, A. Anttila, Phys. Rev. C **15**, 579 (1977)

[11] C. Rolfs, W.S. Rodney, M.H. Shapiro, H. Winkler, Nucl. Phys. A **241**, 460 (1975)

[12] A. Mukhamedzhanov, P. Bém, V. Burjan, C.A. Gagliardi, B.F. Irgaziev, V. Kroha, J. Novak, S.R. Piskor, E. Šimečková, R.E. Tribble et al., *Asymptotic normalization coefficients from the 20 Ne (3 He , d) 21 Na reaction and astrophysical factor for 20 Ne (p , γ) 21* (2006)

[13] S. Lyons, J. Görres, R.J. deBoer, E. Stech, Y. Chen, G. Gilardy, Q. Liu, A.M. Long, M. Moran, D. Robertson et al., Phys. Rev. C **97**, 065802 (2018)

[14] A. Cooper, Ph.D. thesis, University of North Carolina, Chapel Hill, Department of Physics and Astronomy (2019)

[15] J. Karpesky, Ph.D. thesis, Colorado School of Mines (2020)

[16] Best, A., Caciolli, A., Fülüp, Zs., Györky, Gy., Laubenstein, M., Napolitani, E., Rigato, V., Roca, V., Szücs, T., Eur. Phys. J. A **52**, 72 (2016)

[17] F. Ferraro, M.P. Takács, D. Piatti, V. Mossa, M. Aliotta, D. Bemmerer, A. Best, A. Boeltzig, C. Broggini, C.G. Bruno et al., The European Physical Journal A **54**, 44 (2018)

[18] T. Szücs, D. Bemmerer, C. Broggini, A. Caciolli, F. Confortola, P. Corvisiero, Z. Elekes, A. Formicola, Z. Fülöp, G. Gervino et al., Eur. Phys. J. A **44**, 513 (2010)

[19] A. Caciolli, L. Agostino, D. Bemmerer, R. Bonetti, C. Broggini, F. Confortola, P. Corvisiero, H. Costantini, Z. Elekes, A. Formicola et al., Eur. Phys. J. A **39**, 179 (2009)

[20] Mossa, V., Stöckel, K., Cavanna, F., Ferraro, F., Aliotta, M., Barile, F., Bemmerer, D., Best, A., Boeltzig, A., Broggini, C. et al., Eur. Phys. J. A **56**, 144 (2020)

[21] V. Mossa, K. Stöckel, F. Cavanna, F. Ferraro, M. Aliotta, F. Barile, D. Bemmerer, A. Best, A. Boeltzig, C. Broggini et al., Nature **587**, 210 (2020)

[22] The Luna Collaboration, Bemmerer, D., Cavanna, F., Depalo, R., Aliotta, M., Anders, M., Boeltzig, A., Broggini, C., Bruno, C., Caciolli, A. et al., EPL **122**, 52001 (2018)

Measurement and analysis techniques for a study of ^{12}C(p,γ) and ^{13}C(p,γ) deep underground

Jakub Skowronski* for the LUNA collaboration

[1]Università degli Studi di Padova and INFN, Sezione di Padova, Via Francesco Marzolo 8, 35131 Padova, Italy

Abstract. The ^{12}C(p,γ) and ^{13}C(p,γ) reaction cross sections are currently under investigation in the low-background environment of the Laboratory for Underground Nuclear Astrophysics. Both reactions are being studied using different types of solid targets, and employing complementary detection techniques (HPGe spectroscopy, total absorption spectroscopy and activation counting). To reduce systematic uncertainties, targets must be accurately characterized and their degradation monitored under the intense ($\sim 400\ \mu$A) beam of the LUNA-400 accelerator. We present the experimental techniques employed, and the analyses developed for the study of these reactions.

1 Introduction

The reactions ^{12}C$(p,\gamma)^{13}$N and ^{13}C$(p,\gamma)^{14}$N are the first two of the CNO cycle, which is active in the hydrogen burning regions of main sequence, Red Giant Branch (RGB) and Asymptotic Giant Branch stars (AGB). This not only contributes to the energy production, by converting four protons into one helium nucleus, but also governs the abundances of several elements. The ^{13}C is produced by the ^{12}C$(p,\gamma)^{13}$N and the subsequent β^+ decay to ^{13}C. Then it is depleted by the ^{13}C$(p,\gamma)^{14}$N. Hence, both of the reactions play a key role in governing the ^{12}C/^{13}C ratio.

Presently the ^{12}C/^{13}C ratio in the Solar System is estimated to be of about 90 ± 10 [1]. One of the possible origins for the elements in the Solar System are the AGB stars [2], which are remarkably prolific centers of nucleosynthesis. Unfortunately the AGB phase of stellar evolution is marked by intense mixing phenomena, which make the theoretical models particulary difficult to establish [3]. The mixing heavily impacts the atmospheric abundance of both the ^{12}C and ^{13}C, and can arise from several different sources: convective motion inside the star, the angular momentum of the star [4], magnetic buoyancy [5] or gravitional waves [6].

Since the predictions depend on nuclear reaction rates, it is of particular interest to constrain the latter as much as possible. This can help to study the mixing phenomena in more detail and thus to constrain theoretical models. The ^{12}C and ^{13}C are the perfect candidates for this: not only are they the isotopes that burn at lowest energies, but also their ratio can be readily derived from stellar spectra.

A direct measurement of the ^{12}C$(p,\gamma)^{13}$N and ^{13}C$(p,\gamma)^{14}$N cross-sections at astrophysical energies was the goal of a recently conducted experiment at LUNA.

*e-mail: jakub.skowronski@pd.infn.it

2 Current Status

The $^{12}C(p,\gamma)^{13}N$ reaction has been studied by several experiments in the past. The S-factors reported in literature are shown in Fig. 1. Down to 200 keV there are two main data sets [7, 8], which are mostly in agreement and have been confirmed by a later measurement [9]. However, the S-factor is not properly constrained by the data reported below 200 keV. Results by [10] and [11] poorly overlap and are affected by high uncertainties, up to 20% and 41% respectively.

In case of the $^{13}C(p,\gamma)^{14}N$ reaction, which presents transitions to different excited states of ^{14}N, considerable efforts have been done to measure the ground state one, with a reported intensity of ~ 77% [12]. The primary datasets from [7, 12] show good amount of agreement above 250 keV. Nevertheless, the most recent study [13] presents a systematic deviation of 10%. At lower energies, the data start to differ up to 20%. In addition, at the lowest energies the data struggle to agree with the older datasets [14, 15], as can be seen in Fig. 1. The older studies, indeed, deviate by ~ 30% from the most recent ones. For what regards the minor transitions, only one measurement is available [12]. However, the results are affected by large scattering and the statistical uncertainty ranges from 15% to 40%.

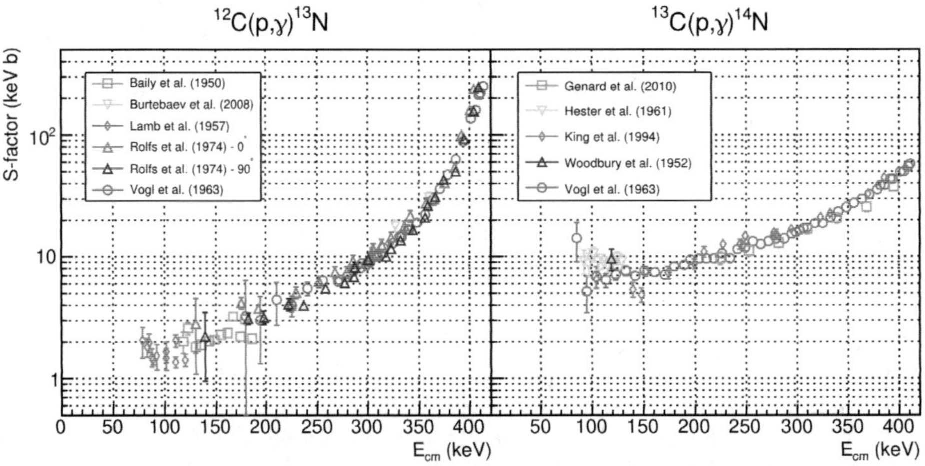

Figure 1: The state of the art for the $^{12}C(p,\gamma)^{13}N$ (on the left) and the $^{13}C(p,\gamma)^{14}N$ (on the right). For the latter, only the transition to the ground state is shown.

3 Experimental Setup

The main objective of the experimental campaign at LUNA was to measure both the $^{12}C(p,\gamma)^{13}N$ and the $^{13}C(p,\gamma)^{14}N$ at astrophysical energies with complementary techniques. By doing this, measurements can be cross-checked and systematic uncertainties reduced. In addition, the deep underground location and high current of LUNA 400 kV accelerator [16] guarantees an environmental background level orders of magnitude lower than above ground [17] enabling high-sensitivity measurements. During the experiment both thin evaporated targets and thick targets were employed to check target stability and normalization.

The first set of measurements was conducted using a HPGe detector (in Fig. 2a) positioned in close geometry (~ 1.35 cm). The advantage of this approach lies in the possibility of performing a complete spectroscopic study for both reactions. Especially in case of the

(a) (b)

Figure 2: Pictures of (a) the HPGe setup and (b) the BGO setup used during the measurements. Both of the setups employed a Pb shielding for the background reduction.

$^{13}C(p,\gamma)^{14}N$, it permits to obtain precise proton capture branchings to each ^{14}N excited state. In addition, the high resolution of the HPGe allowed to study the target degradation with the γ-shape analysis [18]. Finally, the setup allowed to measure both at 0° and 55° and check the the angular distribution of the γ-rays.

The second set of measurements was performed with a large BGO detector [19] (in Fig. 2b). Its close-to-4π geometry and segmentation in 6 crystals are particulary advantageous for measuring very low cross-sections. In case of the $^{13}C(p,\gamma)^{14}N$, the detection in coincidence of all of the energy released as γ-rays allowed to place the region of interest in an almost background-free region [20], owing to the high Q-value of the reaction ($\sim$ 7.5 MeV). For the $^{12}C(p,\gamma)^{13}N$, given the smaller Q-value ($\sim$ 1.9 MeV), different approach had to be used. Since the ^{13}N is β^+ unstable with a half-life of about 10 minutes, the procedure consisted in detecting the two 511 keV γ-rays in coincidence in opposite crystals of the detector. This distinct experimental signature of the positron-electron annihilation grants a significant reduction of the background event rate.

4 Analysis Methods

Given the peculiarity of each measurement, three analysis procedures have been developed:

- For the HPGe data, the γ-shape analysis was performed. The approach consists in a detailed study of the primary γ-peak lines emitted in the two reactions. It allowed both to check the thickness and stoichiometry of the targets. Furthermore, the innovative application of this technique permitted to extract the astrophysical S-factor directly from the minimization, without the need of integrating the γ-peak. An example of the fit can be seen in Fig. 3.
- In case of the ^{13}N activation, an alternative techinque was used as the analysis procedure relied on an iterative fitting procedure: at each step, the differential equation that governs the ^{13}N abundance is solved and the activity calculated. The yield of the reaction is treated as a free parameter and thus the best-fit value can be obtained from likelihood maximization. An example of the fit can be seen in Fig. 3.
- Concerning the total absorption spectroscopy with the BGO detector, the sum γ-peak was parametrized and fitted. In this way, the total counts could be readily extracted from the spectra. This strategy requires to establish the sum-peak efficiency of the detector, which relies on Monte Carlo simulations and calibration measurements, and utilizes the branching information from the HPGe measurement as inputs.

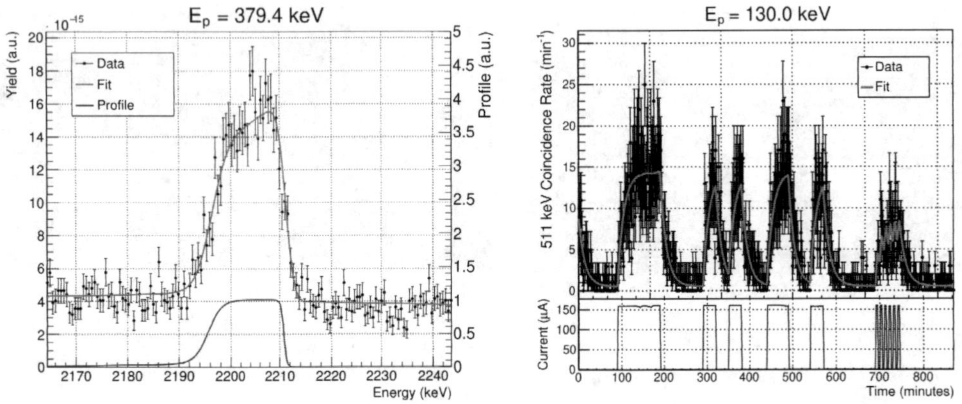

Figure 3: On the left an example of the γ-shape analysis for the HPGe data of the primary transition to the 5691 keV state of ^{14}N. The shape of the γ-peak is modeled both by the exponential drop in the cross-section and by the target profile (in blue). On the right an example of the fit for the ^{12}C$(p,\gamma)^{13}$N activation measurement with the BGO. During the irradiation, the event rate increases and then saturates. Whereas when the beam is stopped, the exponentialy decay is observed.

Finally, the analysis is currently in its final stages and will be presented in a forthcoming publication. The preliminary results show a good agreement among all the different experimental approaches.

References

[1] K. Lodders, H. Palme, H.P. Gail, Landolt-Börnstein **4B**, 712 (2009)
[2] J.M. Trigo-Rodríguez et al., Meteorit. Planet. Sci. **44**, 627 (2009)
[3] F. Herwig, Annu. Rev. Astron. Astrophys. **43**, 435 (2005)
[4] N. Langer et al., Astron. Astrophys. **346**, L37 (1999)
[5] D. Vescovi et al., Astrophys. J. Lett. **897** (2020)
[6] P.A. Denissenkov, C.A. Tout, Mon. Not. R. Astron. Soc. **340**, 722 (2003)
[7] J.L. Vogl, Ph.D. thesis, California Institute of Technology (1963)
[8] C. Rolfs, R. Azuma, Nucl. Phys. A **227**, 291 (1974)
[9] N. Burtebaev et al., Phys. Rev. C **78**, 035802 (2008)
[10] C.L. Bailey, W.R. Stratton, Phys. Rev. **77**, 194 (1950)
[11] W.A.S. Lamb, R.E. Hester, Phys. Rev. **107**, 550 (1957)
[12] J.D. King et al., Nucl. Phys. A **567**, 354 (1994)
[13] G. Genard et al., J. Phys. Conf. Ser. **202**, 012015 (2010)
[14] E.J. Woodbury, W.A. Fowler, Phys. Rev. **85**, 51 (1952)
[15] R.E. Hester, W.A. Lamb, Nucl. Phys. A **121**, 584 (1961)
[16] A. Formicola et al., Nucl. Instrum. Methods Phys. Res. A **507**, 609 (2003)
[17] C. Broggini et al., Prog. Part. Nucl. Phys. **98**, 55–84 (2018)
[18] G.F. Ciani, L. Csedreki et al., Eur. Phys. J. A **56** (2020)
[19] C. Casella et al., Nucl. Instrum. Methods Phys. Res. A **489**, 160 (2002)
[20] A. Boeltzig et al., J. Phys. G **45**, 025203 (2018)

https://doi.org/10.1051/epjconf/202226011009

Fluorine nucleosynthesis and s-processing in AGB stars driven by magnetic-buoyancy mixing

Diego Vescovi[1,*], *Sergio* Cristallo[2,3], *Sara* Palmerini[3,4], *Carlos* Abia[5], and *Maurizio* Busso[3,4]

[1]Goethe University Frankfurt, Max-von-Laue-Strasse 1, Frankfurt am Main 60438, Germany
[2]INAF, Observatory of Abruzzo, Via Mentore Maggini snc, 64100 Teramo, Italy
[3]INFN, Section of Perugia, Via A. Pascoli snc, 06123 Perugia, Italy
[4]Department of Physics and Geology, University of Perugia, via A. Pascoli snc, 06123 Perugia, Italy
[5]University of Granada, Departamento de Fisica Teorica y del Cosmos, 18071 Granada, Spain

Abstract. Asymptotic giant branch (AGB) stars are thought to be among the most important sources of fluorine in our Galaxy. While observations and theory agree at close-to-solar metallicity, stellar models overestimate fluorine production in comparison to heavy elements at lower metallicities. We present predictions for ^{19}F abundance for a set of AGB models with various masses and metallicities, in which magnetic buoyancy induces the formation of the ^{13}C neutron source (the so-called ^{13}C pocket). In our new models, fluorine is mostly created as a consequence of secondary ^{14}N nucleosynthesis during convective thermal pulses, with a minor contribution from the ^{14}N existing in the ^{13}C pocket zone. As a result, AGB stellar models with magnetic-buoyancy-induced mixing show low ^{19}F surface abundances which agree with fluorine spectroscopic observations at both low and near-solar metallicity.

1 Introduction

The cosmic genesis of fluorine is one of the most intriguing problems in nuclear astrophysics. So far, spectroscopic observations of photospheric F enhancements in intrinsic AGB carbon stars [1, 2] and metal-poor extrinsic stars [3, 4] offer the sole direct observation of fluorine production. Its envelope abundance indicates a correlation with those of carbon and *s*-process elements and therefore with the ^{13}C content produced in the interiors of AGB stars [5]. Recently, it was proposed that a process resulting in the creation of an extended ^{13}C pocket and, simultaneously, a tiny quantity of ^{14}N, may solve the issue of ^{19}F overproduction with respect to s-elements in low-mass metal-poor objects [4].

The formation of a ^{13}C pocket in AGB stars requires that some partial mixing of protons from the envelope occurs during a third dredge-up (TDU) phenomenon. In latest FRUITY models [6], the buoyant rise of magnetic flux tubes in the region below the envelope is considered to cause a stable mass upflow which induces a downflow of protons for maintaining mass conservation. As a result of such magnetic mixing, deep profiles of low proton abundances are predicted to be formed. Because protons are nearly completely used for the synthesis of ^{13}C, the low proton concentration significantly reduces the local ^{14}N formation, thereby limiting ^{19}F production as well. Here we examine fluorine nucleosynthesis in low-mass AGB

*e-mail: vescovi@iap.uni-frankfurt.de

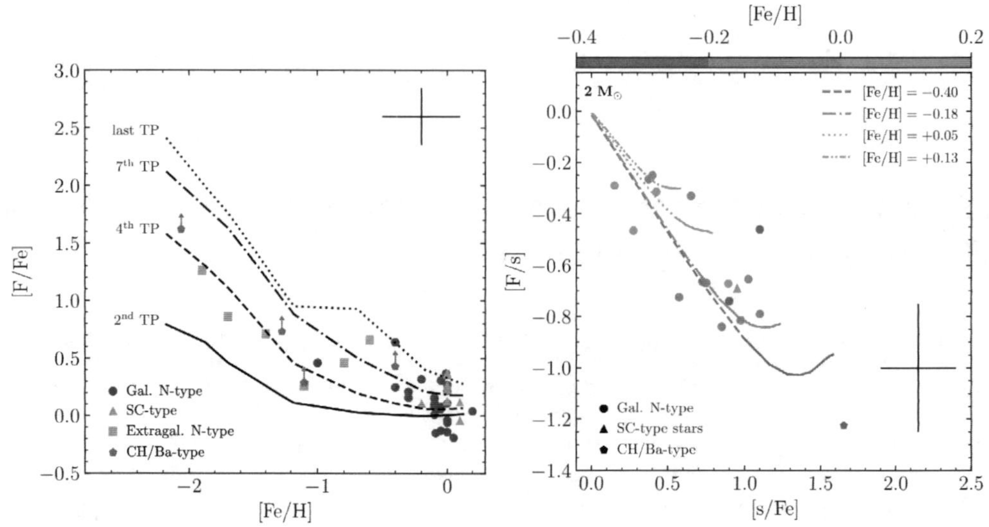

Figure 1. (Left panel) Comparison between observed [F/Fe] ratios as a function of the metallicity and FRUITY magnetic models. Symbols refer to four data groups: circles, galactic (N-type) carbon stars; triangles, SC-type stars; squares, extragalactic carbon stars; pentagons, extrinsic CH/Ba stars. Lines represent theoretical predictions for 2 (for [Fe/H] ≥ −0.7) and 1.5 $M_\odot$ (for [Fe/H] < −0.7) AGB stars at different thermal pulses (TPs). (Right panel) Observed [F/s] vs. average s-element enhancements compared with theoretical predictions for 2 $M_\odot$ and close-to-solar metallicity models. Data points and theoretical lines are color-coded by [Fe/H]. The solid portion of the lines represents the theoretical C-rich phase, while the dashed portion represents the O-rich phase. Typical error bars are indicated.

stars by computing a new series of stellar models accounting for the formation of a magnetic-buoyancy-induced ^{13}C pocket.

2 Comparison with observations

In the following, we compare data for intrinsic AGB carbon stars [2, 4] and extrinsic CH/Ba stars [4, 5] with new FRUITY magnetic models of 1.5 $M_\odot$ and 2 $M_\odot$.

In left panel of Fig. 1 we report the [F/Fe] ratios of the selected sample as a function of [Fe/H]. Within the observational errors, there is a good agreement, confirming the expected F-enhancement trend with the metallicity. The right panel of Fig. 1 shows the comparison between theoretical predictions of FRUITY magnetic models and spectroscopic observations for [F/s] ratios vs. the average s-element enhancement. In this way, the F enhancement for the extrinsic stars is not affected by uncertainties related to the dilution factor and provides a robust tool for comparison. Theoretical expectations can replicate the quasi-linear decreasing trend of [F/s] with the surface s-process enrichment.

In Fig. 2 we perform a similar comparison at low metallicities. We compared model predictions individually since there is no homogeneous sample of stars with both Ba and La. Magnetic models well reproduce the spread observed at different metallicities for both [F/Ba] and [F/La] ratios as a function of total s-process enhancement. Overall, FRUITY magnetic models show a reduction in fluorine production that is consistent with spectroscopic observations of low-metallicity stars. The extended profile and low proton abundance that characterize FRUITY magnetic models [6] have the dual impact of generating large ^{13}C pockets and a small amount of primary ^{14}N. In this scenario, the few available protons make

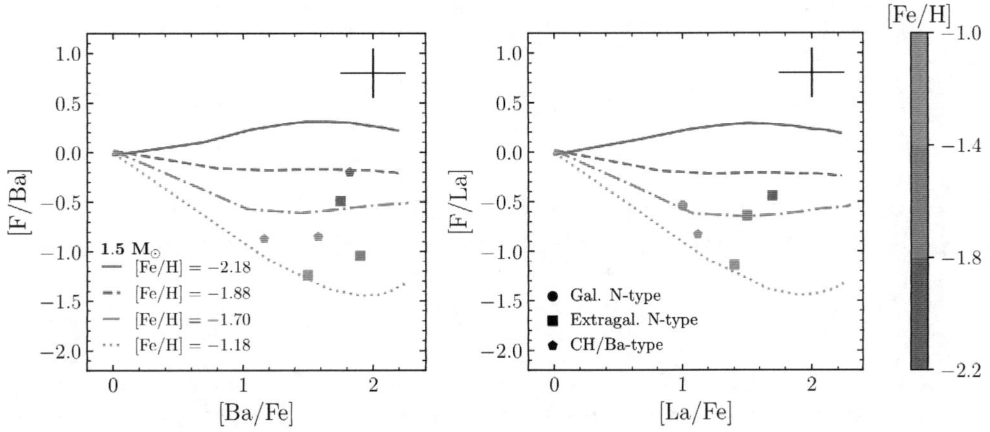

Figure 2. [F/Ba] vs. [Ba/Fe] (left panel) and [F/La] vs. [La/Fe] (right panel) in the sample stars with [Fe/H] ≤ -1.0. Symbols as in Fig. 1. Note that at these low metallicities, theoretical AGB models predict that the star becomes C-rich from first TDU episodes. Data points and theoretical lines are color-coded by [Fe/H]. Typical error bars are indicated.

first ^{13}C through the ^{12}C(p, γ)^{13}C reaction, preventing further proton captures to form ^{14}N, thus inhibiting the nuclear chain ^{14}N(n, p)^{14}C(α, γ)^{18}O(p, α)^{15}N(α, γ)^{19}F. As a consequence, any fluorine appearing in AGB envelopes in these models is of secondary origin, caused by ^{14}N concentrations left over after H-shell burning. As an outcome, FRUITY magnetic models have low fluorine enhancements and large s-enhancements, which are in close agreement with observations in very metal-poor AGB stars [7].

3 Conclusions

The production of fluorine in low-mass AGB stars has been examined in light of new FRUITY stellar models, in which the ^{13}C neutron source is attributed to magnetic-buoyancy-induced phenomena. On the one hand, the new FRUITY magnetic models exhibit a low net ^{19}F production. This is due to the low abundance of ^{14}N in the ^{13}C pocket, which results in negligible fluorine production during ^{13}C radiative burning. The ^{19}F envelope abundance is therefore ascribed only to the amount of the secondary ^{13}C in the H-shell ashes, which depends on the CNO abundances in the star. On the other hand, mixing induced by magnetic buoyancy leads to extended ^{13}C pockets so resulting in large surface s-process enrichments. As a whole, new FRUITY magnetic models simultaneously account for both the observed fluorine and the average s-element enhancements in intrinsic AGB carbon stars and extrinsic CH/Ba stars.

References

[1] A. Jorissen, V.V. Smith, D.L. Lambert, A&A **261**, 164 (1992)
[2] C. Abia, K. Cunha, S. Cristallo, P. de Laverny, A&A **581**, A88 (2015)
[3] S. Lucatello, T. Masseron, J.A. Johnson, M. Pignatari, F. Herwig, ApJ **729**, 40 (2011)
[4] C. Abia, S. Cristallo, K. Cunha, P. de Laverny, V.V. Smith, A&A **625**, A40 (2019)
[5] M. Forestini, S. Goriely, A. Jorissen, M. Arnould, A&A **261**, 157 (1992)
[6] D. Vescovi, S. Cristallo, M. Busso, N. Liu, ApJL **897**, L25 (2020)
[7] D. Vescovi, S. Cristallo, S. Palmerini, C. Abia, M. Busso, A&A **652**, A100 (2021)

Measurement of ^{12}C(n,n') reaction cross section to determine triple-alpha reaction rate in high-density environments

Tatsuya Furuno[1,*], *Takanobu* Doi[2], *Kanako* Himi[1], *Takahiro* Kawabata[1], *Satoshi* Adachi[1], *Hidetoshi* Akimune[3], *Shiyo* Enyo[2], *Yuki* Fujikawa[2], *Yuto* Hijikata[2,4], *Kento* Inaba[5], *Masatoshi* Itoh[6], *Shigeru* Kubono[4], *Yohei* Matsuda[3], *Isao* Murata[7], *Motoki* Murata[8], *Shintaro* Okamoto[2], *Kosuke* Sakanashi[1], and *Shingo* Tamaki[7]

[1]Department of Physics, Osaka University, Toyonaka, Osaka 560-0043, Japan
[2]Department of Physics, Kyoto University, Sakyo, Kyoto 606-8502, Japan
[3]Department of Physics, Konan University, Kobe, Hyogo 658-8501, Japan
[4]RIKEN Nishina Center, Wako, Saitama 351-0198, Japan
[5]Institute for Advanced Synchrotron Light Source, Quantum Scinence and Technology, Sendai, Miyagi 980-8579, Japan
[6]Cyclotron and Radio Isotope Center (CYRIC), Tohoku University, Sendai, Miyagi 980-8578, Japan
[7]Division of Sustainable Energy and Environmental Engineering, Osaka University, Suita, Osaka 565-0871, Japan
[8]Research Center for Nuclear Physics (RCNP), Osaka University, Ibaraki, Osaka 567-0047, Japan

Abstract. The reaction rate of the triple-alpha reaction can be enhanced in hot and dense environments due to the deexcitation of the Hoyle state in ^{12}C by neutrons. The cross section of the deexcitation should be determined for the enhanced reaction rate. We plan to obtain the cross section by measuring the inverse reaction using a neutron beam around 10 MeV and an active target system. In the present paper, we report a proof-of-principle experiment using a neutron beam at 14 MeV. The obtained cross section is consistent with a previous result, demonstrating the validity of our method.

1 Introduction

The triple-alpha reaction is one of the most crucial nucleosynthesis reactions because it is the doorway reaction from $A < 5$ nuclei to heavier elements. In this reaction, the ground state of ^{12}C nucleus is populated from three alpha particles via the two alpha and three alpha resonance states in ^{8}Be and ^{12}C, followed by the gamma decay of the resonance state in ^{12}C. The resonance state in ^{12}C, known as Hoyle state, is located at $E_x = 7.65$ MeV with the spin parity of 0_2^+.

Since the triple-alpha reaction proceeds via the Hoyle state, its branching ratio to the ground state is directly related to the reaction rate. The probability of the gamma decay is experimentally determined as $(4.16\pm0.11)\times10^{-4}$ [1]. Recently, it is pointed out that in hot and dense environments, the branching ratio of the Hoyle state to the ground state can be greatly enhanced due to the deexcitation scattering with the surrounding nuclei [2]. Because the

*e-mail: furuno@ne.phys.sci.osaka-u.ac.jp

neutrons are not obstructed by the Coulomb barrier, the scattering with the neutrons increases the reaction rate the most by a factor of about 40 at the density of 10^6 g/cm^3 and temperature of 10^9 K.

In order to determine the enhanced reaction rate by the neutrons, it is necessary to measure the deexcitation cross section of the Hoyle state and neutrons. However, it is almost impossible to perform such an experiment because the life time of the Hoyle state is extremely short to use as a target. Instead of measuring the direct process, the deexcitation cross section can be obtained on the basis of the detailed balance principle by measuring the inverse reaction, i.e., the inelastic scattering of the ground state in ^{12}C and the neutron beam at about 10 MeV.

We will reconstruct the excitation energy of ^{12}C by detecting the three alpha particles emitted from the Hoyle state, and calculating the invariant mass. The energy of the alpha particles, however, is extremely small ($\sim$100 keV), thus they cannot be detected with a normal detector system. In our experiment, the low-energy alpha particles will be detected using an active target system MAIKo [3]. The MAIKo active target is based on a gaseous time projection chamber (TPC) and capable of recording three-dimensional trajectories of the incident charged particles. In MAIKo, the detection gas of the TPC plays also as the target gas of the scattering. Since the scattering occurs inside the sensitive volume of the TPC, the detection threshold is lowered.

In the present article, we report the proof-of-principle experiment using MAIKo to detect the low-energy alpha particles emitted from the Hoyle state and reconstruct the invariant mass of ^{12}C.

2 Experiment

The experiment was carried out at the intense deuterium-tritium neutron source facility (OKTAVIAN) of Osaka University. The mono-energetic neutron beam at 14 MeV was produced by the DT fusion reaction of a tritium target bombarded by a deuteron beam at 300 keV from a Cockcroft-Walton accelerator. Figure 1 shows the setup of the experiment. The neutron beam was collimated by a polyethylene collimator with a diameter of 20 mm installed inside the concrete wall of 1-m thick. The MAIKo active target was installed just behind the collimator. The number of the incident neutron beam was monitored with a BC501 liquid scintillator. The intensity of the neutron beam at MAIKo was approximately 10k cps in average.

The MAIKo TPC was operated with the iso-C_4H_{10}(10%)+H_2(90%) gas at 100 hPa. The iso-C_4H_{10} gas was used as the carbon target and the H_2 gas was added to suppress the diffusion of the drift electrons and improve the track reconstruction efficiency. The effective length of the target gas was 92.4 mm. When the inelastic scattering of carbon with neutron occurs and the excited carbon nucleus decay into three alpha particles, the alpha particles ionize the TPC gas. The electrons produced in the ionization drift along the electric field formed by the TPC field cage. The drift electrons are multiplied by a gas electron multiplier. The electrons are further amplified and detected by a micro pixel chamber (μ-PIC) [4]. The μ-PIC consists of 256 anode strips and 256 cathode strips with a pitch of 400 μm. The cathode strips are orthogonal to the anode strips. The sensitive area of the μ-PIC is 102.4×102.4 mm^2. Each strip of the μ-PIC is connected to the readout circuits which record the electron drift time. The drift time as a function of anode and cathode strip number provides the two-dimensional projection of the trajectories of the charged particles.

3 Data analysis and results

Figure 2 shows an example of the TPC track of the signal events in which the excited ^{12}C nucleus decay into three alpha particles. The left and right images show the anode and cathode

 https://doi.org/10.1051/epjconf/202226011010

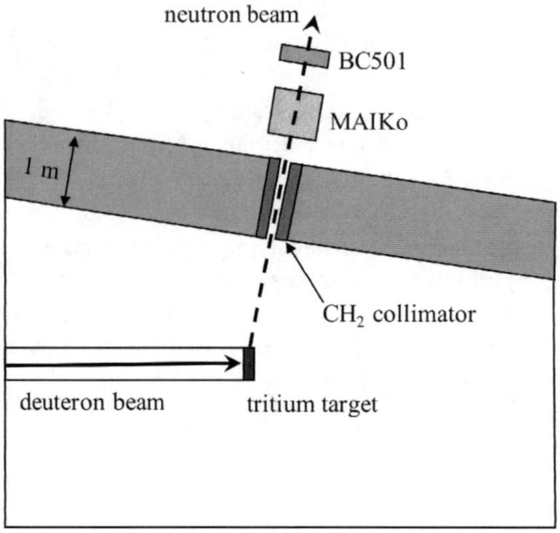

Figure 1. Experimental setup at OKTAVIAN.

projections, respectively. In addition to the signal events, many background events were acquired during the measurement. The background events consist of the elastic scattering with the hydrogen or ^{12}C, and the ^{12}C+n $\rightarrow$^{13}C*$\rightarrow$^{9}Be+^{4}He reaction. The signal events can be distinguished from the background events according to the number of trajectories contained in the TPC images. The signal events contain three trajectories whereas the elastic scattering events contain only one trajectory and the ^{13}C* events contain two trajectories of ^{9}Be and ^{4}He.

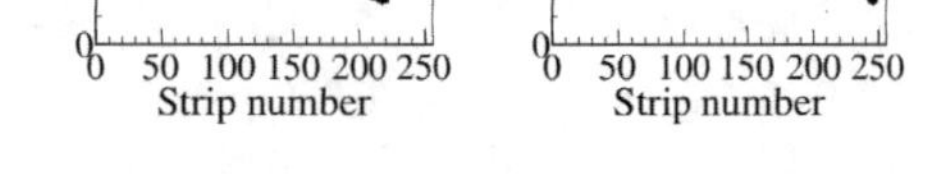

Figure 2. Example of the TPC track of three alpha decay from ^{12}C event. The left and right panels shows the anode and cathode iamges, respectively.

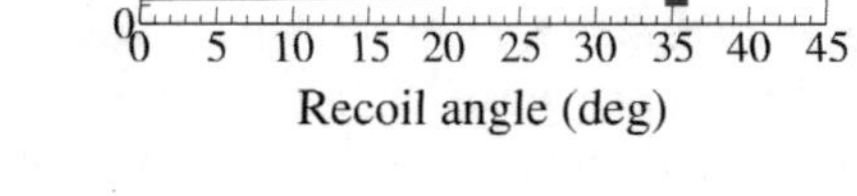

Figure 3. Scatter plot of the kinetic energy and angle of the recoil ^{12}C. The solid line represents the calculated energy and angle for the 0_2^+ state.

We performed an eye scan analysis to categorize the TPC images according to the number of trajectories. About 63,000 events were analyzed by 10 eye scanners. For events that were

Table 1. Summary of the experimental results.

Y	6 ± 2(stat.)
N_b	$[4.5 \pm 0.3(\text{syst.})] \times 10^8$
N_t	9.2×10^{18} /cm^2
ϵ	0.104 ± 0.014(syst.)
σ (Present)	14 ± 6(stat.) $\pm$ 2(syst.) mb
σ (Previous [7])	8.4 mb

categorized as a candidate of the signal event, the eye scanner recorded the coordinates of the vertex position of three alpha trajectories and the end points of the trajectories in the anode and cathode images by mouse clickings. The kinetic energy and momentum of each alpha particle were calculated from the vector between the vertex position and the end point using a computer program SRIM [5]. The kinetic energy, recoil angle, and excitation energy of the ^{12}C nucleus were calculated with the invariant mass spectroscopy method.

Figure 3 represents the correlation between the reconstructed kinetic energy and recoil angle of ^{12}C. The calculated energies and angles for the 0_2^+ state in ^{12}C are plotted with the solid line. The events which lie around the solid line correspond to the inelastic scattering in which the ^{12}C was excited to the 0_2^+ state.

The detection and analysis efficiency of the inelastic scattering events to the 0_2^+ state was estimated by a Monte-Carlo simulation. In the simulation, the inelastic scattering events were generated along the beam region. The decay three alpha particles were then generated from the interaction point. The ionization of the TPC gas by the alpha particles was calculated using SRIM. The electrons due to the ionization were transported towards the μ-PIC plane taking into account for the drift speed and diffusion of the electrons in the gas. The analog signal from the μ-PIC for a single electron was simulated with the computer code Garfield++ [6]. The arrival time of the electrons at the μ-PIC was folded by the analog signal for a single electron. The simulated signals were virtually processed to generate the anode and cathode images. The generated images were analyzed by the eye scanner in the same manner as the experimental data. The detection and analysis efficiency was obtained by the number of events in which three alpha particles stopped inside the sensitive volume of the TPC and the eye scanner identified as the signal event, divided by the number of total generated signal events.

The cross section if the inelastic scattering to the 0_2^+ state was obtained by $\sigma = Y/(N_b N_t \epsilon)$. Y, N_b, N_t, and ϵ are the yield of the signal events reconstructed in the analysis, number of incident neutrons, number of target nuclei for the unit area, and the detection and analysis efficiency, respectively. These values are summarized in Table 1. The cross section determined by the present measurement is, $\sigma = 14 \pm 6(\text{stat.}) \pm 2(\text{syst.})$ mb. This cross section is consistent with the previous one of $\sigma = 8.4$ mb reported by Kondo et al. [7], demonstrating the validity of our method using the MAIKo active target.

4 Future perspectives

We are now constructing a larger active target called MAIKo+ that has a sensitive volume of $300 \times 300 \times 300$ mm^3 to increase the statistics by 10 times from the present MAIKo. The MAIKo+ system will be commissioned at OKTAVIAN in 2022 to measure the same reaction as the present paper. After the commissioning experiment, the physics measurement using the neutron beam at around 10 MeV will be performed at CYRIC, Tohoku University.

References

[1] J. Kelly et al., Nucl. Phys. A **968**, 71 (2017)
[2] M. Beard et al., Phys. Rev. Lett. **119**, 112701 (2017)
[3] T. Furuno et al., Nucl. Instrum. Methods Phys. Res. A **908**, 215 (2018)
[4] A. Ochi et al., Nucl. Instrum. Methods Phys. Res. A **471**, 264 (2001)
[5] J. Ziegler et al., Nucl. Instrum. Methods Phys. Res. B **268**, 1818 (2010)
[6] https://garfieldpp.web.cern.ch/garfieldpp/
[7] K. Kondo et al., J. Nucl. Sci. Tech. **45**, 103 (2008)

Observational properties of peculiar core-collapse supernovae producing r-process elements

Tatsuki Hasegawa[1,*], *Masaomi* Tanaka[1,**], *Nobuya* Nishimura[2,3], *Kyohei* Kawaguchi[4], *Sei* Saito[1], and *Nanae* Domoto[1]

[1]Astronomical Institute, Tohoku University, Aoba, Sendai 980-8578, Japan

[2]Astrophysical Big Bang Laboratory, CPR, RIKEN, Wako, Saitama 351-0198, Japan

[3]RIKEN Nishina Center for Accelerator-Based Science, Wako, Saitama 351-0198, Japan

[4]Institute for Cosmic Ray Research, The University of Tokyo, 5-1-5 Kashiwanoha, Kashiwa, Chiba 277-8582, Japan

Abstract. We investigate the effects of r-process elements on the electromagnetic wave emission from peculiar core-collapse supernovae. We perform a series of radiative transfer simulations based on one-dimensional hydrodynamical explosion models. Varying the mass of ^{56}Ni and r-process nuclei, we examine the effects of r-process elements on the light curves. The ^{56}Ni-powered explosion models with a large amount of r-process-rich material show redder color because the high opacity of r-process elements suppresses the emission at shorter wavelengths. Comparing the results with the color of observed supernovae, we find that the upper limit for the production of r-process elements is $0.1 M_{\odot}$ for ^{56}Ni-powered supernovae.

1 Introduction

The cosmic origin of r-process elements is still under debate. From the recent observations of GW170817 and the relevant emission called kilonova, neutron star mergers are thought to be the plausible candidate. However, it is not yet clear if the neutron star mergers produce the entire r-process elements through the galactic chemical evolution (e.g., [1]). As an alternative scenario, core-collapse supernovae (CCSNe) are regarded as a possible candidate. Typical CCSNe are thought to explode by the neutrino-heating mechanism, which makes r-process nucleosynthesis difficult due to the decrease of neutron-richness (e.g., [2]). However, in recent studies, peculiar CCSNe such as magneto-rotational SNe (e.g. [3]) and the collapsars with the disk wind (e.g., [4]) are suggested to be able to produce r-process elements. In this study, we perform one-dimensional (1D) hydrodynamical calculations and 1D radiative transfer calculations to understand the possible observational properties of peculiar CCSNe with the production of r-process elements.

*e-mail: tatsuki.hasegawa@astr.tohoku.ac.jp
**e-mail: masaomi.tanaka@astr.tohoku.ac.jp

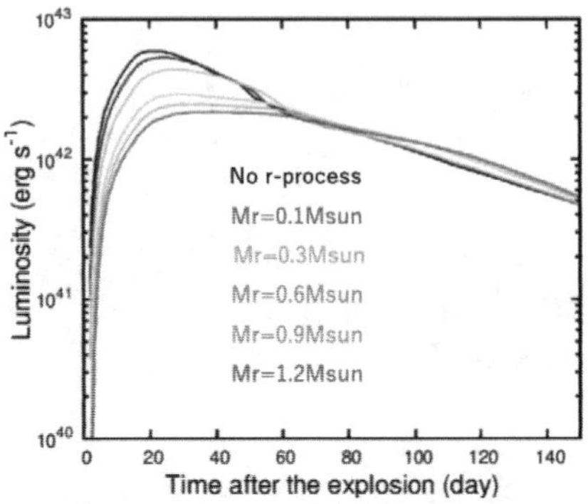

Figure 1. The bolometric luminosities of SNe with different r-process masses.

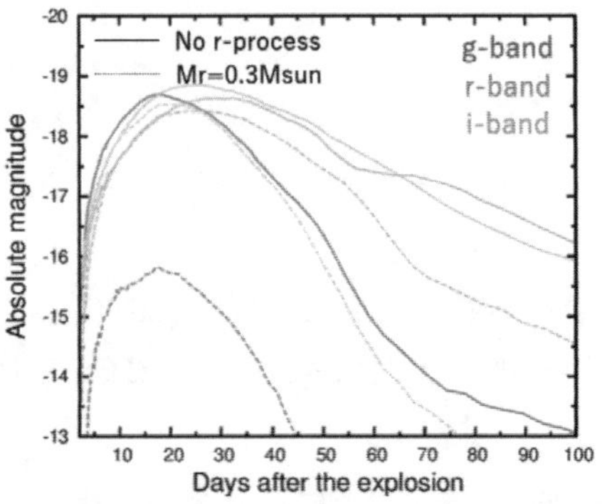

Figure 2. The light curves in each optical band. The solid lines show the model with no r-process elements while the dashed lines show the model with $0.3M_\odot$ of r-process elements.

2 Method

We first perform the 1D hydrodynamical and nucleosynthesis calculations using blcode[1] and nucleosynthesis network with 21 nuclei[2] to derive the ejecta structure and the composition of the elements. We use a progenitor with the zero-age main sequence (ZAMS) mass of $45M_\odot$ (or $13M_\odot$ just before the collapse, s45 [5]). This star is exploded with the kinetic energy of 10^{52} erg, and we track the time evolution of the ejected material. The ^{56}Ni mass in the ejected material is $0.36M_\odot$, which is consistent with ^{56}Ni mass estimated in CCSNe associated with gamma-ray bursts (GRBs).

Using the results of the hydrodynamical calculations, we perform 1D radiative transfer (RT) calculations [6–8]. R-process elements ($Z > 30$) are assumed to be distributed uniformly in the ejecta with the solar r-process abundance pattern. We assume r-process elements are stable as most of the r-process nuclei are close to the stable nuclei in the timescale of interest. Under these assumptions, we calculate the 6 cases with different total mass of r-process elements, i.e., $M_{\mathrm{r-proc}}/M_\odot = 0, 0.1, 0.3, 0.6, 0.9$, and 1.2. We do not consider radioactive decay luminosity of r-process elements as it is subdominant compared with that of ^{56}Ni in our models.

3 Results

Figure 1 shows the light curves of the bolometric luminosity from 1D RT calculations. The different colors indicate the different masses of r-process elements. The model without r-process elements is depicted as the black line. With more r-process elements, the peak luminosity in the bolometric light curve decreases and timescale of the evolution becomes longer.

Figure 2 shows the light curves in the optical band. The solid lines show the model without r-process elements, while the dashed lines show the model with $0.3M_\odot$ of r-process elements. In the model with $0.3M_\odot$ of r-process elements, the light curves in shorter wavelength are remarkably suppressed. The differences in the light curves are attributed to the high opacity of r-process elements, which prevent the photon escaping from the SN ejecta.

[1] https://stellarcollapse.org/index.php/SNEC.html
[2] http://cococubed.asu.edu/code_pages/burn_helium.shtml

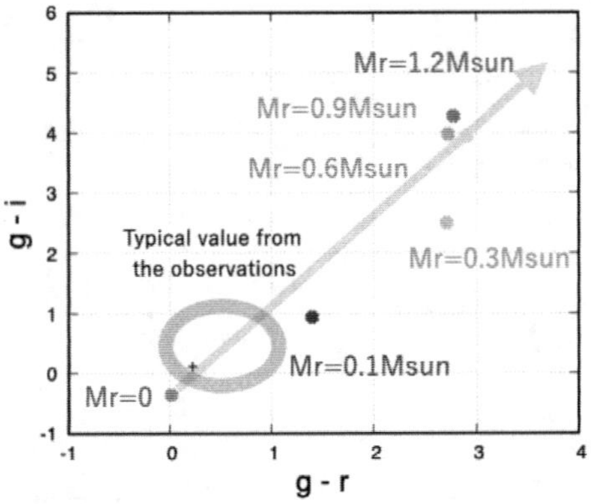

Figure 3. The calculated explosion models on the color-color diagram (g - i vs g - r).

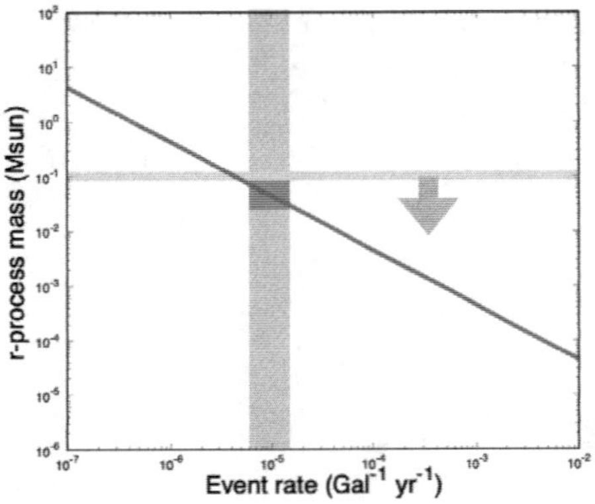

Figure 4. The event rate and the mass of r-process elements produced in one event.

Moreover, the opacity depends on the wavelength, showing a higher opacity toward shorter wavelength [9], which makes the light curve in shorter wavelength more suppressed.

Figure 3 shows the color-color diagram compared with the observational data. The grey region indicates the typical color of observed SNe [10]. From this figure, SNe producing r-process elements larger than $0.1M_\odot$ are deviated from the observed color of typical SNe. This implies that such red SNe have never been observed yet. We can constrain the mass of r-process elements produced in one CCSN, $M_{\rm r-proc} \leq 0.1M_\odot$.

4 Discussion

We discuss the possibility of peculiar SNe such as magneto-rotational SNe as the origin of r-process elements. The red line in Figure 4 shows the mass required to explain the total mass of r-process elements in the Galaxy as a function of the event rate [11]. If the event rate of peculiar SNe is assumed to be similar to that of the SNe with GRB (about 10^{-5} Gal^{-1} yr^{-1} [12]), the required mass of r-process elements per one event to explain the current total mass in the Galaxy is about several $\times 10^{-2} M_\odot$. From the result of this paper, it is found that the mass of r-process elements produced by one SN is less than $0.1M_\odot$. Therefore, the possibility that such peculiar SNe produces r-process elements is not rejected from the observation of typical SNe.

References

[1] Cote, B., Eicher, M. & Arcones, A., et al. 2019, ApJ, 875, 106
[2] Cowan, J. J., Sneden, C., Lawler, J. E., et al. 2021, Reviews of Modern Physics, 93, 15002
[3] Nishimura, N., Sawai, H. & Takiwaki, T., et al. 2015, ApJ, 836, L21
[4] Siegel, D. M., Barnes J. & Metzger, B. D., 2020, Nature, 569, 7755
[5] Sukhbold, T., Ertl, T. & Woosley, S. E., et al. 2016, ApJ, 821, 38
[6] Tanaka, M. & Hotokezaka, K., 2013, ApJ, 775, 113
[7] Kawaguchi, K., Shibata, M. & Tanaka, M., 2018, ApJ, 865, L21
[8] Tanaka, M., Kato, D., Gaigalas, G., et al. 2020, MNRAS, 496, 1369
[9] Tanaka, M., Kato, D., Gaigalas, G., et al. 2018, ApJ, 852, 109

[10] Stritzinger, M. D., Taddia, F. & Burns, C. R., et al. 2015, AA, 609, A135
[11] Hotokezaka, K., Beniamini, P. & Piran, T., 2018, International Journal of Modern Physics D, 27, 1842005
[12] Wanderman, D. & Piran, T., 2010, MNRAS, 406, 1944

https://doi.org/10.1051/epjconf/202226011012

Activation measurements of neutron capture cross sections at various temperatures

Tanja Heftrich[1,*], *Mario* Weigand[1], *Kafa* Al-Khasawleh[1], *Benjamin* Brückner[1], *Sophia* Dellmann[1], *Ozan Can* Dogan[1], *Asmaa* El Mard[1], *Philipp* Erbacher[1], *Fabian* Habermehl[1], *Benedict* Heybeck[1], *Madeleine Margaux* Gail[1], *Kathrin* Göbel[1], *Timo* Kisselbach[1], *Deniz* Kurtulgil[1], *Markus* Reich[1], *René* Reifarth[1], *Silas* Sheriff[1], and *Meiko* Volknandt[1]

[1]Goethe University Frankfurt, Germany

Abstract. About 50% of the elements heavier than iron are produced during the slow neutron capture process. This process occurs in different stellar sites at various energies. To understand the ongoing nucleosynthesis, the probability of a neutron capture for different temperatures and therefore for different stellar sites is essential. Activation experiments using the ^{7}Li(p,n) reaction as neutron source were performed. At a temperature of $k_{\mathrm{B}}T = 25\,\mathrm{keV}$ the cross sections were determined for ^{27}Al, ^{37}Cl and ^{41}K. A new method was developed to perform activation experiments at even lower temperatures. For a proof of principle, the cross section for ^{64}Ni was measured at $k_{\mathrm{B}}T = 25\,\mathrm{keV}$ as well as for $k_{\mathrm{B}}T = 6\,\mathrm{keV}$. To study the impact of isomeric states at higher energies, activations of ^{181}Ta were performed using two different proton energies.

1 Introduction

The slow neutron capture process (s-process) synthesizes about 50% of the heavy elemental abundances. The s-process is divided into two components: main and weak component. The main component refers to thermally pulsing asymptotic giant branch (TP-AGB) stars during the advanced He burning stage. Two neutron sources acting temperature dependent: ^{13}C(α,n) at $k_{\mathrm{B}}T = 5\,\mathrm{keV}$ and ^{22}Ne(α,n) at $k_{\mathrm{B}}T = 30\,\mathrm{keV}$ [1]. The weak component of the s-process refers to He core burning with temperatures of $k_{\mathrm{B}}T = 25\,\mathrm{keV}$ and during the C shell burning at even higher temperatures of $k_{\mathrm{B}}T = 90\,\mathrm{keV}$ [1]. Therefore, depending on astrophysical sites the neutron capture cross sections for a broad energy range of temperatures are essential. The working horse for activation measurements at $k_{\mathrm{B}}T = 25\,\mathrm{keV}$ is the ^{7}Li(p,n) reaction as neutron source. At this temperature measurements for ^{27}Al to study the puzzle of active nucleosynthesis [2] as well as ^{37}Cl and ^{41}K, which act as neutron poisons [1] were performed. In order to investigate cross sections at lower temperatures, a new method for activation measurements at $k_{\mathrm{B}}T = 6\,\mathrm{keV}$ was developed. The principle was proven using a ^{64}Ni ring sample. Isomeric states can have a big impact at the nucleosynthesis path [3]. Many neutron capture measurements are only sensitive to the sum of the cross sections population of the isomeric and ground state. In order to study the neutron capture into the isomeric state at higher temperatures, a measurement of neutron capture on ^{181}Ta into the isomeric and ground state of ^{182}Ta was performed using higher proton energies.

*e-mail: heftrich@iap.uni-frankfurt.de

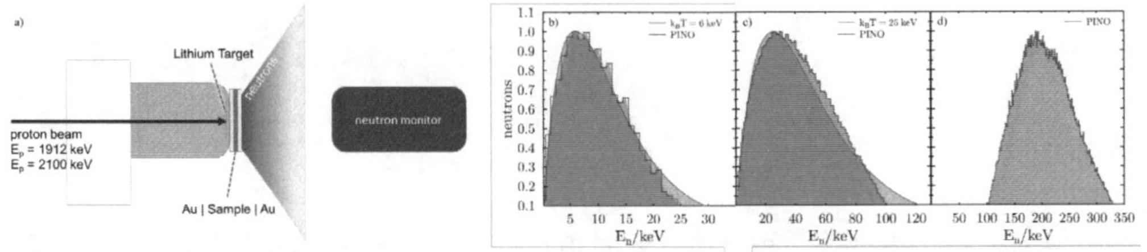

Figure 1. a) Setup for the activation measurements. b)-d) The MD (green) compared with simulated neutron spectrum (blue) for a sample b) ring-shaped. c) circular-shaped. d) circular-shaped with raised proton energy to 2100 keV.

2 Activation setup

Protons were accelerated and impinged on a Li target. Via ^{7}Li(p,n) neutrons are emitted in a cone (Figure 1a). Each sample was positioned between gold foils, which act as monitors for the neutron flux. Neutron flux fluctuations were determined with a Li glass detector. There are two options to perform measurements at different energies: Variation of the impinging proton energy or different sample shapes. For incoming beam energy of 1912 keV, the resulting neutron spectrum for a sample which covers the hole neutron cone follows a Maxwell distribution (MD) at $k_{\mathrm{B}}T = 25$ keV. Figure 1 shows the simulated neutron spectra using PINO [4] which passes different shaped samples (blue) compared to MDs (green). Recently a new method was developed to measure cross sections at lower energies. If the sample is ring-shaped, the angular coverage of the neutron cone represents a MD at various energies depending on the outer and inner radius of the ring and the distance to the Li target. To perform an activation at $k_{\mathrm{B}}T = 6$ keV, an outer radius of 10 mm, an inner radius of 6 mm and a distance to the Li target of 3.2 mm was chosen (Figure 1b). If the proton energy is raised up to $E_P = 2100$ keV, the resulting neutrons which pass the sample are shown in Figure 1d.

3 Analysis

The freshly produced radioactive nuclei were observed by their characteristic γ-rays. A setup consisting of two Broad Energy Germanium Detectors (BEGe) was built up at Goethe University, allowing γ-spectroscopy with high efficiency. The detectors are cylindrical Ge crystals (70x31) mm mounted head to head. The efficiency was determined by measurements and compared with simulations using GEANT. The γ-intensity is included in the simulations as well as the absorption of the γs in the specific samples. The sample is placed in the middle of both detectors. The produced activity is given by: $A = \lambda\left(\frac{C_\gamma}{\epsilon_\gamma f_{\mathrm{m}} f_{\mathrm{dt}}}\right)$, where λ is the decay constant, C_γ are the detected events (Counts), ϵ_γ the efficiency, $f_{\mathrm{m}} = 1 - \exp(-\lambda t_{\mathrm{m}})$ the correction for the decay after the measurement and f_{dt} is the correction for the deadtime of the detection system. The neutron flux was determined by using the gold samples surrounded the target as monitors by: $\Phi_n = \frac{N(^{198}Au)}{N(^{197}Au)\sigma(^{197}Au)}$, where $N(^{198}Au)$ are the gold nuclei which captured a neutron, $N(^{197}Au)$ are the stable gold nuclei of the samples and $\sigma(^{197}Au)$ is spectrum averaged cross section (SACS) for the neutron capture on gold. The activities of produced radioactive gold nuclei were determined using the BEGe setup. The spectrum average cross section for gold were determined using PINO [4] (Table 1). Using the evaluated neutron flux and the activity of the product nuclei and the non-activated nuclei N_{sample}, the cross section is

Table 1. The SACS of gold, measured and literature cross sections.

reaction	k_BT [keV]	this work σ(Au) [mb]	this work σ [mb]	literature σ [mb]
^{64}Ni(n,γ)^{65}Ni	6	1958±35	12±2 [5]	11.8 [8, 9]
^{64}Ni(n,γ)^{65}Ni	25	655±12	4.3±0.3 [5]	9.0[8, 9]
^{27}Al(n,γ)^{28}Al	25	648±9	3.9±0.2 [6]	4.1[8, 10]
^{37}Cl(n,γ)^{38}Cl	25	650±10	2.8±0.2 [7]	2.47[8, 11]
^{41}K(n,γ)^{42}K	25	652±10	19±2 [7]	24.7[8, 12]
^{181}Ta(n,γ)^{182}Ta	25	652±10	771**	863 [8, 13]
^{181}Ta(n,γ)^{182}Ta^m	25	652±10	0.8**	
^{181}Ta(n,γ)^{182}Ta	200*	252±8	197**	
^{181}Ta(n,γ)^{182}Ta^m	200*	252±8	0.51**	

*non MD
** expected uncertainties in the range of 10 %

given by:

$$\sigma = \frac{A}{\lambda N_{\text{sample}} \Phi_n}. \tag{1}$$

4 Summary

Using the standard reaction with a proton energy of 1912 keV and disc-shaped samples, the cross section of ^{27}Al, ^{37}Cl and ^{41}K were measured at $k_BT = 25$ keV. Recently, using a new method to measure cross sections at $k_BT = 6$ keV was developed. The samples were prepared as a ring and therefore the angular-coverage of the neutron cone by the samples refers to to a MD of the neutrons at $k_BT = 6$ keV. At much higher temperatures of 200 keV it was studied the temperature dependency of neutron capture cross section into the isomeric state for the isotope ^{181}Ta. This work has received funding from the Helmholtz International Center for FAIR. We thank M. Dworak, S. Schöffler and P. Ziel for providing excellent beam conditions during these experiments.

References

[1] R. Reifarth, et. al., J. Phys. G: Nucl. Part. Phys. **41** 053101(2014)
[2] R. Diehl, et. al., Nature, **439** 45-47 (2006)
[3] R. Reifarth, et. al., International Journal of Modern Physics A, **33** 1843011 (2018)
[4] R. Reifarth, M. Heil, F. Käppeler, R. Plag, Nucl. Instr. A **608** 139 (2009)
[5] M. Gail, Bachelorthesis (2021)
[6] F. Habermehl, Bachelorthesis (2018)
[7] O.-C. Dogan, Bachelorthesis (2018)
[8] I. Dillmann, et. al., AIP Conf. Proc. **819** 123 (2009)
[9] M. Heil, et. al., Phys. Rev. C **77** 015808 (2008)
[10] K. Wisshak, F. Käppeler, and G. Reffo, Nucl. Sci. Eng. **88** 594 (1984)
[11] K.H. Guber, et.al., Phys. Rev C **65** 058801 (2002)
[12] R. Macklin, Nucl. Sci. Eng. **88** 129 (1984)
[13] K. Wisshak, et.al., Phys. Rev. C **42** 1731 (1990)

Semi-empirical fission model for r-process based on the recent experiments and three-dimensional Langevin approach

Chikako Ishizuka[1,3,*], *Kohsuke* Tsubakihara[2], *Satoshi* Chiba[1], *Yuichiro* Sekiguchi[3], and *Shinya* Wanajo[4]

[1]Institute of Innovative Research, Tokyo Institute of Technology, Tokyo 152-8550, Japan
[2]National Institute of Technology, Asahikawa College, Asahikawa 071-8142, Japan
[3]Department of Physics, Toho University, Funabashi, Chiba 274-8510, Japan
[4]Max Planck Institute for Gravitational Physics (Albert Einstein Institute), Am Mühlenberg 1, Potsdam-Golm, D-14476, Germany

Abstract. Nuclear fission of superheavy elements can affect the r-process nucleosynthesis via a fission cycling process. In that process, the identification of the nuclide of a fission fragment and its abundance is essential for a precise evaluation of their contribution as seed nuclei of the r-process. We have investigated the nuclear fission of SHEs from the proton-rich side to the neutron-rich side using our three- and four-dimensional Langevin models [1, 4]. This model can reproduce the experimental data on nuclear fission of actinides and SHEs, while it can only provide various quantities to each fission fragment, we also developed a semi-empirical charge distribution model based on abundant experimental data of actinides [3]. A new semi-empirical nuclear fission model for the r-process is made by combining these two models.

1 Background and Our Model

Nuclear fission plays an essential role in the rapid neutron capture process (r-process) because it provides seed nuclei for the r-process. After the first observation of the r-elements in neutron star mergers, the fission cycling process in the r-process attracts more attention for a more precise analysis of the r-process nucleosynthesis. However, fission yields strongly depend on nuclear models. Therefore, we need fission yields based on a theory with predictive power. Our semi-empirical fission fragment model Y(Z, A) combines mass yield model Y(A) [2] and the charge distributions evaluated by thousands of experimental data of nuclear fission in actinides [3]. Here, Y(A) [2] is the superposition of three Gaussian functions, which is based on the three-dimensional Langevin calculations with Two Center Shell Model (TCSM) as shown in Figure 1. For details of the charge distributions, please see our previous work [3]. In this paper, we discuss the mass and charge distribution of the present fission model, and evaluate the impact of fission models on the r-process.

$$
\begin{aligned}
Y(Z,A) &\approx Y(A) \\
&\times \frac{1}{\sqrt{2\pi}\sigma(A)} \int_{-0.5}^{0.5} \exp - \frac{(Z - Z_p(A) + t)^2}{2\sigma(A)^2} dt \times \exp - \frac{\Delta E_{sh}(Z,A)}{E_d(A)} .
\end{aligned}
$$

*e-mail: ishizuka.c.aa@m.titech.ac.jp

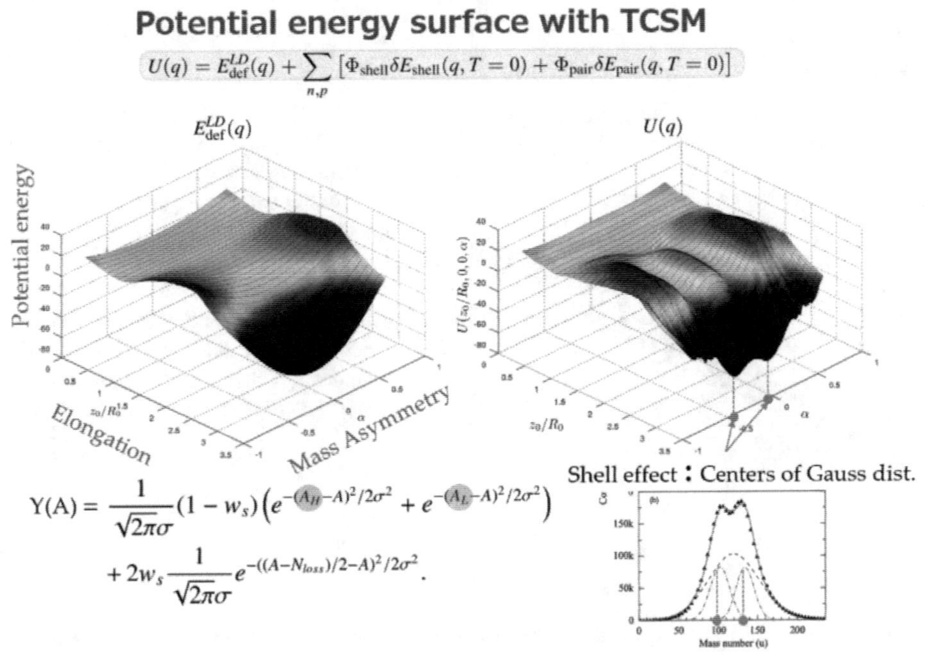

Figure 1. Schematic view of how to construct the mass yield from the potential surface of two center shell model (TCSM). Mass numbers A_H and A_L in the mass yield $Y(A)$ correspond to the valleys of the potential energy surface with respect to the fission fragment mass number.

2 Results and Summary

Figure 2 shows the mass and charge distribution of our Y(Z, A) model based on the three dimensional Langevin model [1] with Two Center Shell Model (TCSM). Our model provides a broader mass distribution with the fragment mass number A=90-165 than Kodama-Takahashi, the standard mass model of the r-process nucleosynthesis. Especially symmetric mass components are more abundant compared to the standard model. Such a trend can also be seen in the charge distributions, as shown in the lower panel in Figure 2.

In Figure 3, we compare three cases, our Y(Z, A) model, Kodam-Takahashi model, and symmetric fission case, with the mass abundance of the r-process nuclides in the solar system. Using our Y(Z, A) model, we obtained an abundance pattern of a single r-process with Y_e = 0.05, corresponding to the electron fraction of a typical black hole and neutron star merger. As a result, we observed significant differences among them, especially around the second peak and the rare-earth peak of the r-process. Thus we found that the fission fragment mass yields can affect the abundance pattern of the r-process via the fission recycling process. Therefore, a more precise fission yield model is necessary to evaluate the contribution of the fission recycling process to the r-process.

For this aim, we have performed the four-dimensional Langevin calculations [4] to more than a hundred nuclides with the proton number $Z = 92 - 122$ from the neutron-deficient side to the neutron-rich side. Figure 3. Compared to the three-dimensional (3D) Langevin model [1], our four-dimensional (4D) Langevin model [4] has an additional degree of freedom concerning nuclear deformation. Therefore, the 4D model can reproduce various nuclear fissions which cannot be well-explained by the 3D model, such as a sudden change

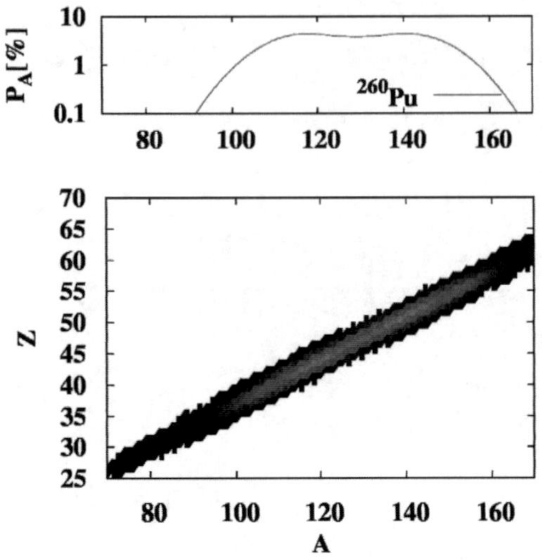

Figure 2. Mass and charge distribution of our semi-empirical yield model Y(Z, A) based on three dimensional Langevin model and recent evaluated yield models using abundant experimental data in actinides. The upper panel shows the mass distribution, while the lower color map displays the distributions of fission fragments on a nuclear chart. The center part of the color map is most abundant.

from asymmetric fission to symmetric fission in Fm isotopes. Our new model based on the 4D model will be discussed in a separate paper.

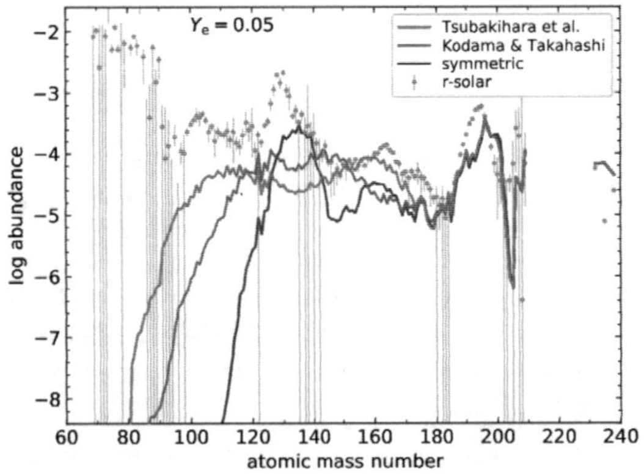

Figure 3. Abundance pattern of a single r-process nucleosynthesis with the electron fraction $Y_e = 0.05$. Red, green, blue lines are results with present model, Kodama-Takahashi model, and symmetric fission model, respectively. Filled grey circles with error bars are the r-elements in the solar abundance.

Acknowledgement

This study is supported by Japan Society for the Promotion of Science (JSPS) KAKENHI Grant No. 18K03642.

References

[1] Y. Aritomo, S. Chiba, Phys. Rev. C **88**, 044614,(2013).

[2] M.Ohta et al., Proceedings of "International Conference on Nuclear Data for Science and Technology 2007", 096 (2008).

[3] K. Tsubakihara, S. Okumura, C. Ishizuka, T. Yoshida, F. Minato, S. Chiba, J. Nucl. Sci. Technol. **58**, 151 (2020).

[4] C. Ishizuka, M.D. Usang, F.A. Ivanyuk, J.A. Maruhn, K. Nishio, S. Chiba, Phys. Rev. C **96**, 064616 (2017).

EPJ Web of Conferences **260**, 11014 (2022) https://doi.org/10.1051/epjconf/202226011014
NIC-XVI

Carbon burning rates on the compound nucleus formation

Masahiko Katsuma[1,2,*]

[1]Advanced Mathematical Institute, Osaka City University, Osaka 558-8585, Japan
[2]Institut d'Astronomie et d'Astrophysique, Université Libre de Bruxelles, B-1050 Brussels, Belgium

Abstract. The ^{12}C+^{12}C reaction rates based on the compound nucleus formation seem to be concordant with the standard rates. The resonant contribution in ^{12}C+^{12}C is also discussed. To put the rates on firm ground, the resonances below $E_{c.m.}$ = 3 MeV will have to be studied further.

1 Introduction

The ^{12}C+^{12}C fusion reaction is one of the key reactions to understand the evolution of massive stars and various explosive scenarios. However, precise measurements of cross sections below $E_{c.m.}$ = 3 MeV are difficult because of the tiny amplitudes caused by the Coulomb barrier. At present, the direct measurements have been performed in $E_{c.m.}$ = 2.1–2.5 MeV [1]. The indirect measurements have been studied with ^{24}Mg(α,α') [2] and Trojan horse method [3]. The derived rates [3] are much faster than CF88 [4], due to the resonant states at $E_{c.m.} \approx 1.5$ MeV, which may have the ^{12}C+^{12}C molecule-like structure. The nuclear fusion for ^{12,13}C+^{13}C have also been discussed experimentally to understand C+C comprehensively [5].

In this presentation, I use a barrier penetration model (BPM), and I show the calculated results of isotope dependence of fusion cross sections and reaction rates in C+C. The transmission coefficients are given by the WKB approximation, semi-classically, and the potentials used in the present work are calculated from a single-folding model [6] with [7, 8]. I also discuss the contribution from the resonances in ^{12}C+^{12}C by comparing the result of BPM with a schematic calculation of the coupled-channels multi-level R-matrix [9].

2 Compound nucleus formation

Before moving on to the results, let me recall the compound nucleus (CN) formation, to understand the reaction mechanism in C+C. The ^{12}C+^{12}C potential obtained from the studies of elastic scattering has predicted the sequences of the rotational excitation in ^{24}Mg [7, 10]. These resonances are the excited states with the ^{12}C+^{12}C molecule-like structure in ^{24}Mg. However, the potential resonances at $E_{c.m.} \approx 0$ are dispersed easily, because of the couplings to reaction channels. Although the inelastic channels are closed at $E_{c.m.}$ = 4.44 MeV, other reaction channels are open, and they work as absorption to the entrance channel. Accordingly, their fragments are distributed around the original energy positions. In fact, many fragments of J^π = 2^+, 4^+ resonances have been observed, in addition to 0^+ [3]. Whereas most of flux are consumed by Coulomb scattering, a small amount of flux is captured into the long-living fragment levels, and exits through the proton, neutron, and α channels after forming a compound nucleus. Under the circumstance, the reactions should be described statistically,

*e-mail: mkatsuma@gmail.com

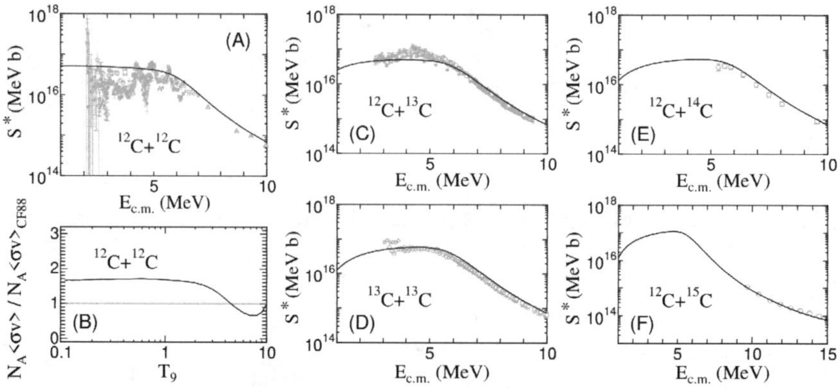

Figure 1. Comparison in S^* between BPM and experimental data: (A) $^{12}C+^{12}C$, (C) $^{12}C+^{13}C$, (D) $^{13}C+^{13}C$, (E) $^{12}C+^{14}C$, and (F) $^{12}C+^{15}C$. The solid curves are the results with BPM. The experimental data are taken from [1, 5, 11, 12]. (B) The derived rates of $^{12}C+^{12}C$ are shown in ratio of CF88 [4].

and the emitted nuclei have to be treated as evaporation products. Therefore, I adopt BPM and R-matrix based on the CN formation in the present study.

In BPM, the energy-averaged fusion cross sections are given by $\sigma_F = (\pi/k^2)\sum_L(2L + 1)\langle|S_{CN}|^2\rangle$. k is the wavenumber; L is the angular momentum between nuclei. $\langle|S_{CN}|^2\rangle$ are the transmission coefficients T_L, given by WKB approximation for $E_{c.m.} < E_B$: $T_L = \{1 + \exp[2\int_{R_1}^{R_2} K(R)\,dR]\}^{-1}$, $K(R) = \{(2\mu)/\hbar^2[\tilde{U}_L(R) - E_{c.m.}]\}^{1/2}$. E_B is the barrier height energy. μ is the reduced mass. R_1 and R_2 are the inner and outer turning points of effective potentials $\tilde{U}_L$. The nuclear potentials in $\tilde{U}_L$ are calculated recursively from the single-folding model [6–8]. The cross sections are also given by R-matrix theory, $\sigma_F = (\pi/k^2)\sum_{cL}(2L+1)\,|\,S^L_{c,0}\,|^2$. $S^L_{c,0}$ is the S-matrix deduced from the R-matrix with the resonance parameters in [3]. The reduced width of $^{12}C+^{12}C$ is statistically assumed to be a constant $\gamma^2_{iL} = 0.001\gamma^2_W$ for all levels, based on the CN formation. γ^2_W is the Wigner limit. In [5], the transmission coefficients are calculated from an approximation using the unitarity relation of S-matrix. To display the cross sections, the S^* factors are defined as $S^* \equiv \sigma_F E_{c.m.} \exp(87.21E_{c.m}^{-1/2}+0.46E_{c.m.})$, e.g. [3].

3 Results & Conclusion

The calculated S^* factors with BPM for $^{12}C+^{12}C$ are shown by the solid curve in Fig. 1(A), and they appear to give the trend of the energy variation in the experimental data [1, 11]. The derived reaction rates are shown in ratio to CF88. (Fig. 1(B)) They seem consistent with CF88. For $^{12,13}C+^{13}C$ and $^{12}C+^{14,15}C$, the present calculations of BPM (solid curves) reproduce the experimental data [5, 12], consistently, as shown in Figs. 1(C) – 1(F).

Figure 2(A) illustrates the isotope dependence of S^* obtained from BPM. In BPM, the $^{12}C+^{12}C$ S^* factors below $E_{c.m.} = 2$ MeV become the largest, so the derived reaction rates are the fastest below $T_9 = 0.8$ (Fig. 2(B)). In addition, the S^* factors are found to be enhanced at the sub-barrier energies as the number of neutrons increases. Particularly, those of $^{12}C+^{15}C$ are enhanced larger. The barrier radius R_B and barrier height energy E_B in the present calculations are shown in Figs. 2(C) and 2(D), as a function of the mass number. R_B (E_B) becomes large (small) as the number of neutrons increases. Especially, R_B suddenly becomes large at ^{15}C. This is caused by the weakly-bound s-wave neutron in ^{15}C. Therefore, the corresponding S^* and reaction rates are expected to be enhanced more by the sharp reduction of E_B.

The resonant contribution in $^{12}C+^{12}C$ is shown in Fig. 3. From the result of the R-matrix calculation by the solid curve in Fig. 3(A), the values of S^* are found to be much smaller than those of [3]. In the present calculation, I include the same 34 levels and four exit channels as

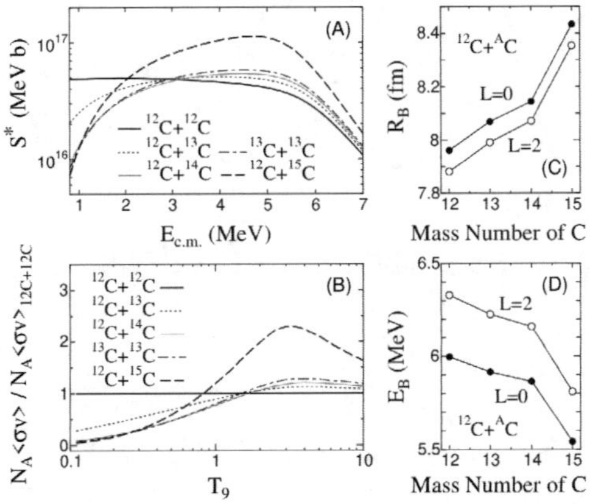

Figure 2. Isotope dependence of (A) S^*, (B) reaction rates, (C) R_B, and (D) E_B, obtained from BPM. (A,B) The curves are the results for the respective system shown in figure. (C,D) R_B and E_B are obtained from $\tilde{U}_L$. The solid and open circles are the values for $L = 0$ and $L = 2$, respectively.

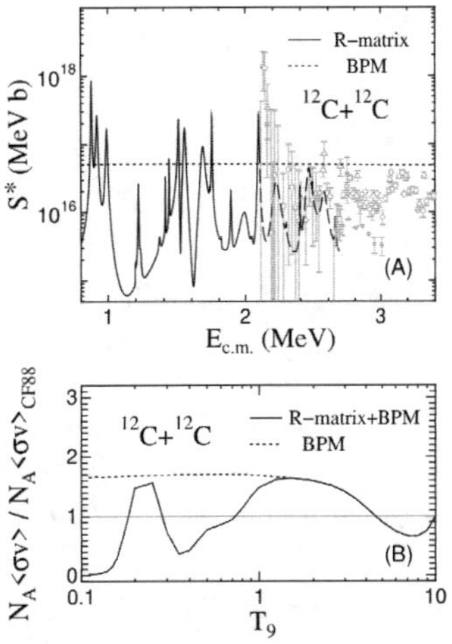

Figure 3. (A) Low-energy extrapolation of S^* and (B) the reaction rates. The solid and dotted curves are the results calculated from R-matrix and BPM. The experimental data are taken from [1, 11].

those in [3]. If $\gamma_{iL}^2 = 0.05\gamma_W^2$ is used as the reduced width of $^{12}C+^{12}C$ at $E_{c.m.} \approx 1.5$ MeV, the reaction rates would increase like those in [3]. The carbon burning rates are sensitive to the reduced width of $^{12}C+^{12}C$. In addition, the reaction rates estimated from the R-matrix extrapolation are confirmed to be reduced from the result of BPM. However, the derived rates at $T_9 = 0.6$ still seem to be consistent with CF88. To put the rates on firm ground, the resonances below $E_{c.m.} = 3$ MeV will have to be scrutinized further.

Acknowledgements

I thank Prof. X.D. Tang for his valuable comments and providing me with the experimental data.

References

[1] W.P. Tan *et al.*, PRL **124**, 192702 (2020); G. Fruet *et al.*, PRL **124**, 192701 (2020); C.L. Jiang *et al.*, PRC **97**, 012801(R) (2018); T. Spillane *et al.*, PRL **98**, 122501 (2007)

[2] T. Kawabata *et al.*, J. Phys. Conf. **436**, 012009 (2013)

[3] A. Tumino *et al.*, Nature **557**, 687 (2018); Springer Proc. Phys. **219**, 17 (2019)

[4] G.R. Caughlan, W.A. Fowler, Atom. Data Nucl. Data Tables **40**, 283 (1988)

[5] M. Notani *et al.*, PRC **85**, 014607 (2012); X.D. Tang, Springer Proc. Phys. **219**, 13 (2019); B.B. Back *et al.*, Rev. Mod. Phys. **86**, 317 (2014)

[6] H. Nishioka *et al.*, Nucl. Phys. A **415**, 230 (1984)

[7] Y. Kondō, M. Katsuma, Nucl. Phys. A **738**, 226 (2004); Y. Kondō, M.E. Brandan, G.R. Satchler, Nucl. Phys. A **637**, 175 (1998)

[8] B.A. Watson *et al.*, Phys. Rev. **182**, 977 (1969)

[9] A.M. Lane, R.G. Thomas, Rev. Mod. Phys. **30**, 257 (1958)

[10] M.E. Brandan, G.R. Satchler, Phys. Rep. **285**, 143 (1997)

[11] E.F. Aguilera *et al.*, PRC **73**, 064601 (2006); D.G. Kovar *et al.*, PRC **20**, 1305 (1979); B. Dasmahapatra *et al.*, Nucl. Phys. A **384**, 257 (1982); M.D. High, B. Čujec, Nucl. Phys. A **282**, 181 (1977); H.W. Becker *et al.*, Z. Phys. A **303**, 305 (1981)

[12] EXFOR, International Network of Nuclear Reaction Data Centres (NNDC); Nuclear Reaction Video (NRV) project supported by Russian Foundation for Basic Research

Measurement of the $^{18}O(\alpha,\gamma)^{22}Ne$ resonances at JUNA

Luohuan Wang[1], *Jun* Su[1,2*], *Yangping* Shen[3], *Jianjun* He[1,2], *Liyong* Zhang[1,2], *Xinyue* Li[1], *Gang* Lian[3], *Bing* Guo[3], *Yaode* Sheng[1], *Yinji* Chen[1], *Hao* Zhang[1], *Ziming* Li[1], *Luyang* Song[1], *Xinzhi* Jiang[1], *Wei* Nan[3], *Weike* Nan[3], *Long* Zhang[3], *Fuqiang* Cao[3], and *Weiping* Liu[3]

[1]Key Laboratory of Beam Technology of Ministry of Education, College of Nuclear Science and Technology, Beijing Normal University, Beijing 100875, China
[2]Beijing Radiation Center, Beijing 100875, China
[3]China Institute of Atomic Energy, P. O. Box 275(10), Beijing 102413, China

Abstract. $^{22}Ne(\alpha,n)^{25}Mg$ is one of the main neutron sources of the *s* process. ^{22}Ne is produced by the $^{14}N(\alpha,\gamma)^{18}F(\beta^+)^{18}O(\alpha,\gamma)^{22}Ne$ reaction chain in the helium burning, thus, the production rate of ^{22}Ne is dominated by $^{14}N(\alpha,\gamma)^{18}F$ and $^{18}O(\alpha,\gamma)^{22}Ne$. At the astrophysical relevant temperatures, the $^{18}O(\alpha,\gamma)^{22}Ne$ reaction rates are determined by several low-energy resonances. In this work, the $^{18}O(\alpha,\gamma)^{22}Ne$ reaction was measured at the 400 kV accelerator of Jinping Underground Nuclear Astrophysics experiment (JUNA). The γ-ray yields of the resonances between 470 to 770 keV were obtained.

1 Introduction

About half of the heavy elements above Fe in our universe are synthesised by the slow neutron capture process (*s*-process) drove by the neutron from the $^{13}C(\alpha,n)^{16}O$ and $^{22}Ne(\alpha,n)^{25}Mg$ reactions [1, 2]. $^{22}Ne(\alpha,n)^{25}Mg$ is considered to be the main neutron source of the *s*-process during convective core helium burning and convective shell carbon burning in massive stars [2], in which the ^{22}Ne is produced through the $^{14}N(\alpha,\gamma)^{18}F(\beta^+\upsilon)^{18}O(\alpha,\gamma)^{22}Ne$ reaction chain in the early phase of core helium burning. Therefore, the $^{14}N(\alpha,\gamma)^{18}F$ and $^{18}O(\alpha,\gamma)^{22}Ne$ reaction rates are the key parameters in the calculation of ^{22}Ne abundance and the subsequent $^{22}Ne(\alpha,n)^{25}Mg$ neutron production rate, which need to be determined precisely.

Several direct measurements [3–5] of the $^{18}O(\alpha,\gamma)^{22}Ne$ resonances have been performed. Trautvetter et al. [3] obtained the resonance strengths of the 662 and 750 keV resonances. Vogelaar et al. [4] measured the precise resonance energies, the primary γ-ray branching ratios, and the resonance strengths of the 662, 750, and 770 keV resonances. And Dababneh et al. [5] investigated the 470 keV resonance by a coincidence method for the first time, however, in which the 470 keV resonance strength was derived by a presumptive branching ratio. The $^{18}O(\alpha,\gamma)^{22}Ne$ reaction needs to be further measured in an underground laboratory with low background to reduce the uncertainties of the low-energy resonance strengths and improve the precision of the reaction rates, which is of great significance to the nucleosynthesis study of *s* process.

*e-mail: sujun@bnu.edu.cn

2 Experimental method

The experiment was performed at the 400 kV accelerator of Jinping Underground Nuclear Astrophysics experiment(JUNA) [6] in the China Jinping Underground Laboratory (CJPL) [7], the $^4He^{2+}$ beam was collimated by two apertures and then bombarded on the ^{18}O enriched isotopic target which was directly water cooling. The average beam current was 500 μA on the target. A cold trap cooled by liquid nitrogen was installed upstream of the target to minimize carbon build-up on the target surface, together with magnetic molecular pumps and dry scroll pumps. Target together with the chamber formed a Faraday cup for measuring the beam charge on the target. A ring electrode applied with a voltage of -300 V was installed between the target and cold trap to suppress secondary electrons emitted from both the target and the last aperture.

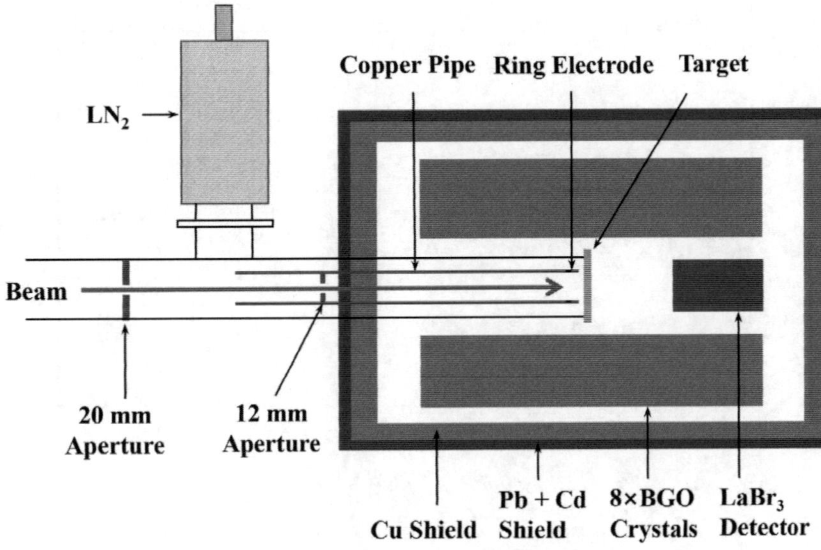

Figure 1: Schematic view of the experimental setup.

The ^{18}O enriched oxygen (98% abundance) gas was used to produce the $Ti^{18}O_x$ (x is the atoms ratio of oxygen to titanium) targets by Filter cathodic vacuum arc (FCVA) technology [8, 9], in which Ta substrates with a diameter of 47.3 mm and a thickness of 0.7 mm was adopted. The targets were able to withstand the high beam loads for several coulombs without noticeable deterioration. The γ rays emitted from $^{18}O(\alpha,\gamma)^{22}Ne$ were detected by a near 4π BGO detector and a $3'' \times 4''$ $LaBr_3$ detector, as shown in Fig. 1. All of the detectors were shielded by 5 mm Cu, 100 mm Pb, and 1 mm Cd materials to further suppress the natural γ-ray background emitted from the rocks and induced by the neutron radioactive capture reactions.

3 Preliminary results

Fig. 2 shows the sum spectra obtained by the BGO detector array of the 470, 662, 750, and 770 keV resonances, respectively, in which the $E_{\rm sum}$ represents the sum of the energies

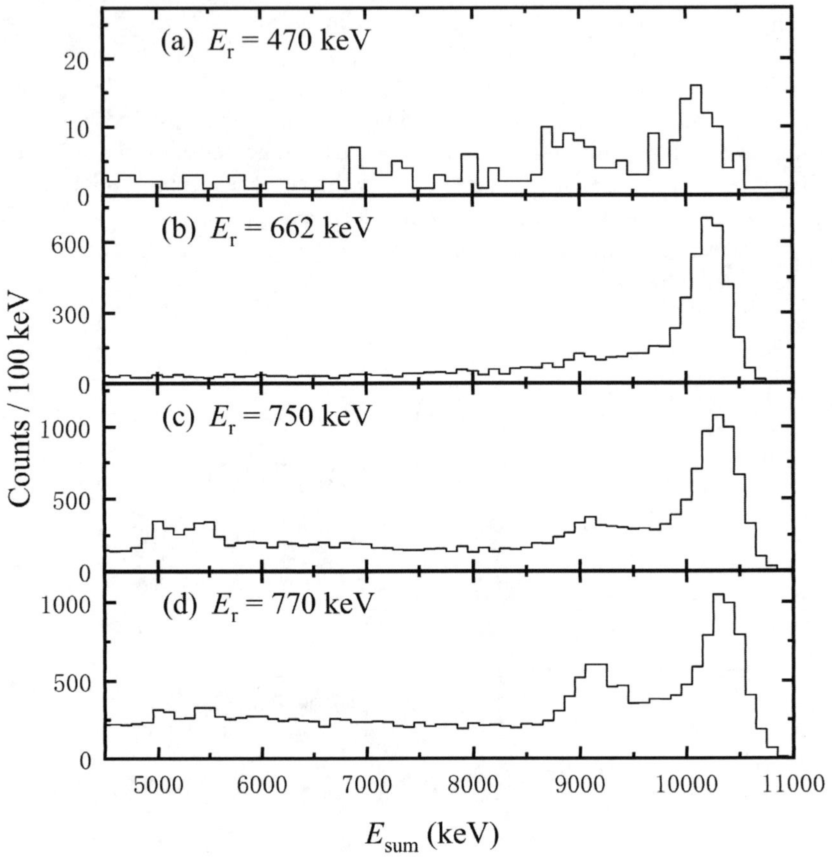

Figure 2: The sum spectra of the $^{18}O(\alpha, \gamma)^{22}Ne$ resonances: (a) at E_r = 470 keV, (b) at E_r = 662 keV, (c) at E_r = 750 keV, (d) at E_r = 770 keV, respectively.

of all eight BGO segments. It can be seen that the sum peaks (E_{sum} = 9500 – 10500 keV) of these resonances are clearly visible in the spectrum.

Based on the counts of the sum peak and the efficiency simulated with GEANT4 program [10], the yield curves of $^{18}O(\alpha, \gamma)^{22}Ne$ were derived, as shown in Fig. 3. In the GEANT4 simulation, the primary γ-ray branching ratios were extracted by fitting the single spectrum obtained of the BGO detector array. The uncertainty of the yields mostly arises from the statistic error, the change of the sum peak range, and the GEANT4 simulation. These yields will be used to extract the resonance strength. Further data analysis is in process and the results will soon be published.

4 Acknowledgments

The authors thank the staff of the CJPL and Yalong River Hydropower Development Company and Tsinghua University for the laboratory support. The present work is supported

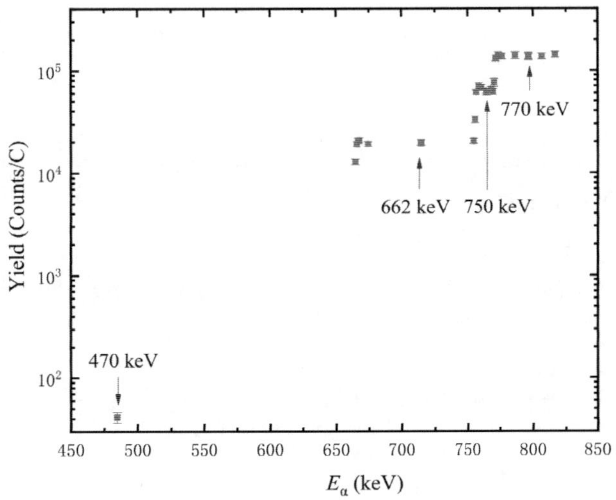

Figure 3: Yield of the 470, 662, 750 and 770 keV resonances of $^{18}O(\alpha,\gamma)^{22}Ne$.

by the National Natural Science Foundation of China (Grant Nos. U1867211, 11490560, 11490562, 11825504, 12075027, 11961141004) and by the Fundamental Research Funds for the Central Universities.

References

[1] F. Käppeler, Prog. Part. Nucl. Phys. **43**, 419 (1999)
[2] F. Käppeler, R. Gallino, S. Bisterzo, W. Aoki, Rev. Mod. Phys. **83**, 157 (2011)
[3] H. Trautvetter, M. Wiescher, K.U. Kettner, C. Rolfs, J. Hammer, Nucl. Phys. **297**, 489 (1978)
[4] R. Vogelaar, T. Wang, S. Kellogg, R. Kavanagh, Phys. Rev. C. **42**, 753 (1990)
[5] S. Dababneh, M. Heil, F. Käppeler, J. Görres, M. Wiescher, R. Reifarth, H. Leiste, Phys. Rev. C. **68**, 025801 (2003)
[6] W. Liu, Z. Li, J. He, X. Tang, G. Lian, Z. An, J. Chang, H. Chen, Q. Chen, X. Chen et al., Sci. China Phys. Mech. Astron. **59**, 642001 (2016)
[7] J. Cheng, K.J. Kang, J. Li, J. Li, Y. Li, Q. Yue, Z. Zeng, Y. Chen, S. Wu, X. Ji et al., Annu. Rev. Nucl. Part. Sci. **67**, 231 (2017)
[8] Z. Zhang, B. Liao, Y. Ou, F. Zhang, X. Zhang, Y. Shen, S. Chen, Q. Hua, G. He, X. Ouyang, Acta Phys. Sin. **69**, 10 (2020)
[9] L. Wang, Y. Shen, J. Su, X. Li, W. Yan, J. He, L. Zhang, B. Liao, Y. Wu, Y. Sheng et al., Nucl. Instr. Meth. B **accepted** (2022)
[10] S. Agostinelli, J. Allison, K.a. Amako, J. Apostolakis, H. Araujo, P. Arce, M. Asai, D. Axen, S. Banerjee, G.. Barrand et al., Nucl. Instr. Meth. A **506**, 250 (2003)

Enhancement of Lithium in Red Clump Stars by Neutrino Magnetic Moments

Kanji Mori[1,2,*], *Motohiko* Kusakabe[3,2], *A. Baha* Balantekin[4,2], *Toshitaka* Kajino[2,3,5], and *Michael A.*Famiano[6,2]

[1]Research Institute of Stellar Explosive Phenomena, Fukuoka University, 8-19-1 Nanakuma, Jonan-ku, Fukuoka-shi, Fukuoka 814-0180, Japan

[2]Division of Science, National Astronomical Observatory of Japan, 2-21-1 Osawa, Mitaka, Tokyo 181-8588, Japan

[3]International Research Center for Big-Bang Cosmology and Element Genesis, and School of Physics, Beihang University, 37 Xueyuan Road, Haidian-qu, Beijing 100083, China

[4]Department of Physics, University of Wisconsin-Madison, Madison, Wisconsin 53706 USA

[5]Graduate School of Science, The University of Tokyo, 7-3-1 Hongo, Bunkyo-ku, Tokyo, 113-0033 Japan

[6]Department of Physics, Western Michigan University, Kalamazoo, MI 49008 USA

Abstract. A recent spectroscopic survey has shown that the lithium abundances in low-mass red-clump stars are higher than theoretical prediction. Motivated by their finding, we implemented effects of additional energy losses induced by neutrino magnetic moments on stellar evolution models. It is found that the lithium abundances in red giants increase because the finite magnetic moments of neutrinos enhance the efficiency of deep mixing and ^{7}Be is conveyed from the hydrogen burning shell to the convective envelope. The results motivate further studies on the effects of non-standard energy losses on the surface compositions of low-mass red giants and red-clump stars.

1 Introduction

The lithium abundances in low-mass stars are long-standing problem in stellar modeling. Giant stars with $A(\mathrm{Li}) = \log N(\mathrm{Li})/N(\mathrm{H}) + 12 > 1.5$ are called lithium-rich giants [1, 2]. Since stellar models predict that lithium in red giants is destroyed by the first dredge-up [3], the mechanism for the lithium enhancement in such giants is an open problem.

Lithium-rich giants with $A(\mathrm{Li}) > 1.5$ account for only $\sim 1\%$ of red giants [e.g. 4]. However, a recent spectroscopic survey [5] has shown that almost all red-clump stars show the lithium abundances of $A(\mathrm{Li}) = 0.71 \pm 0.39$, which are higher than theoretical prediction. This observational result implies missing physical processes in stellar models.

In order to reconcile observations and models, we consider effects of additional energy losses induced by neutrino magnetic moments [NMMs; 6]. It is known that finite NMMs would make a helium core at the tip of red giant branch (RGB) heavier [7, 8]. However, their effect on the stellar surface composition has not been discussed.

*e-mail: kanji.mori@fukuoka-u.ac.jp

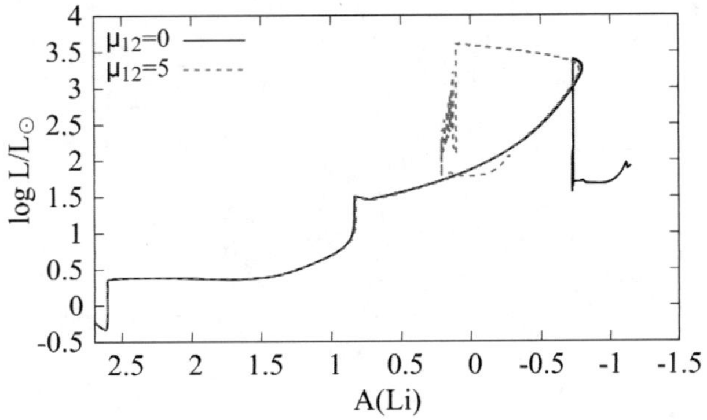

Figure 1. The luminosity L and the surface lithium abundance A(Li) of the models with $\mu_{12} = 0$ and 5.

2 Method

We adopt Modules for Experiments in Stellar Astrophysics [MESA; 9] to calculate stellar models. We fix the initial stellar mass to $1M_{\odot}$ and the metallicity to $Z_{\odot} = 0.0148$ [10]. We adopt the mixing length of $\alpha = 1.6$. We assume thermohaline mixing as the mechanism of deep mixing. The efficiency of thermohaline mixing [11] is fixed to $\alpha_{\rm thm} = 50$.

The energy loss rate induced by NMMs is given as [8]

$$\epsilon^{\mu}_{\rm plas} = 0.318\left(\frac{\omega_{\rm pl}}{10\ {\rm keV}}\right)^{-2}\left(\frac{\mu_{\nu}}{10^{-12}\mu_{\rm B}}\right)^{2}\epsilon_{\rm plas}, \tag{1}$$

where $\omega_{\rm pl}$ is the plasma frequency, μ_{ν} is an NMM, $\mu_{\rm B}$ is the Bohr magneton, and $\epsilon_{\rm plas}$ is the standard plasmon decay rate.

3 Results

We performed simulations of stellar evolution with $\mu_{12} = \mu_{\nu}/(10^{-12}\mu_{\rm B}) = 0 - 5$ [6]. Fig. 1 shows the evolution of the luminosity and A(Li) of the models with $\mu_{12} = 0$ and 5. The star becomes more luminous as it evolves along the red giant branch. In red giants, an inert helium core is formed and its mass gradually grows because of hydrogen shell burning. When the core temperature becomes sufficiently high, helium is ignited in the core. Because the core is degenerate, helium burning at the tip of the red giant branch happens violently. As a result, the core becomes non-degenerate and stable helium burning starts. This evolutionary phase with stable helium core burning is called red clump stars.

As a star evolves in the red giant branch, the lithium abundance decreases. When $L \sim 0.5L_{\odot}$, A(Li) decreases by ~ 1.5 because of the first dredge-up, where the convective envelope penetrates into the stellar interior and lithium is destroyed by higher temperature there. The lithium abundance continues to decrease when $L > 1.5L_{\odot}$. This is because thermohaline mixing conveys the material in the envelope to the hydrogen burning shell. When the star is close to the tip of the red giant branch, A(Li) starts increasing because ^{7}Be created in the hydrogen burning shell is conveyed to the envelope and decays to ^{7}Li [12].

When the effect of NMM is considered, the lithium abundance becomes larger around the tip of the red giant branch. This is because the additional energy loss leads to a smaller core and the density in the envelope becomes lower. The efficiency of thermohaline mixing

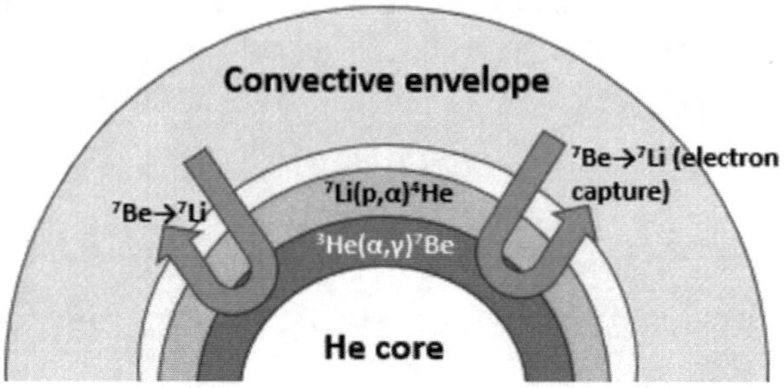

Figure 2. A schematic picture on the lithium enhancement. The figure is not to scale.

then becomes higher. As a result, ^{7}Be is conveyed to the envelope more efficiently. The mechanism of the lithium enhancement is schematically shown in Fig. 2.

Fig. 1 shows that the lithium abundance in red clump stars becomes ~ 10 times higher when $\mu_{12} = 5$ is adopted. The enhancement of lithium may mitigate the lithium problem in red clump stars. Similar effects are expected in the cases of other physical processes such as axions. It is desirable to perform calculations with such additional energy losses induced by new physics to find a solution to the problem. Also, it is important to explore other scenarios including mixing induced by the helium flash [13] and effects of the stellar mass distribution [14].

This work is supported by Research Institute of Stellar Explosive Phenomena at Fukuoka University and JSPS KAKENHI grant number JP21K20369.

References

[1] G. Wallerstein, C. Sneden, Astrophys. J. **255**, 577 (1982)

[2] J.A. Brown, C. Sneden, D.L. Lambert, J. Dutchover, Edward, Astrophys. J. Suppl. Ser. **71**, 293 (1989)

[3] J. Iben, Icko, Astrophys. J. **147**, 624 (1967)

[4] Deepak, B.E. Reddy, Mon. Not. R. Astron. Soc. **484**, 2000 (2019), 1901.03832

[5] Y.B. Kumar, B.E. Reddy, S.W. Campbell, S. Maben, G. Zhao, Y.S. Ting, Nature Astronomy **4**, 1059 (2020), 2007.07045

[6] K. Mori, M. Kusakabe, A.B. Balantekin, T. Kajino, M.A. Famiano, Mon. Not. R. Astron. Soc. **503**, 2746 (2021), 2009.00293

[7] G. Raffelt, A. Weiss, Astron. Astrophys. **264**, 536 (1992)

[8] M. Haft, G. Raffelt, A. Weiss, Astrophys. J. **425**, 222 (1994), astro-ph/9309014

[9] B. Paxton, L. Bildsten, A. Dotter, F. Herwig, P. Lesaffre, F. Timmes, Astrophys. J. Suppl. Ser. **192**, 3 (2011), 1009.1622

[10] K. Lodders, arXiv e-prints arXiv:1912.00844 (2019), 1912.00844

[11] B. Paxton, M. Cantiello, P. Arras, L. Bildsten, E.F. Brown, A. Dotter, C. Mankovich, M.H. Montgomery, D. Stello, F.X. Timmes et al., Astrophys. J. Suppl. Ser. **208**, 4 (2013), 1301.0319

[12] A.G.W. Cameron, W.A. Fowler, Astrophys. J. **164**, 111 (1971)

[13] J. Schwab, Astrophys. J. Lett. **901**, L18 (2020), 2009.01248

[14] J. Chanamé, M.H. Pinsonneault, C. Aguilera-Gómez, J.C. Zinn, arXiv e-prints arXiv:2109.13955 (2021), 2109.13955

EPJ Web of Conferences **260**, 11017 (2022) https://doi.org/10.1051/epjconf/202226011017
NIC-XVI

First direct limit on the 395 keV resonance of the ^{22}Ne(α, γ)^{26}Mg reaction

Eliana Masha[1,2,*] *for the LUNA collaboration*[2]

[1]Università degli studi di Milano and INFN Sezione di Milano
[2]https://luna.lngs.infn.it

Abstract. The ^{22}Ne(α,γ)^{26}Mg reaction (Q = 10614.74 keV) competes with the ^{22}Ne(α, n)^{25}Mg reaction (Q = -478.34 keV) which is the main source of neutrons for the s-process in low-mass Asymptotic Giant Branch and massive stars. The cross sections of these reactions are crucial to fix the cross over temperature at which (α, n) rate exceeds the (α,γ) one. Moreover, the uncertainty on the ^{22}Ne(α,γ)^{26}Mg reaction rate affects also the nucleosynthesis of isotopes between ^{26}Mg and ^{31}P in intermediate-mass stars.
At lower temperatures (T < 300 MK) where the (α,γ) channel is dominant, the rate of the ^{22}Ne(α,γ)^{26}Mg reaction is influenced by several resonances. The present study focuses on the E_α = 395 keV resonance which so far have been studied only indirectly leading to a wide range of possible values for its resonance strength (10^{-15} - 10^{-9} eV).
The experiment has been completed at LUNA using the intense alpha beam of the LUNA 400 kV accelerator and a windowless differential-pumped gas target combined with a high efficiency BGO detector. Experimental details and preliminary results will be shown.

1 Astrophysical motivation and state of the art

The ^{22}Ne(α,γ)^{26}Mg reaction competes with the ^{22}Ne(α, n)^{25}Mg reactions which plays a key role in astrophysics because of its impact on the neutron flux during the s-process. The ^{22}Ne(α, n)^{25}Mg (Q_{value} = -478 keV) reaction is one of the two main neutron sources for the s-process in low-mass Asymptotic Giant Branch (AGB) stars [1] and in massive stars [2]. Its role as a neutron source is affected by the ^{22}Ne(α,γ)^{26}Mg reaction (Q_{value} = 10.6 MeV). This last one can be active during the entire He-burning phase, reducing the amount of the ^{22}Ne before the ^{22}Ne(α, n)^{25}Mg reaction is activated. Therefore, to constrain the role of the ^{22}Ne(α, n)^{25}Mg reaction in the s-process, the rate for both reactions is required. Moreover, recent studies show that the uncertainties of the ^{22}Ne(α,γ)^{26}Mg reaction rate affects also the nucleosynthesis of isotopes between ^{26}Mg and ^{31}P in intermediate-mass AGB stars [3].

The ^{22}Ne(α,γ)^{26}Mg reaction rate at astrophysical temperatures of interest (0.1 GK $\leq$ T $\leq$ 0.4 GK) is dominated by several resonances. In particular, the present work is dedicated to the 395 keV (corresponding to the E = 10949 keV excited level of the ^{26}Mg) resonance which has been studied only indirectly. All these studies, lead to a wide range of reported values for its resonance strength (10^{-14} - 10^{-9} eV). Table 1 summarizes the state of the art

*e-mail: eliana.masha@unimi.it

Table 1: Summary of the literature on the E_α = 395 keV resonance strength.

Lower Limit [eV]	Adopted $\omega\gamma[eV]$	Upper Limit [eV]	Reference
1.4×10^{-14}	1.7×10^{-13}	1.6×10^{-12}	Giesen et al. 1993 [4]
-	4.7×10^{-13}	-	Giesen et al. corrected
-	1.4×10^{-13}	1.3×10^{-12}	NACRE 1999 [5]
-	-	3.6×10^{-9}	Iliadis et al. 2010 [6]
-	-	8.7×10^{-15}	Longland et al. 2012 [7]
-	-	3.6×10^{-9}	STARLIB 2013 [8]
-	-	8.7×10^{-14}	Lotay et al. 2019 [9]
-	9.0×10^{-14}		Jayatissa et al. 2020 [10]
-	-	9.0×10^{-14}	Ota et al. 2020 [11]
-	-	8.7×10^{-15}	Adsley et al. 2021 [12]

for the 395 keV resonance and different adopted values for some recent evaluations of the $^{22}Ne(\alpha,\gamma)^{26}Mg$ reaction rate. To reduce this wide range on the resonance strength and to constrain its role in the $^{22}Ne(\alpha,\gamma)^{26}Mg$ reaction rate we have afforded the first direct study of the 395 keV resonance.

2 Experimental setup and data taking

The study of the 395 keV resonance for the $^{22}Ne(\alpha,\gamma)^{26}Mg$ reaction was performed at LUNA (Laboratory for Underground Nuclear Astrophysics), located at Gran Sasso underground laboratories (LNGS). The Gran Sasso rock overburden of about 1400 m (3800 meters of water equivalent), allows the attenuation of the cosmic-ray muon flux by six orders of magnitude and the neutron component by three orders of magnitude compared with the surface laboratories [13–15]. In the region of interest for the $^{22}Ne(\alpha,\gamma)^{26}Mg$ reaction (E_γ= 9998 – 11286 keV) the muon flux is negligible and the main component of the background is due to neutrons. The study of the 394 keV resonance was performed exploiting the gas target beamline combined with a high-efficiency BGO detection system. The data taking was completed in two subsequent campaigns: Campaign I (Summer 2016) and Campaign II (Spring - Summer 2019).

Specific details from Campaign I can be found in [20]. The present work is focused on the data taking and the results obtained in Campaign II. The setup used in Campaign I was mainly the same used in [16]. Same setup was adopted in Campaign II ([17]) except for some improvements which allowed to refine the results of Campaign I. Figure 1 and Figure 2 show the experimental setup adopted in Campaign I and Campaign II, respectively.

The α beam of 399.9 keV laboratory energy and ~ 250 μA coming from the LUNA 400 kV accelerator [18] was delivered to a dedicated target chamber, filled with 1.0 mbar enriched ^{22}Ne (99.9%). The target chamber was partially occupied by the calorimeter which continuously measures the beam intensity. Both the target chamber and the calorimeter were located inside the borehole of the detector which consists of six optically independent BGO crystals. The beam induced-background generated when the beam reaches particular contaminants in different parts of the setup was monitored delivering 399.9 keV α beam on inert Ar gas at 0.468 mbar. Such pressure was selected to assure the same energy loss as that of ^{22}Ne gas at 1 mbar. At this beam energy, the argon gas is expected to be not reactive and therefore allows to identify contaminants in the target chamber which can be source of background for the experiment. A summary of the data taking of Campaign II is given in Table 2. The

Figure 1: Setup adopted for Campaign I.

Figure 2: Setup improved for Campaign II.

Table 2: Experimental campaign II. Charge Q, running time t in days and beam energy for on-resonance measurements (α+ ^{22}Ne), beam induced background runs (α+ Ar) and for no-beam background are given.

Charge [C]	t [d]	E_α [keV]	Run type
430.5	23.8	399.9	α+ ^{22}Ne
75.6	7.2	399.9	α+ Ar
	40.1	-	Laboratory background

preliminary analysis of Campaign I [20] showed a slight of 0.4σ excess in the net counting rate. Therefore, to confirm a possible resonance detection further improvements on the setup and detailed study of the background sources were required. In particular, at the energies of interest for the 395 keV resonance, the background in the BGO detector at LUNA was mainly given by neutron-induced effects ([21]). To reduce these effects, a 10 cm thick borated (5%) polyethylene (PE-HMW 500 BOR5 by Profilan Kunstofiwerk) shield surrounded the BGO in Campaign II.

Figure 3 shows the comparison of the laboratory background taken with (Campaign II) and without (Campaign I) shielding. The shielding reduced the count rate in the region of interest for the 334 keV resonance by a factor of 3.27 ±0.3.
The acquisition of the list mode data was done using the MC^2 Analyzer software by CAEN which allows to acquire and save information event by event.
The offline analysis was performed using a software program that was able to process the data after the acquisition and generates two types of spectra: the singles sum spectrum obtained by simply summing the individual histograms by each crystal and the add-back spectrum that sums the energies of all crystals.
The detection efficiency was studied combining the experimental measurements with Monte Carlo simulations (Geant3 and Geant4). The low energy efficiency was measured using the standard radioactive sources (^{7}Be ^{60}Co, ^{88}Y, and ^{137}Cs) in different positions along the beam axis inside the target chamber, while the high energy efficiency was completed using the ^{14}N(p, γ)^{15}O reaction at the 278 keV resonance. The adopted detection efficiency in the region of interest for the ^{22}Ne(α, γ)^{26}Mg was of $\sim$ 40%.

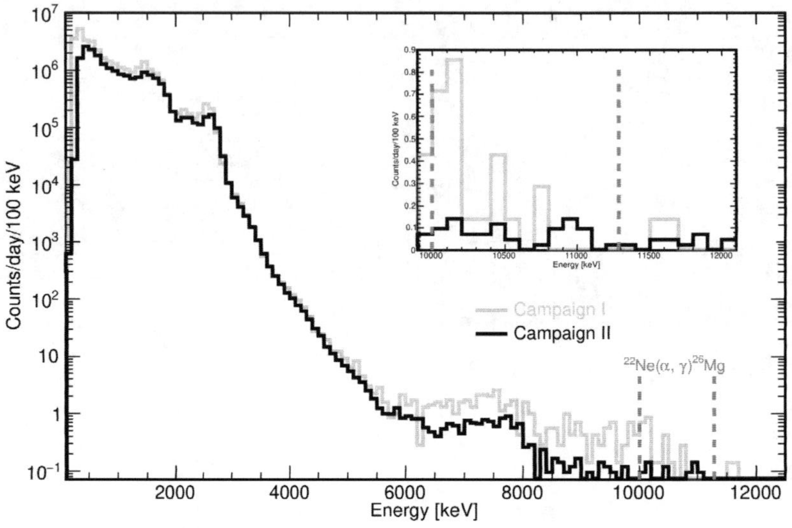

Figure 3: Laboratory background taken during Campaign I (green line) and Campaign II (black line). The dashed magenta lines define the ROI for the 395 keV resonance.

2.1 Results and conclusions

From the analysis of both campaigns no evident signal was detected in the region of interest. The sum of all the corresponding addback spectra for the α + ^{22}Ne runs compared with the

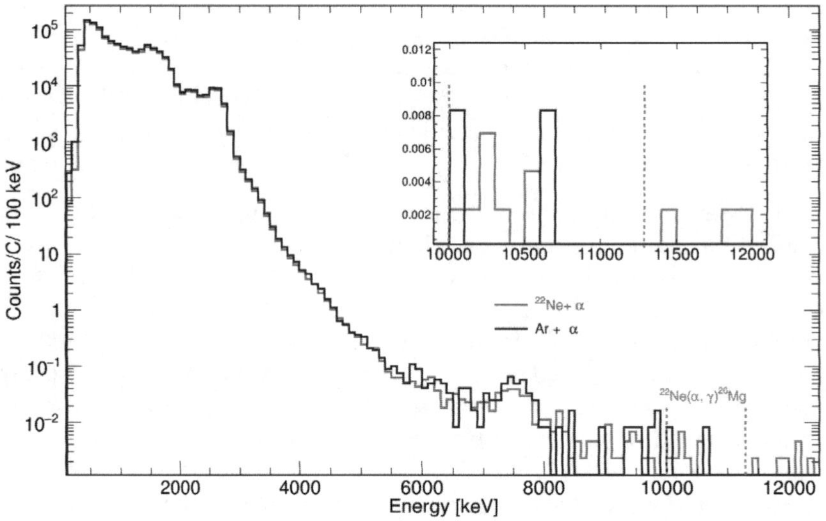

Figure 4: Addback mode spectra. In red the sum of all addback spectra in α+ ^{22}Ne obtained during Campaign II. In blue the sum of the spectra in Ar used for the study of the beam induced background. The dashed magenta lines define the ROI for the 395 keV resonance.

https://doi.org/10.1051/epjconf/202226011017

beam-induced background (α + Ar), and a zoom in the identified region of interest are given in Figure 4.

From the present experiment, an upper limit for the 395 keV resonance strength was determined taking into account several corrections due to the use of the gas target and the effects of the beam passing through the gas. The target characterization is given in [16].
The 395 keV resonance is a narrow resonance, total width much larger than the resonance energy. On the other hand, the target thickness in units of keV (~ 8 keV) is much larger than the total width. Therefore, the connection between the resonance strength $\omega\gamma$ and the experimental yield Y follow the so-called thick-target yield formula.
When the α beam (ΔE_{beam} = 0.1 keV) coming from the accelerator reaches the gas target, the beam broadens due to the energy straggling effects, up to $\sigma E_{straggle}$ = 1.4 keV at the end of the target. Furthermore, the efficiency-corrected density profile deviates from an ideal box shape. The convolution of these two effects introduces a correction factor that takes into account the reduction of the experimental yield due to the finite beam energy width and the target density profile. In the present work, the effect of the energy straggling on the total resonance yield, and hence in the final upper limit result is studied adopting the approach given in [13].
In addition, the interaction with the intense ion beam increases the temperature, and the gas density along the beam path may decrease. This effect, known as the beam heating effect has been studied in different gases and should be taken into account in the energy loss determination. The beam heating correction in neon was studied [22] using the resonance scan technique. The beam heating corrections together with the electron screening effect in the laboratory, which is estimated to be 18% in the adiabatic limit [23] are considered in the determination of the 395 keV resonance upper limit.
The upper limit has been determined by using the Rolke Monte Carlo technique [24]. Taking into account the LUNA result, a new updated ^{22}Ne$(\alpha,\gamma)^{26}$Mg thermonuclear reaction was calculated and its impact on the AGB stars was investigated by studying the abundances of neutron-rich magnesium isotopes in the pulse-driven convective zone. This study was performed by using the COLIBRI code [25] assuming a 5M$_\odot$ AGBs model. The results and their astrophysical impact are under revision and will be published soon.

References

[1] Herwig F. Annual Review of Astronomy and Astrophysics**43** 435 (2005);
[2] Couch R. G. *et al.,* The Astrophysical Journal **190** 95 (1974);
[3] Karakas A. I. *et al.,* The Astrophysical Journal **643** 471 (2006);
[4] Giesen U. *et al.,* Nuclear Physics A **561** 95 (1993);
[5] Angulo C. *et al.,* Nuclear Physics A **656** 3 (1999);
[6] Iliadis C. *et al.,* Nuclear Physics A **841** 31 (2009);
[7] Longland R. *et al.,* Physical Review C **80** 055803 (2009);
[8] Sallaska A. L. *et al.,* The Astrophysical Journal Supplement Series **207** 18 (2013);
[9] Lotay G. *et al.,* The European Physical Journal A **55** 109 (2019);
[10] Jayatissa, G. *et al.,* Phys. Lett. B **802** 135267 (2020),
[11] Ota S. *et al.,* Phys. Lett. B **802** 135256 (2020);
[12] Adsley, P. *et al.,*Physical Review C **103** 015805 (2021);
[13] Bemmerer D. *et al.,*Eur. Phys. J. A**24** 313 (2005);
[14] Caciolli A. *et al.,*Eur. Phys. J. A **34** 179 (2009);
[15] Szucs, T. *et al.,*Eur. Phys. J. A **44** 513 (2010);
[16] Ferraro F. *et al.,* Physical Journal A **54** 44 (2018);

[17] Masha E. EPJ Web of Conferences **227** 02004 (2020);
[18] Formicola A. *et al.,*Nucl. Instr. and Meth. A **507** 609 (2003);
[19] Broggini C. *et al.,* Progress in Particle and Nuclear Physics **98** 55 (2018);
[20] Piatti D. The Study of ^{22}Ne(α,γ)^{26}Mg and ^{6}Li(p, γ)^{7}Be Reactions at LUNA , PhD thesis Padua (2018);
[21] Boeltzig A. *et al.,* Journal of Physics G: Nuclear and Particle Physics **45** 025203 (2018);
[22] Depalo R. The neon-sodium cycle: Study of the ^{22}Ne(p,γ)^{23}Na reaction at astrophysical energies, PhD thesis Padua (2015)
[23] Assenbaum, H. J *et al.,* Zeitschrift für Physik A Atomic Nuclei **327** 4 (1987)
[24] Rolke W., *et al.,* Nucl. Inst. Meth. A **458** 745-758 (2001);
[25] Marigo P., *et al.,* Monthly Notices of the Royal Astronomical Society **434** 488-526 (2013);

New study of the neutron rich ^{136}Te isotope through decay spectroscopy

Min Si[1,*] and *Radomira* Lozeva[1,2,**] for *the ILL3-81-635* collaboration

[1]Univeristé Paris-Saclay, IJCLab, CNRS/IN2P3, F-91405 Orsay, France
[2]Université de Strasbourg, IPHC, 23 rue du Loess, F-67037 Strasbourg, France

Abstract. ^{136}Sb is a neutron rich nucleus with one valence proton and three valence neutrons outside the doubly-magic nucleus ^{132}Sn. It plays an important role in the rapid neutron capture process (r-process) as participates in its path further connected to the nucleosynthesis and the type II supernovae. In this work, a new β decay γ ray data on ^{136}Te is obtained. The experiment is performed at the Institute Laue-Langevin using the Lohengrin spectrometer and the thermal neutron-induced fission reaction on $^{235}_{92}U_{143}$. A specific β-β condition method is investigated for the γ-ray detection.

1 Introduction

The rapid neutron capture process (r-process) represents series of important nuclear reactions preformed at neutron-rich region in nuclear chart, responsible for creating about half of the elements heavier than iron. Nuclei around ^{132}Sn play a significant role in r-process, their nuclear structure properties, such as mass (A), half life ($T_{1/2}$), β delayed neutron emissions P_n and P_{2n} values, are essential to reproduce the r-process abundances and help us to understand the origin of the elements in our universe. Tellurium ($Z = 52$) is found at the second r-process peak ($A \approx 130$) associated with the $N = 82$ neutron shell closure in the solar system r-process distribution and predominantly produced in the main component of the r-process. The ^{136}Te isotopes reveal structure dominated by pairs of valence protons and neutrons above the ^{132}Sn core and nuclei in this neutron-rich region are mostly produced by fission reactions or the β decay of fission products [1]. The previous study about β decay of ^{136}Sb is incomplete [2], the γ ray spectrum of ^{136}Te from β decay of the mother ^{136}Sb isotope is investigated here.

2 Experiment

2.1 Setup

The experiment is performed at the Institute Laue-Langevin by the thermal neutron-induced fission reaction of ^{235}U. The fission productions go through the LOHENGRIN separator and are selected by the magnetic and electric field according to their mass (A) over ionic charge (q) ratio and kinetic energy (E_k) over ionic charge (q). The target nuclei are implanted into a movable tape which surrounded by β detectors and γ detectors. The main configuration

*e-mail: si@ijclab.in2p3.fr
**lozeva@ijclab.in2p3.fr

of the detection system is presented in Fig.1-(a). The β signals are recorded by two plastic detectors of different thickness: thin and thick, placed in front of the detection area of the γ ray detectors, consisting by two clovers and one coaxial detector. The clover detector is made of four high purity germanium crystals, which allows to improve the statistic of high energy γ rays by add-back procedure. To prevent the affect from the long lived contaminants, time chopper signal is used to manage the measurement cycle. A movable tape is used to take away the injected nuclei after each measurement cycle. According to the lifetime of the ^{136}Sb [3], each measurement cycle is divided into three parts: the first 5 s for beam injection, the next 5 s we stop the injection and wait the implanted nuclei to decay and the last 2 s tape is moved to take away the rest decay products (background), see Fig.1-(b).

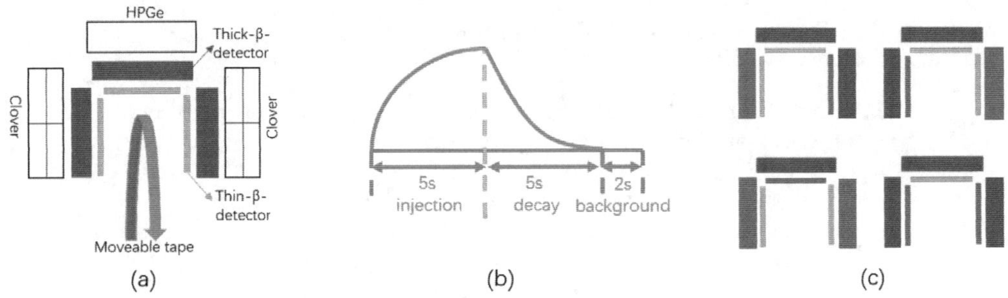

Figure 1. The main configuration of the detection system (a). The time chopper information (b). The schematic drawing of β detector combinations used to trigger the γ-ray spectra (c).

2.2 Data analysis

In order to obtain the γ ray spectrum of ^{136}Te after β decay of ^{136}Sb, we used the time chopper information between 0 s and 10 s. The plastic detectors serve as a β trigger only while their β energy information was not used. Each γ ray is triggered by the combinations of thin and/or the opposite thin and/or thick plastic detector as shown in Fig.1-(c). The schematically drawn combinations represent the used thick/thin β-β coincidence condition. For example, when a set of thick/thin plastic detectors registers a β signal within the same time window (marked in red), this is considered as a β-β coincidence [4] and the γ rays are respectively registered by any of the HpGe detectors. This kind of β-β coincidence condition between two or three β detectors strongly suppresses the background without reducing the statistics of transitions or interest. On the contrary, one can use an anti-coincidence with the β signals to artificially enhance background and exploit background lines.

3 Result

Fig.2 shows part of the energy spectrum after β decay of ^{136}Sb (in blue) obtained in the 0-10 s of each measurement cycle. Adding the β-β coincidence condition by all combinations represents the β-gated spectrum (in red). One can see from the inset, an order of magnitude suppression of background using this condition. In order to find the effect of the β trigger, the β-gated spectrum is compared to the normalised unconditioned γ spectrum. The γ rays from ^{136}Sb β decay to ^{136}Te and β-n decay to ^{135}Te are marked with their energy values. For example, the 606.6 keV and 658.6 keV transitions correspond to the de-excitation of the first excited states in ^{136}Te [2] and ^{135}Te [5], respectively. It is obvious that the application of this β-β coincidence procedure is very significant in the case of weak transitions. To demonstrate this, one can point to such weak transitions in ^{136}Te as the 423.9 keV and 961.5 keV lines, which are not visible before the coincidence condition, while they can be easily identified after the β-β coincidence trigger. In the Fig.2 we can also see the main contaminants coming from the grand-daughter nuclei ^{136}I and ^{136}Xe. The ^{141}Cs is also detected by our system.

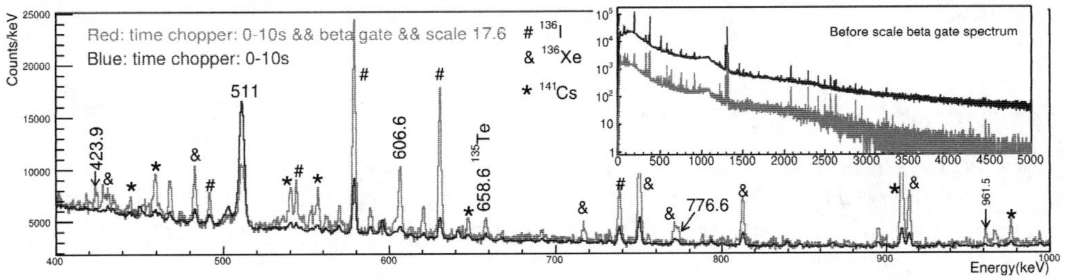

Figure 2. A β-gated (red) γ ray spectrum for ^{136}Te is compared with its un-gated γ ray spectrum (in blue). The transitions in ^{136}Te are marked with their energy values.

The efficiency of the β detectors could be measured by comparing the intensity of γ rays in the Ge spectra with and without condition on β detectors. These measurements could be executed by implanting beam with certain mass number (A) e.g. A =137 into the tape system so, that ^{137}Te and ^{137}I decay quickly to create a pure ^{137}Xe source. Based on the strongest 445 keV (^{137}Cs) [6] and 333 keV (^{136}I) [7] transitions, the results of the efficiency measurements amount to 47(5)% for this experiment. Note that similar value is obtained for the energy range up to about 1.4 MeV.

4 Summary

A specific β-β coincidence trigger is used for the study of β delayed γ ray spectrum. By applying this method to ^{136}Sb β decay experiment, one can conform weak transitions from previous work [2] and identify new transitions. The β efficiency is found to be 47(5)% for Q_β value of around 5 MeV. Note that high energy β rays may also be deposited in the HPGe detectors. Such signals would be added to γ ray detection and may cause an energy shift for these γ rays. As next step one may design a new detector system to reject these kind of synthetic signals as well as to improve the β detection efficiency.

References

[1] M. Mumpower et al., J. Phys. G: Nucl. Part. Phys. **42**, 034027 (2015)
[2] P. Hoff et al.,Phys. Rev. C. **56**, 2865 (1997)
[3] K. Rudstam et al., At. Dat. Nucl. Dat. Tabl. **53**,1-22 (1993)
[4] L. Weissman et al., Nucl. Instr. and Meth. A **423**, 328-336 (1999)
[5] P. Hoff et al., Z. Phys. A - At. Nucl. **332**, 407–411 (1989)
[6] W. R. Western et al., Phys. Rev. C **15**, 1024-1033 (1977)
[7] R. Lozeva et al., Phys. Rev. C **98**, 024323 (2018)

EPJ Web of Conferences **260**, 11019 (2022) https://doi.org/10.1051/epjconf/202226011019
NIC-XVI

The structure of 0^+ states in ^{16}O using real-time evolution method

Hideaki Motoki[1,*] and *Masaaki* Kimura[1,2,3,4,**]

[1]Department of Physics, Hokkaido University, Sapporo 060-0810, Japan
[2]Nuclear Reaction Data Centre (JCPRG), Hokkaido University, Sapporo 060-0810, Japan
[3]Research Center for Nuclear Physics (RCNP), Osaka University, Ibaraki 567-0047, Japan
[4]RIKEN (The Institute of Physical and Chemical Research), Wako, Saitama 351-0198, Japan

Abstract. α + ^{12}C clustering in ^{16}O has been vigorously studied. In the 2000s, a new picture was proposed that the Hoyle state, ^{12}C(0_2^+), is a Bose-Einstein condensate of three α particles by the so-called THSR framework. As a next step, many researchers are interested in 4α condensate state in ^{16}O. In this work, a microscopic calculation named the real-time evolution method (REM) was first applied to a 4α system. As a result, the 0^+ states in ^{16}O up to 4α condensate state were expected to be reproduced simultaneously for the first time.

1 Introduction

At the beginning of this century, a new picture was proposed that the Hoyle state is regarded as the Bose-Einstein condensate of three α particles. It provides us a unique opportunity to study the condensate of bosons composed of four fermions. Up to now, the 3α condensate is almost established, and thus we focus on the 4α system. Several works for ^{16}O, 4α Orthogonality Condition Model (OCM) [1], and extended Tohsaki-Horiuchi-Schuck-Röpke (eTHSR) [2] have been performed. In these studies, the 0_6^+ state was theoretically concluded as a candidate of the 4α condensate [1, 2]. However, the former treats an α particle as a boson and the latter imposes the restrictions on the symmetry of the system resulting in the lack of 0_5^+ state in the 4α OCM. Therefore we employ a microscopic model, REM [3].

2 Theoretical Framework

The Hamiltonian for the $N\alpha$ systems composed of $4N$ nucleons is the same as in the original paper of the REM [3] except the effective nucleon-nucleon 2(3)-body interaction. For evaluating our method with others, we used the Tohsaki No. 1 effective nucleon-nucleon interaction [4]. The intrinsic wave function of the $N\alpha$ system is defined as

$$\Phi(\boldsymbol{Z}_1, \cdots, \boldsymbol{Z}_N) = \mathcal{A}\{\Phi_\alpha(\boldsymbol{Z}_1) \cdots \Phi_\alpha(\boldsymbol{Z}_N)\}, \tag{1}$$

where $\mathcal{A}$ represents the antisymmetrization operator. The $\Phi_\alpha(\boldsymbol{Z})$ denotes the wave packet of the α cluster located at $\boldsymbol{Z}$. The normalized GCM wave function of ^{16}O, $\Psi^{0^+}_{^{16}\mathrm{O}}$, is defined as

*e-mail: motoki@nucl.sci.hokudai.ac.jp
**e-mail: masaaki@nucl.sci.hokudai.ac.jp

$$\Psi^{0^+}_{^{16}\mathrm{O}} = \sum_i \{f_i \hat{P}^{0^+}\Phi_i(\boldsymbol{Z}_1,\cdots,\boldsymbol{Z}_4) + g_i\hat{P}^{0^+}\Phi_i(\boldsymbol{Z}^*_1,\cdots,\boldsymbol{Z}^*_4)\}, \tag{2}$$

where $\hat{P}^{0^+}$ is the angular-momentum (J) and the parity (π) projection operator for $J^\pi = 0^+$. In the REM, the position and momentum of α particles are governed by the equation-of-motion (EOM) of the Gaussian wave packets. In principle, the $\Psi^{0^+}_{^{16}\mathrm{O}}$ can be divided into the $\alpha + {}^{12}\mathrm{C}$ component $\Psi^{0^+}_{\alpha+^{12}\mathrm{C}}$ and the residual part $\Psi^{0^+}_{\mathrm{res}}$. The squared amplitude $|w|^2$, where w is the coefficient of the $\Psi^{0^+}_{\alpha+^{12}\mathrm{C}}$, can be estimated using the method in the Ref. [5].

3 Results

Results were obtained as follows: The intrinsic excitation energy and rebound radius in the REM are 40 MeV and 16.0 fm, respectively. The time of the EOM was evolved until 6000 fm/c by 6 fm/c units. We used the r^2-constraint method [6]. The cutoff value is 7.0 fm.

We first discuss the excitation energy spectra measured from the 4α threshold and r.m.s. charge radii compared with the eTHSR [2] in Fig. 1. As for the bound states, 0^+_1 and 0^+_2, the energies are slightly lower than that of the eTHSR reflecting the size of model space. Table 2 shows the calculated and observed r.m.s. radii and monopole matrix elements from the 0^+_1 state. The monopoles can be understood as an enhancement of the clustering in the excited states. Considering the consistency of the radius and monopole with the OCM, the 0^+_6 state in the REM is expected to be a corresponding state of 0^+_5 state in the OCM. Note that the 0^+_5 state in the OCM is mainly composed of $\alpha + {}^{12}\mathrm{C}(1^-)$ which is not described in the eTHSR due to the size of the model space. The 0^+_V state in the eTHSR is said to be a 4α condensate. As for the 0^+_7 state in the REM, though the radius is almost consistent with that of the eTHSR, the monopole is much larger than that of the eTHSR, which imply that the 0^+_7 state in the REM is a candidate of 4α condensate and has a characteristic property of the clustering enhancement.

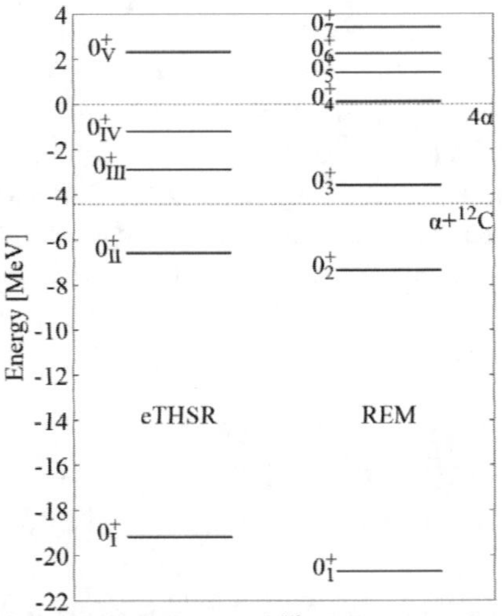

Figure 1. Energy spectra of $^{16}\mathrm{O}(0^+)$ states measured from the 4α threshold of the eTHSR [2] and REM.

Table 1. $\alpha + {}^{12}\mathrm{C}(0^+_1)$ components $|w|^2$, the maximized overlaps in Eq. (3) and the main components.

^{16}O	$\lvert w\rvert^2$	O	Configuration
0^+_1	0.89	0.96	Tetrahedron
0^+_2	0.80	0.64	$\alpha(S) + {}^{12}\mathrm{C}(0^+_1)$
0^+_3	0.16	0.53	
0^+_4	0.02	0.16	
0^+_5	0.42	0.27	$\alpha(S) + {}^{12}\mathrm{C}(0^+_1)$
0^+_6	0.16	0.47	
0^+_7	0.02	0.20	

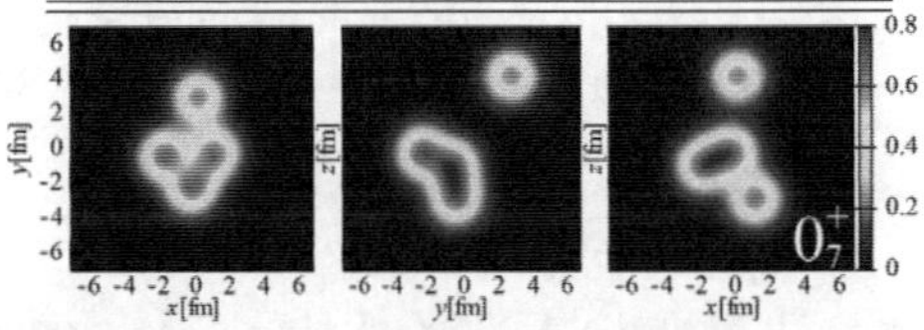

Figure 2. Density distribution of the intrinsic wave function maximized for 0^+_7 state using Eq. (3).

We estimated the squared amplitudes of $\alpha + {}^{12}\mathrm{C}(0^+_1)$ components $|w|^2$, which is shown in Table 1. The $^{12}\mathrm{C}(0^+_1)$ state was made by the same process explained in the original paper of the REM [3]. The maximized overlaps defined in Eq. (3) for each $^{16}\mathrm{O}(0^+)$ state and the configurations are also shown in Table 1.

$$O = |\langle\Psi^{0^+}_{^{16}\mathrm{O}}|P^{0^+}\Phi_{\mathrm{opt}}\rangle|^2/\langle P^{0^+}\Phi_{\mathrm{opt}}|P^{0^+}\Phi_{\mathrm{opt}}\rangle. \tag{3}$$

Table 2. r.m.s. charge radii $R_{\rm rms}$ and monopole matrix elements from the ground state $\mathcal{M}(E0; 0^+_1 \to 0^+_f)$ of the experiment [7] and the 4α OCM [1], eTHSR [2], and REM calculations in ^{16}O.

	$R_{\rm rms}$[fm]				$\mathcal{M}(E0; 0^+_1 \to 0^+_f)[e{\rm fm}^2]$			
REM	EXP.	OCM	eTHSR	REM	EXP.	OCM	eTHSR	REM
0^+_1	2.71(0.02)	2.7	2.7	2.7				
0^+_2		3.0	3.2	3.1	3.55	3.9	5.9	5.8
0^+_3		3.1	3.3	3.2	4.03	2.4	5.7	4.3
0^+_4				5.7				0.9
0^+_5		4.0	4.9	5.2		2.4	0.8	1.1
0^+_6		3.1		3.1	3.3	2.6		2.9
0^+_7		5.6	4.9	4.8		1.0	0.7	2.4

It is easily understood that the 0^+_1 state has a large squared amplitude and the overlap resulting from the tetrahedral configuration. The $0^+_{2,5}$ states have also a rather large squared amplitude, which means that these states are composed of α + ^{12}C(0^+_1). This is consistent with the 4α OCM [1] in which they say that the 0^+_4 state in the 4α OCM is the higher nodal state of the 0^+_2 state. Although the $0^+_{3,6}$ states have overlaps about 50%, the squared amplitude of the α + ^{12}C(0^+_1) component is small, which implies that these states are composed of the other types of configurations; i.g. α + ^{12}C(2^+), α + ^{12}C(1^-). Note that the 0^+_4 state is expected as the continuum state due to the quite large radius. It is expected that the 4α condensate is composed of α + ^{12}C(0^+_2) resulting in the small squared amplitude of the α + ^{12}C(0^+_1) component and small overlap. Therefore the 0^+_7 state is a candidate of the 4α condensate. The density distribution of the intrinsic wave function maximized for the 0^+_7 state is shown in Fig. 2. It looks that The four α particles are loosely interacting with each other.

4 Summary

We focused on the 0^+ states in ^{16}O to investigate the 4α condensate state by adopting the REM analyzing the excitation energy, r.m.s. charge radius and monopole matrix element. Additionally, we showed the squared amplitude of the α + ^{12}C(0^+_1) component As a result, a candidate of the 4α condensate state was obtained. The ^{16}O(0^+) states up to the 4α condensate state were expected to be microscopically described simultaneously for the first time. We will calculate the α + ^{12}C(0^+_2) component to conclude the 0^+_7 state as the 4α condensate state.

Acknowledgments

The numerical calculations were carried out on the high performance computing server at RCNP in Osaka University and on XC40 at YITP in Kyoto University.

References

[1] Y. Funaki, T. Yamada, H. Horiuchi, G. Röpke, P. Schuck, A. Tohsaki, Phys. Rev. Lett. **101**, 082502 (2008)
[2] Y. Funaki, Phys. Rev. C **97**, 021304 (2018)
[3] R. Imai, T. Tada, M. Kimura, Phys. Rev. C **99**, 064327 (2019)
[4] A. Tohsaki, Phys. Rev. C **49**, 1814 (1994)
[5] M. Kimura, Phys. Rev. C **69**, 044319 (2004)
[6] Y. Funaki, H. Horiuchi, A. Tohsaki, Progress of Theoretical Physics **115**, 115 (2006)
[7] F. Ajzenberg-Selove, Nuclear Physics A **460**, 1 (1986)

Three-dimensional supernova simulation of SN 1987A progenitor with implications for multi-messenger signals

Ko Nakamura[1,2,*]

[1]Department of Applied Physics, Fukuoka University, Fukuoka 814-0180, Japan
[2]Research Institute of Stellar Explosive Phenomena, Fukuoka University, Fukuoka 814-0180, Japan

Abstract. Recently a new progenitor model for SN 1987A has been constructed. This progenitor model is based on a slow-merger of 14 $M_\odot$ and 9 $M_\odot$ stars and it satisfies most of the observational constraints such as red-to-blue evolution, lifetime, total mass and position in the Hertzsprung-Russell diagram at collapse, and chemical anomalies. We perform a three-dimensional self-consistent core-collapse supernova simulation using this new progenitor model and find that it successfully presents an explosion and leaves a 1.53 $M_\odot$ neutron star. Assuming a detector sensitivity of Kamiokande-II and the distance to the supernova of 51 kpc, we obtain 16 neutrino detection events in one second. Some characteristic modes in its gravitational wave signal are also discussed in this article.

1 Introduction

SN 1987A emerged in the Large Magellanic Cloud located at a distance of 51.4 ± 1.2 kpc. Multiwavelength studies of SN 1987A have provided unprecedented details of supernova features such as the time evolution of the bolometric luminosity and energy spectrum at all wave bands from infrared to γ-ray. Moreover, a supernova neutrino burst was detected by several neutrino detectors such as Kamiokande-II, and Baksan. Although only about two dozen of the $\sim 10^{28}$ supernova neutrinos that passed through the Earth were detected, they provide us with the first (and ever only one) direct evidence of the supernova driven by the collapsing core of a dying star. This detection implies that there had been a proto-neutron star at least for 10 seconds and it declares the dawn of the neutrino astrophysics.

The light curve and spectra indicate that SN 1987A has roughly typical explosion energy ($\sim 1.2 \times 10^{51}$ erg) and ^{56}Ni mass ($0.07 M_\odot$). On the other hand, SN 1987A is a unique supernova in many senses. First, a blue hot surface of the progenitor star Sk-69°202 found in the pre-explosion images, as well as the light curve without the typical plateau phase of type II-P SNe and the relatively short period of time delay (three hours) between the neutrino burst detection and the shock breakout emission, suggest that the progenitor was a blue-supergiant (BSG). Second, this BSG star is considered that it used to be a RSG at $\sim 2 \times 10^4$ yr since three ring-like nebulae are surrounding the supernova remnant with the high He the CNO abundance ratios and the expansion velocity comparable to the RSG wind velocity.

*e-mail: nakamurako@fukuoka-u.ac.jp

To explain these anomalous features, many pre-explosion evolution scenarios of the progenitor star have been proposed soon after the emergence of SN 1987A: extreme-mass-loss models, helium-enrichment models, low-metallicity models, rapid-rotation models, and restricted-convection models. Also, binary interaction has been considered from early on: accretion models, companion models, and merger models. See [1] for a review of these classical progenitor models.

One of the possible scenarios to explain these features is a slow merger of binary stars where the stars in a close binary evolve into a common envelope phase then the secondary star is gradually being dissolved inside the common envelope in a much longer time scale than the dynamical time scale of the secondary. Recently a new progenitor model based on the slow merger scenario has been constructed [2]. In particular, the progenitor model successfully reproduces most of the anomalous SN 1987A features such as red-to-blue evolution, lifetime, total mass and position in the Hertzsprung-Russell diagram at collapse, and the chemical anomalies.

In this article, we report the results of our self-consistent 3D simulations for core-collapse SN employing this new progenitor model. We solve hydrodynamic evolution and energy-dependent neutrino transport from the center to the outer boundary at 10,000 km. The hydrodynamics code that we employ in this work is essentially the same as in our previous studies [3, 4]. For neutrino transfer calculation we use the state-of-the-art neutrino opacity [5] and solve the neutrino transport taking 20 energy bins with an upper bound of 300 MeV for electron, anti-electron, and heavy-lepton neutrinos.

2 Results

We perform a 3D self-consistent core-collapse supernova simulation using the latest binary progenitor model of SN 1987A. We obtain the explosion energy of 1.5×10^{50} erg and ^{56}Ni mass of $\sim 10^{-2} M_{\odot}$ and these values are still growing at the end of the simulation.

2.1 Shock evolution

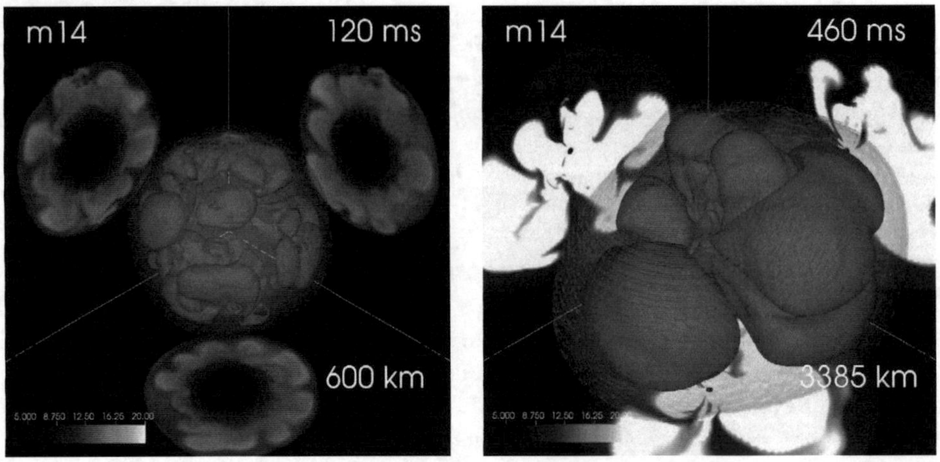

Figure 1. Entropy isosurface within the inner region of the exploding SN 1987A progenitor at 120 ms (left panel) and 460 ms (right panel) after bounce. Note the different box sizes between the panels.

The shock of our 3D model once stalls at $r \sim 150$ km and turns to expand at ~ 220 ms after bounce. The left panel of Figure 1 shows the entropy isosurface of the inner region of our 3D model within a 600 km^3 box at 120 ms after bounce. The matter behind the shock is heated by neutrino and small-scale convective motions are developed. The shock is finally pushed outward via the neutrino heating and at 460 ms after bounce it reaches ~ 1500 km from the center (the right panel of Figure 1). At this later phase, the morphology of the expanding matter is dominated by a few large-scale plumes and low-entropy downflows between them. Although our simulation is limited by the outer boundary at 10,000 km, the shock never recedes after the shock revival and it would successfully make a supernova explosion when it reaches a stellar surface.

2.2 Neutrino signal

Core-collapse SN emits a huge number of neutrinos. We estimate a cumulative neutrino detection event number assuming that a CCSN located at LMC emits neutrinos like our 3D CCSN model and the neutrino emission is nearly isotropic. We observe 14 neutrinos for 0.7 s by Kamiokande-II (K-II) and an appropriate extrapolation gives 16 events in 1 s. These numbers are larger than, but not so far from, the detection events in the case of SN 1987A (from this historical SN, K-II observed 11 neutrinos in ~ 10 s).

2.3 Gravitational wave signal

Gravitational wave from coalescence of compact objects such as black holes and neutron stars has been detected by laser interferometers. Non-spherical motions of dense matter in a CCSN core also produce the gravitational wave and we expect its detection in the future. The gravitational wave from CCSN is characterized by three phases: a prompt signal soon after the core bounce followed by a silent phase, a strong signal stimulated by hydrodynamic instabilities such as convection and Standing Accretion Shock Instability (so-called SASI) lasting for a few hundred ms, and the final explosion phase. Independent of the direction to observers, the spectrogram of our CCSN model presents that the peak frequency of the strong signal shifts from ~ 500 Hz at 150 ms to > 1000 Hz at 500 ms after bounce. This feature is fitted by a turn-over frequency of downflows hitting the central proto-neutron star surface and reflected at a stable region against convection.

3 Acknowledgments

This study was supported in part by Grants-in-Aid for Scientific Research of the Japan Society for the Promotion of Science (JSPS, Nos. JP20K03939, JP21H01121), the Ministry of Education, Science and Culture of Japan (MEXT, Nos. JP20H04748, JP20H05255), by the Central Research Institute of Explosive Stellar Phenomena (REISEP) at Fukuoka University and an associated project (No. 207002), and JICFuS as "Program for Promoting researches on the Supercomputer Fugaku" (Toward a unified view of the universe: from large scale structures to planets).

References

[1] Podsiadlowski, P., PASP, **104**, 717 (1992)
[2] Urushibata, T., Takahashi, K., Umeda, H., et al., MNRAS, **473**, L101 (2018)
[3] Nakamura, K., Takiwaki, T., Kuroda, T., et al., PASJ, **67**, 107 (2015)
[4] Nakamura, K., Takiwaki, T., & Kotake, K., PASJ, **71**, 98 (2019)
[5] Kotake, K., Takiwaki, T., Fischer, T., et al., ApJ, **853**, 170 (2018)

NG-TRAP: Measuring neutron capture cross-sections of short-lived fission fragments

T. Dickel[1,2,*], *I.* Mardor[3,4], *H.* Wilsenach[2], *J.* Ashkenazy[3], *W. R.* Plaß[1,2], *C.* Scheidenberger[1,2,5], and *M. I.* Yavor[6]

[1]GSI Helmholtzzentrum für Schwerionenforschung GmbH, 64291 Darmstadt, Germany
[2]Justus-Liebig-Universität Gießen, 35392 Gießen, Germany
[3]Soreq Nuclear Research Center, Yavne 81800, Israel
[4]Tel Aviv University, Tel Aviv 69978, Israel
[5]Helmholtz Forschungsakademie Hessen für FAIR (HFHF), GSI Helmholtzzentrum für Schwerionenforschung, Campus Gießen, 35392 Gießen, Germany
[6]Institute for Analytical Instrumentation of the Russian Academy of Sciences, 190103 St. Petersburg, Russia

Abstract. We lack significant nuclear physics input to understand the rapid-neutron capture (r-)process fully. The r-process is the source of half the elements heavier than iron and the only way to produce the long-lived actinides we find on earth. This process's key nuclear physics inputs are nuclear masses, cross-sections of (n,γ) and (γ,n), and decay half-lives and branching ratios of neutron-rich isotopes. However, there is currently no method to directly measure neutron-induced reaction rates on short-lived nuclides, so there is no experimental data for the primary nuclear reaction that drives the r-process. We show here a conceptual design of a novel approach to access this information experimentally. The idea is to form a target of short-lived isotopes by confining them as ions in a radio-frequency (RF) trap. Next, they are irradiated with an intense neutron flux, and the reaction products are identified by mass spectrometry. The chosen method is a two-stage process in the presence of high neutron fluxes. The first process is neutron-induced fission in a thin actinide foil to create fission fragments. These fragments are slowed down in a cryogenic stopping cell before being filtered through a radio frequency quadrupole (RFQ) system. The RFQ system selects fission fragments of a specific atomic mass number A and confines them to a small volume in an RF trap, where they are irradiated for a second time in a controlled manner. The resultant $A+1$ isotopes are mass-selectively transported to a multiple-reflection time-of-flight mass spectrometer, where the reaction products are identified and counted.

1 Background

One of the major open questions in physics is the understanding of the abundance distribution of the heaviest elements, and identification of the astrophysical sites of their production

*e-mail: t.dickel@gsi.de

[1]. Astrophysical findings, such as the recent spectroscopic identification of r-process nucleosynthesis in a neutron-star merger [2], provide crucial insights into elemental formation within these newly observed phenomena. Complementarily, measured nuclear physics properties, especially for unstable nuclei, are critical input for network calculations that predict abundance distributions and constrain creation sites [3].

It is possible to measure static properties, such as masses, half-lives, energy levels and radii. However, the measurement of dynamic properties, such as neutron induced cross-sections (reaction rates), on short-lived nuclei (half-lives shorter than a few days) is currently considered impossible [4]. Even though the main motivation for the measurement of these cross-sections is the nucleosynthesis r-process, some of the data are also important for validating nuclear structure and reaction models far away from the valley of stability, as well as for the safe operation of nuclear power plants and the transmutation of nuclear waste.

Indirect methods for these measurements have been proposed. Neutron capture cross-sections may be somewhat constrained by measuring the nuclear level density and the beta-strength function of the resultant nuclei [4]. Presently, the only concepts in the literature for a direct measurement method are combining a storage ring with a nuclear reactor [5], or with an accelerator-based spallation source [6], either which serve as a thermal-neutron "target" that is irradiated by circulating short-lived nuclei. However, technical complexity, safety considerations and high cost make these concepts very challenging.

2 Preparing a target of short-lived fission fragments

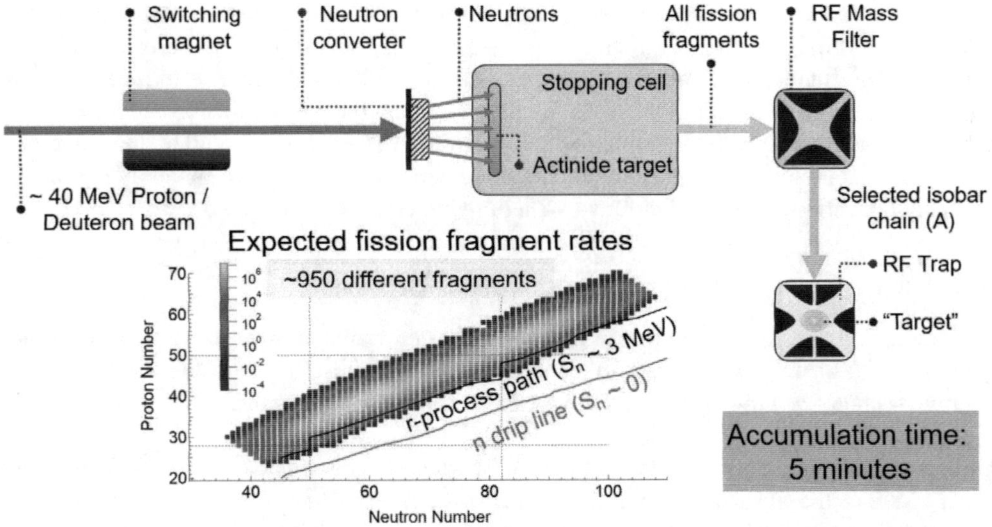

Figure 1. Conceptual layout of the production of short-lived fission fragments, selection of the ions of interest and forming a target of short-lived nuclei in an RF trap. The expected yields (inset) are based on [7], adapted for a gallium-indium neutron converter.

We plan to make use of the intense neutron fluxes that will be available at Soreq Applied Research Accelerator Facility (SARAF-II)[7]. SARAF is a medium-sized high-power accelerator facility that is currently under construction in Soreq Nuclear Research Center at Yavne, Israel, to be completed by 2023. SARAF-II will produce deuteron and proton beams of 40 MeV and 35 MeV, respectively, at continuous wave (CW) currents of up to 5 mA. In combination with world-exclusive liquid targets, based on the existing liquid-lithium target (LiLiT) that was developed at Soreq [8], the full power of the accelerator will be used to generate high-flux neutron beams at various energy ranges. SARAF-II will include at least two high-flux neutron sources. One source, the liquid gallium-indium beam dump [9], will provide a rate in the range of 10^{15} n/sec of high energy neutrons, which will generate the needed rate of fission fragments. In a future upgrade, a second source is planned to produce "cold" neutrons (peaked around 5 meV), with expected fluxes of 10^{10} n/cm^2/sec, which will be used to irradiate the fission fragments and produce a viable measurement rate. The design of the fission fragments production setup has already begun.

We propose to irradiate a thin actinide target (^{232}Th or ^{238}U) by the high-energy neutron flux, generating $\sim 10^{10}$ fissions/s, see Fig. 1. The fission fragments are released from the thin target undergo thermalization in a cryogenic gas-filled stopping cell (CSC). They are subsequently extracted from the CSC using an RF carpet. Such a CSC has already been developed [10] by the Justus-Liebig University (JLU) Gießen, Germany, and major concepts have been tested at GSI, Darmstadt, Germany [11–13]. The fission fragments are then passed through a RF mass filter (selecting the mass number of interest) to an RF trap, where a selected isobar chain will be stored to form a "target". Efficiencies for escape from the target, thermalization in ionic state in helium, extraction from the high density gas, mass number selection and accumulation in an RF trap are based on experience with similar running experiments [11, 14, 15]. From this we expect a total efficiency for the production, separation and accumulation of approximately 8%. If we assume an isotope at the fission peak with an independent fission yield of ~3%, we get a rate of $\sim 6 \times 10^7$ target-nuclei/sec, see inset in Fig. 1. Assuming accumulation of 5 minutes (which enables probing cross-sections down to this half-life range) we can obtain a target of $\sim 2 \times 10^{10}$ ions. We will store ions of all isobars of the selected mass number in the trap and measure their neutron capture cross-sections in parallel. The pseudo potential of the trap needs to be ~500 eV, to be able to recapture ions after they undergo beta decay and neutron capture. Thus, our goal is to store about 2×10^{10} ions at the same time in the RF trap, for the duration of several minuets. The developments (test setup that includes ion optical elements and electronics) for this are currently ongoing at Tel Aviv University, Israel and the Justus-Liebig University Giessen, Germany.

3 Neutron capture reaction and detection

The "target", described in the previous section, will be irradiated by neutrons of a controlled energy spectrum. Following a neutron capture reaction, the resulting nucleus will have recoil energy of about 200 eV, so it will be contained in the trap by the pseudo potential and be thermalized by buffer gas cooling. We now need to select the recoil ions from the large amount of unreacted ones without "destroying" the target. This we do by lowering the voltage on one side of the linear RF trap until the first ions spill out of the trap. The ions, at a rate of 1×10^9 ions/s, will be pushed out of the trap randomly by their thermal motion and space-charge effects, and consequently be cooled in a short RFQ and then accelerated to a kinetic energy of about 1 keV. They are then mass-analyzed in a dipole magnet with a mass resolving power of about 500. This is sufficient to spatially separate recoil (A+1) and unreacted (A) ions. The unreacted ions are guided by electrostatic deflectors back to the linear RF trap (target). The ions spend merely a few milliseconds out of the target, and are then recaptured with high efficiency, as we have demonstrated for a much more challenging case [16], so the number of ions in the target is not depleted significantly.

The reacted ions are guided to the trap of a multiple-reflection time-of-flight mass spectrometer (MR-TOF-MS), which is used for their identification and counting. The MR-TOF-MS can handle more than 10^6 ions per second [17], so the dipole magnet only needs to

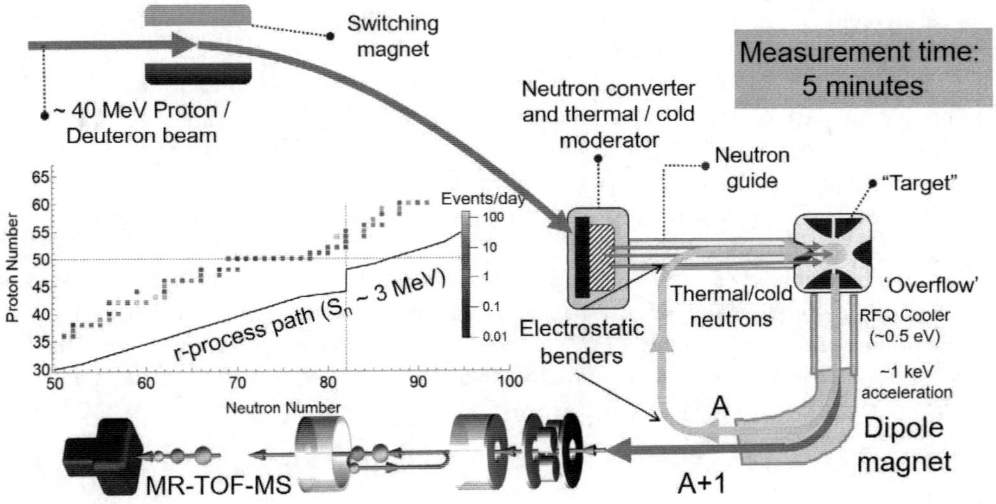

Figure 2. Conceptual layout of the irradiation with cold neutron and detection of reaction products. The inset shows the expected detected event rate per day.

suppress the unreacted ions by 3 orders of magnitude. The counted ions, normalised to the total amount of target nuclei and neutron flux, will infer the desired cross-section. The entire process is depicted in Figure 2. Note that we will measure reaction rates for several nuclei of the selected isobaric-chain simultaneously, which will save on beam time and help to increase statistics. This is possible due to the high mass resolution that can be achieved in modern time-of-flight mass spectrometers [18, 19]. The typical analysis time will be one minute, which is dominated by the average time it takes the ions to leave the target RF trap, much longer than the detection and identification by the MR-TOF-MS, which takes about 10 ms.

In order to measure neutron capture cross-sections of the fission fragments, we intend to irradiate the trapped ions with cold neutrons. The neutron capture cross-section energy (E) dependence is proportional to $\sim 1/E^{1/2}$ for all isotopes (with resonances in the epithermal region, and rarely also in the thermal region) [20]. In order to experimentally constrain the cross-section values at the region of interest for the r-process (10's to 100's of keV), it is sufficient to provide experimental points at the cold or thermal region, preferably at both [21]. Measurements at both points confirm the $\sim 1/E^{1/2}$ trend, whereas a deviation point to a resonance in this region.

Using cold neutrons has two major advantages. First, because of the $1/E^{1/2}$ behaviour, the cross-sections in this energy region are very high ($\sim$1 barn and higher, in some cases by orders of magnitude [22]). Second, cold neutrons can be steered and focused into the target without losses along paths of tens of meters, via usage of so-called "neutron guides". Such guides are available commercially from several companies [23–25] and are used in reactor- and accelerator-based neutron sources (existing and planned [26]). If we assume a cold neutron flux of 10^{10} n/cm^2/sec and a combined ejection and mass measurement efficiency of 50%, we get an overall count rate of a few events per day per barn for neutron capture events. The expected yields for the accessible fission fragments can be found in the inset in Fig. 2. Since this measurement method is accurate and virtually background-free, we will be able to measure cold and/or thermal neutron capture cross-sections of numerous unstable fission fragments with sufficient statistics within a few days.

It is well known that several nuclei have surprisingly high thermal (and thus also cold) neutron capture cross-sections. Veteran examples include stable ^{157}Gd (2.1×10^5 barn) and the famous "reactor poison" ^{135}Xe (2.6×10^6 barn) with $t_{1/2} = 9.14$ hr, and many isotopes with cross-sections in the 10^2, 10^3 and even 10^4 barn range. There is currently no nuclear model to predict which isotopes will exhibit such a high resonance in the thermal region. Even recently, another surprisingly high cross-section was published, following the first thermal (n,γ) measurement of ^{88}Zr ($t_{1/2}$ = 83.4 days), and the result was 8.6×10^5 barn [27], versus the theoretically predicted value of $\sim$10 barns [22]. We thus expect that in our systematic survey of cold and thermal neutron capture cross-sections of fission fragments, we will also discover new "surprises", which may promote theoretical explanations for this phenomenon. Outliers in the range of 10^2 and 10^3 barn will be accessible with our method within a few hours of measurement.

4 Summary and Outlook

We have shown the conceptual design of a novel method to measure neutron capture cross-section of short-lived fission fragments, down to half-lives in the scale of minutes. This is a paradigm change as it is deemed so far as technically impossible to perform such measurements. The measurements will have a significant impact, and will provide invaluable input to astrophysical nucleosynthesis, nuclear structure and reaction models, and nuclear reactor safety and nuclear waste transmutation. Confining short-lived isotopes as ions in a RF trap could also be applied to (γ,n), (n,2n) and (α,n), in addition to (n,γ). Installing a similar system at Extreme Light Infrastructure – Nuclear Physics (ELI-NP) [28] with its unprecedented brilliant gamma beams could allow to directly probe the r-process relevant (γ,n) cross-sections.

5 Acknowledgements

This work was supported by the German Research Foundation (DFG) Grant No. 57564554, by Justus-Liebig-Universität Gießen and GSI under the JLU-GSI strategic Helmholtz partnership agreement, and by the Israel Ministry of Energy, Research Grant No. 220-11-052.

References

[1] A. Arcones, G. Martínez-Pinedo, Phys. Rev. C **83**, 045809 (2011)

[2] E. Pian, P. D'Avanzo, S. Benetti, M. Branchesi, E. Brocato, S. Campana, E. Cappellaro, S. Covino, V. D'Elia, J.P.U. Fynbo et al., Nature **551**, 67 (2017)

[3] M. Mumpower, R. Surman, G. McLaughlin, A. Aprahamian, Progress in Particle and Nuclear Physics **86**, 86 (2016)

[4] A. Larsen, A. Spyrou, S. Liddick, M. Guttormsen, Progress in Particle and Nuclear Physics **107**, 69 (2019)

[5] R. Reifarth, Y.A. Litvinov, Phys. Rev. ST Accel. Beams **17**, 014701 (2014)

[6] R. Reifarth, K. Göbel, T. Heftrich, M. Weigand, B. Jurado, F. Käppeler, Y.A. Litvinov, Phys. Rev. Accel. Beams **20**, 044701 (2017)

[7] I. Mardor, O. Aviv, M. Avrigeanu, D. Berkovits, A. Dahan, T. Dickel, I. Eliyahu, M. Gai, I. Gavish-Segev, S. Halfon et al., The European Physical Journal A **54**, 91 (2018)

[8] S. Halfon, A. Arenshtam, D. Kijel, M. Paul, L. Weissman, O. Aviv, D. Berkovits, O. Dudovitch, Y. Eisen, I. Eliyahu et al., Review of Scientific Instruments **85**, 056105 (2014)

[9] I. Eliyahu, S. Vaintraub, I. Mardor, A. Arenshtam, E. Shvero, E. Reinfeld, O. Ozery, M. Alpha, R. Raz, M. Bukai et al., *High Power Beam Dump and Neutron Sources for SARAF Phase II*, http://www-llb.cea.fr/UCANS8-2019/Eliyahu_poster.pdf/, accessed: 2021-11-25

[10] T. Dickel, W.R. Plaß, H. Geissel, F. Heiße, I. Miskun, S. Purushothman, M.P. Reiter, A.K. Rink, C. Scheidenberger, NIM B **376**, 216 (2016)

[11] W.R. Plaß, T. Dickel, S. Purushothaman, P. Dendooven, H. Geissel, J. Ebert, E. Haettner, C. Jesch, M. Ranjan, M.P. Reiter et al., NIM B **317**, 457 (2013)

[12] M. Ranjan, P. Dendooven, S. Purushothaman, T. Dickel, M. Reiter, S. Ayet San Andrés, E. Haettner, I. Moore, N. Kalantar-Nayestanaki, H. Geissel et al., NIM A **770**, 87 (2015)

[13] Mardor, Israel, Dickel, Timo, Amanbayev, Daler, Ayet San Andrés, Samuel, Beck, Sönke, Benyamin, David, Bergmann, Julian, Constantin, Paul, Cléroux Cuillerier, Alexandre, Geissel, Hans et al., EPJ Web Conf. **239**, 02004 (2020)

[14] W.R. Plaß, T. Dickel, I. Mardor, S. Pietri, H. Geissel, C. Scheidenberger, D. Amanbayev, S. Ayet San Andrés, J. Aysto, D. Balabanski et al., Hyperfine Interactions **240**, 73 (2019)

[15] E. Haettner, W.R. Plaß, U. Czok, T. Dickel, H. Geissel, W. Kinsel, M. Petrick, T. Schäfer, C. Scheidenberger, NIM A **880** (2018)

[16] T. Dickel, W.R. Plaß, W. Lippert, J. Lang, M.I. Yavor, H. Geissel, C. Scheidenberger, J. Am. Soc. Mass Spectr. **28**, 1079 (2017)

[17] T. Dickel, W.R. Plaß, A. Becker, U. Czok, H. Geissel, E. Haettner, C. Jesch, W. Kinsel, M. Petrick, C. Scheidenberger et al., NIM A **777**, 172 (2015)

[18] S. Ayet San Andrés, C. Hornung, J. Ebert, W.R. Plaß, T. Dickel, H. Geissel, C. Scheidenberger, J. Bergmann, F. Greiner, E. Haettner et al., Phys. Rev. C **99**, 064313 (2019)

[19] I. Mardor, S. Ayet San Andrés, T. Dickel, D. Amanbayev, S. Beck, J. Bergmann, H. Geissel, L. Gröf, E. Haettner, C. Hornung et al., Phys. Rev. C **103**, 034319 (2021)

[20] K.S. Krane, *Introductory Nuclear Physics* (Wiley India, 2008), ISBN 9788126517855, `https://books.google.de/books?id=mkZjC1TLXh8C`

[21] R. Reifarth, K. Schwarz, F. Kappeler, The Astrophysical Journal **528**, 573 (2000)

[22] A. Koning, D. Rochman, Nuclear Data Sheets **113**, 2841 (2012)

[23] *Mirrotron*, `https://mirrotron.com/en`, accessed: 2019-08-30

[24] *Swiss Neutronics*, `https://www.swissneutronics.ch/`, accessed: 2019-08-30

[25] *Neutron Guide*, `http://www.neutronguide.com/`, accessed: 2019-08-30

[26] K.H. Andersen, M. Bertelsen, L. Zanini, E.B. Klinkby, T. Schönfeldt, P.M. Bentley, J. Saroun, Journal of Applied Crystallography **51**, 264 (2018)

[27] J.A. Shusterman, N.D. Scielzo, K.J. Thomas, E.B. Norman, S.E. Lapi, C.S. Loveless, N.J. Peters, J.D. Robertson, D.A. Shaughnessy, A.P. Tonchev, Nature **565**, 328 (2019)

[28] N. Zamfir, The European Physical Journal Special Topics **223**, 1221 (2014)

The Dyadic Radionuclide System ^{60}Fe / ^{53}Mn to Distinguish Interstellar from Interplanetary ^{60}Fe

Dominik Koll[1,2,3,*], *Thomas* Faestermann[3], *Gunther* Korschinek[3], *Ingo* Leya[4], *Silke* Merchel[2,5], and *Anton* Wallner[1,2]

[1]NPAA, Research School of Physics, The Australian National University, Canberra, ACT 2601, Australia.
[2]Accelerator Mass Spectrometry & Isotope Research, Helmholtz-Zentrum Dresden-Rossendorf, 01328 Dresden, Germany.
[3]Department of Physics, Technical University of Munich, 85748 Garching, Germany.
[4]Physical Institute - Space Research and Planetology, University of Bern, 3012 Bern, Switzerland.
[5]Isotope Physics, Faculty of Physics, University of Vienna, 1090 Vienna, Austria.

Abstract. The discovery of live ^{60}Fe in a deep-sea crust with proposed interstellar origin followed by evidence for elevated interplanetary ^{3}He in the same crust raised the question on how to unambiguously identify the true production site of the identified ^{60}Fe. Here, we show the implementation of the dyadic radionuclide system ^{60}Fe / ^{53}Mn to serve as a tool for the identification of surplus interstellar ^{60}Fe over interplanetary production. The recent updates in experimental ^{60}Fe and ^{53}Mn data from iron meteorites as well as in production rate models confirm the validity and robustness of this dyadic system for future applications.

1 Introduction

Direct detection of freshly-synthesized supernova nuclides is one of the gold standards to study stars and their impact on the solar system. The radionuclide ^{60}Fe proved to be an excellent candidate because of its relatively long half-life of 2.6 Myr [1, 2], allowing accumulation and detection over several million years.

The first indication of live ^{60}Fe in a deep-sea ferromanganese crust [3] followed by the clear discovery in a 2 Myr old layer [4] raised the question on its astrophysical origin. The proposed supernova (interstellar) origin [4] was subsequently challenged. Elevated levels of ^{3}He in the same crust were used as an argument to propose that the discovered ^{60}Fe was produced by cosmic-ray spallation in (micro-) meteorites and not by stars [5]. Further evidence for a global ^{60}Fe signal [6] was used as an indirect argument for the supernova hypothesis. ^{60}Fe was subsequently identified on timescales of years, thousands of years and millions of years in several archives, see [7] for a recent summary.

The clear identification of the astrophysical production site of any discovered ^{60}Fe requires the exclusion of anthropogenic ^{60}Fe, which was conclusively presented in [8]. Cosmogenic production is governed by cosmic-ray fluxes, target sizes and compositions, and production rates for radionuclides by primary and secondary cosmic rays outside Earth's atmosphere. The most viable candidates to establish cosmic-ray production rates of ^{60}Fe are meteorites, which could be used to distinguish between a rather constant cosmic dust influx plus single large object events such as meteoroids, comets or asteroids and solitary supernova ^{60}Fe influxes.

It is critical to deduce the absolute amount of interplanetary dust influx into a sample. This makes it particularly challenging, because a constant background flux of ^{60}Fe bearing cosmic dust modulated by single large events can mimic supernova influxes. Note, for deep-sea crusts we integrate over several hundred thousand to millions of years, which makes rare events like meteorite falls even over global scales probable. Therefore, a tool to unambiguously identify interstellar over cosmogenic ^{60}Fe is needed. Here, we show that by additionally measuring the long-lived radionuclide ^{53}Mn ($t_{1/2} = 3.7$ Myr), the dyadic radionuclide system ^{60}Fe / ^{53}Mn could be used to overcome these problems and to disentangle interstellar from interplanetary ^{60}Fe influxes. It is worth mentioning that one can estimate the maximum cosmogenic ^{60}Fe concentration in a sample by measuring the stable target element Ni and assuming all of the Ni present is extraterrestrial Ni. This estimation is crude but practical for large ^{60}Fe influxes and low stable Ni abundances and fails for samples with low extraterrestrial but high terrestrial influxes. The hereby proposed system is applicable in any case.

2 Cosmogenic ^{60}Fe and ^{53}Mn from meteorites

In the last few years, we were able to establish a substantial meteorite database for both radionuclides ^{60}Fe and ^{53}Mn using accelerator mass spectrometry (AMS) [9]. AMS is the most sensitive experimental approach to directly determine concentrations of both radionuclides in terrestrial and extraterrestrial samples. Radiochemical neutron activation analysis

*e-mail: d.koll@hzdr.de

and low-level counting have been used in the past to measure both radionuclides in meteorites but are not sensitive enough to search for supernova traces. For ^{60}Fe in particular, one needs to consider waiting times of decades after chemical purification for the in-growth of the daughter nucleus ^{60}Co to achieve secular equilibrium [10]. Note, large databases for the isotope ^{53}Mn exist without ^{60}Fe data [11]. AMS does not require any waiting time since live radionuclides are detected and not their decay. Currently, the only AMS facility in the world capable of measuring both radionuclides at natural levels is the Heavy Ion Accelerator Facility of ANU [2]. The Munich accelerator lab [12], which was pioneering the measurements of both radionuclides using a gas-filled magnet, has been closed in 2020. Clearly, there is the need for existing or new labs to establish new capabilities for ^{60}Fe and ^{53}Mn AMS.

The database with more than 30 single iron meteorites is complemented by an updated production rate model based on the INCL (Liège Intra Nuclear Cascade) code [9]. The predicted production rate ratio is $^{60}\mathrm{Fe}/^{53}\mathrm{Mn} = (2.5 \pm 0.1) \times 10^{-3}$ in units of dpm kg^{-1}[Ni] / dpm kg^{-1}[Fe]. The common units of production rates are specific activities in disintegrations per minute per kg of target element or target mass, in this case stable Ni for ^{60}Fe and stable Fe for ^{53}Mn. The stable element content in terrestrial samples is dominated by terrestrial elements. The specific activity as a measure for the interplanetary influx is therefore not suitable, because the extraterrestrial target element concentration is not quantifiable. Hence, it is necessary to estimate reasonable values of extraterrestrial Ni / Fe in terrestrial samples. Importantly, only the ratio of stable Ni to Fe is needed and not the concentration of both elements. For terrestrial samples, the amount of cosmogenic radionuclides present is much smaller compared to extraterrestrial samples, where intrinsic production is dominant. This relaxes the requirement on the exact knowledge of Ni / Fe.

Using the half-lives of 2.6 Myr and 3.7 Myr, we back-calculate a half-life and target composition independent and directly measurable production rate ratio of $^{60}\mathrm{Fe}/^{53}\mathrm{Mn} = (1.77 \pm 0.07) \times 10^{-3} \times \mathrm{Ni/Fe}$ in units of at / at and wt.%. By factoring out Ni / Fe, the data becomes comparable between samples with different target element concentrations. This allows us to compare the model prediction with extraterrestrial samples (known extraterrestrial Ni / Fe) and with terrestrial archives (unknown extraterrestrial Ni / Fe). The possible values for extraterrestrial Ni / Fe in terrestrial samples are conservatively estimated to be less than 1. Solar system, chondritic and meteorite abundances show the clear trend of $\mathrm{Ni} \ll \mathrm{Fe}$, most commonly Ni / Fe = 0.055 [13]. Iron meteorites, which are metal-rich and elevated in Ni, generally show a higher Ni / Fe of around 0.1. For comparison, Earth's crust has a low Ni to Fe ratio of 0.002, whereas the ratio in sea water is 0.28, but with low concentration of both metals [14]. Using the most common abundances of stable elements in the solar system, the cosmogenic ^{60}Fe to ^{53}Mn atomic ratio obtained in a measurement would be around 10^{-4}.

In contrast to that, the stellar atomic nucleosynthesis ratio is estimated to be 10^{0} - 10^{-2} [15]. The hypothetical interstellar ^{60}Fe would dominate over cosmogenic ^{60}Fe as an excess, whereas any supernova ^{53}Mn is only a fraction of the bulk cosmogenic ^{53}Mn. This makes any accompanied supernova ^{53}Mn challenging to identify over cosmogenic background with only one attempt so far by AMS [16]. Therefore, interstellar ^{60}Fe would appear as an enhanced ^{60}Fe to ^{53}Mn ratio over the cosmogenic baseline, which is dominated by cosmogenic ^{53}Mn.

3 Interstellar ^{60}Fe identification

The very first utilization of this principle happened alongside the first indication of supernova ^{60}Fe in a deep-sea crust [3]. There, it was already estimated, that the atomic ratio of cosmogenic ^{60}Fe to ^{53}Mn should be in the order of 10^{-4}. This strong experimental argument was not adopted in the following studies, possibly due to limited data on extraterrestrial production rates and meteorite samples. Seventeen years later in 2016, this dyadic radionuclide system yielded evidence for supernova ^{60}Fe on the surface of the Moon [17]. Lunar regolith was analyzed for ^{60}Fe and ^{53}Mn and compared to four different meteorites before the database was significantly extended. The specific activity ratio was calculated, since the lunar surface itself is the target for the cosmic ray interactions. The meteorite activity ratio was determined to be $^{60}\mathrm{Fe}/^{53}\mathrm{Mn} = (2.68 \pm 0.35) \times 10^{-3}$ in units of dpm kg^{-1}[Ni] / dpm kg^{-1}[Fe]. This agrees perfectly with the current model value of $(2.5 \pm 0.1) \times 10^{-3}$ and the meteorite database value of 2.6×10^{-3}. The lunar surface ratio was determined to be $(10 - 50) \times 10^{-3}$, much higher than for meteorites. The measured atomic $^{60}\mathrm{Fe}/^{53}\mathrm{Mn}$ of lunar material would be misleading because of the low lunar Ni / Fe of 0.001 - 0.01, which is similar to the ratio on Earth and significantly lower than for meteorites. The specific activity reveals the clear enhancement, which shows that for lunar material the knowledge of Ni / Fe is crucial.

More recently, ^{60}Fe was discovered in Antarctic surface snow [8]. Besides ruling out anthropogenic production, the identification of its astrophysical origin was important since the Antarctic snow covered only the most recent 20 years of terrestrial history and could therefore reveal an influx right at this moment. Here, the specific activity ratio approach was not possible, because the stable elements in the sample are of terrestrial origin. The same meteorite data from [17] was converted to atomic ratios with a value of $^{60}\mathrm{Fe}/^{53}\mathrm{Mn} = (1.9 \pm 0.2) \times 10^{-3} \times \mathrm{Ni/Fe}$ in units of at / at and wt.%. Similarly, this again agrees perfectly with the outlined model value of $(1.77 \pm 0.07) \times 10^{-3} \times \mathrm{Ni/Fe}$ and the meteorite database value of $1.8 \times 10^{-3} \times \mathrm{Ni/Fe}$. The measured ^{60}Fe to ^{53}Mn ratio in Antarctic snow was around 17×10^{-3}. The extraterrestrial Ni / Fe was conservatively estimated to be less than 1. This yields a more than two orders of magnitude enhancement over cosmogenic background for chondritic composition Ni / Fe = 0.055. There is still a clear enhancement even for the extreme case of Ni / Fe = 1. The interstellar origin of the discovered ^{60}Fe becomes now even clearer with the updated model and the significantly extended meteorite database (Figure 1).

Table 1: Summary of $^{60}Fe\,/\,^{53}Mn$ ratios from the theoretical model, the meteorite database and the samples lunar regolith as well as Antarctic snow. Depending on the origin of the analyzed sample, one has to compare any measured data with the first or the second $^{60}Fe\,/\,^{53}Mn$ ratio for extraterrestrial or terrestrial samples, respectively. The unit dpm / kg X refers to the specific activity ratio with X being the respective target element for the production of ^{60}Fe or ^{53}Mn. In particular for lunar regolith, the measured atomic ratio is misleading because of the low intrinsic Ni / Fe. $^{60}Fe\,/\,^{53}Mn$ ratios around 10^{-3} indicate cosmogenic production, whereas higher ratios indicate a surplus of interstellar ^{60}Fe.

	Sample	**Ni / Fe**	^{60}**Fe** / ^{53}**Mn** [dpm / kg X]	^{60}**Fe** / ^{53}**Mn** [at/at] × Ni / Fe
Leya *et al.* [9]	Model	. . .	$(2.5 \pm 0.1) \times 10^{-3}$	. . .
This work	based on [9]	. . .	. . .	$(1.77 \pm 0.07) \times 10^{-3}$
Knie *et al.* [18]	Dermbach Met.	0.91	2.3×10^{-3}	1.6×10^{-3}
Knie *et al.* [18]	Emery Met.	0.09	2.6×10^{-3}	1.8×10^{-3}
Ott *et al.* [19]	Gebel Kamil Met.	0.24	2.7×10^{-3}	1.9×10^{-3}
Fimiani *et al.* [17]	NWA 6369 Met.	0.07	2.2×10^{-3}	1.5×10^{-3}
Leya *et al.* [9]	Database (27 Met.)	0.1	2.6×10^{-3}	1.8×10^{-3}
				[at/at]
Fimiani *et al.* [17]	Lunar regolith	0.001 - 0.01	$(10 - 50) \times 10^{-3}$	$10^{-5} - 10^{-4}$
Koll *et al.* [8]	Antarctic snow	. . .	. . .	17×10^{-3}

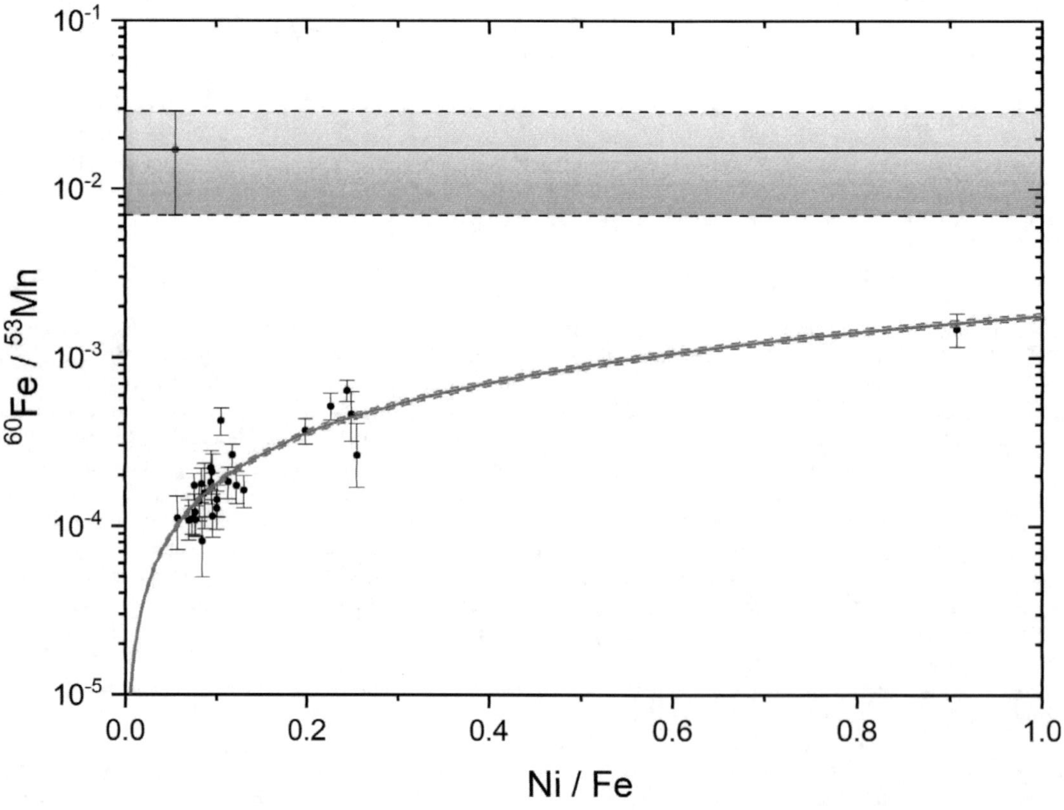

Figure 1: Antarctic snow data [8] in green with the updated, independent meteorite model in red and available meteorite data in black [9]. The enhancement of ^{60}Fe over the expected cosmogenic ratio is clearly visible. The datapoint is exemplarily displayed for the most likely stable element abundance Ni / Fe = 0.055.

By establishing a significant meteorite database for both radionuclides and having an independent model, the dyadic radionuclide system $^{60}Fe/^{53}Mn$ is now robust enough to be widely applied for the search of interstellar ^{60}Fe in any terrestrial or extraterrestrial archive. The model values for the production rate ratios are in very good agreement with now 31 single meteorite measurements, see Table 1. This system allows to unambiguously identify supernova produced ^{60}Fe over cosmogenic background.

Acknowledgments

The work at TUM was supported by the DFG Cluster of Excellence "Origin and Structure of the Universe". The support for the Heavy Ion Accelerator Facility at ANU by NCRIS is highly appreciated. AW acknowledges funding from the Australian Research Council's Discovery scheme, project numbers DP180100495 und DP180100496. DK was supported by an AINSE Ltd. Postgraduate Research Award (PGRA).

References

[1] G. Rugel, T. Faestermann, K. Knie *et al.*, Phys. Rev. Lett. 103, 072502 (2009).
[2] A. Wallner, M. Bichler, K. Buczak *et al.*, Phys. Rev. Lett. 114, 041101 (2015).
[3] K. Knie, G. Korschinek, T. Faestermann *et al.*, Phys. Rev. Lett. 83, 18 (1999).
[4] K. Knie, G. Korschinek, T. Faestermann *et al.*, Phys. Rev. Lett. 93, 171103 (2004).
[5] S. Basu, F. M. Stuart, C. Schnabel *et al.*, Phys. Rev. Lett. 98, 141103 (2007).
[6] A. Wallner, J. Feige, N. Kinoshita *et al.*, Nature 532, 69 (2016).
[7] D. Koll, T. Faestermann, G. Korschinek, *et al.*, EPJ Web Conf. 232, 02001 (2020).
[8] D. Koll, G. Korschinek, T. Faestermann *et al.*, Phys. Rev. Lett. 123, 072701 (2019).
[9] I. Leya, J.C. David, T. Faestermann *et al.*, Meteorit. Planet. Sci. 55, 4 (2020).
[10] K. Nishiizumi and M. Honda, Meteorit. Planet. Sci. 42, 118 (2007).
[11] K. Nishiizumi, Nucl. Tracks Radiat. Meas. 13, 4 (1987).
[12] D. Koll, C. Busser, T. Faestermann *et al.*, Nucl. Instrum. Meth. B 438, 180 (2019).
[13] H. Palme, K. Lodders, and A. Jones, in Treatise on Geochemistry, Elsevier, Oxford (2014), pp. 15–36.
[14] CRC Handbook of Chemistry and Physics, 97th edition (2016–2017), pp. 14-17.
[15] S.E. Woosley and T.A. Weaver, Astrophys. J. Suppl. Ser. 101, 181 (1995).
[16] G. Korschinek, T. Faestermann, M. Poutivtsev *et al.*, Phys. Rev. Lett. 125, 031101 (2020).
[17] L. Fimiani, D.L. Cook, T. Faestermann *et al.*, Phys. Rev. Lett. 116, 151104 (2016).
[18] K. Knie, S. Merchel, G. Korschinek *et al.*, Meteorit. Planet. Sci. 34, (1999).
[19] U. Ott, S. Merchel, S. Herrmann *et al.*, Meteorit. Planet. Sci. 49, 8 (2014).

Regulated NiCu Cycles with the New ^{57}Cu(p,γ)^{58}Zn Reaction Rate and the Influence on Type-I X-Ray Bursts: GS 1826−24 Clocked Burster

Yi Hua Lam[1,2,*], *Ning* Lu[1,2,3,**], *Alexander* Heger[4,5,6,7,***], *Adam Michael* Jacobs[7,8], *Nadezda A.* Smirnova[9], *Teresa Kurtukian* Nieto[9], *Zac* Johnston[7,8], and *Shigeru* Kubono[10,11]

[1]Institute of Modern Physics, Chinese Academy of Sciences, Lanzhou 730000, China
[2]School of Nuclear Science and Technology, University of Chinese Academy of Sciences, Beijing 100049, China
[3]School of Nuclear Science and Technology, Lanzhou University, Lanzhou 730000, China
[4]School of Physics and Astronomy, Monash University, Victoria 3800, Australia
[5]OzGrav-Monash – MOCA, School of Physics and Astronomy, Monash University, VIC 3800, Australia
[6]Center of Excellence for Astrophysics in Three Dimensions (ASTRO-3D), Australia
[7]The Joint Institute for Nuclear Astrophysics, Michigan State University, East Lansing, MI 48824, USA
[8]Department of Physics and Astronomy, Michigan State University, East Lansing, MI 48824, USA
[9]CENBG, CNRS/IN2P3 and University of Bordeaux, Chemin du Solarium, 33175 Gradignan cedex, France
[10]RIKEN Nishina Center, 2-1 Hirosawa, Wako, Saitama 351-0198, Japan
[11]Center for Nuclear Study, University of Tokyo, 2-1 Hirosawa, Wako, Saitama 351-0198, Japan

Abstract. In Type-I X-ray bursts (XRBs), the rapid-proton capture (rp-) process passes through the NiCu and ZnGa cycles before reaching the region above Ge and Se isotopes that hydrogen burning actively powers the XRBs. The sensitivity study performed by Cyburt *et al.* [1] shows that the ^{57}Cu(p,γ)^{58}Zn reaction in the NiCu cycles is the fifth most important rp-reaction influencing the burst light curves. Langer *et al.* [2] precisely measured some low-lying energy levels of ^{58}Zn to deduce the ^{57}Cu(p,γ)^{58}Zn reaction rate. Nevertheless, the order of the 1^+_1 and 2^+_3 resonance states that dominate at $0.2 \lesssim T(\mathrm{GK}) \lesssim 0.8$ is not confirmed. The 1^+_2 resonance state, which dominates at the XRB sensitive temperature regime $0.8 \lesssim T(\mathrm{GK}) \lesssim 2$ was not detected. Using isobaric-multiplet-mass equation (IMME), we estimate the order of the 1^+_1 and 2^+_3 resonance states and estimate the lower limit of the 1^+_2 resonance energy. We then determine the ^{57}Cu(p,γ)^{58}Zn reaction rate using the full *pf*-model space shell model calculations. The new rate is up to a factor of four lower than the Forstner *et al.* [3] rate recommended by JINA REACLIBv2.2. Using the present ^{57}Cu(p,γ)^{58}Zn, the latest ^{56}Ni(p,γ)^{57}Cu and ^{55}Ni(p,γ)^{56}Cu reaction rates, and 1D implicit hydrodynamic Kepler code, we model the thermonuclear XRBs of the clocked burster GS 1826−24. We find that the new rates regulate the reaction flow in the NiCu cycles and strongly influence the burst-ash composition. The ^{59}Cu(p,α)^{56}Ni and ^{59}Cu(p,γ)^{60}Zn reactions suppress the influence of the ^{57}Cu(p,γ)^{58}Zn reaction. They strongly diminish the impact of the nuclear reaction flow that bypasses the ^{56}Ni waiting point induced by the ^{55}Ni(p,γ)^{56}Cu reaction on burst light curve.

The ^{57}Cu(p,γ)^{58}Zn reaction is the fifth most influential (p,γ) reaction that affects the light curve of GS 1826−24, the clocked burster [4], as found by Cyburt *et al.* [1]. Langer *et al.*

*e-mail: lamyihua@impcas.ac.cn
**Presenter
***e-mail: alexander.heger@monash.edu

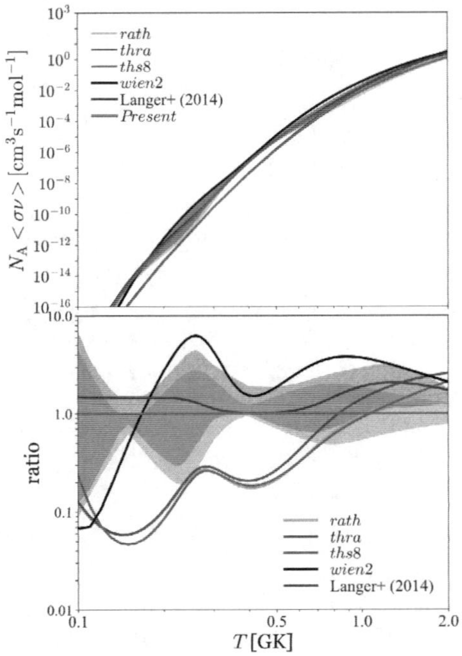

Figure 1. *Top Panel*: The *Present* and Langer *et al.* [2] ^{57}Cu(p,γ)^{58}Zn thermonuclear reaction rates, and available reaction rates compiled in to JINA REACLIB v2.2: *rath*, *thra*, *ths8*, and *wien2* [3], the rate recommended by JINA REACLIB v2.2. All available rates of REACLIB v2.2 use the ^{58}Zn proton threshold, $S_p(^{58}$Zn) = 2.277 MeV. *Bottom Panel*: The comparison of the *Present* rate with Langer *et al.* and all available reaction rates of REACLIB v2.2. The uncertainty of the *Present* rate (*red zone*) folds the uncertainties from the $S_p(^{58}$Zn) and nuclear structure. The uncertainty of Langer *et al.* rate is indicated as *blue zone*. The rates of *rath*, *thra*, and *ths8* based on the Hauser-Feshbach statistical model are very close to one another from 0.1 to 2.0 GK, and they are lower than the *Present* rate up to one order of magnitude at temperature $T \lesssim 0.9$ GK. The *Present* rate is up to a factor of two lower than Langer *et al.* rate from 0.8 to 2 GK covering the typical maximum temperature of accreted envelope of the GS 1826−24 burster, and up to a factor of four lower than the *wien2* rate [3].

[2] experimentally confirmed some low-lying excited states of ^{58}Zn, which are resonance states dominantly contribute to the ^{57}Cu(p,γ)^{58}Zn reaction rate. The order of the dominant resonance states, 1^+_1 and 2^+_3, that they precisely measured was, however, unconfirmed. The 1^+_2 resonance state, one of the dominant resonances sensitive to the clocked burst temperature regime, $0.8 \lesssim T$(GK) $\lesssim 2$, was not detected.

In the present study, we first compare the theoretical and experimental IMME c coefficients for the $A = 58$, isospin $I = 1$ multiplets to obtain the root-mean-square (rms) deviation value 22 keV as the theoretical uncertainty. Then, we study and propose the most plausible order of the 2^+_3 and 1^+_1 states based on the $A = 58$, $I = 1$ multiplets of ^{58}Zn, ^{58}Cu, and ^{58}Ni. We then use IMME to estimate the lowest limit energy of the 1^+_2 state (3.664 ± 0.022 MeV). The corresponding resonance energy is 1.384 MeV. This is 329 keV higher than the direct estimation of shell-model calculation by Langer *et al.* [2]. The contribution of the 1^+_2 resonance state reduces by about one order of magnitude, reducing the total reaction rate in the temperature regime $0.8 \lesssim T$(GK) $\lesssim 2$. With the presently deduced nuclear structure information, we use the full *pf*-model space shell model calculations with GXPF1A interaction [5] to construct the *Present* ^{57}Cu(p,γ)^{58}Zn reaction rate for the typical XRB temperature range, in particular, the temperature regime of $0.8 \lesssim T$(GK) $\lesssim 1.6$ relevant to the GS 1826−24 burster. The comparison of the *Present* rate with Langer *et al.* rate and with other reaction rates compiled into JINA REACLIB v2.2 [6] is shown in Fig. 1.

Using the GS 1826−24 model obtained from Kepler, we construct three XRB models based on each combination of reaction rates: **(1)** the *Present* ^{57}Cu(p,γ)^{58}Zn, *et al.* Kahl [7] ^{56}Ni(p,γ)^{57}Cu, and Valverde [8] ^{55}Ni(p,γ)^{56}Cu, **(2)** Langer [2] ^{57}Cu(p,γ)^{58}Zn, Kahl [7] ^{56}Ni(p,γ)^{57}Cu, and Valverde [8] ^{55}Ni(p,γ)^{56}Cu, and **(3)** *wien2* [3] ^{57}Cu(p,γ)^{58}Zn. These **(1)**, **(2)**, and **(3)** combinations are labeled as *Present*§, *Present*$^{\heartsuit}$, and *baseline* models, respectively.

We remark that the observed burst tail end of *Epoch Jun 1998* of the GS 1826−24 burster is closely reproduced by the *Present*§, *Present*$^{\heartsuit}$, and *baseline* models (Fig. 5 of Ref. [9]). Figure 2 shows the burst-ash composition at the burst tail end produced by three models. Using

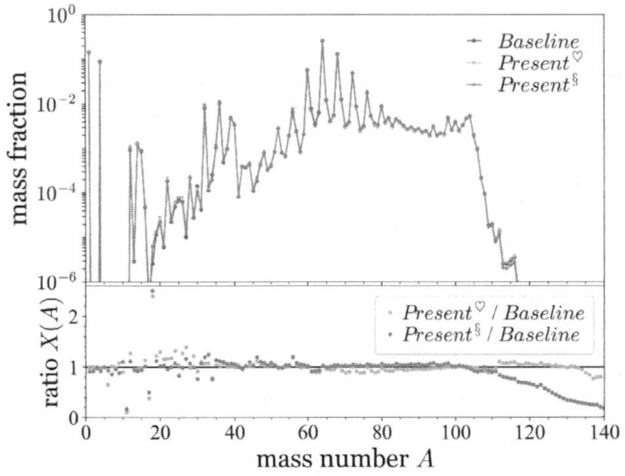

Figure 2. The averaged mass fractions for each mass number at burst tail end when $t \approx 180$ s after the burst peak. The GS 1826−24 model from Jacobs *et al.* [10] is the adjusted to reproduce the recurrence time of *Epoch Jun 1998*. The model uses JINA REACLIB v2.2. and is named as *baseline* model. The *Present*§ (or *Present*$^{\heartsuit}$) model adopts the same astrophysical configurations of *baseline* but implement the *Present* (or Langer *et al.*) ^{57}Cu(p,γ)^{58}Zn, Kahl *et al.* [7] ^{56}Ni(p,γ)^{57}Cu and Valverde *et al.* [8] ^{55}Ni(p,γ)^{56}Cu rates, see text.

the *Present* ^{57}Cu(p,γ)^{58}Zn rate, the productions of ^{12}C is reduced by a factor of 0.2, the nuclei with A = 17 and 18 breaking out from the hot CNO cycle are affected up to about a factor of 0.5 and 2.5, respectively. The abundances of the daughters of SiP, SCl, and ArK cycles are reduced up to a factor of 0.7. The overall production of ^{56}Ni and its remnant increases up to a factor of 1.2 due to the *correlated influence* between the new ^{56}Ni(p,γ)^{57}Cu, ^{57}Cu(p,γ)^{58}Zn, and ^{55}Ni(p,γ)^{56}Cu rates. The abundances of nuclei with $A = 64 - 104$ produced by *Present*§ are closer to *baseline* than the ones produced by *Present*$^{\heartsuit}$. Furthermore, the abundances of the nuclei with $A = 105 - 140$ are decreased by up to a factor of 0.2 (red dots in the bottom panel of Fig. 2). Notably, the Langer *et al.* ^{57}Cu(p,γ)^{58}Zn reaction rate produces a different set of burst-ash composition compared to the *Present* ^{57}Cu(p,γ)^{58}Zn rate, especially the nuclei with $A = 20 - 34$; the abundance of the nuclei with $A = 65 - 84$, is reduced by up to a factor of 0.9; and the abundance of the nuclei with $A = 100 - 134$ is closer to *baseline* than the impact suggested by the change to the *Present* ^{57}Cu(p,γ)^{58}Zn reaction rate.

The impact of *Present* ^{57}Cu(p,γ)^{58}Zn, new ^{56}Ni(p,γ)^{57}Cu and ^{55}Ni(p,γ)^{56}Cu rates on burst ash composition is due to the cumulative effect of redistributing and reassembling of reaction flows in the NiCu cycles that eventually affects the nucleosyntheses of ^{56}Ni, ^{57}Cu, ^{58}Zn, and nuclei in the ZnGa cycles. A detail analysis of the evolution of the episode of clocked bursts is presented in Ref. [9].

This work was financially supported by the Strategic Priority Research Program of Chinese Academy of Sciences (CAS, Grant Nos. XDB34000000 and XDB34020204) and National Natural Science Foundation of China (No. 11775277), Chinese Academy of Sciences President's International Fellowship Initiative (No. 2019FYM0002), Australian Research Council Centre of Excellence for Gravitational Wave Discovery (OzGrav, No. CE170100004) and for All Sky Astrophysics in 3 Dimensions (ASTRO 3D, No. CE170100013), IN2P3/CNRS, France, Master Project – "Exotic nuclei, fundamental interactions and astrophysics".

References

[1] R. H. Cyburt, A. M. Amthor, A. Heger *et al.*, Astrophys. J. **55**, 830 (2016).
[2] C. Langer, F. Montes, A. Aprahamian *et al.*, Phys. Rev. Lett. **113**, 032502 (2014).
[3] O. Forstner, H. Herndl, H. Oberhummer *et al.*, Phys. Rev. C. **64**, 045801 (2001).
[4] F. Makino, IAU. Circ., No. 4653 (1988).
[5] M. Honma, T. Otsuka, B. A. Brown *et al.*, Phys. Rev. C. **69**, 034335 (2004).
[6] R. H. Cyburt, A. M. Amthor, R. Ferguson *et al.*, Astrophys. J. Supp. **189**, 240 (2010).
[7] D. Kahl, P. J. Woods, T. Poxon-Pearson, *et al.*, Phys. Lett. B. **797**, 134803 (2019).
[8] A. A. Valverde, M. Brodeur, G. Bollen *et al.*, Phys. Rev. Lett. **123**, 239905E (2019).
[9] Y. H. Lam, N. Lu, A. Heger *et al.*, arXiv:2107.11552.
[10] J. M. Jacobs, A. Heger *et al.*, *Burst Environment, Reactions and Numerical Modelling Workshop* (BERN18), 11 – 15 Jun, 2018, Monash Prato Centre, Tuscany, Italy.

Cooling of neutron stars with quark-hadron continuity

Tsuneo Noda[1,*], *Nobutoshi* Yasutake[2], *Masa-aki* Hashimoto[3], *Toshiki* Maruyama[4], and *Toshitaka* Tatsumi[5]

[1]Kurume Institute of Technology, 2228-66 Kamitsu-machi, Kurume, Fukuoka 830-0052, Japan
[2]Chiba Institute of Technology, Chiba 275-0023, Japan
[3]Department of Physics, Kyushu University, Fukuoka 819-0395, Japan
[4]Advanced Science Research Centre, Japan Atomic Energy Agency, Ibaraki 319-1195, Japan
[5]Osaka Sangyo University, Osaka 574-8530, Japan

Abstract. Neutron stars are high-density objects formed by the gravitational collapse of massive stars, and the whole star can be likened to a giant nucleus. The interior of a neutron star is considered to contain exotic particles and states which do not appear in a normal nucleus. The internal states are constrained by observations of masses and radii via the equation of state of highly dense nuclear matter. Within these constraints, a variety of exotic states have been discussed. The internal state of neutron stars is closely related to its neutrino emission process, which cools the star from the inside. This effect can be compared with observations of the surface temperature of neutron stars. However, despite the wide range of observations of neutron stars, the nature of the neutron star matter remains uncertain. We consider quark matter as an exotic state and perform cooling calculations for neutron stars, incorporating the effects of nucleon superfluidity and quark colour superconductivity. We take into account the "quark-hadron continuity", in which the neutron superfluidity is succeeded by the d quark pairing. Furthermore, we obtained the range of the neutron star cooling curve, taking into account the difference in surface temperature due to the composition of the surface layer. We found that the existence of quark matter causes strong neutrino emission from quarks, which is moderately suppressed by superfluidity and superconductivity, and can explain the cold surface temperature of neutron stars.

1 Introduction

Neutron stars are extremely dense objects, which can be likened to giant atomic nuclei. They are born at high temperature during supernova explosions, but have no internal heat source, and are cooled by neutrino and X-ray radiation processes. The dominant cooling process in neutron star cooling is neutrino emission in the early stages, and X-ray emission from the surface in the later stages. Neutrino emission is strongly dependent on the state of the dense nuclear matter inside, and there is a strong relationship between the cooling of the neutron star and the state of the dense nuclear matter.

The state of dense nuclear matter constituting a neutron star can be exotic and different from that of normal nuclear matter. Such exotic states have a different neutrino emission

*e-mail: noda@kurume-it.ac.jp

mechanism to that of normal matter and have a significant impact on the thermal evolution of neutron stars. Exotic states are explored in the terrestrial experiments, but the internal states of neutron stars are relatively cold compared to their density, and it is very difficult to reproduce them in heavy-ion collisions. The most effective method to explore the internal states of neutron stars is to compare theoretical calculations with observations.

The states of the dense nuclear matter in neutron stars must be common to all neutron stars, and therefore the equation of state (EoS) describing the state of matter must also be common. Recent observations of neutron star masses of $2M_\odot$ have resulted in a very strong constraint on the EoS. The maximum mass determined by the EoS and the TOV equation must exceed $2M_\odot$. Furthermore, the EoS corresponds to the state of internal matter and is directly linked to the neutrino emission process.

The simplest neutrino emission process is the β decay of neutrons and their inverse reaction, known as the Direct URCA (DU), but these processes are forbidden in degenerate neutron stars with a low proton fraction ($< 1/9$) because they violate the momentum conservation. The Modified URCA (MU) process, which solves the momentum conservation problem by involving other nucleons, is thought to be the dominant neutrino emission process in neutron stars. However, the MU process has a much lower emissivity than the DU process and cannot explain the temperatures of some cold neutron stars. To account for the observed temperatures of neutron stars, we would need a strong neutrino emission process, such as the DU process, or an exotic state with similar strong cooling which are only valid for some stars. Such a state above the critical density would lead to weak neutrino emission from normal nuclear matter in lighter neutron stars, where the central density does not reach the critical density, and to strong neutrino emission in heavier stars. This way can explain the observed surface temperatures under a unified EoS.

We investigate the thermal evolution of neutron stars by constructing a model of the neutron star cooling, considering the EoS in which nucleons decay into quark matter at high densities. The model takes into account the effects of nucleon superfluidity and quark colour superconductivity (CSC), as well as the quark-hadron continuum in superfluidity. And we investigate the effects of these "superstates".

2 Cooling Models

We consider that the neutron stars consist the standard hadronic matter at low density, and deconfined quark matter at high density. The hadronic matter transits into the nucleon superfluidity state below the critical temperature, and quark matter transits into CSC state. The effects of superfluidity/superconductivity included to the neutrino emissivities. We take account of ${}^1S_0$ and ${}^3P_2$ states for neutrons and ${}^1S_0$ state for protons in the hadronic phase. For quark phase, we include the effects of Colour-Flavour-Locking (CFL) paring, Two-Flavour-Superconductivity (2SC) pairing or "2SC+X" phase.

The superfluid/superconducting phases have two effects on neutrino emissivities. One is that they cause strong neutrino emission during the transition from the normal phases[1], and the another one is that they suppress other neutrino emission process after the transition. The suppressed neutrino emissivities can be obtained with the gap energy Δ and the original emissivity ε_ν as $\varepsilon_\nu \exp(-\Delta/k_{\mathrm{B}}T)$, here k_B denotes the Boltzmann constant and T is the temperature of the layer. We construct the models of the critical temperature as parametrized functions which depend on the density as Figure 1 for nucleon superfluidity.In this calculation, four models were created by varying the maximum critical temperature and fixing the density at which the critical temperature is maximum. The models were labelled A to D in ascending order of the critical temperature.

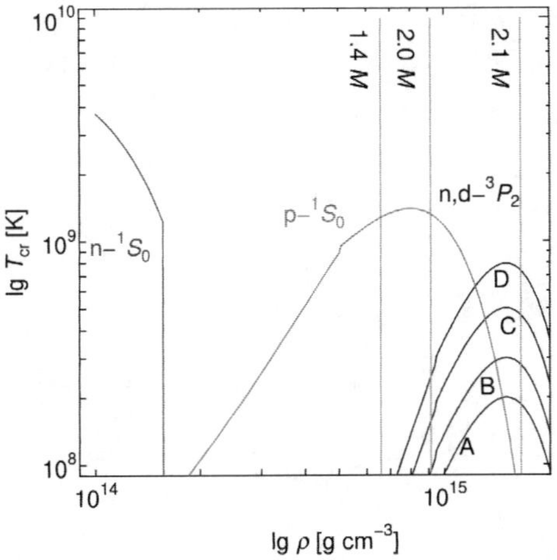

Figure 1. Superfluid temperatures of each state. Black lines denote the difference of neutron $^3P_2$ model.

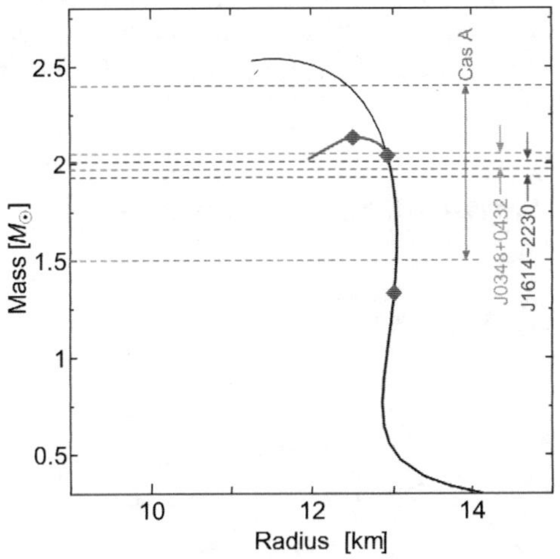

Figure 2. *M–R* relation of the adopted EoS. Green dots denotes the calculation models.

For the CSC, we assume the gap Δ is large enough comparing with the temperature T (ie. $\Delta > 10$ MeV). All colours/flavours of quarks are in superconducting state in the CFL state, and 2 of 3 colours are in superconducting state in 2SC state. The CFL state strongly suppresses the neutrino emissivity of quarks, and the 2SC state also suppresses it in about 1/3. We focus on the concept of "quark hadron continuity" of superfluidity. In this concept, the neutron $^3P_2$ superfluidity is taken over by the d-quark $^3P_2$ pairing during the crossover from hadronic matter to quark matter, and this d-quark paring can coexist with the 2SC CSC of other quarks[2]. We assume that the continuity occurs during the transition and it carries the superfluid effect on the neutrino emission. We also assume that the density dependence of the critical temperature of the d-quark $^3P_2$ paring is connected with that of the neutron $^3P_2$ superfluidity. We include the quark-hadron continuity effect to our calculation, as the name of "2SC+X" for the comparison with other CSC parings (2SC, CFL). 2SC+X state are combination of 2SC state and d-quark $^3P_2$ state which is carried from the neutron $^3P_2$ state. We treat the 2SC component as pure 2SC state, and the $^3P_2$ component as the extension of neutron $^3P_2$ state. We calculate the neutrino emissivity in each layer of stars.

We adopt the EoS which includes the first order phase transition between the hadronic phase and the quark phase[3]. We assume the dimensional structure of mixed phase at the transition. The maximum mass with this EoS is $2.13M_\odot$ which satisfies recent $2M_\odot$ observations[4, 5], even including the quark phase. The *M–R* relation is shown in Figure 2. In the mixed phase, the dimensional structure is determined by Wigner-Seitz approximation, which fixes the cell size and the symetricity. The volume fraction of each phase can be calculated by the cell size, and it can determine the bulk neutrino emissivity. Also, the centre of the maximum mass star is constructed by mixed phase, it causes the neutrino emission processes of both of hadronic phase and quark phase are activated. In the hadronic phase, the proton fraction can exceed 1/9 at high density region, the strong neutrino emission by the DU is activated.

We choose three masses of stars as calculation models (1.41, 2.04, and $2.13M_\odot$). And four kinds of neutron ${}^3P_2$ superfluid models and quark CSC paring (CFL, 2SC, and 2SC+X) as parameters. First, we set the surface composition as ^{56}Fe, and calculate the cooling curves. And then we show the effect of light element surface composition for one fixed superfluid model.

3 Results and discussions

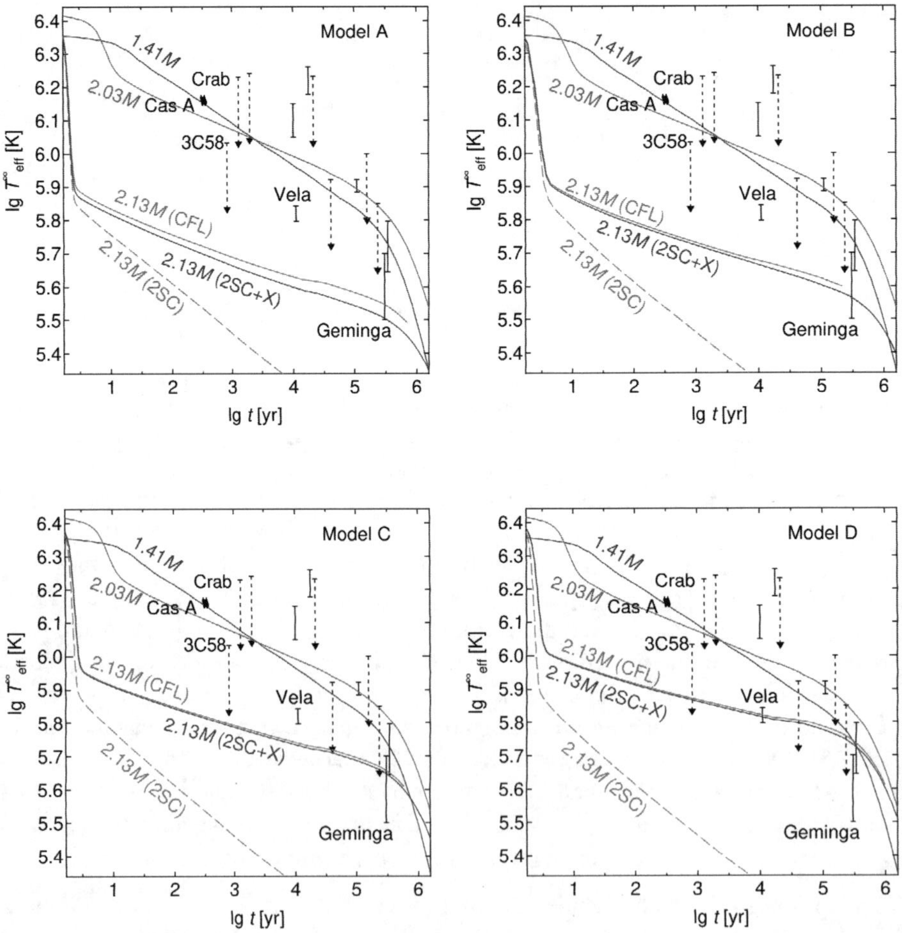

Figure 3. Cooling curves with various superfluid models. Blue (green) line denotes $1.41M_\odot$ ($2.03M_\odot$), respectively. Red solid (dashed) line denotes $2.13M_\odot$ with CSC paring of CFL (2SC), respectively. Purple line denotes $2.13M_\odot$ with 2SC+X model. Each superfluid models are shown on the right-top corners of each panels.

We show the cooling curves of neutron stars with superfluid model A, B, C, and D with ^{56}Fe surface in Figure 3, and the curves with ^{4}He surface in Figure 4. The curves of $2.13M_\odot$ show the difference of CSC: The red solid curves for CFL, the red dashed curves for 2SC, and the purple solid curves for 2SC+X. In each models, heavy stars cool faster than lighter stars in early stage ($t < 10^3$ yr), and the medium mass star stays warm in late stage($t > 10^3$ yr). Also, except the 2SC cases, the curves of $2.13M_\odot$ are sensitive to the superfluid model of neutron $^3P_2$ and they can explain cold neutron stars. The CSC paring affects to the cooling curves for the heaviest star model. With 2SC paring, the cooling curves locate low temperature region and they cannot match with observational data with the lower limit. With 2SC+X, the cooling curves are similar behaviour with CFL cases. Focusing on the observational data of Vela, the cooling curves of CFL and 2SC+X can cross the data, but 2SC model cannot reach the lower limit value.

The results show that the CSC model has an impact on the cooling of neutron stars. If the CSC state realized in neutron stars is the 2SC state, heavy stars with quarks show very fast cooling, which cannot explain the currently observed isolated neutron stars. In contrast, considering CFL state, the neutrino emission by quarks is strongly suppressed, and the DU process of hadronic phase becomes dominant. The cooling curves of CFL state, can pass the low temperature observations in the cooling diagram. In the case of CFL and 2SC+X, the superfluidity of neutrons $^3P_2$ has a significant influence on the cooling of neutron stars.

Considering CSC, there is one problem. The CFL paring suits for the cooling results which shows moderate strong cooling, but the appearing density is thought to be much higher density. On the other hand, the 2SC paring suits for the appearing density, but the cooling result shows too strong cooling for explaining the observation. The 2SC+X phase can solve this problem. The appearing density of 2SC+X suits for the density of neutron star, and it suppresses strong neutrino emission by quarks in all colours.

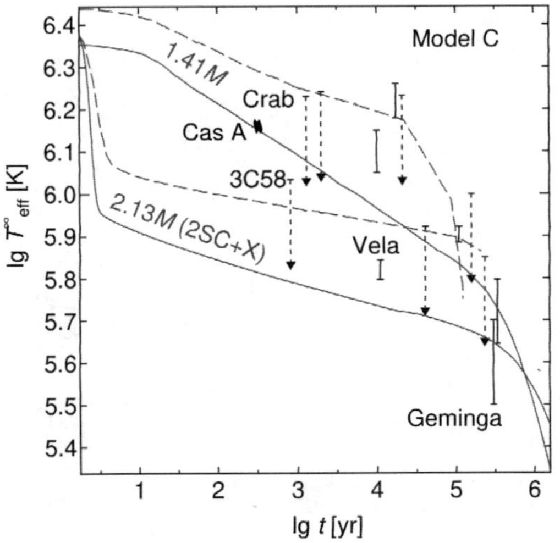

Figure 4. Cooling curves of $1.41M_\odot$ and $2.13M_\odot$ (2SC+X) with the light element surface, and with the Model C. Solid (dashed) curves denote ^{56}Fe (^{4}He) surface composition, respectively.

4 Summary

We have demonstrated the neutron star cooling with quark hadron continuity. Considering the continuity, the 2SC+X state can explain the cooling of neutron stars. Also we notice that the neutron $^3P_2$ superfluidity which is continued by the d-quarks affects the thermal history of neutron star cooling. This trend does not change even if the surface composition is modified.

We have calculated the cooling curves with light surface elements with the same model of the model C shown in Figure 4.

By the results of our calculation, the cold neutron star can be explained by CSC quark matter which causes strong neutrino emission. The neutrino emissivity is strong enough to drop the star temperature, but the appearance of CSC phase suppresses the emission rate and the cooling curves can pass the observed values. To consider neutron star cooling with quark phase, it is important to consider three "superstates" of neutrons, protons, and quarks. Also, the exotic states which we have not included, such as hyperon mixing or meson condensation, are possible candidates of strong cooling of neutron stars. Considering these states also require superfluid effect of corresponding particles. The study of the exotic states which realized in neutron stars should be discussed in the both of nuclear physics and astrophysics continuously.

References

[1] D. Page *et al.*, Astrophys. J. Suppl. **155**, 623 (2004).
[2] Y. Fujimoto *et al.*, Phys. Rev. D. **101**, 094009 (2020).
[3] N. Yasutake *et al.*, J. Phys. Conf. Ser. **665**, 012068 (2016).
[4] P. B. Demorest *et al.*, Nature **467**, 1081 (2010).
[5] J. Antoniadis *et al.*, Science **340**, 448 (2013).

Effect of the Nuclear Equation of State and Relativistic Turbulence on Core-Collapse Supernovae

Luca Boccioli[1,*], *Grant* Mathews[1], and *Evan* O'Connor[2]

[1]Center for Astrophysics, Department of Physics & Astronomy, University of Notre Dame, 225 Nieuwland Science Hall, Notre Dame, IN 46556, USA
[2]The Oskar Klein Centre, Department of Astronomy, Stockholm University, AlbaNova, SE-106 91 Stockholm, Sweden

Abstract. The nuclear Equation of State (EoS) is an important component in the evolution and subsequent explosion of core collapse supernovae. We make a survey of various equations of state that can be found in the literature and analyze their effect on the explosion. To simulate the supernovae, we use the general relativistic spherically-symmetric code GR1D, modified to take into account the effects of three-dimensional turbulence through a new mixing length theory approach (STIR). We show that the viability of the explosion is quite EoS dependent and that the strength of explosions correlate best with the central entropy density right after bounce and the onset of turbulent mixing in the proto-neutron star.

1 Introduction

Core-collapse supernovae have been one of the main focuses of computational astrophysics for more than 50 years. The tremendous advance in CPU technology enabled very detailed three-dimensional simulations to be developed, which greatly improved our knowledge of the explosion mechanism, but there are still many challenges left. One of the main uncertainties affecting the physics of supernovae is the Equation of State (EOS) of nuclear matter. Core-collapse supernovae generate thermodynamic conditions that can go from very low densities in the mantle (10^3 g cm^{-3}) to very high densities in the core (10^{15} g cm^{-3}) as well as all the values in between. At the same time, temperatures can also vary from 1 MeV to 100 MeV, while electron fractions can be as low as ~ 0.01 and as large as ~ 0.6. To adequately describe the explosion, a detailed knowledge of the EOS of matter across all these thermodynamic regimes is required. However, at high densities the EOS is still poorly constrained, especially when considering finite temperature effects [1–4].

Other important challenges that need to be addressed concern neutrinos and their interactions with matter. It is now quite well established that neutrino-delayed heating [5] is the main mechanism responsible for the explosion of massive stars. Therefore, a detailed knowledge of the thermodynamic dependence of neutrino-matter interactions, alongside realistic algorithms that describe the transport of neutrinos across the star, are required [6–8]. In this proceeding, however, we will only focus on the EOS.

*e-mail: lbocciol@nd.edu

EOS	Type	n_0 (fm^{-3})	J (MeV baryon^{-1})	K_0 (MeV baryon^{-1})	m_n^*/m_n -	$\mathrm{M_{max}}$ $\mathrm{M_\odot}$
SFHo	RMF	0.1583	31.57	245.40	0.7608	2.05
DD2	RMF	0.1491	32.73	242.70	0.5625	2.14
HShen	RMF	0.1455	36.95	281.00	0.6340	2.21
LS220	SRO	0.1549	28.61	219.85	1.0000	2.04
APR	SRO	0.1600	32.59	266.00	0.6985	2.19
KDE0v1	SRO	0.1646	34.58	227.53	0.7442	1.97
SLy4	SRO	0.1595	32.00	229.90	0.6952	2.05

Table 1. The different columns represent the nuclear saturation density (n_0) the symmetry energy at saturation density (J), the nuclear incompressibility (K_0), the nucleon effective mass (m_n^*) and the maximum mass of a neutron star ($\mathrm{M_{max}}$). The first four equations of state are calculated using RMF theory, while the last three were generated using the SRO code from [31].

In this work, we use the spherically symmetric model in full general relativity described in [9] to study the impact of the EOS on the explosion. This model can achieve an explosion by means of the inclusion of turbulent convection using Mixing-Length Theory (MLT). Specifically, it is a general relativistic version of STIR, first developed by [10]. In the past, several studies were carried out, analyzing the effect of the EOS on the dynamics of CCSNe (e.g. [11–29]). Here, we employ 7 different EOS, based on different microscopic models, and with those simulate four different progenitor models.

Our work was inspired by two recent papers [26, 28], that study the effect of the EOS on the structure of the PNS and on the explosion. Both papers focus on EOSs that were calculated using Skyrme energy density functionals, whereas our study also includes EOSs generated using Relativistic Mean Field (RMF) models. In these two studies, by varying several Skyrme parameters, they observed that the effective nucleon mass m_n^* is the parameter that best correlates with the strength of the explosion. In this work, however, we will study how EOSs generated using different microscopical theories affect the explosion. Natural units will be adopted throughout the manuscript, i.e. $G = c = \mathrm{M_\odot} = 1$.

2 Methods

2.1 Nuclear Equations of State

We used 7 different EOS to simulate 4 different progenitors with masses of 9, 15, 20 and 25 $\mathrm{M_\odot}$ from [30]. The EOS we chose are listed in table 1. In previous studies [26, 28] some of the nuclear parameters that can more significantly impact the explosion have been identified. In particular, the effective nucleon mass in symmetric matter at saturation density m_n^* was found to be the one that best correlates with the strength of the explosion. The details of the EOSs used in this study is summarized below:

- The SFHo and DD2 EOSs are taken from [32], which uses a Relativistic Mean Field (RMF) approach, with the parametrization of nucleon interactions from [15] and [33], respectively. Nuclei near saturation density are treated using an NSE approach, with several thousands nuclei taken into account.

- The HShen EOS from [34] is based on RMF theory. The difference with the SFHo and DD2 EOS is that the HShen uses the Single Nucleus Approximation (SNA), based upon a Thomas-Fermi model, to describe nuclei near nuclear saturation density.
- The APR EOS [35, 36]: At high densities it uses realistic nuclear potentials resulting from fits to nucleon-nucleon scattering. It also includes a phase transition to a neutral pion condensate. It is then connected to a network of ~3300 nuclei in Nuclear Statistical Equilibrium (NSE) at low densities using the SRO code [31].
- The LS220 [11], KDE0v1 [37] and SLy4 [38]: These are based on Skyrme parameterizations as described in [31]. Additionally, nuclei at densities near nuclear saturation density are described using the compressible liquid drop theory, in the same fashion as [11]. These EOS are then connected to a network of ~3300 nuclei in Nuclear Statistical Equilibrium (NSE) at low densities using the SRO code [31].

The last four EOSs were taken from the SROEOS code of [31]. Hence, at densities below nuclear saturation density, they are described using a compressible liquid-drop model at finite temperature in which heavy nuclei are described using SNA. We therefore divide these EOSs in two categories: RMF-type (the first three) and SRO-type (the last four). The gravitational mass versus radius relationships for cold neutron can be seen in Figure 1, together with some observational constraints on the mass and radius of Neutron Stars. Despite the fact that some of them are not consistent with observation constraints, these EOSs represent a reasonable catalogue of possible equations of state.

2.2 Supernova model

One of the mechanisms that can more significantly impact the delayed-neutrino heating mechanism is neutrino-driven turbulent convection, as many previous studies have shown [46–49]. The reason is that material behind the expanding shock is heated by neutrinos escaping the proto-neutron star, and turbulent convection can increase the time spent by the material in the gain region, increasing the probability of interaction with neutrinos. Hence, neutrino-driven turbulence is a key component of the explosion mechanism of supernovae.

To analyze how different EOSs affect the explosion, we use the model described in [9], which uses a general relativistic version STIR [10] to include neutrino-driven turbulent convection in the spherically symmetric code `GR1D`. STIR includes the effects of turbulent convection in spherically symmetric simulations using a time-dependent Mixing-Length Theory (MLT) approach. The energy coming from turbulent convection can therefore trigger an explosion even in spherically symmetric simulations. We now briefly summarize the main features of STIR, but more details can be found in [49], [10] and [9]. The metric used in `GR1D` is:

$$
\begin{aligned}
ds^2 &= g_{\mu\nu} x^\mu x^\nu \\
&= -\alpha(r,t)^2 dt^2 + X(r,t)^2 dr^2 + r^2 d\Omega^2 \,,
\end{aligned} \tag{1}
$$

where α is the lapse function and X is a function of the gravitational mass at radius r.

STIR adds an extra equation to the standard hydrodynamic ones solved in `GR1D`, to account for the time evolution of the turbulent energy:

$$
\begin{aligned}
&\frac{\partial D v_{\rm turb}^2}{\partial t} + \frac{1}{r^2}\frac{\partial}{\partial r}\left[\frac{\alpha r^2}{X} D (v_{\rm turb}^2 v - D_{\rm K} \nabla v_{\rm turb}^2)\right] \\
&= -\alpha X \left(\rho v_{\rm turb}^2 \frac{\partial v}{\partial r} + \rho v_{\rm turb} \omega_{\rm BV}^2 \Lambda_{\rm mix} - \rho \frac{v_{\rm turb}^3}{\Lambda_{\rm mix}} \right) ,
\end{aligned} \tag{2}
$$

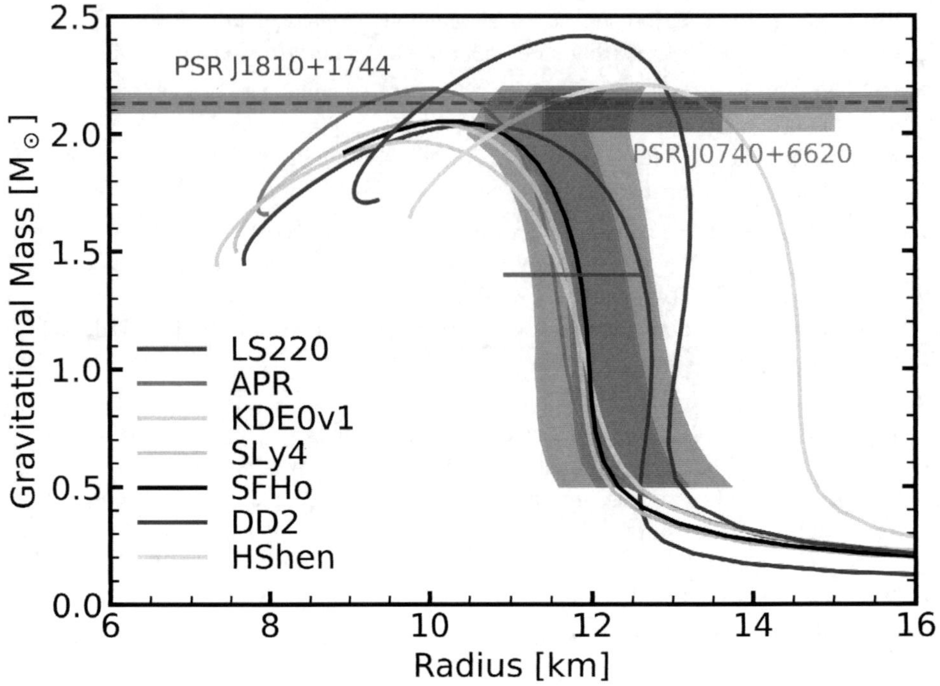

Figure 1. Gravitational mass vs radius, from the various equations of state adopted in the present work. The green shaded region shows the NS mass-radius constraints from model A of [39]. The grey band represents the mass and radius of the millisecond pulsar PSR J0740+6620. The mass has been measured by [40] and [41], with a value of 2.08 ± 0.07 $M_\odot$. The radius has been independently estimated by [42] and [43] using NICER data. Here we show the latter, where the lighter grey region represents a 95% confidence level, corresponding to a radius of $12.39^{+2.63}_{-1.68}$ km, while the darker region shows the 1σ confidence level, corresponding to a radius of $12.39^{+1.30}_{-0.98}$ km. The blue dotted line is the mass of PSR J1810+1744 measured by [44], and the shaded blue region represents the 1σ confidence level. The brown horizontal band shows the constraint on the radius of a 1.4 $M_\odot$ neutron star from [45].

where $D = X\rho W$ is the conserved density, $W = (1 - v^2)^{-1/2}$ is the Lorentz factor and $v_{\rm turb}$ is the turbulent velocity. The velocity is defined as $v = Xv^r$, where v_r is the coordinate velocity. Finally, $D_{\rm K} = \alpha_{\rm K} v_{\rm turb} \Lambda_{\rm mix}$ is the diffusion coefficient. The most important quantities of this model are the mixing length

$$\Lambda_{\rm mix} = \alpha_{\rm MLT} \frac{P}{\rho \, {\rm d}\phi/{\rm d}r} , \tag{3}$$

and the Brunt-Väisälä frequency

$$\omega_{\rm BV}^2 = \frac{\alpha^2}{\rho h X^2} \left(\frac{{\rm d}\phi}{{\rm d}r} - v \frac{\partial v}{\partial r} \right) \left(\frac{\partial \rho (1+\epsilon)}{\partial r} - \frac{1}{c_s^2} \frac{\partial P}{\partial r} \right) , \tag{4}$$

where $\phi = \ln \alpha$ reduces to the Newtonian potential in the non-relativistic limit and c_s is the sound speed. The quantity $h = 1 + \epsilon + P/\rho$ represents the relativistic enthalpy, where P is the pressure and ϵ is the internal energy.

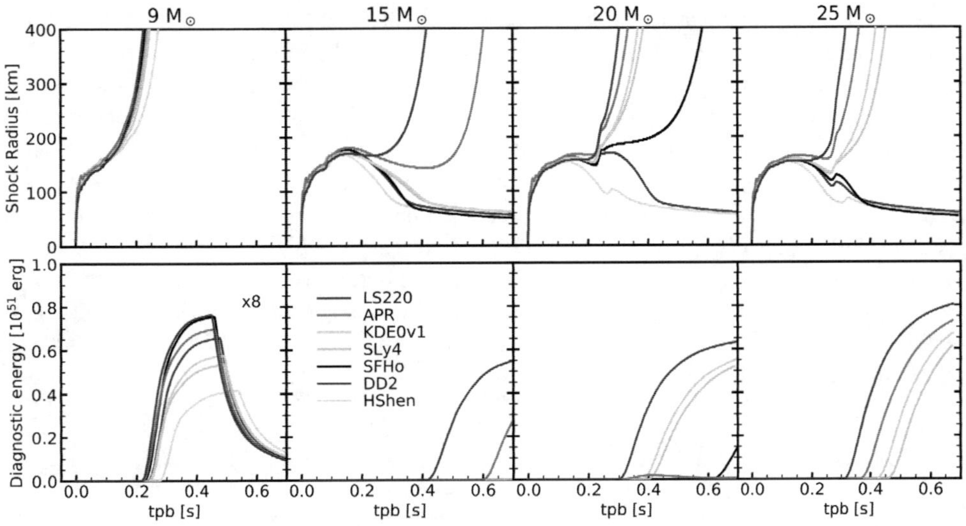

Figure 2. The upper panels display shock radius vs time for different progenitors and EOSs, while the lower panels display the diagnostic energy vs time. Notice that for the 9 $M_{\odot}$ progenitor the diagnostic energies have been multiplied by a factor of 8 so that they could be more clearly shown on the same scale used for the other progenitors.

The parameters controlling the strength of the explosion are: $\alpha_{\rm MLT}$ and $\alpha_{\rm K}$. The first one controls the magnitude of the mixing length, whereas the second one controls the magnitude of the diffusion coefficients $D_{\rm K}$. We have not shown all the hydrodynamic equations here, but it's important to point out that other diffusion coefficients, analogous to $D_{\rm K}$, are added to the evolution equations for neutrino energy density, internal energy and electron fraction. The total number of parameters in the model is therefore 5. However, we fix $\alpha_{\rm K}$ and the other parameters governing diffusion to 1/6, consistently with the choices of [50] and [10]. The only parameter we vary is $\alpha_{\rm MLT}$.

The simulations carried out for this work were run using a modified version of `GR1D` [51, 52], which is based on spherically symmetric hydrodynamics and radiation transport in full General Relativity (GR). Neutrino opacitiy tables were generated using the open-source code `NuLib` [52].

3 Results

Given the high densities reached inside the PNS, the dynamics of supernovae can be greatly impacted by the EOS. To illustrate this, we show some of the most significant thermodynamic quantities at the center of the PNS in Figure 3, for different progenitors and EOSs. As one can clearly see, central density and temperature can vary significantly as a function of EOS, but they do not correlate with the strength of explosion. On the contrary, the central entropy in the first 50 ms after bounce correlates extremely well with the strength of the explosion, even though the range spanned by different EOSs is not as large as for the other two quantities. One can then conclude that the hierarchy of central entropies follows the hierarchy of explosion strength, with the only exception of the SLy4 and KDE0v1 EOSs, which will be discussed below.

A similar correlation was found by the studies of [26] and [28]. In both papers, they only focused on one type of EOS which, at high densities, is based on Skyrme interactions between nucleons. By doing this, they were able to vary different nuclear parameters — such as the effective mass, the symmetry energy, the incompressibility and others — while keeping all the other ones fixed. This approach allowed them to observe the impact of different properties of the EOS on the explosion. Both analyses revealed that the quantity responsible for the greatest change in PNS structure is the effective mass of nuclei at saturation density m_n^*. This translates into a change in the dynamics of explosion, which depends on how fast the PNS contraction is [28], since it changes the location and thermodynamic properties of the neutrinospheres for all flavors [26]. As a side note, they also pointed out that a change in the effective mass corresponds to a change in the central entropy: larger effective masses lead to larger central entropies.

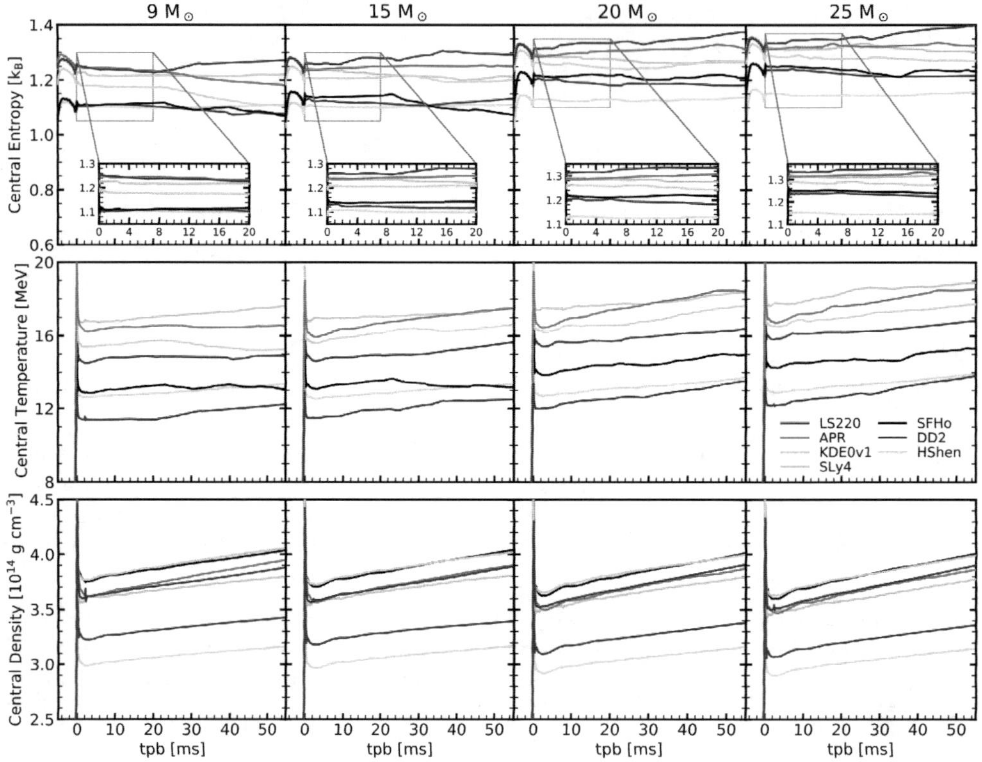

Figure 3. The first three rows show the main central thermodynamic variables for different progenitors (columns) and EOSs (colors) as a function of time post bounce. The last row shows the ratio of the nucleon effective mass to the nucleon mass calculated for the central values of densities and temperatures as a function of time post bounce. The central entropy at early times tpb $\lesssim$ 5 ms is larger for EOSs that yield stronger explosions. See the text for a more detailed discussion.

We complement these studies by including not only EOSs based on Skyrme interactions, such as the LS220, KDE0v1, and SLy4, but also the APR EOS as well as EOSs based on RMF theory. Since the finite-temperature part of the APR EOS was constructed using the SRO-code [31], which was also used to generate the Skyrme-type EOSs, they share the same properties near saturation density. Hence, we consider them to be "SRO-type". The difference between the APR and the Skyrme-type EOSs is that the nuclear potentials for the Skyrme-type EOSs are derived from the energy density of nuclear matter, whereas for the APR they are derived from nucleon-nucleon scattering. The EOSs calculated using RMF theory are the SFHo, DD2 and HShen. The first two use an NSE treatment of nuclei near saturation density, and the HShen uses a Thomas-Fermi description of nuclei instead. In the following paragraphs we will investigate the differences among these frameworks used to calculate the nuclear EOS.

We begin our analysis by comparing the EOSs calculated using Skyrme interactions (i.e. LS220, KDE0v1 and SLy4). One can easily notice that the effective mass correlates with the strength of explosion. If however one compares all of the EOSs considered in this study, this correlation breaks down. Also, despite the fact that the APR uses the same treatment of nuclei near saturation density as the Skyrme-type EOSs, it does not follow the correlation between strength of explosion and effective mass. This, however, is not completely unexpected since the effective mass in the APR EOS has a more complicated density dependence [36]. Overall, this shows that the effective mass is not the only parameter that affects the strength of explosion. Interestingly, the only quantity that correlates with the strength of the explosion, regardless of the framework used to calculate the nuclear EOS, is the central entropy.

As mentioned above, KDE0v1 and SLy4 don't follow this correlation between strength of explosion and central entropy. From Figure 2, one can see that shock trajectories and explosion energies for these two EOSs are very similar across progenitors, with KDE0v1 yielding slightly stronger explosions. From the top panel of Figure 3, however, one can see that KDE0v1 gives a slightly smaller central entropy than SLy4. Hence, this shows that nuclear parameters other than m_n^* can also change the dynamics of the explosion. This can be seen by considering a Fermi liquid theory, in which the central entropy is given by $S_c \sim m_n^* T_c/\rho_c^{2/3}$ [53], where ρ_c, T_c and S_c are the central density, temperature and entropy, respectively. KDE0v1 has a larger saturation density, which increases the central density, and therefore lowers the central entropy. At the same time, it has a larger nucleon effective mass m_n^*, which increases the central entropy. These two competing effects slightly disrupt the hierarchy of central entropy. However, in this study the EOSs differ for several nuclear parameters, and therefore one doesn't expect a perfect correlation.

4 Conclusions

In this proceeding we have shown how the EOS affects the explosion of CCSNe using spherically symmetric, general relativistic simulations with a parametric treatment of turbulent convection.

We highlight a remarkable correlation between the central entropy immediately after bounce the strength of the explosion. This is in agreement with previous studies by [26] and [28], carried out by considering only Skyrme-type EOSs. In these works, they concluded that the effective nucleon mass is the nuclear parameter that most strongly correlates with the explosion. Our results using the three Skyrme EOSs considered in this work confirm this.

In addition to Skyrme-type EOSs we have also taken into account the APR EOS and three RMF-type EOSs. This breaks down the correlation between strength of explosion and effective nucleon mass. Because of that, the correlation between effective nucleon mass and central entropy reported in those studies [26, 28] is also no longer present in our simulations. However, we do obtain a remarkable correlation between central entropy and strength of the explosion. One can therefore conclude that the central entropy immediately after bounce is the best indicator of explodability, and it therefore plays an important role in determining the strength of the explosion. Finally, since SRO-type and RMF-type EOSs do not follow the same correlation between effective mass and strength of explosion, we conclude that the approach used to calculate the EOS, and more specifically the treatment of nuclear matter near saturation density, has a significant impact on the explosion of CCSNe.

Acknowledgments

The authors would like to thank Andre da Silva Schneider for his help in calculating the symmetry energy of the EOSs adopted. Work at the University of Notre Dame supported by the U.S. Department of Energy under Nuclear Theory Grant DE-FG02-95-ER40934. EOC would like to acknowledge Vetenskapsrådet (the Swedish Research Council) for supporting this work under award numbers 2018-04575 and 2020-00452.

References

[1] T. Klähn, D. Blaschke, S. Typel, E.N.E. van Dalen, A. Faessler, C. Fuchs, T. Gaitanos, H. Grigorian, A. Ho, E.E. Kolomeitsev et al., PhRvC **74**, 035802 (2006)
[2] K. Hebeler, J.M. Lattimer, C.J. Pethick, A. Schwenk, ApJ **773**, 11 (2013), 1303.4662
[3] N.B. Zhang, B.A. Li, J. Xu, ApJ **859**, 90 (2018), 1801.06855
[4] G.F. Burgio, H.J. Schulze, I. Vidaña, J.B. Wei, Progress in Particle and Nuclear Physics **120**, 103879 (2021), 2105.03747
[5] H.A. Bethe, J.R. Wilson, ApJ **295**, 14 (1985)
[6] M. Liebendörfer, M. Rampp, H.T. Janka, A. Mezzacappa, ApJ **620**, 840 (2005), astro-ph/0310662
[7] C. Kato, H. Nagakura, Y. Hori, S. Yamada, ApJ **897**, 43 (2020), 2001.11148
[8] A. Mezzacappa, E. Endeve, O.E.B. Messer, S.W. Bruenn, Living Reviews in Computational Astrophysics **6**, 4 (2020), 2010.09013
[9] L. Boccioli, G.J. Mathews, E.P. O'Connor, ApJ **912**, 29 (2021), 2102.06767
[10] S.M. Couch, M.L. Warren, E.P. O'Connor, The Astrophysical Journal **890**, 127 (2020)
[11] J.M. Lattimer, D.F. Swesty, NuPhA **535**, 331 (1991)
[12] E. O'Connor, C.D. Ott, ApJ **730**, 70 (2011), 1010.5550
[13] M. Hempel, T. Fischer, J. Schaffner-Bielich, M. Liebendörfer, ApJ **748**, 70 (2012), 1108.0848
[14] H.T. Janka, Annual Review of Nuclear and Particle Science **62**, 407 (2012), 1206.2503
[15] A.W. Steiner, M. Hempel, T. Fischer, ApJ **774**, 17 (2013), 1207.2184
[16] S.M. Couch, ApJ **765**, 29 (2013), 1206.4724
[17] Y. Suwa, T. Takiwaki, K. Kotake, T. Fischer, M. Liebendörfer, K. Sato, ApJ **764**, 99 (2013), 1206.6101
[18] T. Fischer, M. Hempel, I. Sagert, Y. Suwa, J. Schaffner-Bielich, European Physical Journal A **50**, 46 (2014), 1307.6190
[19] P. Char, S. Banik, D. Bandyopadhyay, ApJ **809**, 116 (2015), 1508.01854
[20] J.P. Olson, M. Warren, M. Meixner, G.J. Mathews, N.Q. Lan, H.E. Dalhed, arXiv e-prints arXiv:1612.08992 (2016), 1612.08992
[21] S. Furusawa, H. Togashi, H. Nagakura, K. Sumiyoshi, S. Yamada, H. Suzuki, M. Takano, Journal of Physics G Nuclear Physics **44**, 094001 (2017), 1707.06410
[22] S. Richers, C.D. Ott, E. Abdikamalov, E. O'Connor, C. Sullivan, PhRvD **95**, 063019 (2017), 1701.02752
[23] H. Nagakura, W. Iwakami, S. Furusawa, H. Okawa, A. Harada, K. Sumiyoshi, S. Yamada, H. Matsufuru, A. Imakura, ApJ **854**, 136 (2018), 1702.01752

[24] V. Morozova, D. Radice, A. Burrows, D. Vartanyan, ApJ **861**, 10 (2018), 1801.01914
[25] A. Burrows, D. Vartanyan, J.C. Dolence, M.A. Skinner, D. Radice, SSRv **214**, 33 (2018), 1611.05859
[26] A.S. Schneider, L.F. Roberts, C.D. Ott, E. O'Connor, Phys. Rev. C **100**, 055802 (2019)
[27] A. Harada, H. Nagakura, W. Iwakami, H. Okawa, S. Furusawa, K. Sumiyoshi, H. Matsufuru, S. Yamada, ApJ **902**, 150 (2020), 2003.08630
[28] H. Yasin, S. Schäfer, A. Arcones, A. Schwenk, PhRvL **124**, 092701 (2020)
[29] S. Ghosh, N. Wolfe, C. Fröhlich, arXiv e-prints arXiv:2107.13016 (2021), 2107.13016
[30] T. Sukhbold, T. Ertl, S.E. Woosley, J.M. Brown, H.T. Janka, The Astrophysical Journal **821**, 38 (2016)
[31] A.S. Schneider, L.F. Roberts, C.D. Ott, Phys. Rev. C **96**, 065802 (2017)
[32] M. Hempel, J. Schaffner-Bielich, NuPhA **837**, 210 (2010), 0911.4073
[33] S. Typel, G. Röpke, T. Klähn, D. Blaschke, H.H. Wolter, PhRvC **81**, 015803 (2010), 0908.2344
[34] H. Shen, H. Toki, K. Oyamatsu, K. Sumiyoshi, ApJS **197**, 20 (2011), 1105.1666
[35] A. Akmal, V.R. Pandharipande, D.G. Ravenhall, PhRvC **58**, 1804 (1998), nucl-th/9804027
[36] A.S. Schneider, C. Constantinou, B. Muccioli, M. Prakash, PhRvC **100**, 025803 (2019), 1901.09652
[37] B.K. Agrawal, S. Shlomo, V.K. Au, PhRvC **72**, 014310 (2005), nucl-th/0505071
[38] E. Chabanat, P. Bonche, P. Haensel, J. Meyer, R. Schaeffer, NuPhA **627**, 710 (1997)
[39] J. Nättilä, A.W. Steiner, J.J.E. Kajava, V.F. Suleimanov, J. Poutanen, A&A **591**, A25 (2016), 1509.06561
[40] H.T. Cromartie, E. Fonseca, S.M. Ransom, P.B. Demorest, Z. Arzoumanian, H. Blumer, P.R. Brook, M.E. DeCesar, T. Dolch, J.A. Ellis et al., Nature Astronomy **4**, 72 (2020), 1904.06759
[41] E. Fonseca, H.T. Cromartie, T.T. Pennucci, P.S. Ray, A.Y. Kirichenko, S.M. Ransom, P.B. Demorest, I.H. Stairs, Z. Arzoumanian, L. Guillemot et al., ApJL **915**, L12 (2021), 2104.00880
[42] M.C. Miller, F.K. Lamb, A.J. Dittmann, S. Bogdanov, Z. Arzoumanian, K.C. Gendreau, S. Guillot, W.C.G. Ho, J.M. Lattimer, M. Loewenstein et al., arXiv e-prints arXiv:2105.06979 (2021), 2105.06979
[43] T.E. Riley, A.L. Watts, P.S. Ray, S. Bogdanov, S. Guillot, S.M. Morsink, A.V. Bilous, Z. Arzoumanian, D. Choudhury, J.S. Deneva et al., ApJL **918**, L27 (2021), 2105.06980
[44] R.W. Romani, D. Kandel, A.V. Filippenko, T.G. Brink, W. Zheng, ApJL **908**, L46 (2021), 2101.09822
[45] T. Dietrich, M.W. Coughlin, P.T.H. Pang, M. Bulla, J. Heinzel, L. Issa, I. Tews, S. Antier, Science **370**, 1450 (2020), 2002.11355
[46] S.M. Couch, C.D. Ott, ApJ **799**, 5 (2015), 1408.1399
[47] D. Radice, C.D. Ott, E. Abdikamalov, S.M. Couch, R. Haas, E. Schnetter, ApJ **820**, 76 (2016), 1510.05022
[48] D. Radice, E. Abdikamalov, C.D. Ott, P. Mösta, S.M. Couch, L.F. Roberts, Journal of Physics G Nuclear Physics **45**, 053003 (2018), 1710.01282
[49] Q.A. Mabanta, J.W. Murphy, ApJ **856**, 22 (2018), 1706.00072
[50] B. Müller, M. Viallet, A. Heger, H.T. Janka, ApJ **833**, 124 (2016), 1605.01393
[51] E. O'Connor, C.D. Ott, Classical and Quantum Gravity **27**, 114103 (2010), 0912.2393
[52] E. O'Connor, ApJS **219**, 24 (2015), 1411.7058

[53] G. Baym, C. Pethick, *Landau Fermi-Liquid Theory: Concepts and Applications* (John Wiley & Sons, Ltd, 1991)

EPJ Web of Conferences **260**, 11026 (2022) https://doi.org/10.1051/epjconf/202226011026
NIC-XVI

First ^{80}Se(n,γ) cross section measurement with high resolution in the full stellar energy range 1 eV - 100 keV and its astrophysical implications for the s-process

V. Babiano-Suarez[1], J. Balibrea-Correa[1], L. Caballero-Ontanaya[1], C. Domingo-Pardo[1], I. Ladarescu[1], J. Lerendegui-Marco[2], J. L. Tain[1], F. Calviño[18], A. Casanovas[18], A. Tarifeño-Saldivia[18], C. Guerrero[2], O. Aberle[3], V. Alcayne[4], S. Amaducci[5,6], J. Andrzejewski[7], L. Audouin[8], M. Bacak[3,9,10], M. Barbagallo[3,11], S. Bennett[12], E. Berthoumieux[10], J. Billowes[12], D. Bosnar[13], A. Brown[14], M. Busso[15,16], M. Caamaño[17], M. Calviani[3], D. Cano-Ott[4], F. Cerutti[3], E. Chiaveri[3,12], N. Colonna[11], G. Cortés[18], M. A. Cortés-Giraldo[2], L. Cosentino[5], S. Cristallo[15,19], L. A. Damone[11,20], P. J. Davies[12], M. Diakaki[21,3], M. Dietz[22], R. Dressler[23], Q. Ducasse[24], E. Dupont[10], I. Durán[17], Z. Eleme[25], B. Fernández-Domínguez[17], A. Ferrari[3], P. Finocchiaro[5], V. Furman[26], K. Göbel[27], R. Garg[22], A. Gawlik-Ramięga[7], S. Gilardoni[3], I. F. Gonçalves[28], E. González-Romero[4], F. Gunsing[10], H. Harada[29], S. Heinitz[23], J. Heyse[30], D. G. Jenkins[14], A. Junghans[31], F. Käppeler[32], Y. Kadi[3], A. Kimura[29], I. Knapová[33], M. Kokkoris[21], Y. Kopatch[26], M. Krtička[33], D. Kurtulgil[27], C. Lederer-Woods[22], H. Leeb[9], S. J. Lonsdale[22], D. Macina[3], A. Manna[34,35], T. Martínez[4], A. Masi[3], C. Massimi[34,35], P. Mastinu[36], M. Mastromarco[3], E. A. Maugeri[23], A. Mazzone[11,37], E. Mendoza[4], A. Mengoni[38], V. Michalopoulou[21,3], P. M. Milazzo[39], F. Mingrone[3], J. Moreno-Soto[10], A. Musumarra[5,40], A. Negret[41], R. Nolte[24], F. Ogállar[42], A. Oprea[41], N. Patronis[25], A. Pavlik[43], J. Perkowski[7], L. Persanti[11,15,19], C. Petrone[41], E. Pirovano[24], I. Porras[42], J. Praena[42], J. M. Quesada[2], D. Ramos-Doval[8], T. Rauscher[44,45], R. Reifarth[27], D. Rochman[23], Y. Romanets[28], C. Rubbia[3], M. Sabaté-Gilarte[2,3], A. Saxena[46], P. Schillebeeckx[30], D. Schumann[23], A. Sekhar[12], A. G. Smith[12], N. V. Sosnin[12], P. Sprung[23], A. Stamatopoulos[21], G. Tagliente[11], L. Tassan-Got[3,21,8], Th. Thomas[27], P. Torres-Sánchez[42], A. Tsinganis[3], J. Ulrich[23], S. Urlass[31,3], S. Valenta[33], G. Vannini[34,35], V. Variale[11], P. Vaz[28], A. Ventura[34], D. Vescovi[15], V. Vlachoudis[3], R. Vlastou[21], A. Wallner[47], P. J. Woods[22], T. Wright[12], and P. Žugec[13]
and the n_TOF Collaboration

[1]Instituto de Física Corpuscular, CSIC - Universidad de Valencia, Spain
[2]Universidad de Sevilla, Spain
[3]European Organization for Nuclear Research (CERN), Switzerland
[4]Centro de Investigaciones Energéticas Medioambientales y Tecnológicas (CIEMAT), Spain
[5]INFN Laboratori Nazionali del Sud, Catania, Italy
[6]Dipartimento di Fisica e Astronomia, Università di Catania, Italy
[7]University of Lodz, Poland
[8]Institut de Physique Nucléaire, CNRS-IN2P3, Univ. Paris-Sud, Université Paris-Saclay, F-91406 Orsay Cedex, France
[9]TU Wien, Atominstitut, Stadionallee 2, 1020 Wien, Austria
[10]CEA Irfu, Université Paris-Saclay, F-91191 Gif-sur-Yvette, France
[11]Istituto Nazionale di Fisica Nucleare, Sezione di Bari, Italy
[12]University of Manchester, United Kingdom
[13]Department of Physics, Faculty of Science, University of Zagreb, Zagreb, Croatia

[14]University of York, United Kingdom
[15]Istituto Nazionale di Fisica Nucleare, Sezione di Perugia, Italy
[16]Dipartimento di Fisica e Geologia, Università di Perugia, Italy
[17]University of Santiago de Compostela, Spain
[18]Universitat Politècnica de Catalunya, Spain
[19]Istituto Nazionale di Astrofisica - Osservatorio Astronomico di Teramo, Italy
[20]Dipartimento Interateneo di Fisica, Università degli Studi di Bari, Italy
[21]National Technical University of Athens, Greece
[22]School of Physics and Astronomy, University of Edinburgh, United Kingdom
[23]Paul Scherrer Institut (PSI), Villigen, Switzerland
[24]Physikalisch-Technische Bundesanstalt (PTB), Bundesallee 100, 38116 Braunschweig, Germany
[25]University of Ioannina, Greece
[26]Joint Institute for Nuclear Research (JINR), Dubna, Russia
[27]Goethe University Frankfurt, Germany
[28]Instituto Superior Técnico, Lisbon, Portugal
[29]Japan Atomic Energy Agency (JAEA), Tokai-Mura, Japan
[30]European Commission, Joint Research Centre (JRC), Geel, Retieseweg 111, B-2440 Geel, Belgium
[31]Helmholtz-Zentrum Dresden-Rossendorf, Germany
[32]Karlsruhe Institute of Technology, Campus North, IKP, 76021 Karlsruhe, Germany
[33]Charles University, Prague, Czech Republic
[34]Istituto Nazionale di Fisica Nucleare, Sezione di Bologna, Italy
[35]Dipartimento di Fisica e Astronomia, Università di Bologna, Italy
[36]Istituto Nazionale di Fisica Nucleare, Sezione di Legnaro, Italy
[37]Consiglio Nazionale delle Ricerche, Bari, Italy
[38]Agenzia nazionale per le nuove tecnologie (ENEA), Bologna, Italy
[39]Istituto Nazionale di Fisica Nucleare, Sezione di Trieste, Italy
[40]Dipartimento di Fisica e Astronomia, Università di Catania, Italy
[41]Horia Hulubei National Institute of Physics and Nuclear Engineering, Romania
[42]University of Granada, Spain
[43]University of Vienna, Faculty of Physics, Vienna, Austria
[44]Department of Physics, University of Basel, Switzerland
[45]Centre for Astrophysics Research, University of Hertfordshire, United Kingdom
[46]Bhabha Atomic Research Centre (BARC), India
[47]Australian National University, Canberra, Australia

Abstract. Most elements heavier than iron have been generated in the stellar media by means of neutron capture reactions, approximately half are produced by the slow neutron capture or *s*-process. Radiative neutron capture cross section measurements are of fundamental importance for the study of this mechanism. In this contribution we present a brief summary on the measurement and results for the ^{80}Se(n,γ) cross-section. The experiment was carried out at CERN n_TOF EAR1 via the time of flight (ToF) technique, using four C_6D_6 scintillation detectors with very fast response. More than a hundred new resonances have been analyzed for the first time with a high accuracy. The MACS obtained at kT = 8 keV is 36% smaller than the recommended value in KADoNiS. Some of the astrophysical implications of this result are elucidated in this contribution.

1 Introduction

The *s*-process is one of the main nucleosynthesis mechanisms proposed to explain the abundance curve of elements [1]. For the nucleosynthesis of elements with mass $A > 90$, the *s*-process is ascribed to low mass stars during their AGB evolutionary phase, in which neutrons are released by the reactions ^{22}Ne(α,n)^{25}Mg and ^{13}C(α,n)^{16}O corresponding to He-flash and ^{13}C-pocket, respectively [2]. For the mass region $60 < A < 90$ the weak *s*-process was introduced [1]. This additional component takes place in massive stars where neutrons are released by the reactions ^{22}Ne(α,n)^{25}Mg and ^{12}C(^{12}C,n)^{23}Mg during the core-He and shell-C burning stages [2]. For both main and weak *s*-process components the relatively low neutron flux allows unstable nuclei produced to β-decay before capturing another neutron. In some specific cases where the beta-decay half-life spans between weeks and several years, the unstable nucleus can capture another neutron before decaying, inducing a split in the *s*-process path. In these branching-point nuclei temperature and neutron density play an important role in shaping the final isotopic-abundance pattern. Because of the branching sensitivity to the varying physical conditions and evolutionary time-scales, stellar models [3–7] are required for a realistic assessment of the *s*-process production yields, based on the neutron capture cross sections and β-decay rates of all isotopes involved.

^{79}Se (3.27×10^5 y [8]) is one of the most relevant *s*-process branching [9] isotopes, for it allows to constrain the thermal conditions during the last evolutionary stage of massive-stars, before they explode as supernova. The thermal sensitivity of the nucleosynthesis around ^{79}Se arises from a few quantum levels at low excitation energy that are populated in the stellar media. The comparison between the abundances observed for the *s*-only isotopes of Kr [10], and those predicted by the stellar models, will give information about the thermal conditions of the stellar media. To this aim one needs to know both, the cross-section of the branching nucleus ^{79}Se, whose measurement is scheduled for 2022 [11], and also the cross sections of the neighbouring stable nuclei. In this context, we carried out the neutron capture cross section measurement of the ^{80}Se isotope using the ToF technique, at CERN n_TOF [12], under the framework of the ERC-funded project HYMNS [13]. Experimentally, there existed only one previous measurement on ^{80}Se using the ToF technique [14]. However, the latter had a poor neutron-energy resolution and a rather limited neutron energy range. These drawbacks have been remarkably improved with the present high resolution measurement, which covers the entire energy range of astrophysical interest between 1 eV and 100 keV.

2 Experimental setup and measuring technique

At CERN n_TOF [15] neutrons are produced via spallation reactions of a proton beam impinging on a lead block. 7×10^{12} protons reach the spallation target every proton-pulse, thereby generating $\sim$ 300 neutrons per incident proton. The outcoming neutrons pass through a layer of borated water which moderates their energy leading to a white neutron flux covering the range from meV to GeV in each single shot. Charged particles are removed from the beam using a magnet, whereas the rest of particles are collimated into a beam by two different collimators along the beam line. Neutrons travel $\sim$185 m to reach the experimental area EAR1, where the pulsed beam shows a Gaussian profile of 18 mm FWHM at the sample position [16]. In addition to neutrons, a flux of γ-rays generated by spallation- and neutron captures in the moderator, reaches the experimental hall. The prompt γ-rays is called γ-flash and saturates the response of all radiation detectors for few μs with a very steep rise-edge. This is utilized to determine the time reference for determining the ToF of each neutron bunch. Therefore, the kinetic energy in eV of a neutron with a measured ToF t_{ToF} (μs) is given by Eq. 1, in which L is the flight distance in m. The short bunch width of 6 ns RMS and the long flight distance ensure an excellent neutron-energy resolution.

$$E_n = \left(\frac{72.2977L}{t_{ToF}}\right)^2 \tag{1}$$

Table 1. Disk shaped samples of 20 mm diameter used during the ^{80}Se(n,γ) ToF experiment. Dummy is not included.

Sample	Pureness (%)	Mass (g)	Thickness (mm)
^{80}Se	99.87(10)	2.965(5)	3.50(1)
^{197}Au	99.99	0.600(5)	0.10(1)
^{nat}Pb	100	7.300(5)	2.10(1)

Fig. 1 shows a photograph of the experimental setup consisting in four C_6D_6 scintillation detectors [17] placed at 125° with respect to the beam line, and their faces pointing at the capture sample from ~10 cm distance. These detectors are built in carbon fiber to minimize contaminant neutron captures in structural materials, an effect known as intrinsic neutron sensitivity. Each detector contains approximately 1 l volume of C_6D_6 scintillation liquid, which is coupled to a traditional PMT to give a very fast response. Their low efficiency allows one to apply the Pulse-Height Weighting Technique [18, 19].

Figure 1. General view of experimental setup mounted in EAR1 for the measurement of the ^{80}Se(n,γ) cross section.

The samples used for this experiment are listed in Tab. 1. Apart from the selenium sample under study, gold was measured to normalize the obtained ^{80}Se(n,γ) yield by means of the Saturated Resonance Method (SRM) [20]. The ancillary lead sample was measured to estimate the background contribution corresponding to neutrons scattered by the sample and captured in the structural material of the experimental hall [21]. In addition, a replica of the ^{80}Se sample without the sample itself, called empty sample, was measured to estimate the background component unrelated to the sample and caused by neutrons captured in the surroundings.

3 Data analysis

The measured count-rate as a function of the ToF was transformed into a neutron capture yield by means of Eq. 2. N^w stands for the weighted number of counts already background subtracted, the neutron flux is indicated by N_n and the energy of the capture cascade corresponding is $E_C \approx S_n + E_n$.

$$Y(E_n) = \frac{N^w(E_n)}{N_n(E_n)E_C(E_n)} \tag{2}$$

Before calculating the capture yield, the performance of all C_6D_6 detectors along the experiment was studied in terms of gain and count-rate stability to filter-out possible accidental errors or systematic bias in the data taking. The detectors were calibrated using radioactive sources of ^{137}Cs, ^{88}Y and AmBe together with the ^{197}Au(n,γ) deposited energy spectra, ensuring a valid range between 662 keV and 6.512 MeV. Additionally, a series of Monte Carlo simulations of the entire experimental setup were performed to calculate the weighting function, which was applied following the methodology described in Ref. [19]. The latter includes yield-correction factors to account for the threshold-effects in the detectors, as well as γ-ray summing and conversion-electron effects. The neutron flux available during the experiment was determined from the evaluated version measured with high precision at n_TOF [22].

Disagreements between the determined and real neutron flux, or between simulated and true experimental setup were taken into account using the aforementioned SRM [20]. After subtracting all backgrounds present in the measurement [21], the capture yield was obtained using Eq. 2.

Once the yield was calculated, a resonance analysis was carried out using the SAMMY code [23]. This code is based on the R-matrix theory that gives a phenomenological description of the neutron induced reactions [24, 25]. SAMMY performs a Bayesian fit to the experimental data using an initial set of R-matrix resonance parameters and employing, by default, the recommended Reich-Moore approximation. Experimental effects such as Doppler broadening or neutron multiple scattering effects are taken into account in SAMMY by providing experimental data as temperature or sample thickness. Fig. 2 shows the final capture yield obtained in this work and fitted with SAMMY. In summary, 113 resonances have been analyzed, 98 of them for the first time and only 15 were previously known from transmission [26, 27]. There was no resonance data available from the previous ToF measurement [14]. The high resolution and high statistics of the measurement led us to propose a spin-parity change in the first resonance at 1.47 keV, and to individually analyze the multiple scattering component in the second resonance at 1.98 keV of neutron energy.

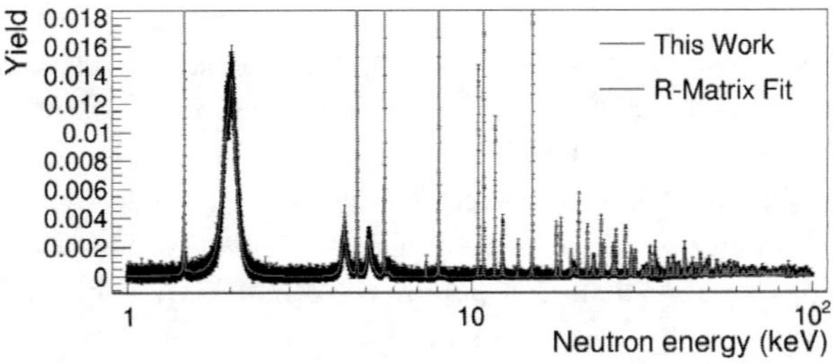

Figure 2. Measured ^{80}Se(n,γ) yield between 1 keV and 100 keV (red data-points) and R-matrix analysis fit (red curve).

4 MACS

During the *s*-process, neutrons in the stars are thermalized following the Maxwell distribution of velocities. Hence, the MACS is the input quantity required in stellar models to study this nucleosynthesis mechanism. At a temperature T, the MACS is given by Eq. 3.

$$\mathrm{MACS} = \frac{\langle \sigma_v \rangle}{v_T} = \frac{2}{(kT)^2\sqrt{\pi}} \int_{E_i=0}^{E_f=\infty} \sigma_\gamma(E) E e^{-E/kT} dE. \quad (3)$$

The MACS for ^{80}Se was determined from the measured capture yield (Fig. 2) by means of the SAMMY code. Fig. 3 shows the MACS of ^{80}Se(n,γ) determined in this work and compared to that obtained from resonance parameters fom the JEFF3.3 evaluated library [28]. Also the MACS from KADoNiS [29] and that calculated in the previous measurement [14] are displayed for comparison.

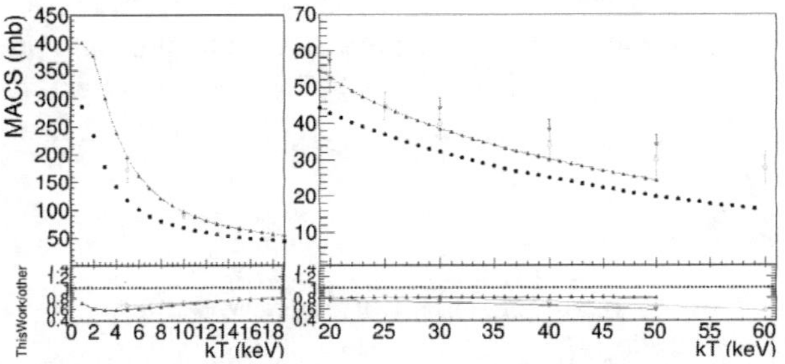

Figure 3. MACS of ^{80}Se(n,γ) determined in this work compared to evaluated libraries and previous measurement.

The ratio between this work and the rest of contributions is displayed in the bottom panel of the figure. On average, our MACS is between 20% and 40% lower than previous values.

Along the TP-AGB evolutionary stage, thermal energies of $kT \sim 8$ keV and $kT \sim 23$ keV are reached. In massive stars neutrons are produced at energies of $KT \sim 26$ keV and $kT \sim 90$ keV during core He-burning and shell C-burning, respectively [2]. Differences in MACS at these temperatures are more influential on the astrophysical aspects. Tab. 2 displays the different results at energies close to these points, in which the compared models have available data. At $kT = 10$ keV, the MACS obtained from the JEFF3.3 evaluated library is compatible with that available in KADoNiS, but both are 40% overestimated compared to the new result at this energy. Relative differences decrease at 30 keV, where the new MACS is smaller by 30%. Finally, following this trend, the new value for the MACS at 90 keV is 25% smaller than that obtained from JEFF3.3. However, the MACS in KADoNiS is overestimated by a factor of 2.5, whereas no value is available at this point from the previous measurement [14].

Table 2. Value of the MACS at three different temperatures. Only statistical uncertainties are included in these results.

	MACS (mb)		
	at 10 keV	at 30 keV	at 90 keV
JEFF 3.3	96.9	38.7	11.9
KADoNis	93.6(6.8)	39.8(4.1)	23.7(4.3)
Walter et al.	–	44(3)	–
This Work	**67.6(4)**	**32.2(3)**	**9.5(1)**

5 Summary and outlook

A high resolution measurement of the ^{80}Se(n,γ) cross section has been carried out at CERN n_TOF covering for the first time a very broad neutron-energy range, from meV up to 100 keV. The MACS could be accurately determined for the relevant stellar temperatures, significantly improving previous statistical and systematic uncertainties. As next steps we intend to investigate the impact of the new results on weak s-process nucleosynthesis and massive stars [30]. This results will be required for a full interpretatio of the ^{79}Se branching, whose measurement is scheduled at CERN n_TOF for 2022 [11].

6 Acknowledgement

We acknowledge support from from the European Research Council (ERC) under the European Union's Horizon 2020 research and innovation program (grant agreement No. 681740), and the Spanish Science Ministry for funding projects FPA2017-83946-C2-1-P and PID2019-104714GB-C21. This work is part of the PhD Thesis of V. Babiano-Suarez.

References

[1] E.M. Burbidge, G.R. Burbidge, W.A. Fowler, F. Hoyle, Reviews of Modern Physics **29**, 547 (1957)

[2] F. Käppeler, Progress in Particle and Nuclear Physics **66**, 390 (2011)

[3] S. Degl'Innocenti, P.G. Prada Moroni, M. Marconi, A. Ruoppo, Astrophysics and Space Science **316**, 25 (2008)

[4] B. Paxton, L. Bildsten, A. Dotter, F. Herwig, P. Lesaffre, F. Timmes, Astrophysical Journal Supplement **192**, 3 (2011)

[5] M. Pignatari, F. Herwig, R. Hirschi, M. Bennett, G. Rockefeller, C. Fryer, F.X. Timmes, C. Ritter, A. Heger, S. Jones et al., Astrophysical Journal Supplement **225**, 24 (2016)

[6] R. Gallino, C. Arlandini, M. Busso, M. Lugaro, C. Travaglio, O. Straniero, A. Chieffi, M. Limongi, The Astrophysical Journal **497**, 388 (1998)

[7] S. Cristallo, L. Piersanti, O. Straniero, *The FRUITY database on AGB stars: past, present and future*, in *Journal of Physics Conference Series* (2016), Vol. 665 of *Journal of Physics Conference Series*, p. 012019, 1405.3392

[8] G. Jörg, R. Bühnemann, S. Hollas, N. Kivel, K. Kossert, S. Van Winckel, C.L.v. Gostomski, Applied Radiation and Isotopes **68**, 2339 (2010)

[9] F. Käppeler, Progress in Particle and Nuclear Physics **43**, 419 (1999)

[10] U. Ott, F. Begemann, J. Yang, S. Epstein, Nature **332**, 700 (1988)

[11] V. Babiano-Suárez, J. Balibrea-Correa, L. Caballero, F. Calviño, D. Cano-Ott, A. Casanovas, N. Colonna, S. Cristallo, C. Domingo-Pardo, C. Guerrero et al., Tech. Rep. CERN-INTC-2020-065, INTC-P-580, CERN, Geneva (2020), http://cds.cern.ch/record/2731962

[12] V. Babiano-Suárez, L. Caballero, C. Domingo-Pardo, C. Guerrero, A. Tarifeño-Saldivia, Tech. Rep. CERN-INTC-2018-005, INTC-P-536, CERN, Geneva (2018), http://cds.cern.ch/record/2299660

[13] High-sensitivitY Measurements of key stellar Nucleo-Synthesis reactions (HYMNS), ERC-consolidator grant agreement no. 681740, PI: C. Domingo Pardo.

[14] G. Walter, H. Beer, F. Kaeppeler, G. Reffo, F. Fabbri, Astronomy and Astrophysics **167**, 186 (1986)

[15] C. Guerrero, A. Tsinganis, E. Berthoumieux, M. Barbagallo, F. Belloni, F. Gunsing, C. Weiß, E. Chiaveri, M. Calviani, V. Vlachoudis et al., European Physical Journal A **49**, 27 (2013)

[16] S. Andriamonje, M. Calviani, Y. Kadi, R. Losito, V. Vlachoudis, E. Berthoumieux, F. Gunsing, Y. Giomataris, T. Papaevangelou, C. Guerrero et al., Journal of Korean Physical Society **59**, 1601 (2011)

[17] P.F. Mastinu, R. Baccomi, E. Berthoumieux, D. Cano-Ott, F. Gramegna, C. Guerrero, C. Massimi, P.M. Milazzo, F. Mingrone, J. Praena et al., Tech. Rep. CERN-INTC-2013-002, CERN (2013)

[18] R.L. Macklin, J.H. Gibbons, Physical Review **159**, 1007 (1967)

[19] U. Abbondanno, G. Aerts, H. Alvarez, S. Andriamonje, A. Angelopoulos, P. Assimakopoulos, C.O. Bacri, G. Badurek, P. Baumann, F. Bečvář et al., Nuclear Instruments and Methods in Physics Research A **521**, 454 (2004)

[20] R.L. Macklin, J. Halperin, R.R. Winters, Nuclear Instruments and Methods **164**, 213 (1979)

[21] P. Žugec, N. Colonna, D. Bosnar, S. Altstadt, J. Andrzejewski, L. Audouin, M. Barbagallo, V. Bécares, F. Bečvář, F. Belloni et al., Nuclear Instruments and Methods in Physics Research A **760**, 57 (2014), 1406.6865

[22] M. Barbagallo, C. Guerrero, A. Tsinganis, D. Tarrío, S. Altstadt, S. Andriamonje, J. Andrzejewski, L. Audouin, V. Bécares, F. Bečvář et al., European Physical Journal A **49**, 156 (2013)

[23] N.M. Larson, ORNL/TM-9179/R8, Oak Ridge National Laboratory, Oak Ridge, TN (2008)

[24] E.P. Wigner, L. Eisenbud, Physical Review **72**, 29 (1947)

[25] A.M. Lane, R.G. Thomas, Reviews of Modern Physics **30**, 257 (1958)

[26] G.M. Novoselov, V.G. Krivenko, L.L. Litvinskij, I.M. Simonov, Jadernye Konstanty (Nuclear Constants) **1-2**, 3 (1995)

[27] S.F. Mughabghab, D.I. Garber, *Neutron cross sections. Volume I. Resonance parameters* (Academic Press Inc, United States, 1973)

[28] A.J.M. Plompen, O. Cabellos, C. De Saint Jean, M. Fleming, A. Algora, M. Angelone, P. Archier, E. Bauge, O. Bersillon, A. Blokhin et al., European Physical Journal A **56**, 181 (2020)

[29] I. Dillmann, T. Szücs, R. Plag, Z. Fülöp, F. Käppeler, A. Mengoni, T. Rauscher, Nuclear Data Sheets **120**, 171 (2014)

[30] M. Pignatari, R. Gallino, M. Heil, M. Wiescher, F. Käppeler, F. Herwig, S. Bisterzo, The Astrophysical Journal **710**, 1557 (2010)

Underground measurement at LUNA found no evidence for a low-energy resonance in the ^{6}Li(p, γ)^{7}Be reaction

Denise Piatti[1,*] for the LUNA Collaboration

[1]Univerisità degli Studi di Padova and INFN Division of Padova, Via Marzolo 8 35136 Padova (Italy)

Abstract. The ^{6}Li(p,γ)^{7}Be reaction is involved in all three main nucleosynthesis scenarios: Big Bang Nucleosynthesis, the interaction of cosmic rays with interstellar matter, and stellar nucleosynthesis.
Conflicting experimental results have been reported in literature for the ^{6}Li(p,γ)^{7}Be reaction cross section trend at astrophysical energies. A recent direct measurement found a resonance-like structure at $E_{\mathrm{c.m.}}$ = 195 keV, corresponding to an excited state at $E_{\mathrm{x}} \sim$ 5800 keV in ^{7}Be which, however, has not been confirmed by either theoretical calculations or other direct measurements. In order to clarify the existence of this resonance, a new experiment was performed at the Laboratory for Underground Nuclear Astrophysics, located deep underground at Laboratori Nazionali del Gran Sasso (Italy). The ^{6}Li(p,γ)^{7}Be cross section was measured in the energy range $E_{\mathrm{c.m.}}$ = 60-350 keV with unprecedented sensitivity and no evidence for the alleged resonance was found.

1 Introduction

According to simulations of the Galaxy chemical evolution most of the solar system lithium was provided by low-mass stars [1] while less than half of it was produced by Big Bang Nucleosynthesis (BBN) [2, 3] or Galactic cosmic rays interacting with interstellar matter.

The predicted BBN ^{6}Li/^{7}Li isotopic ratio is $\sim 10^{-5}$ [4], significantly lower than the solar system value of 0.08 [5]. Very low ^{6}Li/^{7}Li values are expected for neutrino nucleosynthesis [6] and for stellar sources as well. In contrast, in case of Galactic or structure formation cosmic rays the ^{6}Li/^{7}Li production ratio is close to unity [7].

The ^{6}Li/^{7}Li isotopic ratio has been indeed proposed as a tool to constrain non-standard lithium production mechanisms [8] and pollution of stellar atmospheres [9] in the context of the cosmological lithium problem,

The ^{6}Li(p,γ)^{7}Be reaction plays a key role in determining the stellar ^{6}Li/^{7}Li. The ^{6}Li(p,γ)^{7}Be reaction may indeed not only deplete ^{6}Li but also convert some of it to ^{7}Li, through ^{7}Be radioactive decay.

Measurements of the ^{6}Li(p,γ)^{7}Be reaction cross section at low energies have reported inconsistent results on the slope of the astrophysical S-factor [10, 11]. Moreover, the positive slope reported by [12] was interpreted as a new resonance at $E_{\mathrm{c.m.}}$ = 195 keV, corresponding to an excited level at $E_x \approx$ 5800 keV with $J^\pi = (1/2^+, 3/2^+)$ and $\Gamma_p \approx$ 50 keV. No evidence for such a resonance was found in the ^{3}He(^{4}He,γ)^{7}Be reaction at $E_{\mathrm{c.m.}}$ = 4210 keV as reported in recent comprehensive study [13].

*e-mail: denise.piatti@pd.infn.it

https://doi.org/10.1051/epjconf/202226011027

None of the theoretical calculations of the ^{6}Li(p,γ)^{7}Be S-factor are able to reproduce the newly-reported resonance [14, 15, and references therein], unless this is added *ad-hoc* to reproduce the experimental data [16].

2 Experimental Setup

We performed a new experiment [17] at the Laboratory for Underground Nuclear Astrophysics (LUNA), located deep underground at Laboratori Nazionali del Gran Sasso (Italy) [18].

A schematic view of the experimental setup is shown in Fig. 1. The LUNA-400 accelerator

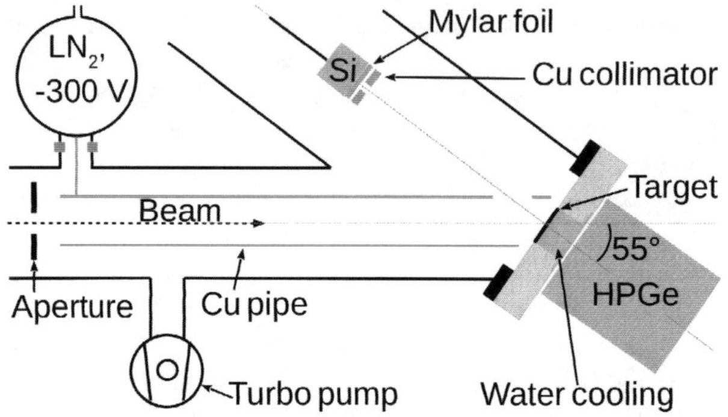

Figure 1: Sketch of the experimental setup used for the measurement of the ^{6}Li(p,γ)^{7}Be cross section at LUNA [17].

[19] high-intensity proton beam was collimated by a 3 mm diameter aperture and delivered through a copper pipe to the target, mounted at 55° with respect to the beam direction. The Cu tube was used both as a cold trap and for secondary electron suppression. Three evaporated targets (thicknesses 100 – 200 μg/cm^2) were made from $^{6}Li_2WO_4$ powder and one (thickness 20 μg/cm^2) was made using $^{6}Li_2O$ powder. The ^{6}Li isotopic enrichment level was 95% for all targets, which were water cooled during irradiation in order to limit target degradation [17].

A High-Purity Germanium (HPGe) detector positioned in close geometry to the target and at 55° with respect to the beam direction was used to detect ^{6}Li(p,γ)^{7}Be reaction γ-rays. To detect the α and ^{3}He particles from the ^{6}Li(p,α)^{3}He reaction concurrently with the gamma rays from the ^{6}Li(p,γ)^{7}Be reaction, a silicon detector was installed at 125° from the beam direction. Efficiencies for both detectors were obtained using GEANT simulations, fine tuned through the comparison with experimental results [17].

3 Analysis and Results

To make consistency checks and verify results are unaffected by systematic effects, a measurement of the ^{6}Li(p,γ)^{7}Be and ^{6}Li(p,α)^{3}He excitation functions was performed for each target in the whole dynamic range of the LUNA-400 accelerator [17].

The ^{6}Li(p,γ)^{7}Be experimental yield was calculated from the sum of the contributions from the direct capture to the ground state (γ_0) and to the 429 keV excited state of ^{7}Be (γ_1).

For the calculation of the ^{6}Li(p,γ)^{7}Be reaction S-factor, we adopted a relative approach [17]: the (p,γ) yield was normalized at each energy to the (p,α) yield. This ratio can be expressed in terms of the (p,γ) and (p,α) S-factors. We adopted for the ^{6}Li(p,α)^{3}He reaction the S-factor parametrization from [20]. For the (p,α) channel, the angular distribution coefficients A_k and related uncertainties were taken from [21, and references therein]. For the (p,γ) channel we adopted the theoretical angular distribution described in [14]. Finally the measured S-factor was corrected for electron screening using the adiabatic approximation [22] with screening potential U_e = 273 eV [20].

Our S-factor data have a monotonic dependence on the energy and show no evidence of the resonance reported by [12], see Fig.2. The ^{6}Li(p,γ)^{7}Be reaction cross section was measured in the energy range 60 – 350 keV with ≤2% statistical and 12% systematic uncertainty. An R-matrix fit of our data and the data from [23] was performed and used to calculate a new ^{6}Li(p,γ)^{7}Be reaction rate. The proposed reaction rate is 9% lower than NACRE [24] and 33% higher than reported in NACREII [16] at 2 MK, relevant for ^{6}Li depletion in pre-main sequence stars, and the reaction rate uncertainty has been significantly reduced [17], see Fig.3.

The result of a recent indirect study supports LUNA extrapolation for the ^{6}Li(p,γ)^{7}Be S-factor [25].

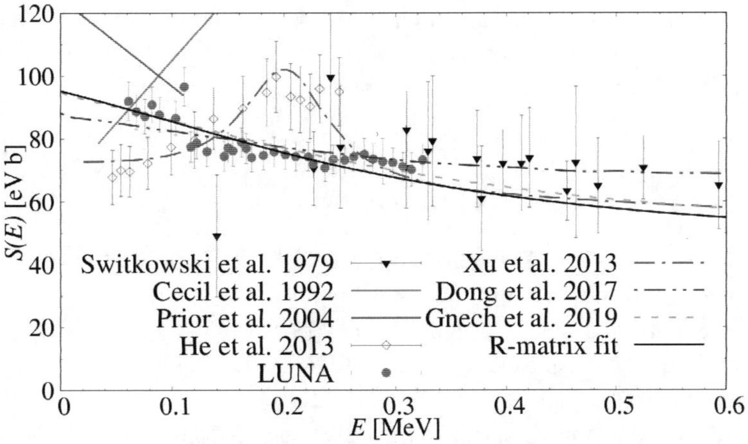

Figure 2: Astrophysical S-factor for the ^{6}Li(p,γ)^{7}Be reaction as obtained by LUNA in red [17]. Previous experimental data and theoretical evaluations are also shown for comparison. The solid black line represents an R-matrix fit of LUNA data and data from [23].

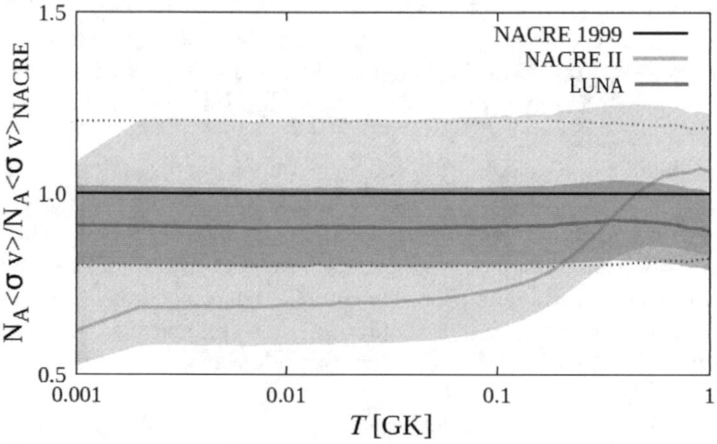

Figure 3: Reaction rate for the ^{6}Li(p,γ)^{7}Be reaction, normalized to the NACRE rate [24]. The NACRE II rate [16] is also shown for comparison. Dashed lines represent the uncertainty on the NACRE rate, while shaded areas represent the uncertainties from LUNA experiment (red) and from NACRE II (grey).

References

[1] N. Prantzos *et al.*, Astronomy and Astrophysics **542**, A67 (2012)
[2] C. Pitrou *et al.*, Physics Reports **754**, 1 (2018)
[3] B. D. Fields *et al.*,Journal of Cosmology and Astroparticle Physics **2020**, 010 (2020)
[4] M. Anders *et al.*, Physical Review Letters **113**, 042501 (2014)
[5] K. Lodders, Astrophysical Journal **591**, 1220 (2003)
[6] A. Sieverding *et al.*, Astrophysical Journal **876**, 151 (2019)
[7] B. D. Fields and T. Prodanović, Astrophysical Journal **623**, 877 (2005)
[8] J. C. Howk *et al.*, Nature **489**, 121 (2012)
[9] G. Harutyunyan *et al.*, Astronomy and Astrophysics **618**, A16 (2018)
[10] F. Cecil *et al.*, Nuclear Physics A **539**, 75 (1992)
[11] R. M. Prior *et al.*, Physical Review C **70**, 055801 (2004)
[12] J. He *et al.*, Physics Letters B **725**, 287 (2013)
[13] T. Szücs *et al.*, Physical Review C **99**, 055804 (2019)
[14] A. Gnech and L. E. Marcucci, Nuclear Physics A **987**, 1 (2019)
[15] G. X. Dong *et al.*, Journal Physics G Nuclear and Particle Physics **44**, 045201 (2017)
[16] J. Huang *et al.*, Atomic Data and Nuclear Data Table **96**, 824 (2010)
[17] D. Piatti *et al.*, Physical Review C **102**, (2020)
[18] F. Cavanna and P. Prati, International Journal of Modern Physics A **33**, 1843010 (2018)
[19] A. Formicola *et al.*, Nuclear Instruments and Methods in Physics Research **507**, 609 (2003)
[20] J. Cruz *et al.*, Physics Letters B **624**, 181 (2005)
[21] C. Brune *et al.*, Nuclear Instruments and Methods in Physics Research A **389**, 421 (1997)
[22] H. J. Assenbaum, K. Langanke and C. Rolfs, Zeitschrift für Physik A Atomic Nuclei **327**, 461 (1987)

[23] Z. Switkowski *et al.*, Nuclear Physics A **331**, 50 (1979)
[24] C. Angulo *et al.*, Nuclear Physics A **656**, 3 (1999)
[25] G.G. Kiss *et al.*, Physical Review C **104**, 015807 (2021)

Sensitivity Study of Nuclear Reactions Influencing Photospheric Radius Expansion X-Ray Bursts

Yi Hua Lam[1,2,*], *Alexander* Heger[3,4,5,6,**], *Zac* Johnston[6], and *Adelle Jane* Goodwin[7]

[1]Institute of Modern Physics, Chinese Academy of Sciences, Lanzhou 730000, China
[2]School of Nuclear Science and Technology, University of Chinese Academy of Sciences, Beijing 100049, China
[3]School of Physics and Astronomy, Monash University, Victoria 3800, Australia
[4]OzGrav-Monash – MOCA, School of Physics and Astronomy, Monash University, VIC 3800, Australia
[5]Center of Excellence for Astrophysics in Three Dimensions (ASTRO-3D), Australia
[6]Joint Institute for Nuclear Astrophysics, Michigan State University, East Lansing, MI 48824, USA
[7]International Centre for Radio Astronomy Research – Curtin University, Perth, WA 6845, Australia

Abstract. Up to now, more than 62 of the 115 X-ray sources of low-mass-X-ray binaries have been identified as photospheric radius expansion (PRE) bursters [1]. Galloway and collaborators expect more PRE bursters in their near future analysis [2]. Although more than half of the discovered X-ray sources are PRE bursters, the bursting mechanism of PRE burster is still not adequately understood. This is because of the complicated hydrodynamics and variable accretion rates. An example is the accretion-powered millisecond pulsar SAX J1808.4−3658 [3, 4] that powered up the brightest Type-I X-ray burst (XRB) recorded by NICER in recent history [5]. The first 1D multi-zone model of SAX J1808.4−3658 was recently constructed [6, 7]. The pioneering model offers a first concurrent and direct comparison with the observed light curves, fluences, and recurrence times. With the three observables, a comparison between theory and observations could be more sensitive than the previous studies of the clocked burster and post-processing models. We perform a sensitivity study on (α, p), (α, γ), (p, α), and (p, γ) reactions with a total up to ~1,500 reactions. Our current result indicates that the observables are more sensitive to the competition between the reactions involving alpha-capture, e.g., the $^{22}\mathrm{Mg}(\alpha, \mathrm{p})$ and $^{22}\mathrm{Mg}(\mathrm{p}, \gamma)$ reactions competing at the ^{22}Mg branch point [8].

Generally, due to less extensive rapid-proton capture (rp-) process but mainly powered by alpha-proton (αp-) process, the light curves of PRE bursts commence with rapid onsets ($\lesssim$1 s), extend with wide plateau ($\approx$10 s), and end with short burst tail ($\lesssim$30 s) [9–11]. A close match of modeled and observed PRE burst permit us to probe the complicated burst mechanism of PRE bursters. Johnston *et al.* [6] recently constructed the first PRE burst model quantitatively describing the accretion-powered millisecond pulsar SAX J1808.4−3658, whereas the other simplified model built by Goodwin *et al.* [7] provides an intuitive grasp of the observables of SAX J1808.4−3658 burster. The host neutron star of this PRE burster spins up to millisecond periods by the accretion of stellar matter from a low-mass companion star in its X-ray binary system [3]. The essential properties of an epoch of this PRE burster includes the burst *light curves*, *fluences* (integral of flux over time), and *recurrence times*. This PRE burster offers

*e-mail: lamyihua@impcas.ac.cn
**e-mail: alexander.heger@monash.edu

more concurrent observables for us to compare with models. Therefore, we expect the comparison of these PRE-burster observables is more sensitive than merely the comparison of light curves and recurrence times, e.g., the comparison of light curves and recurrence times of the GS 1826−24 clocked burster. Recently, only the PRE models were used to quantitatively match with these three observables. Table 1 shows the comparison of available models, their characteristics, and XRB observables that had been studied with published results.

Table 1. XRB models and the respective observables that have been studied.

Models	self-consistent hydrodynamics	modeled light curve	matched with observation: light curve	recurrence time	fluence
0D One-zone post-processing	No	Yes	–	–	–
0D One-zone	No	Yes	–	–	–
1D Multi-zone GS 1826−24 [12, 13]	Yes	Yes	Yes	Yes	–
1D Multi-zone SAX J1808.4−3658 [6, 7]	Yes	Yes	Yes	Yes	Yes

Note: The models' features that have not been used to quantitatively compare with the observable are indicated as dash lines.

Concurrently comparing light curves, fluences, and recurrence times permits us to scrutinize the constructed XRB models, e.g., the assumed hydrodynamics, to simulate and to study the extreme astrophysical environment. Moreover, the sensitivity study on these three observables helps us to identify important reactions, and then to constrain and to improve the pioneer PRE XRB model, e.g., the H/He/metaliticiy ratios and accretion rate. With a more constrained PRE XRB model, this opens new opportunities for us to study neutron-star compactness, mass, radius, and cooling processes of SAX J1808.4−3658 burster of which the host neutron star is extremely cold, exhibiting rapid and active cooling processes [14]. In additions, the hydrodynamic data generated from an 1D multi-zone SAX J1808.4−3658 XRB model could be beneficial for one-zone post-processing and one-zone models to obtain the nuclear energy generation (XRB flux) consistent with the input hydrodynamic snapshots.

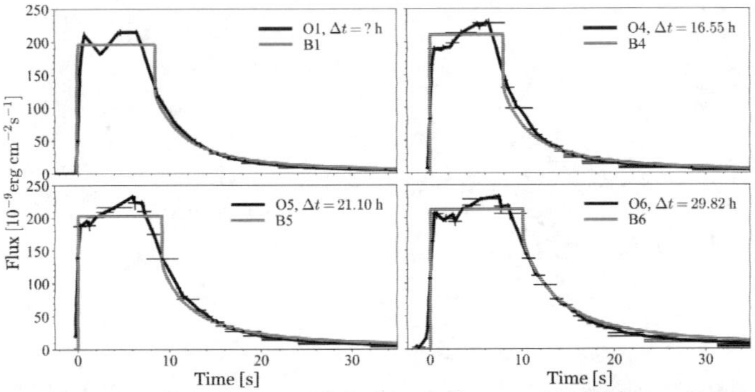

Figure 1. SAX J1808.4−3658 PRE bursts of *Epoch Oct 2002*. The observed bursts O1, O4, O5, and O6 (black lines), and the corresponding modeled bursts B1, B4, B5, and B6 (red lines) based on the restored Johnston *et al.* [6] model using Hu *et al.* ^{22}Mg(α, p) reaction rate [8] with mass fractions $X_{\mathrm{H}}/X_{\mathrm{He}} = 0.413/0.567$ and time dependent accretion rate.

Figure 1 shows the modeled and observed PRE burst light curves of SAX J1808.4−3658. We reproduce the observed PRE bursts of SAX J1808.4−3658 (*Epoch Oct. 2002*) using Johnston *et al.* [6] and Goodwin *et al.* [7] models with Non-Smoker ^{22}Mg(α, p) reaction rate (yellow diamonds and purple stars in Fig. 2). Nevertheless, with implementing the most updated ^{22}Mg(α, p) reaction rate experimentally deduced by Hu *et al.* [8] with the lowest uncertainty compared to all currently available ^{22}Mg(α, p) rates, an extra burst is produced and the overall recurrence times of all PRE bursts are shifted (blue open squares in Fig. 2). This most updated ^{22}Mg(α, p) rate is in fact almost one order of magnitude lower than Non-Smoker ^{22}Mg(α, p) rate. The primary reason for producing an extra burst is due to the competition between the ^{22}Mg(α, p) and ^{22}Mg(p, γ) reactions at the ^{22}Mg branch point within the accreting envelope maximum temperature range of PRE burst which lasts for around 10 s and is 1/3 of the duration of a PRE burst. The process of restoring the PRE model permits us to constrain the mass

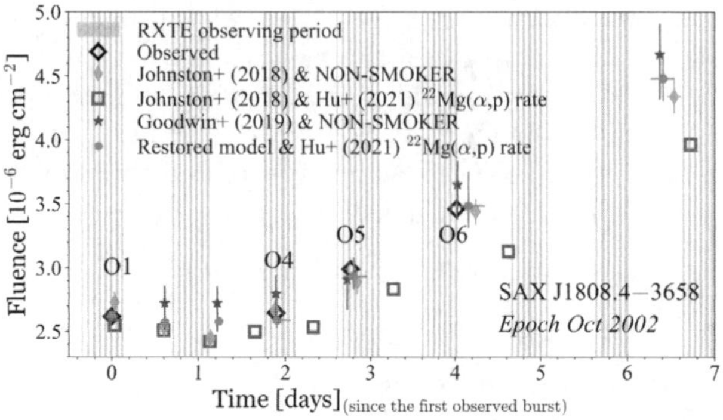

Figure 2. The bursts' fluences and times of the SAX J1808.4−3658 PRE burster, recorded by the Rossi X-ray Timing Explorer (RXTE) [1], based on Johnston *et al.* [6] and Goodwin *et al.* [7] models with NON-SMOKER ^{22}Mg(α, p) reaction rate, and the restored Johnston *et al.* model with Hu *et al.* ^{22}Mg(α, p) reaction rate [8]. The light curves of observed (and modeled) bursts O1 (B1), O4 (B4), O5 (B5), and O6 (B6) are plotted in Fig. 1.

fractions of H and He in the accreting envelope of the PRE model to be $X_{\rm H} = 41.3 \pm 0.3$ % and $X_{\rm He} = 56.7 \pm 0.3$ %. We find that the comparison of fluences-recurrence times due to a varied selected reaction is more sensitive than merely comparing the light curves of XRB, e.g., the light curve of GS 1826−24 clocked burster. This finding motivates us to explore the sensitivity of ~1,500 reactions along the nuclear reaction flow path during an episode of PRE bursts. These ~1,500 reactions consist of (α, p), (α, γ), (p, α), and (p, γ) reactions.

Our preliminary results indicate that these three observables are more sensitive to the competition between the reactions involving alpha-capture, e.g., the ^{22}Mg(α, p) and ^{22}Mg(p, γ) as shown above. Contrarily to the influence of ^{22}Mg(α, p) reaction, if the ^{34}Ar(α, p) reaction is one order of magnitude lower than its NON-SMOKER counterpart, the PRE model produces one PRE burst fewer than the model implementing NON-SMOKER ^{34}Ar(α, p) rate. We anticipate that a list of important reactions of PRE burst is different from what was found by Cyburt *et al.* [15] based on the GS 1826−24 model. A set of more precisely determined important reaction rates definitely allows us to further constrain the pioneering PRE models. Further results will be made available and published elsewhere [16].

This work was financially supported by the Strategic Priority Research Program of Chinese Academy of Sciences (CAS, Grant Nos. XDB34000000) and National Natural Science Foundation of China (No. 11775277), the Chinese Academy of Sciences President's International Fellowship Initiative (No. 2019FYM0002), the Australian Research Council Centre of Excellence for Gravitational Wave Discovery (OzGrav, No. CE170100004) and for All Sky Astrophysics in 3 Dimensions (ASTRO 3D, No. CE170100013), Australian Research Council's Discovery Projects funding scheme (DP200102471). We are appreciative of the computing resource provided by the Institute of Physics (PHYS_T3 cluster) and the ASGC (Academia Sinica Grid-computing Center) Distributed Cloud resources (QDR4 and FDR5 clusters) of Academia Sinica, Taiwan.

References

[1] D. K. Galloway, J. in 't Zand, J. Chenevez *et al.*, Astrophys. J. Suppl. **249**, 32 (2020).
[2] D. K. Galloway *private communication*.
[3] R. Wijnands, M. van der Klis, Nature **394**, 344 (1998).
[4] D. Chakrabarty, E. Morgan, M. Muno *et al.*, Nature **424**, 42 (2003)
[5] P. Bult, G. K. Jaisawal, T. Güver *et al.*, Astrophys. J. Lett. **885**, L1 (2019).
[6] Z. Johnston, A. Heger, D. K. Galloway, Mon. Not. R. Astron. Soc. **477**, 2112 (2018).
[7] A. J. Goodwin, D. K. Galloway, A. Heger *et al.*, Mon. Not. R. Astron. Soc. **490**, 2228 (2019).
[8] J. Hu, H. Yamaguchi, Y. H. Lam *et al.*, Phys. Rev. Lett. **127**, 172701 (2021).
[9] W. H. G. Lewin, W. D. Vacca, E. M. Basinska, Astrophys. J. **277**, L57 (1984).
[10] Y. Tawara, T. Kii, S. Hayakawa *et al.*, Astrophys. J. **276**, L41 (1984).
[11] T. Strohmayer, L. Bildsten, *New Views of Thermonuclear Bursts* in *Compact Stellar X-ray Sources* (Cambridge University Press, Cambridge, UK, 2010) p. 113.
[12] A. Heger, A. Cumming, D. K. Galloway *et al.*, Astrophys. J. Lett. **671**, L141 (2007).
[13] B. Paxton, P. Marchant, J. Schwab *et al.*, Astrophys. J. Suppl. Ser. **220**, 15 (2015).
[14] C. O. Heinke, P. G. Jonker, R. Wijnands *et al.*, Astrophys. J. **691**, 1035 (2009).
[15] R. H. Cyburt, A. M. Amthor, A. Heger *et al.*, Astrophys. J. **830**, 55 (2016).
[16] Y. H. Lam, A. Heger, Z. Johnston, A. J. Goodwin *et al.*, *in preparation*.

Neutrino and Antineutrino pair-Emission in Strong Magnetic Field in Relativistic Quantum Approach

Tomoyuki Maruyama[1,*], *A.Baha* Balantekin[4], *Myung-Ki* Cheoun[3,], *Toshitaka* Kajino[2,], and *Grand J.* Mathews[5]

[1]College of Bioresource Sciences, Nihon University, Fujisawa 252-8510, Japan
[2]Department of Physics, University of Wisconsin, Madison, WI 53706, USA
[3]Beihang University, School of Physics, Int. Center for Big-Bang Cosmology and Element Genesis, Beijing 100083, China
[4]Department of Physics, Soongsil University, Seoul, 156-743, Korea
[5]Center of Astrophysics, Univ. of Notre Dame, Notre Dame, IN 46556, USA

Abstract. We study the $\nu\bar{\nu}$-pair emission from electrons and protons in a relativistic quantum approach. In this work we calculate the luminosity of the $\nu\bar{\nu}$-pairs emitted from neutron-star-matter with a strong magnetic field, and find that this luminosity is much larger than that in the modified Urca process. The $\nu\bar{\nu}$-pair emission processes in strong magnetic fields significantly contribute to the cooling of the magnetars.

1 Introduction

Magnetic fields in neutron stars play important roles in the interpretation of many observed phenomena. Magnetars, which are associated with super strong magnetic fields [1, 2], have properties different from normal neutron stars. Thus, phenomena related magnetars can provide a lot of information about the physics of the magnetic field.

Many authors have paid attention into cooling processes of neutron stars (NS) because it gives important information on neutron star structure [3]. Neutron stars are cooled by neutrino emission, and a magnetic field is expected to affect the emission mechanism largely because a strong magnetic field can supply energy and momentum into cooling processes.

In this work we study two kinds of the processes, the $\nu\bar{\nu}$-pair synchrotron radiation $p(e^-) \rightarrow p(e^-) + \nu + \bar{\nu}$ by calculating it through the transition between the different Landau levels for protons and electrons [4].

2 Formalism

In this work we assume the neutron-star matter composing of proton, neutron and electron, a uniform magnetic field along the z-direction, $\boldsymbol{B} = (0, 0, B)$. We take the electro-magnetic vector potential A^μ to be $A = (0, 0, xB, 0)$ at the position $\boldsymbol{r} \equiv (x, y, z)$. The relativistic proton (electron) wave function ψ is obtained from the following Dirac equation:

$$\left[\alpha_z p_z - i\alpha_x \partial_x + \alpha_y (p_y - eB_x) + (M - U_s)\beta - \frac{\kappa}{M} B\Sigma_z + U_0\right] \psi(x, p_z, s) = E\psi(x, p_z, s), \quad (1)$$

*e-mail: maruyama.tomoyuki@nihon-u.ac.jp

where M is the proton (electron) mass, κ is the AMM, e is the particle charge, and E is the single particle energy written as

$$E(n, p_z, s) = \sqrt{p_z^2 + (\sqrt{2|e|Bn + M^2} - s\kappa B/M)^2} + U_0 \tag{2}$$

with n being the Landau level number. In the above equations, U_s and U_0 are the scalar and vector mean-fields which are given by the relativistic mean-field approach with the parameter-set PM1 [5].

The detailed expressions to calculate the neutrino emissivity are written in Ref. [4]. Energy intervals between two states with different Landau numbers are much larger than the temperature, so that we cannot use the Sommerfeld expansion at the low temperature limit.

3 Results

In Fig. 1a, we show the density dependence of the luminosities in the $\nu\bar{\nu}$-pair emission by using the parameter-set, PM1-SF1 for the mean-fields. For comparison, we give the $\bar{\nu}$-luminosities from the modified Urca (MU) process [6].

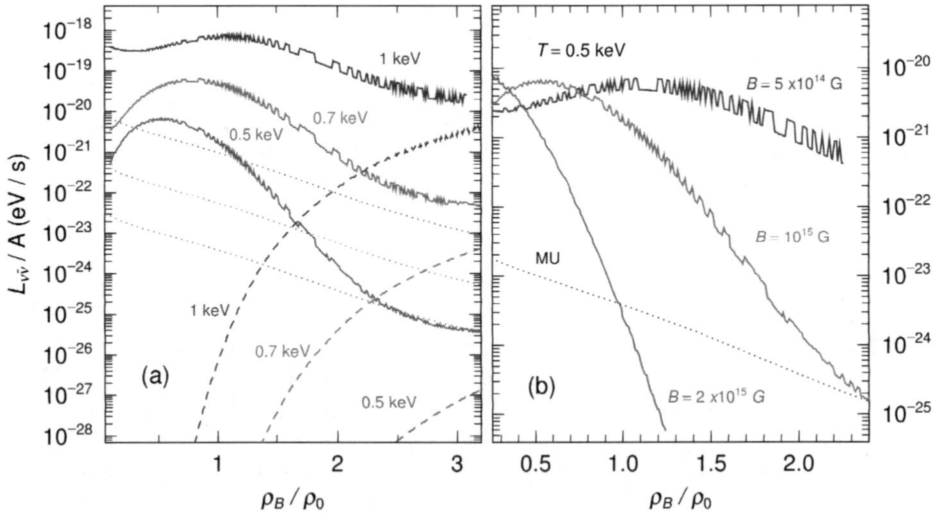

Figure 1. (a) Density dependence of the $\nu\bar{\nu}$-pair emission luminosity per nucleon. In the left panel (a) The solid and dashed lines represent the contributions from protons and electrons for $B = 10^{15}$G at $T = 0.5$, 0.7 and 1 keV (from bottom to top). The dotted lines indicate the neutrino luminosities from the MU process. In the right panel (b) solid lines show them for $B = 5 \times 10^{14}$ G, 10^{15} G and 2×10^{15} G at $T = 0.5$ keV. The dotted lines indicate the neutrino luminosity from the MU process.

We see that in the strong magnetic field, the neutrino luminosities become larger than those from the MU process particularly in low density regions. The calculation results include fluctuations. The density dependence of the factor $f(E_i)[1 - f(E_f)]$ does not smoothly vary for strong magnetic fields because the energy intervals between the initial and final states are larger than the temperature.

In Fig. 1b we show the density dependence of the $\nu\bar{\nu}$-luminosities at $B = 5 \times 10^{14}$ G, $B = 10^{15}$ G and $B = 2 \times 10^{15}$ G. We see that as the magnetic field strength decreases, the luminosity increases in the density region, $\rho_B/\rho_0 \gtrsim 1$. The $\nu\bar{\nu}$-pairs are produced via the transition of the proton and electron between different Landau levels. As the magnetic field strength increases, the momentum transfer from the magnetic field becomes larger, and the energy interval between the initial and final states is also larger. The former effect enhances the emission rate, but the latter effect suppresses it. In the present calculations the latter effect dominates in the higher density region.

4 Summary

We have used a relativistic quantum approach to study neutrino and anti-neutrino emission in the $\nu\bar{\nu}$-pair production and DU processes from NS matter with strong magnetic fields. We see that the magnetic field has a role to enhance the neutrino emission in the both processes largely.

The $\nu\bar{\nu}$-pair emission process has a much larger effect than that of the MU process in strong magnetic fields. We can conclude that the $\nu\bar{\nu}$-pair emission process is dominant in the low density region, $\rho_B \lesssim \rho_0$, for a cooling process of magnetars whose magnetic field strength is $10^{14} - 10^{15}$ G. Therefore, our results suggest that one needs to introduce the $\nu\bar{\nu}$-pair emission process when calculating the cooling rate of magnetars.

The present results demonstrate that the magnetic field very largely contributes to the neutrino emission processes, and that the quantum calculations are necessary to describe the momentum transfer from the magnetic field exactly. In the high density region, $\rho_B \gtrsim 3\rho_0$, the direct Urca (DU) process must appear, and its contribution is much larger than those of the $\nu\bar{\nu}$-pair emission and MU processes. The strong magnetic field must also contribute to the neutrino emission from the DU process [7]

This work was supported in part by the Grants-in-Aid for the Scientific Research from the Ministry of Education, Science and Culture of Japan (JP19K03833).

References

[1] B. Paczyński, Acta. Astron. **41**, 145 (1992).
[2] For a review, G. Chanmugam, Annu. Rev. Astron. Astrophys. **30**, 143 (1992).
[3] D.G. Yakolev and C.J. Pethick, Ann. Rev. Astron. Astrophys, **42**, 169 (2004).
[4] T. Maruyama, A B. Balantekin, M.-K. Cheoun, T. Kajino and G.J. Mathews, Phys. Lett. **B805**, 135413 (2020).
[5] T. Maruyama, H. Fujii, T. Muto and T. Tatsumi, Phys. Lett. **B337**, 19 (1994).
[6] O.V. Maxwell, ApJ **319**, 691 (1987); D.G. Yakolev, K.P. Levenfish A&A **297**, 717 (1995).
[7] T. Maruyama, A. B. Balantekin, M-K. Cheoun, T. Kajino, M. Kusakabe, G.J. Mathews, Phys. Lett. **B824**, 136813 (2022).

EPJ Web of Conferences **260**, 11030 (2022) https://doi.org/10.1051/epjconf/202226011030
NIC-XVI

Radiative proton–capture reactions with ^{112,114}Cd at astrophysically important energies

Polytimos Vasileiou[1,*], *Theo J.* Mertzimekis[1], *Achment* Chalil[1,**], *Christina* Fakiola[1], *Ioannis* Karakasis[1], *Anastasia* Kotsovolou[1], *Stefanos* Pelonis[1], and *Aikaterini* Zyriliou[1]

[1]National & Kapodistrian University of Athens, Zografou Campus, GR–15784, Athens, Greece

Abstract. The reaction network in the neutron-deficient part of the nuclear chart around $A \approx 100$ contains several nuclei of importance to astrophysical processes, such as the p–process. This work reports on the results from recent experimental studies of the radiative proton–capture reactions ^{112,114}Cd$(p,\gamma)^{113,115}$In. Experimental cross sections for the reactions have been measured for proton beam energies residing inside the respective Gamow windows for each reaction, using isotopically enriched ^{112}Cd and ^{114}Cd targets. Two different techniques, the in–beam γ–ray spectroscopy and the activation method have been implemented, where the latter is considered mandatory to account for the presence of low–lying isomers in ^{113}In ($E \approx 392$ keV, $t_{1/2} \approx 100$ min), and ^{115}In ($E \approx 336$ keV, $t_{1/2} \approx 4.5$ h). Following the measurement of the cross sections, the astrophysical S factors have been subsequently deduced.

1 Introduction

The origin of a group of 35 neutron–deficient nuclides, spanning from ^{74}Se to ^{196}Hg, commonly referred to as p nuclei, has been a long–standing puzzle in nuclear astrophysics. The solar abundances of these nuclei are one to two orders of magnitude lower than those of the respective s and r nuclides in the same mass regime. Various astrophysical sites and nucleosynthesis scenarios have been proposed to account for the production of these nuclei, with the main production mechanism being referred to as the p–process [1].

Due to the vastness of the p–process reaction network, the Hasuser–Feshbach statistical model [2] is often employed to estimate the reaction rates for the majority of these reactions. Experimental input is critical in the attempt to constrain the model parameters, and better understand the driving mechanisms behind the p–process. Such is the aim of this work, which reports on cross section measurements of the radiative proton–capture reactions with ^{112,114}Cd, for energies inside the astrophysical important Gamow window [1] (1.6–4.8 MeV for both reactions [3]). These reactions lead to the production of ^{113}In, nowadays widely accepted not to be a "pure" p nucleus, having non–negligible s– and r–process components [4], and the slightly heavier ^{115}In, a nucleus also studied for its medical applications [5].

*e-mail: polvasil@phys.uoa.gr
**Present address: IRFU, CEA, Université Paris-Saclay, France

https://doi.org/10.1051/epjconf/202226011030

2 Experimental details

The proton beams, ranging from E_p = 3.0–4.0 MeV, were provided by the 5.5 MV T11 Tandem Accelerator of NCSR "Demokritos". The detector setup consisted of 3 HPGe detectors, angled at 55°, 90° and 165° with respect to the beam direction (a picture of the setup is shown in Fig. 1). Isotopically enriched (99.7%) ^{112}Cd and ^{114}Cd targets were used for the study of the radiative proton–capture reactions ^{112,114}Cd$(p,\gamma)^{113,115}$In, using two different techniques, namely, in–beam γ–spectroscopy (described in detail in Refs. [3, 7, 8]) and the activation method [3, 8, 9], where the latter was deemed necessary to account for the presence of low–lying isomers in ^{113,115}In [6].

Figure 1: Picture of the detector setup used in the experiments.

3 Analysis and Results

Transitions feeding the ground state that were observed in the in–beam spectra of our experiments are marked with *'s in Fig. 2, and were used to determine the in–beam ground state cross section (σ_{gs}). The same procedure was followed for the in–beam measurement of the isomeric transitions, by measuring the photopeaks marked with #'s in the in–beam spectra (Fig. 2).

The full contributions of the isomeric transitions were determined by means of the activation technique, described in detail in Refs. [3, 8, 9]. Finally, the total cross sections were determined as:

$$\sigma_T = \sigma_{gs} + \sigma_{is} \tag{1}$$

where σ_T, σ_{gs}, σ_{is} are the total, ground state and isomeric cross sections, respectively. Subsequently, the astrophysical S factors were deduced through:

$$S(E) = E\sigma_T(E)\exp(-2\pi\eta) \tag{2}$$

where η is the Sommerfeld parameter.

The experimental results were compared with data existing in the literature, as well as with theoretical calculations with the TALYS code [10], using all the available combinations of Optical Potentials (OMP), Nuclear Level Densities (NLD) and γ–Strength Functions

 https://doi.org/10.1051/epjconf/202226011030

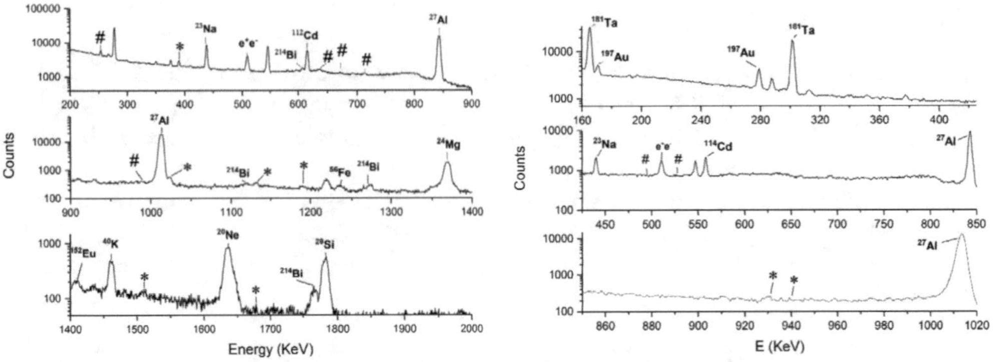

Figure 2: (a) Horizontal split–view (0.2–2.0 MeV) of a typical in–beam spectrum for the reaction $p+^{112}$Cd. Transitions to the gs of ^{113}In are marked with an (∗). Transitions to the isomeric state are marked with (#). Additional photopeaks arising from background radiation or other beam induced reactions are also marked. (b) Horizontal split–view (0.16–1.02 MeV) of a typical in–beam spectrum for the reaction $p+^{114}$Cd.

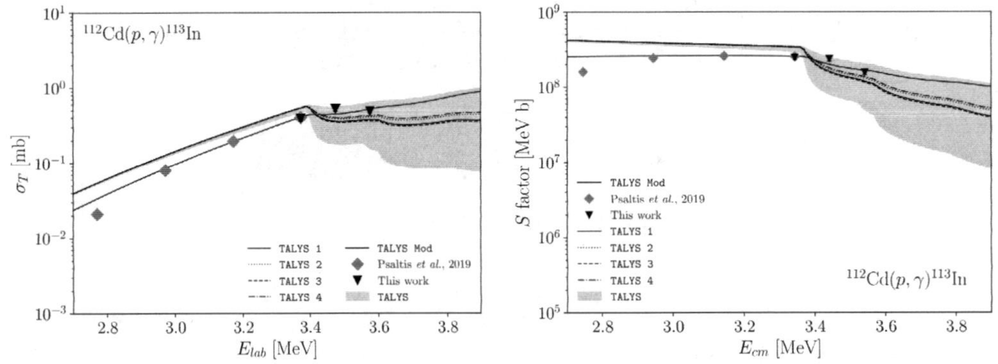

Figure 3: Total cross sections and astrophysical *S* factors for the reaction ^{112}Cd$(p,\gamma)^{113}$In (for details see text).

(γSF), with their default parameters, reaching a total of 96 combinations. The minimum and maximum value for each energy step was used to create the green shaded area in Figs. 3 and 4. The best data matching combinations of models, using their default parameters are also shown in these figures, labeled TALYS 1–4. The set of modified model parameters, achieving the best simultaneous agreement with all of the experimental data, is also plotted (labeled TALYS Mod. in Figs. 3, 4).

4 Conclusions

The experimental data presented in this work for the reaction ^{112}Cd$(p,\gamma)^{113}$In are in excellent agreement with those of the earlier work of Psaltis et al. [8], for the 3.4 MeV energy point,

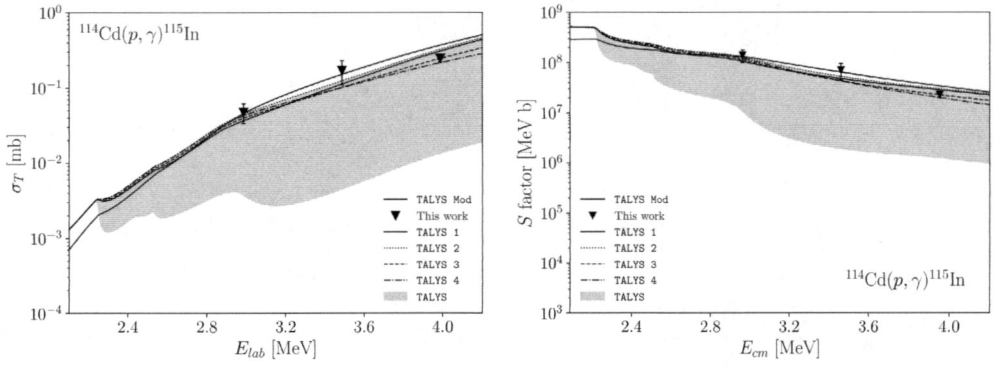

(a) (b)

Figure 4: Same as in Fig. 3, for the reaction ^{114}Cd$(p,\gamma)^{115}$In.

extending the measurements above the neutron emission threshold for the reaction (E_{th} = 3.397 MeV [6]), while still inside the Gamow energy window of astrophysical importance.

At the same time, the experimental data for both reactions studied in this work are in good agreement with the theoretical predictions of the TALYS v1.95 code, with the TALYS Mod. combination leading to a significantly improved agreement between theory and experiment for all of the studied channels, in a simultaneous fashion.

Further experimental and theoretical work is required for a firm insight at the driving mechanisms behind the *p*–process nucleosynthesis, in an energy region where the experimental data, even for stable nuclei, remain scarce.

Acknowledgements

AZ and AC acknowledge support by the Hellenic Foundation for Research and Innovation (HFRI) and the General Secretariat for Research and Technology (GSRT) under the HFRI PhD Fellowship grant (GA. No. 101742/2019 and 74117/2017, respectively).

H.F.R.I. Hellenic Foundation for Research & Innovation

References

[1] C. Iliadis, *Nuclear Physics of Stars* (John Whiley & Sons, 2015)
[2] W. Hauser and H. Feshbach, Phys. Rev. **87**, 366 (1952)
[3] P. Vasileiou et al., Nucl. Phys. A **1015**, 122298 (2021)
[4] M. Pignatari et al., Int. J. Mod. Phys. B **25 (04)**, 1630003 (2016)
[5] F. Tárkányi et al., Radiochim. Acta **93**, 561–569 (2005)
[6] Nudat3, National Nuclear Data Center, www.nndc.bnl.gov/nudat3/
[7] A. Khaliel et al., Phys. Rev. C **96**, 035806 (2017)
[8] A. Psaltis et al., Phys. Rev. C **99**, 065807 (2019)
[9] G. Gyürky et al., Eur. Phys. J. A **55**, 41 (2019)
[10] A. Koning et al., TALYS manual, v1.95 Edition, www.talys.eu (2020)

SHADES

$^{22}Ne(\alpha, n)^{25}Mg$ reaction rate in the Gamow window

David Rapagnani[1], *Chemseddine* Ananna[1], *Antonino* Di Leva[1], *Gianluca* Imbriani[1], *Matthias* Junker[2], *Marco* Pignatari[3,4,5,6], and *Andreas* Best[1,*]

[1]Università di Napoli "Federico II" and INFN - Sezione di Napoli, 80126 Napoli, Italy
[2]INFN, Laboratori Nazionali del Gran Sasso (LNGS), 67100 Assergi, Italy
[3]Konkoly Observatory, Research Centre for Astronomy and Earth Sciences, Hungarian Academy of Sciences, Konkoly Thege Miklos ut 15-17, H-1121 Budapest, Hungary
[4]E.A. Milne Centre for Astrophysics, Department of Physics and Mathematics, University of Hull, HU6 7RX, United Kingdom
[5]Joint Institute for Nuclear Astrophysics, Center for the Evolution of the Elements, Michigan State University, 640 South Shaw Lane, East Lansing, MI 48824, USA
[6]NuGrid Collaboration, nugridstars.org

Abstract. Neutron capture reactions are the main contributors to the synthesis of the heavy elements through the s-process. Together with $^{13}C(\alpha, n)^{16}O$, which has recently been measured by the LUNA collaboration in an energy region inside the Gamow peak, $^{22}Ne(\alpha, n)^{25}Mg$ is the other main neutron source in stars. Its cross section is mostly unknown in the relevant stellar energy (450 keV < E_{cm} < 750 keV), where only upper limits from direct experiments and highly uncertain estimates from indirect sources exist. The ERC project SHADES (UniNa/INFN) aims to provide for the first time direct cross section data in this region and to reduce the uncertainties of higher energy resonance parameters. High sensitivity measurements will be performed with the new LUNA-MV accelerator at the INFN-LNGS laboratory in Italy: the energy sensitivity of the SHADES hybrid neutron detector, together with the low background environment of the LNGS and the high beam current of the new accelerator promises to improve the sensitivity by over 2 orders of magnitude over the state of the art, allowing to finally probe the unexplored low-energy cross section. Here we present an overview of the project and first results on the setup characterization.

1 Introduction

In massive stars ($M>8M_{\odot}$) $^{22}Ne(\alpha, n)^{25}Mg$ is the main neutron source for the synthesis of the isotopes made by the slow neutron capture process (s-process) in the mass region A 60-90, between iron and the first neutron magic peak (N=50) [1]; together with $^{13}C(\alpha, n)^{16}O$, which has recently measured with high precision by LUNA collaboration ([2] and references therein) it is also responsible for the synthesis of s-process elements with mass $A \sim 90-209$ in Asymptotic Giant Branch (AGB) stars [3–6]. Despite of the many efforts in measuring $^{22}Ne(\alpha, n)^{25}Mg$ during the last fifty years (e.g. [7–11]), present uncertainty on reaction rate

*e-mail: andreas.best@na.infn.it

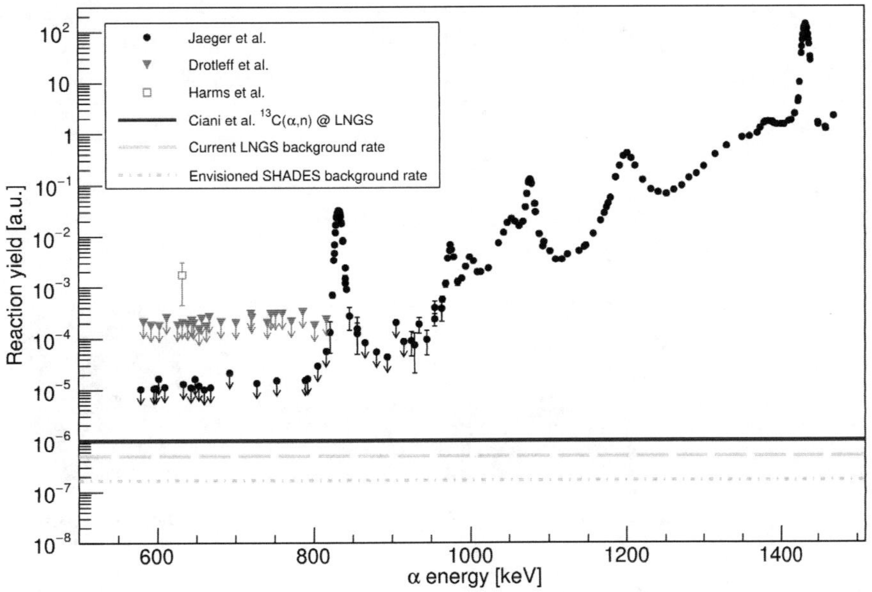

Figure 1. Reaction yield data from Jaeger et al. ([8], black circles), Drotleff et al. ([9], red triangles) and Harms et al. ([10], green rectangle). The background level estimated for the $^{13}C(\alpha,n)^{16}O$ LUNA measurement campaign [2] and the one envisioned for SHADES are also shown. See the text for more details.

estimate reaches a value of about 500% at temperature around 0.2-0.3 GK, the one of interest for helium burning. In addition, only upper limit values are known at the moment for the cross section, whereas the knowledge of its actual value is necessary for stellar models. For this the European Research Council founded project SHADES (Scintillator-He3 Array for Deep-underground Experiments on the S-process) is developing an innovative low background measurement campaign which will be described in the following sections.

2 SHADES

The challenging experimental condition limited the sensitivity of measurements performed so far. As it is clear from Figure 1, only upper limits for the reaction cross section at energies below $E_\alpha \lesssim 800$ keV are currently known. Since the efforts of Jeager et al. [8], which reduced by one order of magnitude the upper limit, no progress was made on $^{22}Ne(\alpha,n)^{25}Mg$ while a greater sensitivity is still required to measure actual values of its cross section. The SHADES project goal is to improve the experimental sensitivity by at least two order of magnitude. This increase in sensitivity will be obtained by (1) performing the reaction cross section measurement underground at Gran Sasso National Laboratory (LNGS) of the Italian Institute of Nuclear Physics (INFN), (2) employing the new high current LUNA-MV ion beam accelerator, (3) with the use of an innovative detection array which grants a selective discrimination of reaction neutrons and (4) with an high purity ^{22}Ne gas target system. This efforts will allow to perform (5) precise astrophysical estimates of s-process influenced elemental abundance, solving a longstanding issue on this important astrophysical process.

 https://doi.org/10.1051/epjconf/202226011031

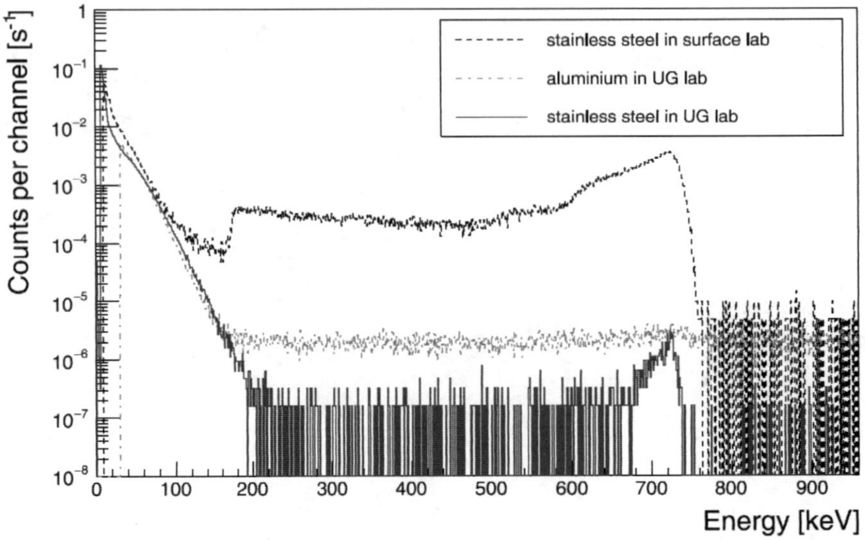

Figure 2. The spectra, as presented in [12], for surface measurements with stainless steel He3 proportional counters (black dots) and underground measurements with both aluminum (red dots) and stainless-steel (blue line) He3 proportional counters. See also [13] for more details.

2.1 Underground Laboratory

Being deep underground, LNGS offer a unique environment for low neutron background measurements. In this respect, Figure 2 shows the spectra of stainless steel and aluminum He3 proportional counters tested on surface laboratory and at LNGS: the natural neutron flux suppression is so high underground that the intrinsic activity of aluminum becomes the main source of background.

2.2 LUNA-MV

The new LUNA-MV accelerator, currently under installation at LNGS, is a 3.5 MV Singletron Accelerator [14] with an ECR source capable of deliver up to 500 μA of high energy resolved alpha particles beams. It has also an extremely high stability over time, a feature desirable since the long measurement runs expected for SHADES.

2.3 Array

The SHADES array is composed by 12 EJ-309 liquid scintillators and 18 He3 counters held by a borated polyethylene support which acts also as a shield for environmental neutrons (Figure 3). A reaction neutron will be recorded only after being moderation inside the scintillators and capture in the He3 proportional counters. Together with a Pulse Shape Discrimination analysis [15], this procedure allows a further background suppression.

2.4 Gas Target

In order to reach the sensitivity necessary to measure ^{22}Ne$(\alpha, n)^{25}$Mg at the lower energies, beam induced background events must be kept at minimum. Carbon and boron contamination, due to their higher (α, n) cross section, can be a formidable source of background

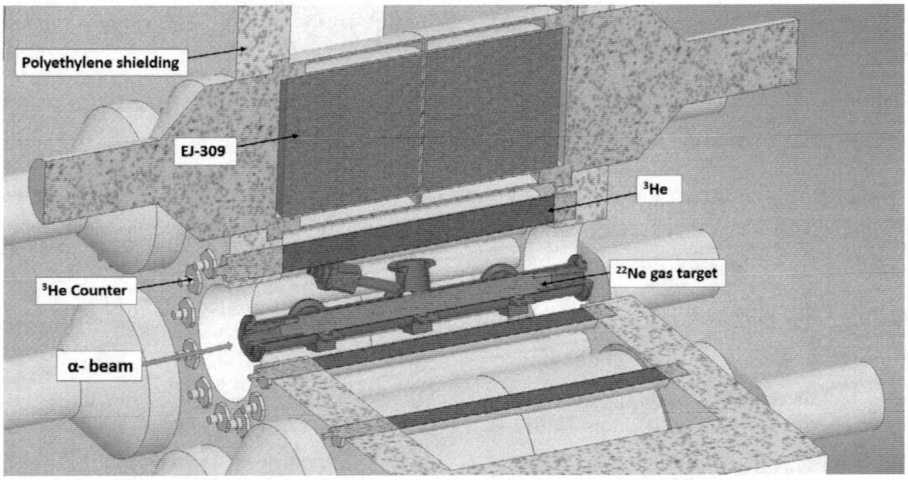

Figure 3. Cut-view of the SHADES detectors array and gas target setup.

neutrons. For this, a new recirculating high purity 30 keV thick Ne22 extended gas target was designed and is presently under commissioning in the Nuclear Astrophysics Laboratory of University of Naples "Federico II". A combination of high flow and high vacuum pumps in different pumping stage and differential configuration allow to reach the high vacuum conditions of LUNA-MV of $\lesssim 10^{-7}$ mbar from target region pressure of about 5 mbar in a few centimeters of beam line. Indeed, to optimize detection efficiency, the gas target dimensions has been kept as low as possible. Carbon-free pumps, gaskets and sealings and a gas purifier were also adopted to reduce carbon contaminant from vacuum elements degassing and air leaks. A detector at 165° with respect the ion beam direction will allow to measure back-scattered α-particles for ion beam current monitoring. Several pressure and temperature gauges will serve as stability monitor for gas target during the long reaction cross section measurement runs.

2.5 Astrophysical Evaluations

After an R-Matrix analysis of the cross section data, the astrophysical impact of SHADES experimental outcomes will be evaluated using the NuGrid codes and their stellar library for a complete set of stellar masses and metallicities [16–19].

3 Outlook

The ERC-funded project "Scintillator-He3 Array for Deep-underground Experiments on the S-process" (SHADES) aims to measure ^{22}Ne$(\alpha, n)^{25}$Mg down to neutron threshold energy with an innovative detection array and a low-background environment granted by the LNGS underground laboratories and an extremely pure gas target system. The detectors array and gas target design is already complete, while the array characterization is ongoing on surface laboratory in the University of Naples "Federico II" and underground at LNGS. The gas target is currently under installation and commissioning and will enter soon in the characterization phase through Ion Beam Analysis measurements to be performed at the Tandem Accelerator laboratory of the University of Campania in Caserta during winter 2021. The preparation of underground measurements is expected for Spring-Winter 2022 while the ^{22}Ne$(\alpha, n)^{25}$Mg rate measurements is planned for the whole 2023. Finally R-Matrix analysis and nucleosynthesys calculations will be performed during 2024.

Acknowledgement

The authors acknowledge funding from the European Research Council (ERC-StG 2019 #852016). MP acknowledges support to NuGrid from STFC (through the University of Hull's Consolidated Grant ST/R000840/1), and access to VIPER, the University of Hull HPC Facility. MP also Acknowledges the National Science Foundation (NSF, USA) under grant No. PHY-1430152 (JINA Center for the Evolution of the Elements), the "Lendulet-2014" Program of the Hungarian Academy of Sciences (Hungary), the ERC Consolidator Grant funding scheme (Project RADIOSTAR, G.A. n. 724560, Hungary), the ChETEC COST Action (CA16117), supported by the European Cooperation in Science and Technology, the ChETEC-INFRA project funded from the European Union's Horizon 2020 research and innovation programme (grant agreement No 101008324), and the IReNA network supported by NSF AccelNet.

References

[1] Woosley S.E. and Weaver T.A., Astrophys. J. Suppl. Ser. **101**, 181 (1995).
[2] Ciani G. et al. PRL **127**, 152701 (2021).
[3] Gallino R. et al., Astrophys. J. **333**, L25 (1988).
[4] Busso M., Gallino R., and Wasserburg G.J., Annual Review of Astronomy and Astrophysics **37**:239-309 (1999).
[5] Falk H., Annual Review of Astronomy and Astrophysics **43**:435-479 (2005)
[6] Karkas A.I. and Lattanzio J.C., Publications of the Astronomical Society of Australia **31** (2014).
[7] Haas F.X. and Bair J.K., Phys. Rev. C **7**, 2432 (1973).
[8] Jeager M. et al., Phys. Rev. Lett. **87**, 20 (2001).
[9] Drotleff H.W. et al., APJ **414**, 735-739 (1993).
[10] Harms V. et al., PRC **43**, 6 (1991).
[11] Massimi C. et al., PLB **768**, 1-6 (2017).
[12] Csedreki L. et al., NIM A **994**, 165081 (2021).
[13] A. Best et al., NIM B **812**, 1-6 (2016).
[14] Sen G. et al., NIM B 450 (2019).
[15] Balibrea-Correa J. et al., NIM A **906**, 103-109 (2018).
[16] Pignatari M. et al., The Astrophysical Journal Supplement Series **225**, Number 2 (2016).
[17] Ritter C. et al., Monthly Notices of the Royal Astronomical Society **480**, Issue 1, p.538-571 (2018).
[18] Battino U. et al., Monthly Notices of the Royal Astronomical Society **489**, Issue 1, p.1082-1098 (2019).
[19] Andrews S. et al., The Astrophysical Journal **890**, Issue 1, id.35, 14 pp. (2020). Monthly Notices of the Royal Astronomical Society, Volume 480, Issue 1, p.538-571

Description of weak-interaction rates within the relativistic energy density functional theory

Ante Ravlić[1,*], *Esra* Yüksel[2], *Yifei* Niu[3], *Nils* Paar[1], *Gianluca* Colò[4], and *Elias* Khan[5]

[1]Department of Physics, Faculty of Science, University of Zagreb, Bijenička c. 32, 10000 Zagreb, Croatia
[2]Physics Department, Faculty of Science and Letters, Yildiz Technical University, TR-34220, Esenler/Istanbul, Turkey
[3]School of Nuclear Science and Technology, Lanzhou University, Lanzhou 730000, China
[4]Dipartimento di Fisica, Università degli Studi di Milano, Milano, Italy and INFN, Sezione di Milano, Via Celoria 16, 20133 Milano, Italy
[5]Institut de Physique Nuclèaire, Universitè Paris-Sud, IN2P3-CNRS, Universitè Paris-Saclay, F-91406 Orsay Cedex, France

Abstract. A new theoretical framework has been established and applied in the calculation of electron capture (EC) and β-decay rates in stellar environment, characterized by high density and temperature. For the description of the nuclear properties, the finite-temperature Hartree Bardeen-Cooper-Schrieffer (FT-HBCS) theory based on the relativistic derivative-coupling D3C* interaction is employed. In order to describe the charge-exchange transitions, the finite-temperature proton-neutron quasi-particle random-phase approximation is developed (FT-PNRQRPA) which includes both temperature and pairing correlations. In the FT-HBCS calculations, only the isovector pairing is included, while in the residual interaction of the FT-PNRQRPA both the isovector and isoscalar pairing contribute. In this work, results for EC and β-decay rates are presented in the temperature interval $T = 0$–1.5 MeV and stellar density $\rho Y_e = 10^7$ and 10^9 g/cm^3. Both allowed 0^+, 1^+ and first-forbidden transitions 0^-, 1^- and 2^- are included in the calculations. It is shown that interplay between pairing correlations and finite-temperature effects can lead to significant changes in rates. It is also important to include de-excitations, i.e. transitions with negative Q-value, that become increasingly significant at higher temperatures especially for pf-shell nuclei.

1 Introduction

Reactions mediated by the weak force are of significant importance for nuclear and particle physics, as well as nuclear astrophysics, where they specify the dynamics of core-collapse supernovae (CCSNe) and the time-scale of r-process. Electron capture (EC) plays a prominent role in the dynamics of CCSNe, which are mainly determined by two parameters: (i) electron-to-baryon ratio (Y_e) and (ii) the core entropy. An exploding star in the last stages of its life maintains a delicate balance between gravitational pull and degeneracy pressure originating from the Pauli blocking between electrons. The EC reactions remove electrons from

*e-mail: aravlic@phy.hr

the system, thus lowering Y_e, while outgoing neutrinos carry out entropy from the core. At certain stages in the collapse, β-decay can compete with EC when temperature is in the range $T_9(K) = 1$–10 ($T_9(K)$ denoting temperature in 10^9 K) and product of stellar density (ρ) and Y_e is $\rho Y_e \geq 10^7$ g/cm^3 [1]. Therefore, for a complete understanding of the CCSNe dynamics, we need to study reaction rates of both EC and β-decay in stellar medium. Since the conditions that occur in CCSNe are beyond the experimental reach, simulations of CCSNe have to rely on theoretical predictions. Therefore, it is necessary to develop a robust theoretical framework for the description of key weak force reactions under extreme conditions of density and temperature. First steps toward this direction were made some time ago by Fuller, Fowler and Newman in Refs. [2–5] using the independent particle model. They have recognized the key role played by the Gamow-Teller (GT) resonance. Presently, the theoretical methods can be split into two main categories: (i) based on the shell-model (SM) calculations [6–8] and (ii) based on the random-phase approximation (RPA). Although shell-model calculations can reproduce experimental strength functions with success, they are computationally infeasible for nuclei with $A > 70$ due to the huge configuration space required for the description of nuclear excited states. Compared to the shell model calculations, the RPA method based on the nuclear energy density functionals (EDF) has a solid microscopic basis and the advantage of calculating main nuclear properties and excitations across the nuclide chart. Previously, the finite-temperature RPA (FT-RPA) method based on the relativistic [9] and non relativistic [10, 11] energy density functionals was used in the calculation of the EC rates and cross sections. It was shown that the unblocking effect of temperature in the charge-exchange transitions has a non-negligible effect on the EC results. Unblocking of the Gamow-Teller excitations at finite temperatures and their consequences on the EC and β-decay rates were also studied in the framework of the thermal quasi-particle RPA (TQRPA) which treats thermal excitations in a thermodynamically consistent way [12–14].

Recently, we have developed a theoretical framework for the calculation of the nuclear charge-exchange transitions in open-shell nuclei at finite temperatures, based on the relativistic nuclear EDFs. The proton-neutron relativistic quasi-particle RPA (FT-PNRQRPA) has been applied on top of the FT-HBCS to calculate the relevant charge-exchange transitions alongside the EC and β-decay rates [15–17]. In this work, we also use the same model to study the temperature dependence of EC and β-decay rates of selected Ti and Fe nuclei.

2 Theoretical formalism

For the calculation of the nuclear properties, we employ the RMF theory based on the meson-exchange relativistic energy density functional with derivative coupling D3C* [18]. Within this framework, nucleons interact via the exchange of isoscalar-scalar σ meson, isoscalar-vector ω meson and isovector-vector ρ meson, together with the electromagnetic (EM) interaction. The model is formulated starting from the Lagrangian density [18, 19]

$$\mathcal{L} = \mathcal{L}_N + \mathcal{L}_m + \mathcal{L}_{int}, \tag{1}$$

where $\mathcal{L}_N$ is the Dirac Lagrangian for free nucleons, $\mathcal{L}_m$ is the free meson Lagrangian and $\mathcal{L}_{int}$ is the interaction term (see Refs. [18–20] for details). We treat both the mean-field and pairing correlations at finite-temperature within the FT-HBCS theory [21]. At finite temperature, occupation probabilities of single-particle states are given by

$$n_k = v_k^2(1 - f_k) + u_k^2 f_k, \tag{2}$$

where u_k, v_k are the FT-HBCS amplitudes and f_k is the Fermi-Dirac factor, $f_k = (1 + e^{-\beta E_k})$, $\beta = 1/k_B T$, T denoting the temperature and k_B is the Boltzmann constant. The quasi-particle

Table 1. The monopole pairing strengths for neutrons(protons) $G_{n(p)}$ (in units MeV/A where A is the nuclear mass number), adjusted to reproduce the empirical neutron(proton) pairing gaps $\Delta^{5p}_{n(p)}$ calculated using the five-point formula [22] for ^{52}Ti and ^{58}Fe. Calculations are performed using the FT-HBCS with D3C* interaction.

Nucleus	Δ^{5p}_n [MeV]	Δ^{5p}_p [MeV]	G_n [MeV/A]	G_p [MeV/A]
^{52}Ti	0.95	1.63	19.40	30.60
^{58}Fe	1.45	1.73	21.60	30.60

(q.p.) energies are given by $E_k = \sqrt{(\varepsilon_k - \lambda)^2 + \Delta_k^2}$, where ε_k is the single-particle energy, λ is the chemical potential for either protons or neutrons and Δ_k is the state-dependent pairing gap obtained by solving the gap equation [21]. In this work, we use a monopole pairing interaction characterized by strength parameters $G_{n(p)}$ for neutrons(protons) which are determined according to the five-point formula [22]. Corresponding parameters $G_{n(p)}$ for ^{52}Ti and ^{58}Fe can be found in Table 1, adjusted to reproduce the five-point neutron(proton) pairing gaps $\Delta^{5p}_{n(p)}$.

The FT-PNRQRPA consists in a small-amplitude limit of the time-dependent FT-HBCS theory. Recently, it was developed and applied to study the charge-exchange transitions in Ref. [15]. The particle-hole (ph) part of the residual interaction is derived self-consistently from the same EDF used in the ground-state calculation. The strength parameter of the Landau-Migdal interaction $g' = 0.76$ is determined by reproducing the experimental excitation energy of GT$^-$ resonance in ^{208}Pb [18]. The particle-particle (pp) part of the residual interaction has both isovector ($T = 1, S = 0$) and isoscalar ($T = 0, S = 1$) components. The isovector component assumes the pairing part of the Gogny D1S interaction [16], while for the isoscalar component a formulation consisting of a sum of two Gaussians is employed

$$V_{12} = V_0^{is} \sum_{j=1}^{2} g_j e^{-r_{12}^2/\mu_j^2} \prod_{S=1,T=0}, \tag{3}$$

where $\prod_{S=1,T=0}$ denotes projector on $T = 0, S = 1$ states, r_{12} is the relative coordinate, $\mu_1 = 1.2$ fm, $\mu_2 = 0.7$ fm and the strength parameters are set to $g_1 = 1$ and $g_2 = -2$ [23]. The isoscalar pairing strength is constrained using the following functional [24]

$$V_0^{is} = V_L + \frac{V_D}{1 + e^{a+b(N-Z)}}, \tag{4}$$

with values defined in Ref. [17] and adjusted to reproduce all experimental data on β-decay half-lives in the range $8 \leq Z \leq 82$.

The FT-PNRQRPA equation can be written in the matrix form for a well defined angular momentum and parity J^π as [15, 25]

$$\begin{pmatrix} \tilde{C} & \tilde{a} & \tilde{b} & \tilde{D} \\ \tilde{a}^\dagger & \tilde{A} & \tilde{B} & \tilde{b}^T \\ -\tilde{b}^\dagger & -\tilde{B}^* & -\tilde{A}^* & -\tilde{a}^T \\ -\tilde{D}^* & -\tilde{b}^* & -\tilde{a}^* & -\tilde{C}^* \end{pmatrix} \begin{pmatrix} \tilde{P} \\ \tilde{X} \\ \tilde{Y} \\ \tilde{Q} \end{pmatrix} = E_n \begin{pmatrix} \tilde{P} \\ \tilde{X} \\ \tilde{Y} \\ \tilde{Q} \end{pmatrix}, \tag{5}$$

where the excited state energy is determined by the eigenvalue E_n, while the set of eigenvectors $\begin{pmatrix} \tilde{P} & \tilde{X} & \tilde{Y} & \tilde{Q} \end{pmatrix}^T$ is used to calculate the transition strength. Explicit forms of the submatrices $\tilde{A}, \tilde{B}, \tilde{C}, \tilde{D}, \tilde{a}, \tilde{b}, \tilde{a}^T, \tilde{b}^T, \tilde{a}^\dagger, \tilde{b}^\dagger$ can be found in Ref. [25]. Calculations can be performed for both directions of the isospin projection $\Delta T_z = \pm 1$. Finally, the strength function for the charge-exchange external field operator $\hat{F}$ in the $\Delta T_z = -1$ direction $(n \rightarrow p)$ can be written as

$$B(\hat{F})_n = \frac{1}{1 - e^{-\beta(E_n - \lambda_{np})}} \left| \sum_{\pi > \nu} \left[(\tilde{X}^n_{\pi\nu} u_\pi v_\nu + \tilde{Y}^n_{\pi\nu} v_\pi u_\nu) \sqrt{1 - f_\pi - f_\nu} \right.\right.$$
$$\left.\left. + (\tilde{P}^n_{\pi\nu} u_\pi u_\nu - \tilde{Q}^n_{\pi\nu} v_\pi v_\nu) \sqrt{f_\pi - f_\nu} \right] \langle \pi || \hat{F} || \nu \rangle \right|^2, \tag{6}$$

where $\pi(\nu)$ denote quasi-proton(neutron) states, $\lambda_{np} = \lambda_n - \lambda_p$ is the difference between the neutron and proton chemical potential and summation is performed over states with $E_\pi > E_\nu$. The exponential prefactor $(1 - e^{-\beta(E_n - \lambda_{np})})^{-1}$ appears to relate the FT-PNRQRPA strength function to the physical strength function [26]. This implies that not only positive energy transitions with $E > \lambda_{np}$ should be included in the calculation of the strength function but also the negative energy transitions with $E < \lambda_{np}$, i.e. de-excitations. The latter are calculated using the strength function in the $\Delta T_z = +1$ channel (see Refs. [17, 26] for further details).

3 Description of electron capture rates

In presupernova environment, nuclei are assumed to be fully ionized and immersed in electron plasma, described by the Fermi-Dirac distribution of electrons. Such a system is characterized by its temperature T and stellar density ρY_e. To calculate the EC rates including both allowed $(0^+, 1^+)$ and first-forbidden transitions $(0^-, 1^-, 2^-)$ we use the formalism given in Refs. [27, 28] where the current-current form of the weak interaction Hamiltonian is assumed. By following Refs. [27, 28] the EC cross section σ_{ec} can be expressed in terms of multipole transition operators $\hat{\mathcal{M}}_J, \hat{\mathcal{L}}_J, \hat{\mathcal{T}}^{mag}_J, \hat{\mathcal{T}}^{el}_J$, for a well defined J^π. The coupling constant g_A of the axial-vector part of multipole transition operators is quenched from the free-nucleon value $g_A = -1.26$ to $g_A = -1.0$ [16]. Electron capture rates are calculated by folding the cross section σ_{ec} with a Fermi-Dirac distribution of electrons

$$\lambda_{ec} = \frac{(m_e c^2)^3}{\pi^2 \hbar^3} \int_{W_0}^{\infty} p W \sigma_{ec}(W) f_e(W) dW, \tag{7}$$

where the dimensionless electron energy is defined as $W = E_e/(m_e c^2)$, m_e is the electron mass and $p = p_e/(m_e c)$ is the electron momentum. The minimum electron energy W_0 is imposed by the condition on outgoing neutrino energy $E_\nu > 0$. The Fermi-Dirac distribution is defined by the electron chemical potential μ_e and temperature T

$$f(W, \mu_e, T) = \frac{1}{\exp\left(\frac{W m_e c^2 - \mu_e}{k_B T}\right) + 1}. \tag{8}$$

The electron chemical potential μ_e is determined by the charge-neutrality condition [16, 29]. While in Ref. [16] we have performed the sensitivity study of EC rates on the isoscalar pairing strength V_0^{is}, here we constrain it by using the functional form of Eq. (4), with parameters determined by reproducing the experimental β-decay half-lives [17]. In Figure 1, we show the temperature evolution of EC rates for two pf-shell nuclei, ^{52}Ti and ^{58}Fe, at stellar densities $\rho Y_e = 10^7$ and 10^9 g/cm^3. The FT-HBCS + FT-PNRQRPA calculations with D3C* interaction (blue dashed line) are compared to the large-scale shell model (LSSM) calculations of Langanke *et. al.* (red squares) [6] and shell-model calculations based on the pf-GXPF1J interaction (green triangles) [7, 30]. It is seen that EC rates increase with temperature and the stellar density ρY_e. The former fact is explained by thermal unblocking of q.p. levels at finite temperature and contribution of de-excitations (see Ref. [17] for details), while the latter is a consequence of increasing the lepton phase-space, thus including more transition strength in the EC rate integral of Eq. (7). The present model somewhat overestimates the shell-model rates, however, the overall trends are reasonably reproduced.

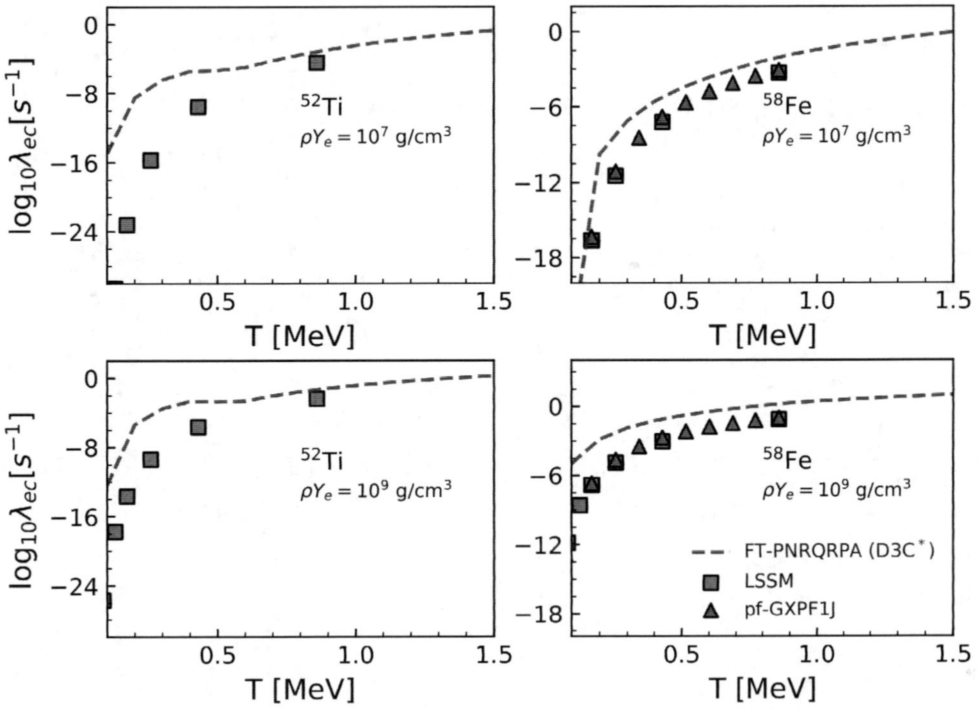

Figure 1. Electron capture rates λ_{ec} for ^{52}Ti and ^{58}Fe for stellar densities $\rho Y_e = 10^7$ and 10^9 g/cm^3. The FT-PNRQRPA results (blue dashed line) with D3C* interaction are shown together with the LSSM results (red squares) [6], and the shell-model results based on the pf-GXPF1J interaction (green triangles) [7, 30].

4 Description of β-decay rates

Same assumptions of presupernova environment are also applied in the study of stellar β-decay rates. Both allowed and first-forbidden transitions are included in the calculations. The general form of the β-decay rate in stellar conditions is given by [29, 31]

$$\lambda_\beta = \frac{\ln 2}{K} \int_0^{p_0} p^2 (W_0 - W)^2 F(Z, W) C(W)[1 - f(W, \mu_e, T)] dp, \tag{9}$$

where W_0 is the maximum electron energy given by the difference between initial and final nuclear mass, and integration is performed up to a maximal electron momentum p_0. The Fermi function $F(Z, W)$ takes into account distortion of electron wavefunctions [29]. The shape-factor $C(W)$ depends on the quantum numbers of considered transitions, and its functional form can be found in Ref. [31]. The constant K is measured in super-allowed β-decay

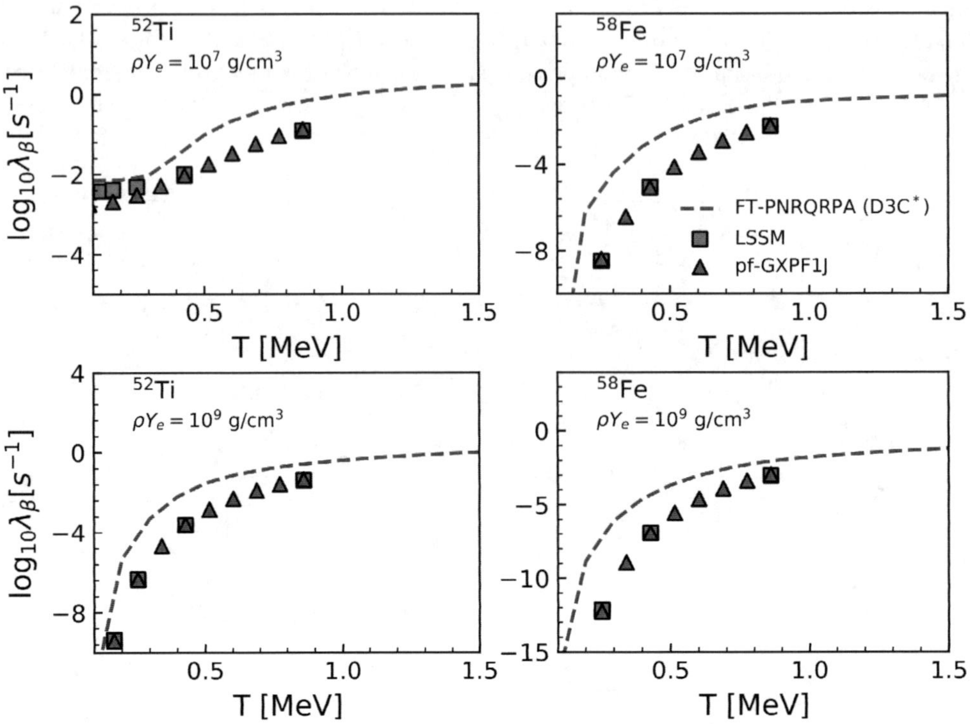

Figure 2. The same as Fig. 1 but for β-decay rate λ_β.

to be $K = 6144 \pm 2$ s [32]. It is important to note that within our model, the same parameters used in EC calculations are applied to the β-decay calculation. Thus, the axial-vector coupling is quenched to $g_A = -1.0$. In Figure 2, the results calculated with the FT-HBCS+FT-PNRQRPA model using D3C* interaction (blue dashed line) are compared to the LSSM results (red squares) [6], and the shell-model results based on the pf-GXPF1J interaction (green triangles) [7, 30] at $\rho Y_e = 10^7$ (upper panels) and $\rho Y_e = 10^9$ g/cm^3 (lower panels). Although the shell-model results include only the allowed GT transitions, influence of first-forbidden transitions for pf-shell nuclei is negligible up to $T < 1.5$ MeV [17]. In the case of β-decay, it is observed that rates decrease with increasing ρY_e. This effect is related to the phase-space blocking with increasing electron chemical-potential. As the temperature increases, electron chemical potential is decreased, thus allowing for an increase in the β-decay rate. Also, there are two main reasons for the increase of β-decay rate with temperature: (i) vanishing of pairing correlations and (ii) contribution of de-excitations (check Ref. [17] for more details on de-excitations). With vanishing pairing correlations, low lying strength in the Q_β window gets redistributed to lower excitation energies, thus increasing the rate. Furthermore, with increasing temperature, negative energy transitions (de-excitations) become unblocked and have a dominant contribution to the β-decay rate, especially for pf-shell nuclei near the valley of stability [17]. It is observed that the FT-HBCS+FT-PNRQRPA results have good agreement with two shell-model results, meaning that main trends of the temperature evolution of β-decay rates are reproduced.

5 Conclusions

Recently developed FT-HBCS+FT-PNRQRPA based on the relativistic D3C* derivative coupling interaction has been applied to study both the EC and β-decay rates in presupernova conditions. Our model allows us to study the interplay between finite-temperature and pairing correlation effects, as well as their influence on reactions mediated by the weak force. Rates are characterized by their dependence on temperature T and stellar density ρY_e. In stellar environment, nuclei can be found in highly-excited states, thus the proper treatment of negative energy transitions, i.e. de-excitations is of crucial importance. Model calculations were performed for two pf-shell nuclei ^{52}Ti and ^{58}Fe for EC and β-decay stellar rates and compared with respective shell-model calculations. It is shown that both EC and β-decay rates increase with increasing temperature, however, due to different reasons. The EC rates mainly increase because of the thermal unblocking of q.p. states while the β-decay rates are considerably influenced by the negative energy transitions. For both EC and β-decay rates, vanishing of pairing correlations at critical temperature influences the rates. It is demonstrated that rates based on the FT-HBCS+FT-PNRQRPA model can reproduce the main trends of shell-model rates, with considerably less computational resources. Therefore, our model is instrumental to provide large-scale rate tables of EC and β-decay to be used in astrophysical simulations of interest. Extension of the model from spherical to axially-deformed nuclei is left for future work.

6 Acknowledgments

This work is supported by the "QuantiXLie Centre of Excellence", a project co-financed by the Croatian Government and European Union through the European Regional Development Fund, the Competitiveness and Cohesion Operational Programme (KK.01.1.1.01.0004). Y. F. N. acknowledges the support from National Natural Science Foundation of China under Grant No. 12075104.

References

[1] G. Martinez-Pinedo, K. Langanke, D.J. Dean, The Astrophysical Journal Supplement Series **126**, 493 (2000)
[2] G. M. Fuller, W. A. Fowler, M. J. Newman, The Astrophysical Journal Supplement Series **42**, 447 (1980)
[3] G. M. Fuller, W. A. Fowler, M. J. Newman, The Astrophysical Journal **252**, 715 (1981)
[4] G. M. Fuller, W. A. Fowler, M. J. Newman, The Astrophysical Journal Supplement Series **48**, 279 (1982)
[5] G. M. Fuller, W. A. Fowler, M. J. Newman, The Astrophysical Journal **293**, 1 (1985)
[6] K. Langanke, G. Martínez-Pinedo, Atomic Data and Nuclear Data Tables **79**, 1 (2001)
[7] K. Mori, M.A. Famiano, T. Kajino, T. Suzuki, J. Hidaka, M. Honma, K. Iwamoto, K. Nomoto, T. Otsuka, The Astrophysical Journal **833**, 179 (2016)
[8] T. Suzuki, M. Honma, H. Mao, T. Otsuka, T. Kajino, Phys. Rev. C **83**, 044619 (2011)
[9] Y.F. Niu, N. Paar, D. Vretenar, J. Meng, Phys. Rev. C **83**, 045807 (2011)
[10] A.F. Fantina, E. Khan, G. Colò, N. Paar, D. Vretenar, Phys. Rev. C **86**, 035805 (2012)
[11] N. Paar, G. Colò, E. Khan, D. Vretenar, Phys. Rev. C **80**, 055801 (2009)
[12] A.A. Dzhioev, A.I. Vdovin, V.Y. Ponomarev, J. Wambach, K. Langanke, G. Martínez-Pinedo, Phys. Rev. C **81**, 015804 (2010)

[13] A.A. Dzhioev, K. Langanke, G. Martínez-Pinedo, A.I. Vdovin, C. Stoyanov, Phys. Rev. C **101**, 025805 (2020)
[14] A.A. Dzhioev, A.I. Vdovin, V.Y. Ponomarev, J. Wambach, Bulletin of the Russian Academy of Sciences: Physics **72**, 269 (2008)
[15] E. Yüksel, N. Paar, G. Colò, E. Khan, Y.F. Niu, Phys. Rev. C **101**, 044305 (2020)
[16] A. Ravlić, E. Yüksel, Y.F. Niu, G. Colò, E. Khan, N. Paar, Phys. Rev. C **102**, 065804 (2020)
[17] A. Ravlić, E. Yüksel, Y.F. Niu, N. Paar, Phys. Rev. C **104**, 054318 (2021)
[18] T. Marketin, D. Vretenar, P. Ring, Phys. Rev. C **75**, 024304 (2007)
[19] S. Typel, Phys. Rev. C **71**, 064301 (2005)
[20] T. Nikšić, N. Paar, D. Vretenar, P. Ring, Computer Physics Communications **185**, 1808 (2014)
[21] A.L. Goodman, Nuclear Physics A **352**, 30 (1981)
[22] M. Bender, K. Rutz, P.G. Reinhard, J.A. Maruhn, Eur. Phys. J. A **8**, 59 (2000)
[23] N. Paar, T. Nikšić, D. Vretenar, P. Ring, Phys. Rev. C **69**, 054303 (2004)
[24] Z. Niu, Y. Niu, H. Liang, W. Long, T. Nikšić, D. Vretenar, J. Meng, Physics Letters B **723**, 172 (2013)
[25] E. Yüksel, G. Colò, E. Khan, Y.F. Niu, K. Bozkurt, Phys. Rev. C **96**, 024303 (2017)
[26] E.M. Ney, A. Ravlić, J. Engel, N. Paar (2021), unpublished
[27] J. Walecka, *Theoretical Nuclear and Subnuclear Physics*, Theoretical Nuclear and Subnuclear Physics (Imperial College Press, London, 2004), ISBN 9789812388988
[28] J.S. O'Connell, T.W. Donnelly, J.D. Walecka, Phys. Rev. C **6**, 719 (1972)
[29] G. Martínez-Pinedo, Y.H. Lam, K. Langanke, R.G.T. Zegers, C. Sullivan, Phys. Rev. C **89**, 045806 (2014)
[30] T. Suzuki, private communication
[31] T. Marketin, L. Huther, G. Martínez-Pinedo, Phys. Rev. C **93**, 025805 (2016)
[32] J.C. Hardy, I.S. Towner, Phys. Rev. C **79**, 055502 (2009)

Red Dwarfs as Sources of Cosmic Rays

Vera Yu. Sinitsyna[1,*], *Vera G.* Sinitsyna[1,**], and *Yurii I.* Stozhkov[1]

[1]P.N. Lebedev Physical Institute, Moscow, Russia

Abstract. The Pamela, Fermi, AMS-02 spectrometer experimental data cannot be explained using the diffusive models of propagation of cosmic-rays accelerated at the supernova shocks and require the existence of nearby sources of cosmic rays at the distances less than one kpc. These sources could explain the growth of the ratio of galactic positrons to electrons with an energy increase, the complex dependence of the exponent of the proton and alpha spectra on the energy, the cosmic ray anomaly component origin. We consider active dwarf stars as possible sources of galactic cosmic rays in the energy range up to $\sim 10^{14}$ eV. The generation of high-energy cosmic rays should be accompanied by high-energy gamma-ray emission, which may be detected. The TeV gamma-ray emission mostly of flaring type from active red dwarf stars V388 Cas, V547 Cas, V780 Tau, V962 Tau, V1589 Cyg, GJ 3684, GJ 1078 and GL 851.1 was detected recently with SHALON long-term observations. This result confirms that active dwarf stars are also the sources of high-energy galactic cosmic rays.

1 Introduction

Supernova explosions are considered as sources of Galactic cosmic rays. Indeed, investigations of last years show that the Supernova Remnants (SNRs) are powerful accelerators of cosmic-ray particles up to $\sim 10^{17}$ eV. But, the proton and helium spectra at high and very high energies measured by PAMELA and AMS-02 can't be explained using the diffusive models of propagation of supernova shock born cosmic-rays. Also, there is no explanation of the excess of 10 - 300 GeV positron flux detected by PAMELA and AMS-02 experiments (Fig. 1). An interpretation of all these observations hints at a presence of nearby sources of cosmic rays at the distances less than 1 kpc from the solar system. We consider active dwarf stars as possible sources of galactic cosmic rays with energies up to $\sim 10^{14}$ eV whose generation is accompanied by the X-ray and high-energy γ-ray emission including TeV-energy range.

2 On red dwarf stars

The dwarf stars belong to the G - M classes of the main sequence of stars and are more than 90% of all 2×10^{11} stars in our Galaxy. The nearest of the dwarf stars are at several parsecs from Earth and are uniformly distributed inside of the galactic disc [1]. These stars produce powerful stellar flares, which sometimes occur several times per day, but its flare frequency is uncertain. The total energy release estimates as $10^{34} - 10^{36}$ ergs [1, 2]. The total energy of cosmic rays produced by the dwarf star flares in the Galaxy is estimated as $\approx 10^{51} - 10^{53}$ ergs [3]. This could provide the amount of energy of charged cosmic rays in our Galaxy compared with ones from SN explosions. If active dwarf stars produce cosmic rays with $E \leq 10^{14}$ eV, it can explain the anomalies detected by PAMELA, AMS-02, CALET, DAMPE [4, 5].

*e-mail: sinicinavy@lebedev.ru
**e-mail: sinits@sci.lebedev.ru

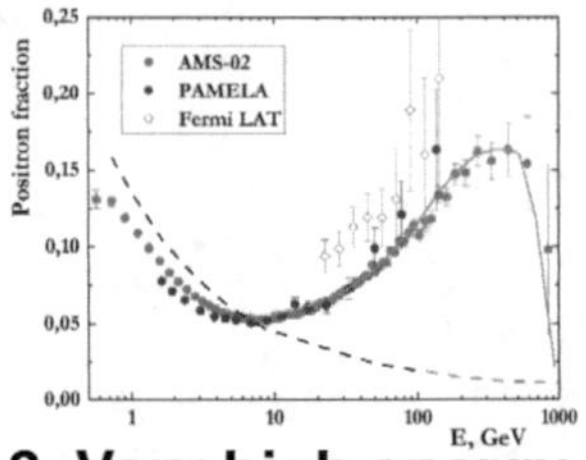

Figure 1. The ratio of e^+ flux to the sum of e^- and e^+ fluxes versus energy of these particles. Blue points are the PAMELA data and red points are the AMS-02 data. Dashed line shows model predictions from [10]. Blue line shows the result of calculations of ratio fluxes of e^+ to $(e^+ + e^-)$ versus energy if the probability of positron release from stellar flare site is taken into account.

3 Very high energy emission of red dwarf stars

TeV γ-ray emission from the M-stars was detected in SHALON experiment [3] on the level of $> 5\sigma$ and the average fluxes shown in Table 1. During the long-term observations of red dwarfs the flux variations of at energies up to 10 TeV were detected. Both, average (△) and flaring (▲) spectral energy distributions (SED) are presented in Fig. 2. The energy spectrum shapes at $E > 800$ GeV can be fitted with a soft power law: $dF/dE \propto E^k$, the details are listed in Table 1. The duration of high flux periods corresponding to the flare spectra was found to befrom 2 to 5 days [3]. The detected TeV γ-ray fluxes show that these objects produce flares with a total energy of $\sim 10^{34} - 10^{35}$ ergs. But, the most powerful solar flare is 10^{32} ergs.

The searches for counterparts for the detected at TeV energy M-stars in X-rays and MeV-GeV energies were performed. X-ray emission from the position of these objects was detected by ROSAT . In Fig. 2 the emission maps of listed red dwarfs by SHALON in comparison with data from X-ray data are presented. The X-ray emission (0.1 - 2.4 keV) from the position of discussed red dwarfs (error $\sim 10''$) was detected with ROSAT with energy fluxes from 3.9×10^{-13} to 2.5×10^{-12} erg cm^{-2} s^{-1} and indexes of ~ -2. It corresponds to the estimated luminosity of from 5×10^{27} to 1×10^{29} erg s^{-1}. High-energy emission from unassociated sources at distances of $< 0.6°$ from V962 Tau, V1589 Cyg, GJ 1078 and GL 851.1 were also detected by Fermi LAT. The spectra of each of unassociated objects are presented in Fig. 2 with bow tie and is consistent with SHALON data for these red dwarfs at TeV energies. Unfortunately, V388 Cas, V547 Cas, V780 Tau and GJ 3684 were not active enough after 2008 to produce a significant detection by Fermi LAT launched in 2008. But, V1589 Cyg, V962 Tau, GJ 1078, GL 851.1 were quite active during the whole period of observations and the nearby Fermi LAT unassociated sources may actually be these red dwarfs.

4 Discussion

The discovery of flaring red dwarf stars at $E \approx 1 \div 20$ TeV by SHALON, confirms, that these stars are also the sources of high energy cosmic rays and can explain a number phenomenon viewed in space experiments. The space experiments in the energy range from ~ 50 GeV up to several hundreds of GeV show that the spectra of protons and α-particles do not follow single power law with constant exponent γ [5, 7]. If galactic cosmic rays are produced by shock-waves from supernova explosions and cosmic ray life-time in Galaxy is about $10^6 - 10^8$ years the diffusion process smooths the spectra of these particles. As there are many active dwarf stars nearby the solar system and their stellar activity is quasi-periodic, one can expect

Table 1. Red stars of Flare star catalogue [1] viewed by SHALON.
Here ^{a} Integral flux at energy > 800 GeV in units of $10^{-12} cm^{-2} s^{-1}$ and
^{b} Index in fitting of TeV differential spectrum in form of power low $dF/dE \propto E^{k_\gamma}$ and $dF/dE \propto E^{k_{Flare}}$)

Source	Distance, pc	Spectral type	$F(> 0.8$ TeV$)^a$	k_γ^b	k_{Flare}^b
V388 Cas	10.5	M5V	0.84 ± 0.20	-2.51 ± 0.15	-2.91 ± 0.18
V547 Cas	10.1	dM2.5e/M2V	0.14 ± 0.03	-2.41 ± 0.29	-2.12 ± 0.32
V780 Tau	10.4	m/M5.5 or M7V	0.23 ± 0.03	-2.52 ± 0.15	-3.01 ± 0.21
V962 Tau	-	-	0.39 ± 0.04	-2.54 ± 0.16	-2.95 ± 0.22
V1589 Cyg	23 - 32	dM4.5e	0.13 ± 0.02	-2.91 ± 0.18	-3.15 ± 0.29
GJ 1078	20.6	M4.5V	0.38 ± 0.14	-2.46 ± 0.48	-3.11 ± 0.49
GJ 3684	16.5	m/M4V	0.33 ± 0.11	-2.64 ± 0.27	-3.04 ± 0.26
GL 851.1	21.7	dM0e or K5V	0.55 ± 0.09	-2.54 ± 0.42	-2.32 ± 0.21

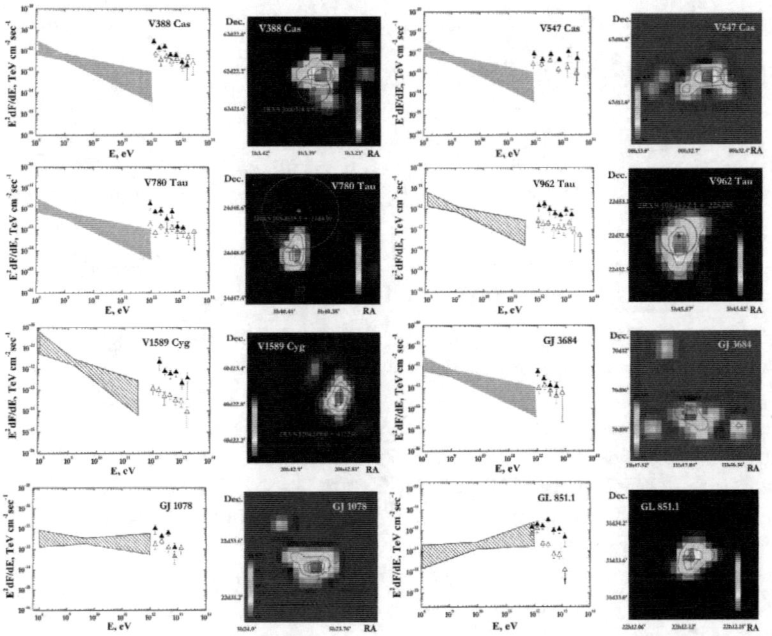

Figure 2. SEDs of red dwarfs by SHALON and Fermi LAT; Emission maps of red dwarfs by SHALON.

that the value of exponent γ may change if one or several such stellar objects are at the distances of several pc from the Sun.

Another problem could be explained with nearby cosmic ray sources is so-called anomaly component consisting of single ionized atoms like He, N, O, Ne, C, Mg, Si and others [8, 9] with energies from ~ 5 to 100 MeV/nucleon. The anomaly component has a positive radial gradient in the heliosphere and undergoes a solar modulation. The mechanism of anomaly component production was proposed in [9] but the spacecraft Voyager-1, 2 did not confirm it. These single ionized atoms could be formed in flares of active dwarf stars if they are near the solar system. These single ionized atoms will be totally stripped if the ionization cross-section of $\sigma \approx 10^{-18}$ cm^2 and the amount of interstellar matter between their source and the solar system is ≥ 1. The interstellar medium density near Sun is about $\rho \sim 0.02$ atom/cm^3. Thus, the distance between the source and the solar system is $L \leq 1/(\rho\sigma)$, i.e. $L \leq 30$ pc, and there are about 560 flaring red dwarf stars within the sphere of the ~ 30 pc radius.

An important effect, discovered by the PAMELA and confirmed by AMS-02, is positron excess or unexpected positron fraction raise instead of decrease [10]. The observed positron fraction is starting to increase at ~ 5 GeV and then monotonically grows up to ~ 350 GeV, then at energies > 500 GeV it falls down. A possible interpretation of this effect may involve the active dwarf stars as the sources of cosmic rays up to $E \sim 10^{13}$ eV [6] This new type sources of cosmic rays provides essential addition positron flux to one produced in the interstellar medium in interactions of cosmic rays with interstellar matter. (see line in Fig. 1).

References

[1] R.E. Gershberg et al., Astron. Astrophys. Suppl. Ser. **139**, 555 (1999)
[2] S. Ohm & C. Hoischen, MNRAS **474**, 1335-1341 (2018)
[3] V.G. Sinitsyna, V.Y. Sinitsyna, Y.I.Stozhkov, Adv. Space Res. **64(12)**, 2585-2594 (2019)
[4] O. Adriani, G.C. Barbarino, G.A. Bazilevskaya et al., Nature **458(7238)**, 607-609 (2009)
[5] M. Aguilar, D. Aisa, B. Alpat et al., Phys. Rev. Lett. **114**, 171103 (2015)
[6] Yu.I. Stozhkov, Bull. of RAS ser. Physics **75(3)**, 323-326 (2011)
[7] O. Adriani et al., Rivista del Nuovo Cimento **40(10)**, 473-522 (2017)
[8] M. Garcia-Munoz, G.M. Mason, J.A. Simpson, Astrophys. Journ. Lett. **182**, L81 (1973)
[9] L. Fisk, B. Kozlovsky, R. Ramaty, Astrophys. Journ. Lett. **190**, L35 (1974)
[10] I.V. Moskalenko & A.W. Strong , Astrophys. J. **493**, 694-707 (1998)

Measuring the ^{15}O(α,γ) ^{19}Ne Reaction in Type I X-ray Bursts using the GADGET II TPC: Software

Ruchi Mahajan[1,*], *A.* Adams[1,2], *J.* Allmond[3], *H.* Alvarez Pol[4], *E.* Argo[1,2], *Y.* Ayyad[4], *D.* Bardayan[5], *D.* Bazin[1,2], *T.* Budner[1,2], *A.* Chen[6], *K.* Chipps[3], *B.* Davids[7], *J.* Dopfer[1,2], *M.* Friedman[8], *H.* Fynbo[9], *R.* Grzywacz[10], *J.* Jose[11], *J.* Liang[6], *S.* Pain[3], *D.* Perez-Loureiro[12], *E.* Pollacco[13], *A.* Psaltis[14], *S.* Ravishankar[15], *A.* Rogers[16], *L.* Schaedig[1,2], *L. J.* Sun[1], *J.* Surbrook[1,2], *T.* Wheeler[1,2], *L.* Weghorn[1,2], and *C.* Wrede[1,2]

[1]Facility for Rare Isotope Beams, Michigan State University, East Lansing, Michigan 48824, USA
[2]Department of Physics and Astronomy, Michigan State University, East Lansing, Michigan 48824, USA
[3]Oak Ridge National Laboratory, Oak Ridge, Tennessee 37830, USA
[4]Instituto Galego de Física de Altas Enerxías (USC), Spain.
[5]Department of Physics, University of Notre Dame, Notre Dame, Indiana 46556, USA
[6]Department of Physics and Astronomy, McMaster University, Hamilton, Canada.
[7]TRIUMF, Vancouver, British Columbia, Canada.
[8]The RACAH institute of Physics, Hebrew University of Jerusalem, Israel.
[9]Aarhus University, Aarhus C, Denmark
[10]University of Tennessee, Knoxville, USA.
[11]Universitat Politècnica de Catalunya, Barcelona, Spain.
[12]Canadian Nuclear Laboratories, Canada.
[13]IRFU, CEA Saclay, Gif-sur-Ivette, France.
[14]Technical University of Darmstadt,Germany.
[15]Department of Computational Mathematics, Science and Engineering, Michigan State University, East Lansing, Michigan 48824, USA.
[16]University of Massachusetts Lowell, USA.

Abstract. ^{15}O(α,γ) ^{19}Ne is regarded as one of the most important thermonuclear reactions in type I X-ray bursts. For studying the properties of the key resonance in this reaction using β decay, the existing Proton Detector component of the Gaseous Detector with Germanium Tagging (GADGET) assembly is being upgraded to operate as a time projection chamber (TPC) at FRIB. This upgrade includes the associated hardware as well as software and this paper mainly focusses on the software upgrade. The full detector set up is simulated using the ATTPCROOTv2 data analysis framework for ^{20}Mg and ^{241}Am.

1 Introduction

A type I X-ray burst can occur when a binary system consisting of a neutron star and an ordinary star is gravitationally bound so that hydrogen-rich material is accreted onto the surface of the neutron star [1]. The peak temperature in these bursts is of the order of 1-2 GK, enabling breakout from the CNO cycles and leading to the synthesis of heavy elements with mass number up to 100 [1]. There are a few reaction bottlenecks in the study of type I X-ray bursts whose unknown reaction rates can have large effects on simulated X-ray burst profiles [2]. The ^{15}O(α,γ) ^{19}Ne reaction in particular has been singled out as the most important reaction rate out of all currently unknown rates to measure [3]. At breakout temperatures ($\approx$0.5 GK), this reaction is strongly dominated by a single resonance with center of mass energy 506 keV corresponding to a ^{19}Ne state having excitation energy of 4034 keV [4]. It is technically not feasible to measure the strength of this resonance directly with current facilities due to the need for a very intense and low-energy beam of ^{15}O. However, it is possible to determine the resonance strength indirectly from the spin, lifetime and branching ratio (Γ_α/Γ)

*e-mail: mahajan@frib.msu.edu

of 4034 keV state. Since the spin and lifetime are well known one only needs to measure Γ_α/Γ to determine the strength [5]. Attempts have been made to measure Γ_α/Γ using direct reactions to populate the 4034 keV state and search for the α-particle emission, but they have effectively only led to strong upper limits of $(2.9\pm2.1)\times10^{-4}$ [6]. With this strong motivation, we have proposed an experiment to measure a finite value for Γ_α/Γ and hence the reaction rate at FRIB using an upgraded version of GADGET [7]. These measurements will proceed via the ^{20}Mg(βpα)^{15}O decay sequence using β decay of ^{20}Mg followed by a proton-emission to the 4034 keV ^{19}Ne state. We are currently upgrading GADGET's Proton Detector into a TPC with the goal of resolving and identifying the proton and the α-particle tracks. A high granularity MICROMEGAS board with 1024 (2.2×2.2 mm^2) pads and high-density GET electronics [8] has been installed to accommodate the large number of electronics channels. In order to test the functionality of TPC, a ^{241}Am source is placed inside the detector near the upstream end. The TPC has been simulated using the ATTPCROOTv2 data analysis framework based on the FairRoot package for ^{20}Mg and ^{241}Am decay events. LISE++ simulations were also performed in order to model the FRIB beam properties and purity. The paper is organized as follows: Sect. 2 illustrates in detail the ATTPCROOTv2 framework and Sect. 3 highlights the result of these simulations and the conclusions.

2 ATTPCROOTv2 Simulations

ATTPCROOTv2 is a ROOT based framework which requires external libraries (FairSoft and Fair Root) and a set of physics generators to simulate and analyze data from Active Target Time Projection Chamber (AT-TPC) and its prototype detectors [9]. Using this framework, a user can unpack as well as analyze the data and also create a customized geometry of interest for performing realistic simulations on an event-by-event basis using a virtual Monte Carlo package. Each generated event can correspond to a particular decay sequence including ^{20}Mg(βpα)^{15}O and ^{241}Am α-decay for the present case. Once the simulated data is created, pulse shape analysis is used to process the readout from the pad planes. Pattern recognition algorithms can be used to evaluate each event, and the trajectories of the particles in the detector can then be tracked. The Monte Carlo simulations are performed using the Geant4 toolkit and the HDF5 library is needed for data formatting. More information about this simulation package and digitization can be found in Ayyad *et al.* [9].

3 Results and Conclusion

Using the method described in Sect. 2, simulated tracks were analyzed. The ^{20}Mg(βpα)^{15}O events of interest will have a unique 3D topology in the TPC. The proton energies will be roughly 1.2 MeV based on detailed Doppler broadening analysis of the 4034-keV γ-ray peak [10]. For these simulations, P10 gas mixture (90% Ar and 10% CH_4) at atmospheric pressure was considered. The protons will be identified by their characteristic Bragg curves and have a range of 2-3 cm. The α particles will deposit 506 keV of total energy in very short (<4mm), dense tracks at the point of proton emission. The ranges can be extended if desired by lowering the gas pressure at the expense of beam stopping efficiency. Figure 1 (a) shows a simulated event of interest inside the detector geometry and the projection of a proton-α track on the pad plane is depicted in Figure 1 (b). The scattered points between 10-30 mm in Figure 1 (b) refers to the β-particles from the decay of ^{20}Mg to ^{20}Na. ^{241}Am α-decay events were also simulated as shown in Figure 1 (c) and Figure 1 (d), respectively. Based on these simulations, a machine learning algorithm is being developed that will be integrated with the ATTPCROOTv2 analysis framework to identify candidates in the data. Transfer learning will be used to refine the machine-learned models after the experiment using real in-situ data on ^{20}Mg(βpα)^{15}O events for single protons and ^{20}Mg($\beta\alpha$)^{16}O daughter-decay events for single alphas.

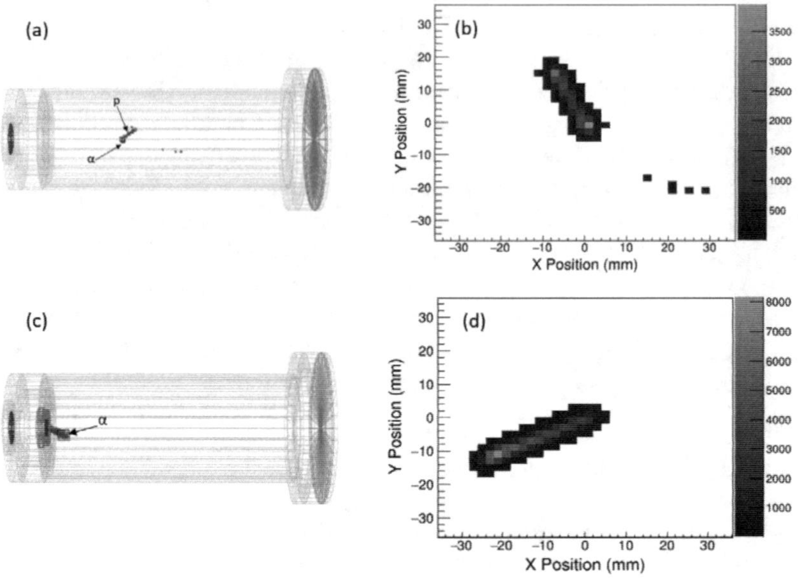

Figure 1. Panel (a) and (b)-ATTPCROOTv2 simulation of GADGET's TPC for ^{20}Mg(βpα)^{15}O, 3D render and 2D projection. Panel (c) and (d)-ATTPCROOTv2 simulation of GADGET's TPC for ^{241}Am α-decay sequence, 3D render and 2D projection.

LISE++ simulations were performed using narrow slit settings in the pre separator stage to reduce the momentum acceptance to 0.16% for ^{36}Ar primary beam with an energy of 200 MeV/A, which produces a ^{20}Mg secondary beam having energy of 107.2 MeV/A. The simulation predicts a beam purity of 99.985% with ^{15}N and ^{16}O as stable contaminants.

These simulations will be useful in selecting events of interest based on their unique signature to provide background free measurements.

This work has been supported by the U. S. Department of Energy under award no: DE-SC0016052 and the U. S. National Science Foundation under award no: 1565546 and 1913554.

References

[1] H. Schatz *et al.*, Phys. Rev. Lett. **86**, 3471 (2001).
[2] R. H. Cyburt *et al.*, Astrophys. J. **830**, 55 (2016).
[3] M. Wiescher *et al.*, J. Phys. G **25**, R 133 (1999).
[4] B. Davids *et al.*, Astrophys. J. **735**, 40 (2011).
[5] S. Mythili *et al.*, Phys. Rev. C **77**, 035803 (2008).
[6] W. Tan *et al.*, Phys. Rev. Lett. **98**, 242503 (2007).
[7] M. Friedman *et al.*, Nuclear Instrum. Methods in Physics Research A **940**, 93 (2019).
[8] E. Pollacco *et al.*, Nuclear Instrum. Methods in Physics Research A **887**, 81 (2018).
[9] Y. Ayyad *et al.* Nuclear Instrum. Methods in Physics Research A **954**, 161341 (2020).
[10] B. Glassman *et al.* Phys. Rev. C **99**, 065801 (2019).

Reactor activations to constrain astrophysically relevant cross sections

Sophia Dellmann[1], *René* Reifarth[1], *Mario* Weigand[1], *Klaus* Eberhardt[2], *Ruchi* Garg[3,4], *Christopher* Geppert[2], *Tanja* Heftrich[1], *Deniz* Kurtulgil[1], and *Claudia* Lederer-Woods[3]

[1]Goethe University, Frankfurt, Germany
[2]Johannes Gutenberg University, Mainz, Germany
[3]University of Edinburgh, United Kingdom
[4]Michigan State University, USA

Abstract. The determination of astrophysically relevant neutron-induced cross sections is particularly difficult when the involved isotopes are radioactive or the cross sections are very small. Activation experiments at reactors offer the possibility to overcome these limitations with high neutron fluxes. The flux determination is typically based on the activation of two monitors with known cross sections to separate the different flux components. The usually applied cadmium difference method allows a distinction between the thermal and the epithermal part. By a combination of two linear functions representing both monitors the neutron flux components can be determined. However, if more than two monitors are used, the linear system of equations is overdetermined, which allows the identification of a probability distribution. In this proceeding, the feasibility and relevance of this method is demonstrated.

1 Reactor activations

Reactor activations offer the possibility to determine neutron-induced cross sections of astrophysical interest. The difficulties encountered involving small cross sections or radioactive isotopes can be overcome by high neutron fluxes in the reactor. In particular, experiments with epithermal neutrons providing the resonance integral (RI) allow constraints on the effective cross sections in the astrophysically relevant keV range. The neutron flux in a reactor can be divided into a thermal, an epithermal component and a component of fast neutrons. The (n,γ) effective cross section decreases rapidly with increasing energy, so that the fast neutrons can be neglected. The thermal component can be suppressed with cadmium because cadmium has a high effective cross section for thermal neutrons [1]. Thus, both the thermal and the epithermal part of the neutron flux can be determined by applying the cadmium difference method.

2 Analytical determination of the neutron flux

To determine the neutron flux analytically, the assumption $R_A = \sigma_A^{th}\Phi^{th} + \sigma_A^{epi}\Phi^{epi}$, with $R_A = \frac{N_{A+1}}{N_A}$ for an isotope with mass A can be used. σ_A^{th} and σ_A^{epi} denote the thermal neutron capture cross section and the RI respectively. The thermal flux Φ^{th} and epithermal flux Φ^{epi}

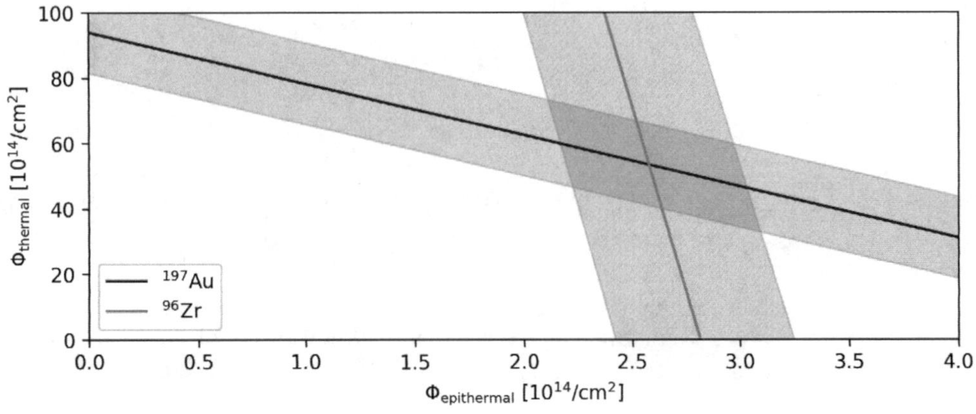

Figure 1. Relation between the thermal neutron flux Φ^{th} and the epithermal component Φ^{epi} determined by a system of linear equations for two monitors.

components can only be disentangled if at least two monitors with different ratios $\sigma_A^{epi}/\sigma_A^{th}$ and $\sigma_B^{epi}/\sigma_B^{th}$ are used. If only two monitor isotopes are used, the corresponding equations result in a system of linear equations (SLE) and can be analytically solved for Φ^{th} and Φ^{epi}:

$$\Phi^{th} = \frac{\sigma_B^{epi} R_A - \sigma_A^{epi} R_B}{\sigma_B^{epi}\sigma_A^{th} - \sigma_A^{epi}\sigma_B^{th}}, \Phi^{epi} = \frac{\sigma_A^{th} R_B - \sigma_B^{th} R_A}{\sigma_B^{epi}\sigma_A^{th} - \sigma_A^{epi}\sigma_B^{th}}$$

Figure 1 illustrates the analytical determination by combining two monitors. From the figure it is already obvious that large uncertainties may result from this method.

3 Probability distribution of the neutron flux

If more than two monitors are used, the SLE is overdetermined. The idea is now to determine the most likely pair of Φth and Φ^{epi}. It is therefore necessary to construct the probability distribution for all monitor combinations and each isotope. Thus, the uncertainty for the production ratio R_A for each flux pair has to be determined:

$$\mathrm{d}R_A = \sqrt{(\Phi^{th}\mathrm{d}\sigma_A^{th})^2 + (\Phi^{epi}\mathrm{d}\sigma_A^{epi})^2}$$

For the probability distribution follows:

$$P_A^{\Phi} = C_A \exp\left(-\frac{(R_A^{\Phi} - R_A^{exp})^2}{2\left((\mathrm{d}R_A^{\Phi})^2 + (\mathrm{d}R_A^{exp})^2\right)}\right)$$

The normalization constant c_A has to be determined such that the integral is 1. The probability distributions for all monitor pairs need to be combined:

$$P^{\Phi} = c \prod c_A P_A^{\Phi}$$

The global normalization constant c has to be determined such that the integral is 1. By increasing the number of monitors, the flux values can now be determined more precisely.

 https://doi.org/10.1051/epjconf/202226011035

4 Cadmium-encapsulated activation at TRIGA reactor Mainz

The neutron flux determination can be illustrated by the example of a reactor activation at the TRIGA reactor in Mainz. Four monitors ^{96}Zr, ^{96}Zr, ^{45}Sc and ^{197}Au covered with a 1 mm thick cadmium layer were activated for three hours. Figure 2 shows graphically the linear equations for the neutron fluxes in this activation. The probabilities for the respective neutron fluxes are plotted on the split axes.

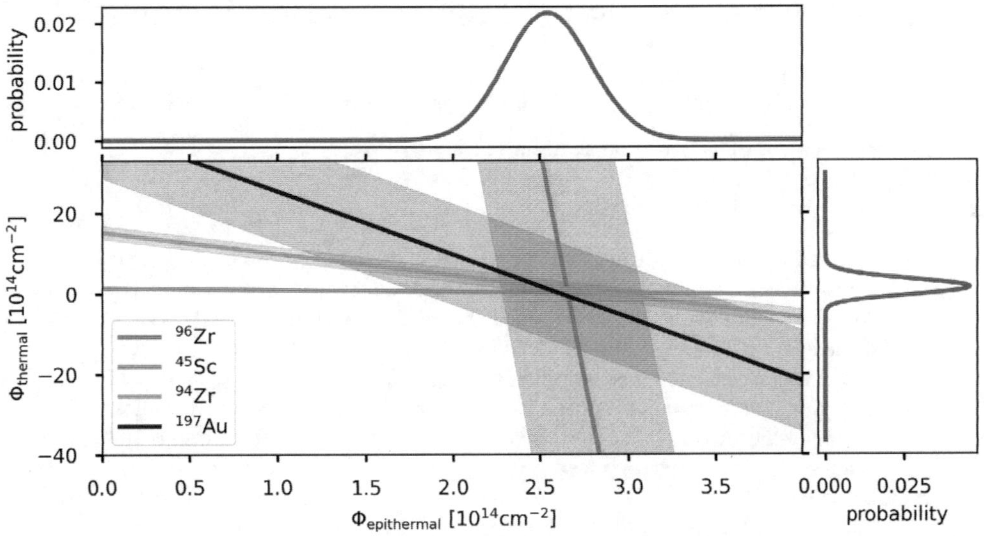

Figure 2. Relation between the thermal Φ^{th} the epithermal neutron flux Φ^{epi} determined by four monitors. The shared axes show projections of the probability distribution.

Graphically, the fluxes determined by the different methods agree with each other. The data necessary for the calculation are the production rates of the monitors used ^{96}Zr, ^{94}Zr, ^{45}Sc, and ^{197}Au, shown in table 1. The neutron fluxes obtained are shown in table 2. It lists the values of the neutron fluxes displayed in figure 2. For the analytical determination the monitor pair ^{45}Sc and ^{197}Au was chosen exemplarily. The uncertainty especially for the epithermal part has been reduced drastically.

$R_{^{95}Zr^{94}Zr}$	$R_{^{97}Zr^{96}Zr}$	$R_{^{46}Sc^{45}Sc}$	$R_{^{198}Au^{197}Au}$
$(0.73 \pm 0.02) \cdot 10^{-10}$	$(1.4 \pm 0.002) \cdot 10^{-9}$	$(3.89 \pm 0.08) \cdot 10^{-9}$	$(3.99 \pm 0.2) \cdot 10^{-7}$

Table 1. Values of the production rates of the four monitors.

	Φ^{th}[cm^{-2}]	Φ^{epi}[cm^{-2}]
Analytical determination (^{45}Sc and ^{197}Au)	$(3.65 \pm 0.59) \cdot 10^{13}$	$(2.41 \pm 0.05) \cdot 10^{14}$
Most probable neutron flux	$(2.89 \pm 0.51) \cdot 10^{13}$	$(2.52 \pm 0.04) \cdot 10^{14}$

Table 2. Values for the thermal Φ^{th} and the epithermal Φ^{epi} neutron flux component determined analyticaly for the monitor pair ^{45}Sc and ^{197}Au and second by using the probability approch.

References

[1] Frankle, C., Physical Review C **Volume 50**, p. 2774 (1994)

A moderately-sized nuclear network to assist multi-D hydrodynamic simulations of supernova explosions

Axel Sanz[1,*], *Rubén* Cabezón[2,**], and *Domingo* García-Senz[1,3,***]

[1]Departament de Física. Universitat Politècnica de Catalunya (Spain)
[2]Scientific Computing Center (sciCORE). Universität Basel (Switzerland)
[3]Institut d'Estudis Espacials de Catalunya

Abstract. A key ingredient in any numerical study of supernova explosions is the nuclear network routine that is coupled with the hydrodynamic simulation code. When these studies are performed in more than one dimension, the size of the network is severely limited by computational issues. In this work, we propose a nuclear network (*net87*) which is close to one hundred nuclei and could be appropriate to simulate supernova explosions in multidimensional studies. One relevant feature is that electron and positron captures on free protons and neutrons have been incorporated to the network. Such addition allows for a better track of both, the neutronized species and the gas pressure. A second important feature is that the reactions are implicitly coupled with the temperature, which enhances the stability in the nuclear statistical equilibrium (NSE) regime. Here we analyze the performance of *net87* in light of both: the computational overhead of the algorithm and the outcome in terms of the released nuclear energy and produced yields in typical Type Ia Supernova conditions.

1 Introduction

The disrupting mechanism in Type Ia Supernova (SN Ia) explosions is of thermonuclear origin and, therefore, a strong feedback between the hydrodynamics and the released nuclear energy is expected [1]. Any hydrodynamic code aiming to simulate these explosions in more than one dimension has to incorporate good enough nuclear routines, although not too time-consuming in computing terms.

Pioneering multidimensional simulations of SN Ia incorporated around 10 nuclei with the aim of having, at best, a reasonable depiction of the released nuclear energy. For example, Benz et al. [2] implemented an α-network to keep track of the nuclear evolution during the collision of two white dwarfs. Timmes et al. [3] analyzed the performance of two networks, with 7 and 14 species, concluding that they are able to account for the nuclear generation rate within 20% precision. The use of these small networks has been commonplace in multi-D simulations of supernova explosions. See [4–6], as an example of works connected with different SN Ia explosion scenarios. At present, multidimensional simulations incorporate $\simeq$ 50 nuclei, being able to reproduce the released nuclear energy within a narrow deviation, $<$ 5% with respect a larger network [7].

*e-mail: axel.sanz@estudiantat.upc.edu
**e-mail: ruben.cabezon@unibas.ch
***e-mail: domingo.garcia@upc.edu

In this work, we describe a network of 87 species, *net87*, which can be used in explosive astrophysical scenarios. In particular, we applied it to assist with the explosion of a massive white dwarf (WD) made of ^{12}C+^{16}O with central density $\rho_c \simeq 2\times10^9$ g cm^{-3} in the so-called Chandrasekhar-mass explosion models. It is also adequate to track the detonation of a tiny helium shell at densities $\simeq 5\times10^5$ g cm^{-3} located on top of a moderately massive WD, which is a necessary condition to study the SubChandrasekhar-mass route to SN Ia.

Our proposal has two important features. First, the molar fractions of the species, Y_i, and the temperature, T, are found jointly, after implicitly solving the system of nuclear reactions coupled with the energy equation. It is well known that the implicit coupling of $\{Y_i, T\}$ leads to a much stable behavior around the NSE regime [8, 9] and recent studies have identified the implicit coupling between the released nuclear energy and the nuclear reaction rates as a key point in simulating high-density combustion [10]. The second novelty of this work is that our moderately-sized network includes the capture reactions $e^- + p \rightarrow n + \nu_e$ and $e^+ + n \rightarrow p + \bar{\nu}$. As far as we know, this is the first time that reactions of this kind are taken into account in a reduced network routine belonging to a multi-D hydrodynamic code dealing with SN Ia explosions. As we show below, the benefits of including these reactions -namely a more accurate depiction of the electronic pressure and neutronized yields- come at no relevant computational cost.

2 Calculations and results

net87 adds a cluster of proton and neutron reactions in between each node of a standard $\alpha-$network from ^{20}Ne up to ^{60}Zn (see Fig. 1). The variation of the molar fractions is integrated jointly with the temperature equation with an implicit Newton-Raphson scheme. We also included e^- and e^+ captures on free protons and neutrons. They have well-known rates [11] and may account for up to 60% of bulk captures for Chandrasekhar-mass SN explosions.

Figure 2 shows the results of a ^{12}C+^{16}O burning test with initial temperature $T_9 = 1$ and density $\rho_9 = 2$ (both in units of 10^9). An artificial decrease of density, simulating an adiabatic expansion, was switched on around iteration 400, well after NSE is established, to reach the freeze-out. Note the decrease in pressure and temperature when the endoenergetic e$^-$-captures become relevant, around iteration 300. Such reduction in pressure will have dynamical consequences in multi-D simulations of SN Ia. Additionally, the NSE plateau is well reproduced, with the abundances showing no oscillations.

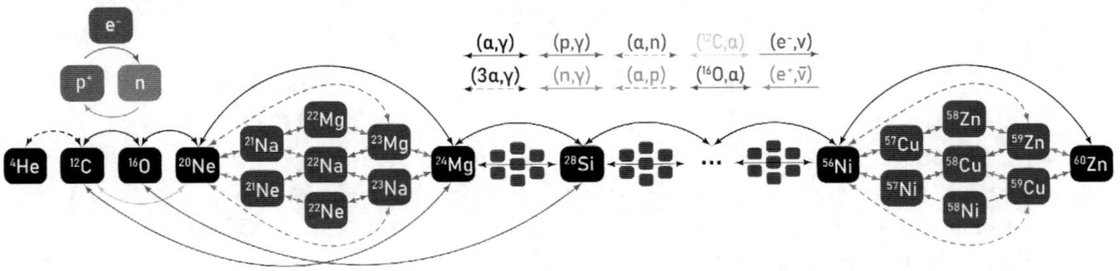

Figure 1. Schematics of the *net87* network, showing the α, p, and n reactions in the first and last cluster nodes (in purple) between α-chain elements (in black). Three heavy-ion reactions are also included as they are the main seeds for α particles. Finally, the e^- and e^+ captures are depicted on the top-left.

Table 1 shows the computational performance of *net87*. The routine scales approximately as $N \log N$ with the number of species as going from *net14* (with 14 species) to *net87* rises the computing time in a factor $\simeq 15$. The overload of including e^- and e^+ captures is negligible.

Table 1. Wall-clock time, average number of Newton-Raphson iterations per global iteration, total number of global iterations, and average wall-clock time per single iteration for the $^{12}C+^{16}O$ test and for each nuclear network (*net86* is *net87* without e^-, e^+-captures)

Network	$\bar{t}$ (s)	$\overline{\text{NR It.}}$	It.	$\bar{t}$/It. ($\times 10^{-4}$ s)
net14	0.152 ± 0.009	4.23	1792	0.20
net86	1.92 ± 0.13	4.89	1265	3.10
net87	1.78 ± 0.08	4.87	1164	3.08

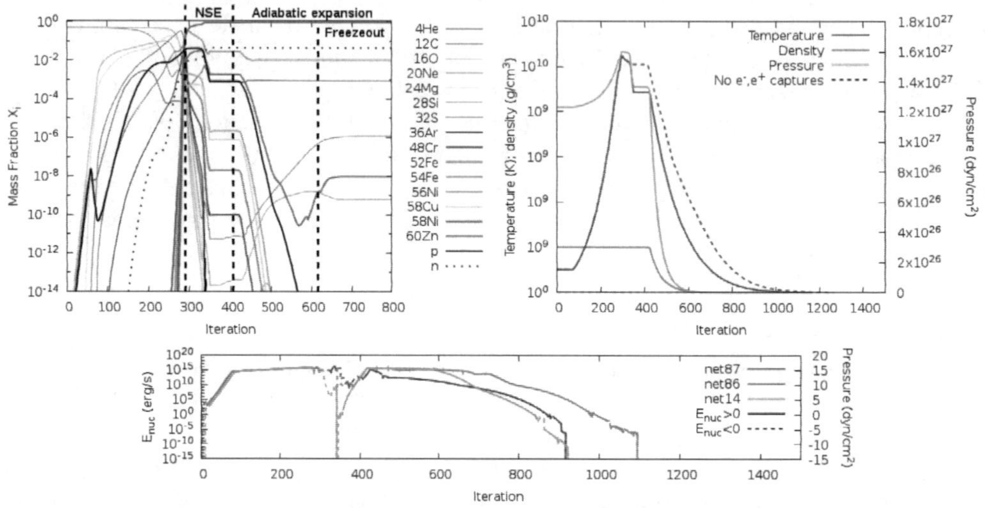

Figure 2. Evolution of the most relevant quantities in the $^{12}C+^{16}O$ test. Top-left: nuclear abundances for *net87*; top-right: temperature, density, and pressure for *net87* and *net86*; bottom: nuclear energy generation rate for *net87, net86* and *net14*.

References

[1] Hillebrandt, W. & Niemeyer, J. C., ARA&A 38, 191 (2000)
[2] Benz, W., Hills, J. G., & Thielemann, F. K., ApJ 342, 986 (1989)
[3] Timmes, F. X., Hoffman, R. D., & Woosley, S. E., ApJS 129, 377 (2000)
[4] Reinecke, M., Hillebrandt, W., & Niemeyer, J. C., A&A 347, 739 (1999)
[5] García-Senz, D., Bravo, E., & Woosley, S. E., A&A 349, 177 (1999)
[6] Plewa, T., Calder, A. C., & Lamb, D. Q., ApJL 612, L37 (2004)
[7] Gronow, S., Collins, C. E., Sim, S. A., & Röpke, F. K., A&A 649, A155 (2021)
[8] Mueller, E., A&A, 162, 103 (1986)
[9] Cabezón, R. M., García-Senz, D., & Bravo, E., ApJS 151, 345 (2004)
[10] Zingale, M., Katz, M. P., Willcox, D. E., & Harpole, A., Research Notes of the American Astronomical Society 5, 71 (2021)
[11] Fuller, G. M., Fowler, W. A., & Newman, M. J., ApJS 42, 447 (1980)

Impact of the decay width in Breit-Wigner formula on Maxwellian-averaged cross section for neutron capture on ^{16}O

Si-Zhe Xu and *Shi-Sheng* Zhang

[1]School of Physics, and International Research Center for Big-Bang Cosmology and Element Genesis, Beihang University, Beijing 100191, P. R. China

Abstract. Neutron capture on ^{16}O may serve as a crucial neutron poison reaction in weak s-process occurred in massive stars. In our previous study [Zhang, S. S., Xu, S. Z., He, M. et al. Eur. Phys. J. A **57**, 114 (2021)], we found that the contribution from low-lying resonances to Maxwellian-averaged cross section (MACS) progressively increases as the energy goes beyond 70 keV. In Breit-Wigner formula for resonant cross section, the decay width is a decisive quantity. In this paper, we discuss the impact of three kinds of decay widths, i.e. a constant width and two energy-dependent widths, on reaction cross sections and MACSs. The penetration factor adopts semi-classical WKB approximation and the asymptotic solution, respectively. We clarify that energy-dependent width are necessary for a reasonable behavior of resonance cross section around resonance peak and low-energy region far from the peak. The difference of two energy-dependent widths decreases from 3.7 to 1.5 with the energy increasing from 0.01 keV to 1000 keV. It results in similar behavior of resonance cross sections, but a slight difference by 1% for MACSs when E > 50 keV.

1 Introduction

Light nuclei play an important role in the studies of neutron capture nucleosynthesis. Some of them may absorb a large amount of neutrons, as so called "neutron poisons", which have an important impact on the synthesis of heavy elements. ^{16}O may be one of the candidates due to its high abundance and large reaction rate of neutron capture process. It has been shown that the difference of the $^{16}O(n,\gamma)^{17}O$ rates apparently influence the weak s-process in massive stars [1], which trigger us to study the properties of $^{16}O(n,\gamma)^{17}O$, especially for its cross section.

The cross section of neutron capture on ^{16}O mainly consists of three parts: direct capture, resonance, and the interference between the direct capture and the resonance. We have estimated these terms separately and found that the contribution from low-lying resonance to Maxwellian-averaged cross section (MACS) progressively increases as the energy goes beyond 70 keV, in which the resonant cross section takes the form of Breit-Wigner formula with an energy-dependent decay width [2, 3].

Usually, the decay width of a resonant state is defined as a constant in textbooks and some references [4, 5]. Except for a constant width, an energy-dependent width is also utilized in some works [6–8], including our former attempts [2, 3]. There are two ways commonly used to obtain an energy-dependent width. One way is to calculate the penetration factor by using semi-classical WKB approximation, then decay width as a function of energy can be evaluated. The other way is to express penetration factor by the asymptotic solution (AS) of Bessel function, which is adopted in our former work [3]. Here, we aim at clarifying the effectiveness and rationality among these three kinds of decay widths, and showing their impacts on resonant cross sections and MACS.

This paper is made up of three sections. In section 2, we simply review the theoretical framework to achieve penetration factors in two ways, i.e. WKB approximation and asymptotic solution. In section 3, we show the distinct impact of constant and energy-dependent decay widths on resonant cross sections and MACSs. Finally, we draw a simple summary in section 4.

2 Theoretical framework

2.1 Maxwellian-averaged cross section

As key inputs of nucleosynthesis simulations, the reaction rates $N_A < \sigma v >$ is equivalent to MACS $< \sigma >_{k_BT}$ except for a factor, i.e., $N_A < \sigma v >= N_A v_{k_BT} < \sigma >_{k_BT}$, where N_A is the Avogadro number, and v_{k_BT} is the thermal velocity. The expression of MACS reads,

$$< \sigma >_{k_BT}= \frac{2}{\sqrt{\pi}} \frac{1}{(k_BT)^2} \int_0^{\infty} \sigma(E)\, E\, exp(-\frac{E}{k_BT})\, dE, \tag{1}$$

which is the convolution of the reaction cross section $\sigma(E)$, the incident energy E, and exponential term with Boltzmann constant k_B and temperature T.

2.2 Breit-Wigner formula for resonant cross section

The reaction cross section mainly consists of three parts: direct capture, resonance, and the interference between the direct capture and the resonance. The Breit-Wigner formula is commonly utilized for the estimation of the resonant cross section, written as

$$\sigma_R(E) = \frac{\pi\hbar^2}{2\mu E} \frac{2J+1}{(2J_a+1)(2J_X+1)} \frac{\Gamma_a\Gamma_b}{(E-E_r)^2+(\Gamma/2)^2}, \tag{2}$$

in which μ is the reduced mass, defined as $\mu = m_a m_X/(m_a + m_X)$, E_r is the resonance energy, and E is the incident energy of the projectile; J, J_a and J_X refer to total angular momentums of the compound nucleus, the projectile a, and the target X, respectively; Γ_a and Γ_b are the decay widths of incoming particle a and outgoing particle b, and total width Γ is the summation of them. As shown in Breit-Wigner formula, the decay width of incoming particle a is one of important resonance parameters, but lacks of detailed description in the textbooks. Usually, one regards decay width as a constant in the calculations. We will show the difference aroused by constant and energy-dependence widths in the following.

2.3 The decay width

The decay width usually increases with incident particle energy E as the barrier decreases. An energy-dependence width takes the form as shown in Ref. [9]

$$\Gamma_l = \frac{3\hbar}{R}\sqrt{\frac{2E}{\mu}} P_l \theta_l^2, \tag{3}$$

where P_l is the penetration factor, and θ_l^2 is the dimensionless reduced width, defined as $\theta_l^2 = \chi_l^*(R)\chi_l(R)R/3$. In this definition, Γ_l is the function of the particle energy E, proportional to the penetration factor P_l and $\sqrt{E}$.

2.4 Penetration factor

Penetration factor P_l is defined as the ratio of the probabilities of particles appearing at infinity to those appearing at R, which can be expressed by the radial wave function. For $r > R$, The solution is a linear combination of regular Coulomb wave function $F_l(r)$ and irregular Coulomb wave function $G_l(r)$. Therefore, it reads

$$P_l = \frac{\chi_l^*(\infty)\chi_l(\infty)}{\chi_l^*(R)\chi_l(R)} = \frac{1}{F_l^2(R) + G_l^2(R)}. \tag{4}$$

2.4.1 The asymptotic solution (AS)

Without Coulomb interaction, the radial wave function can be simplified as a linear combination of spherical Bessel function $j_l(r)$ and Neumann function $n_l(r)$, i.e. so called the asymptotic solution (AS). Hence, the penetration factor can be rewritten as:

$$P_l = \frac{1}{\rho^2(j_l^2(R) + n_l^2(R))}, \tag{5}$$

where $\rho = kR$ with wave number k and matter radius of the nucleus R. In Table 1, we list the penetration factors P_l and level shift S_l for different reaction channels with angular momentum $l = 0, 1, 2, ...$ Here, we mainly focus on the low-lying $2p_{3/2}$ resonant state, in which $l = 1$.

Table 1. Penetration factors P_l and level shift S_l for different reaction channels without Coulomb interaction [10].

l	P_l	S_l
0	1	0
1	$\rho^2/(1+\rho^2)$	$1/(1+\rho^2)$
2	$\rho^4/(9+3\rho^2+\rho^4)$	$-(18+3\rho^2)/(9+3\rho^2+\rho^4)$
l	$\frac{\rho^2 P_{l-1}}{(l-S_{l-1})^2+\rho^2P_{l-1}^2}$	$\frac{\rho^2(l-S_{l-1})}{(l-S_{l-1})^2+\rho^2P_{l-1}^2} - l$

2.4.2 The WKB approximation

The WKB approximation is commonly used for barrier tunneling penetration problems, which is the approximated solution of a differential equation like $y'' = -f(x)y$. Here, we use $f(r) = 2\mu[E - V_l(r)]/\hbar$, so that the penetration factor follows [9]:

$$P_l(E) = \frac{\chi_l^*(\infty)\chi_l(\infty)}{\chi_l^*(R)\chi_l(R)} = \left[\frac{V_l(R)}{E} - 1\right]^{1/2} exp\left[-\frac{2\sqrt{2\mu}}{\hbar}\int_R^{R_0}\sqrt{V_l(r) - E}dr\right], \tag{6}$$

where R denotes the nuclear matter radius, and R_0 is the classical turning point satisfied with $V_l(R_0) = E$. Since there is no Coulomb interaction for neutron capture process, it can be reduced to

$$\int_R^{R_0} \sqrt{V_l(r) - E}dr = R\left[\sqrt{E_l}ln\left(\sqrt{\frac{E_l}{E}} + \sqrt{\frac{E_l}{E} - 1}\right) - \sqrt{E_l - E}\right]. \quad (7)$$

Later on, we will discuss the impact of penetration factors on the decay widths, the impact of the energy-dependent widths on the resonant cross sections and MACSs.

3 Results and discussions

3.1 Decay widths

In last section, we introduce two ways to obtain penetration factors P_l, i.e. the AS and WKB approximation. Given the penetration factors, we can calculate the decay widths with Eq. 3. The dimensionless reduced width θ_l^2 is also required. With measured resonance energy E_R and width $\Gamma_n(E_R)$ from experiments, θ_l^2 is easy to be determined by the Eq. 3. For the case of $^{16}O(n,\gamma)^{17}O$, the decay width $\Gamma_n = 40$ keV and the resonant energy $E_R = 411$ keV, are lately measured [11]. Then, θ_l^2 can be fixed to 5.128×10^{-2} with P_l^{AS} in Eq. 3. The matter radius of ^{16}O takes the value of 2.6 fm, and the angular momentum $l = 1$ corresponding to p-orbital valence neutron.

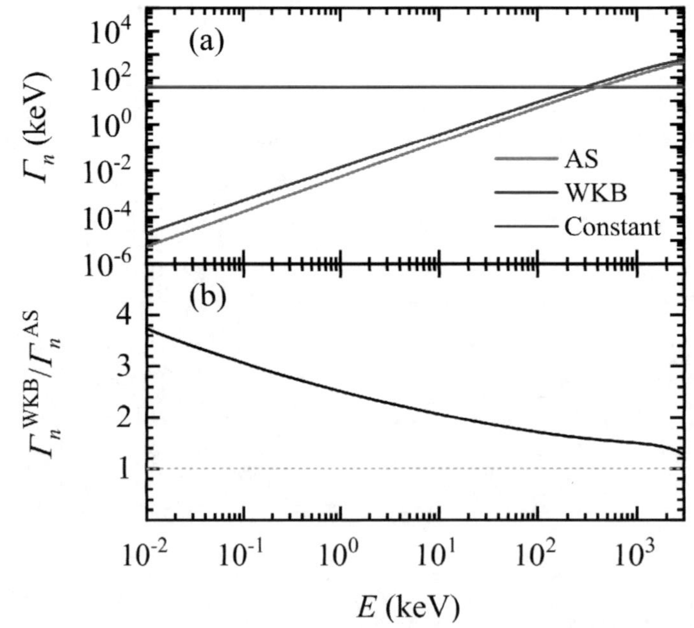

Figure 1. (a) A constant decay width (black line), and two decay widths $\Gamma_n(E)$ as a function of energy E, based on the penetration factors from the asymptotic solution (red line) and the WKB approximation (blue line), for $^{16}O(n,\gamma)^{17}O$, respectively. (b) ratio of Γ_n^{WKB} to Γ_n^{AS} as a function of energy E.

We plot the decay widths $\Gamma_n(E)$ as a function of energy E in Figure 1(a). The ratio of $\Gamma_n^{WKB}(E)$ to $\Gamma_n^{AS}(E)$ is also shown in Figure 1(b). In low-energy region, Γ_n^{WKB} is larger than Γ_n^{AS}, with a ratio about 4 at $E = 0.01$ keV. The difference becomes smaller as the energy increases.

3.2 Resonant cross sections

With the energy-dependent decay width $\Gamma_n(E)$, we achieve the resonant cross section $\sigma_R(E)$ from Breit-Wigner formula. In Figure 2, we compare the resonant cross sections $\sigma_R(E)$ with a constant width, and those based on two energy-dependent decay widths. It can be clearly seen that the resonance cross sections with a constant width show an abnormal increase in the low-energy region far from the resonance peak. It can be analyzed from From Eq. 2. When E approaches 0 keV, $(E - E_r)^2 + (\Gamma/2)^2$ becomes a constant, $\sigma(E) \propto 1/E$ in low-energy region. This is completely opposite to the trend of the resonance cross sections evaluated by energy-dependent widths, which becomes smaller as neutron energy decreases. It is more reasonable because lower particle energy away from resonance peak leads to smaller reaction probability, in accordance with lower cross section.

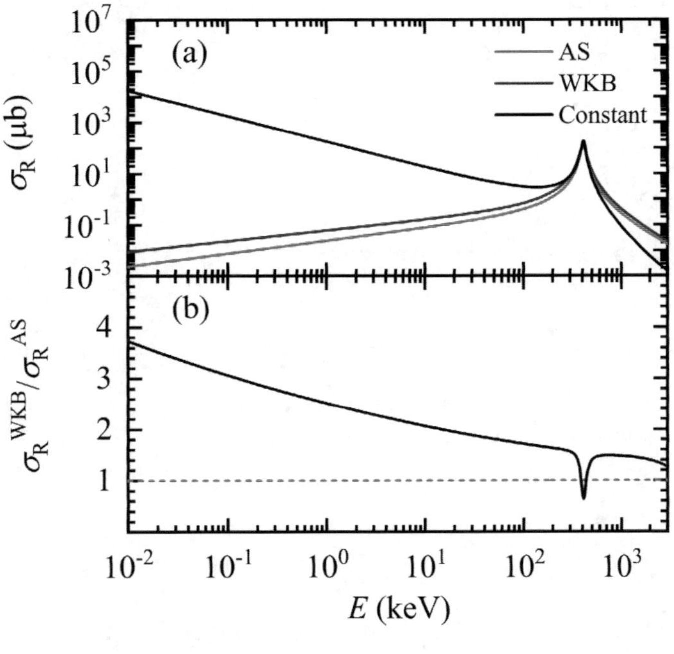

Figure 2. (a) Resonant cross section of 411 keV resonance of $^{16}O(n,\gamma)^{17}O$, with a constant decay width Γ_n = 40 keV (black line), and two energy-dependent $\Gamma_n(E)$ based the asymptotic solution (red line) and the WKB approximation (blue line). (b) ratio of σ_R^{WKB} to σ_R^{AS} as a function of energy E.

The ratio of σ_R^{WKB} to σ_R^{AS} generally decreases with E increasing, except for the region nearby resonance energy. This can be seen from Eq. (2) that the ratio $\sigma_R^{WKB}/\sigma_R^{AS}$ is proportional to the ratio $\Gamma_n^{WKB}/\Gamma_n^{AS}$ when E is far away from resonance energy E_r. However, the ratio of resonant cross sections rapidly declines around E_r. This is because the ratio $\sigma_R^{WKB}/\sigma_R^{AS}$ is proportional to the ratio $\Gamma_n^{AS}/\Gamma_n^{WKB}$, when $\Gamma_\gamma \ll \Gamma_n$ and $E \to E_r$. Therefore, compared to Γ_n^{AS}, the larger Γ_n^{WKB} causes smaller resonant cross section around the resonance energy.

3.3 Maxwellian-averaged cross sections

Figure 3 displays the MACSs as a function of k_BT, based on three decay widths: a constant, Γ_l^{WKB}, and Γ_l^{AS}, respectively. The MACSs with a constant width deviate a lot from the other two, especially for low-energy region, which is consistent with abnormal increase of the resonant cross section shown in Figure 2 (a). For energy-dependent widths $\Gamma(E)$, the ratio $\Gamma_n^{WKB}/\Gamma_n^{AS}$ becomes close to 1.0 (less than 1%) when $E > 50$ keV. From this sense, the WKB approximation is also a good way to evaluate MACSs.

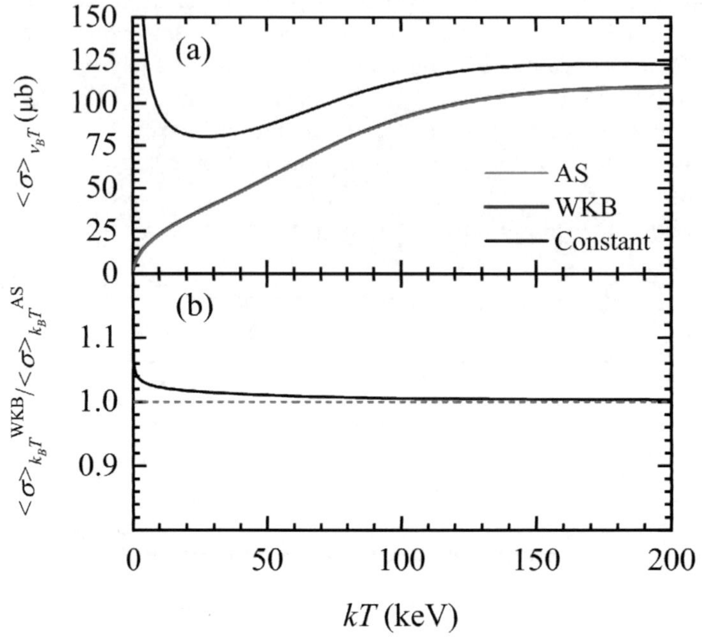

Figure 3. (a) MACSs for $^{16}O(n,\gamma)^{17}O$ based on three decay widths: constant (black dashed line), Γ_l^{WKB} (blue line), and Γ_l^{AS} (red line), respectively. (b) ratio of $<\sigma>_{k_BT}^{WKB}$ to $<\sigma>_{k_BT}^{AS}$ as a function of k_BT.

4 Summary

In this paper, we study the impacts of three kinds of decay widths in Breit-Wigner formula on resonant cross sections and MACSs. We found that a constant decay width shows abnormal behavior of the resonant cross sections especially for the low-energy region far away from the resonance peak. Therefore, energy-dependent widths are strongly recommended. Via comparing energy-dependent widths from commonly used WKB approximation and the asymptotic solution, we see that the deviations decrease with the energy goes up. Correspondingly, the ratio of the resonant cross sections shows similar behavior except for the region around the resonance peak. Interestingly, the differences between these two ways shrunk quickly and turn out to be less than 1% with increasing energy beyond 50 keV. Therefore, we make a conclusion that WKB approximation is also a good way to be used for neutron capture process.

References

[1] P. Mohr, C. Heinz, M. Pignatari, I. Dillmann, A. Mengoni, F. Käppeler, The Astrophysical Journal **827**, 29 (2016)
[2] M. He, S.S. Zhang, M. Kusakabe, S. Xu, T. Kajino, The Astrophysical Journal **899**, 133 (2020)

[3] S.S. Zhang, S.Z. Xu, M. He, M.S. Smith, The European Physical Journal A **57**, 1 (2021)
[4] C. Bertulani, Computer Physics Communications **156**, 123 (2003)
[5] J.J. Sakurai, E.D. Commins, *Modern quantum mechanics, revised edition* (1995)
[6] M. Caamano, D. Cortina-Gil, W. Mittig, H. Savajols, M. Chartier, C. Demonchy, B. Fernández, M.G. Hornillos, A. Gillibert, B. Jurado et al., Physical review letters **99**, 062502 (2007)
[7] S. Ceci, M. Korolija, B. Zauner, Physical review letters **111**, 112004 (2013)
[8] I. Hamamoto, Physical Review C **77**, 054311 (2008)
[9] D.D. Clayton, *Principles of stellar evolution and nucleosynthesis* (University of Chicago press, 1983)
[10] A. Trkov, M. Herman, D. Brown et al., Brookhaven National Laboratory **80** (2012)
[11] Y. Nagai, M. Kinoshita, M. Igashira, Y. Nobuhara, H. Makii, K. Mishima, T. Shima, A. Mengoni, Physical Review C **102**, 044616 (2020)

 https://doi.org/10.1051/epjconf/202226011038

Three-dimensional hydrodynamics simulations of shell burning in Si/O-rich layer of pre-collapse massive stars

Takashi Yoshida[1,*], *Tomoya* Takiwaki[2], *Kei* Kotake[3], *Koh* Takahashi[4], *Ko* Nakamura[3], *Hideyuki* Umeda[5], *David R.* Aguilera-Dena[6], and *Norbert* Langer[7,8]

[1]Yukawa Institute for Theoretical Physics, Kyoto University, Kyoto 606-8502, Japan
[2]Division of Science, National Astronomical Observatory of Japan, Tokyo 181-8588, Japan
[3]Department of Applied Physics & Research Institute of Stellar Explosive Phenomena, Fukuoka University, Fukuoka 814-0180, Japan
[4]Max-Planck-Institute for Gravitational Physics, D-14476 Potsdam, Germany
[5]Department of Astronomy, Graduate School of Science, The University of Tokyo, Tokyo 113-0033, Japan
[6]Institute of Astrophysics, FORTH, Department of Physics, University of Crete, GR-71003 Heraklion, Greece
[7]Argelander-Institut für Astronomie, Universität Bonn, D-53121 Bonn, Germany
[8]Max-Planck-Institute für Radioastronomie, D-53121 Bonn, Germany

Abstract. We perform three-dimensional (3D) hydrodynamics simulations of shell burning in the silicon- and oxygen-rich layers in pre-collapse massive stars. We adopt a non-rotating 27 $M_\odot$ star having an extended O/Si/Ne layer and a fast-rotating 38 $M_\odot$ star having a Si/O layer, that has experienced chemically homogeneous evolution. Both pre-collapse stars show large-scale turbulent motion with a maximum Mach number of ~0.1 in the convective layers activated by neon and oxygen shell burning. The radial proflie of the angle-averaged mass fraction distribution in O/Si/Ne layer is more homogeneous in the 3D simulation compared to the 1D evolution for the 27 $M_\odot$ star. The angle-averaged specific angular momentum in the Si/O layer of the fast-rotating 38 $M_\odot$ star tends to become roughly constant in the convective layer of the 3D simulation, which is not considered in the 1D evolution.

1 Introduction

Multi-dimensional (multi-D) hydrodynamics simulations of shell burning in convective layers during advanced evolution of massive stars have been performed to investigate the convection properties in evolved massive stars (e.g., [1, 2]). Recently, it has been pointed out that asphericity of wide silicon- and oxygen-rich (Si/O) convective layers, such as large-scale convective flows, in pre-collapse stars gives a favorable condition to energetic supernova explosion [3, 4]. Then, three-dimensional (3D) hydrodynamics simulations of the Si/O convective layers of pre-collapse massive stars have been performed for hundreds seconds and showed a large-scale and high-velocity turbulent motion in these layers (e.g., [5–8]). We perform 3D hydrodynamics simulations of shell burning in wide Si/O-rich convective layers of

*e-mail: yoshida@yukawa.kyoto-u.ac.jp

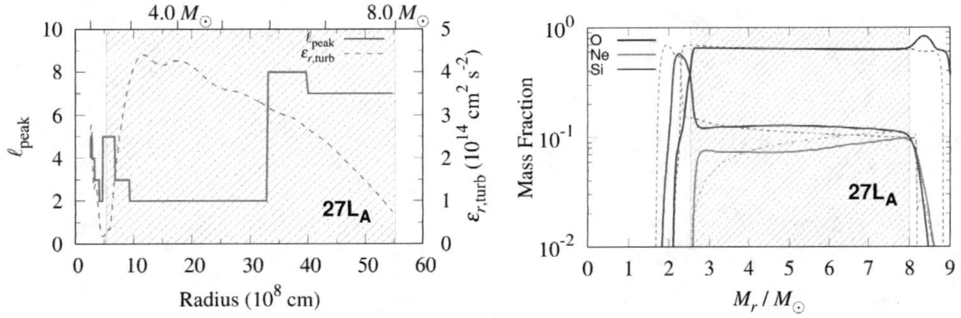

Figure 1. Left panel: The radial profiles of the peak mode $\ell_{\rm peak}$ obtained by the spectrum analysis of the radial turbulent velocity (solid curve) and the radial turbulent kinetic energy (dashed curve) of the 27 $M_\odot$ star. Right panel: the radial profiles of the oxygen (blue), neon (green), and silicon (magenta) mass fractions. Solid and dashed curves correspond to the last and initial profiles of the 3D simulation.

pre-collapse massive stars for 100–200 s. We investigate convection properties of the O/Si/Ne layer induced by the neon shell burning in a non-rotating 27 $M_\odot$ star and rotation effects of the oxygen-shell burning in the Si/O layer of a fast-rotating 38 $M_\odot$ star.

2 Models

We adopt two pre-collapse massive star models for 3D hydrodynamics simulations of shell burning. The one is a non-rotating 27 $M_\odot$ star (model 27L$_{\rm A}$ [7, 9]) with the solar metallicity, having the convective O/Si/Ne layer with the range of (5.8–60) $\times 10^8$ cm. The other is a fast-rotating 38 $M_\odot$ star with $Z{\sim}1/50\ Z_\odot$ and the initial rotation velocity of 600 km s^{-1} [10, 11], having the Si/O convective layer in the range of (4.7–17) $\times 10^8$ cm. This star has experienced chemically homogeneous evolution. We perform 3D hydrodynamics simulations of the 27 $M_\odot$ star and the fast-rotating 38 $M_\odot$ star for 218.5 s and 91.6 s, respectively, until the central temperature becomes 10^{10} K using multi-dimensional hydrodynamics code 3DnSEV [7].

3 Results

First, we show convection properties of the neon shell burning in the O/Si/Ne layer of the 27 $M_\odot$ star [9]. We obtained the angle-averaged turbulent Mach number of $\langle Ma^2 \rangle^{1/2} \sim 0.1$. The angle-averaged turbulent velocity reaches 4.2×10^7 cm s^{-1} (see also left panel of Figure 1). This high-velocity turbulence forms large-scale turbulence eddies in the convective layer. In the power spectrum analysis of the radial turbulent velocity, the peak mode of $\ell_{\rm peak} = 3$ is obtained in a wide range of the convective layer (left panel of Figure 1). The averaged value of the peak mode in the O/Si/Ne layer is $\langle \ell_{\rm peak} \rangle = 3.56$. The turbulent motion is activated by neon shell burning during the contraction and extends to the outer boundary.

The turbulent mixing in the 3D simulation causes the radial profile of the angle-averaged neon and silicon mass fractions different from that in the 1D stellar evolution. The radial gradient of the neon and silicon mass fractions in the O/Si/Ne layer in the 3D simulation becomes less steep compared to the 1D evolution profile (right panel of Figure 1). The mixing time scale of the 3D simulation is compatible to the advection time scale of the turbulent flow and shorter than the diffusion time scale deduced using the mixing length theory.

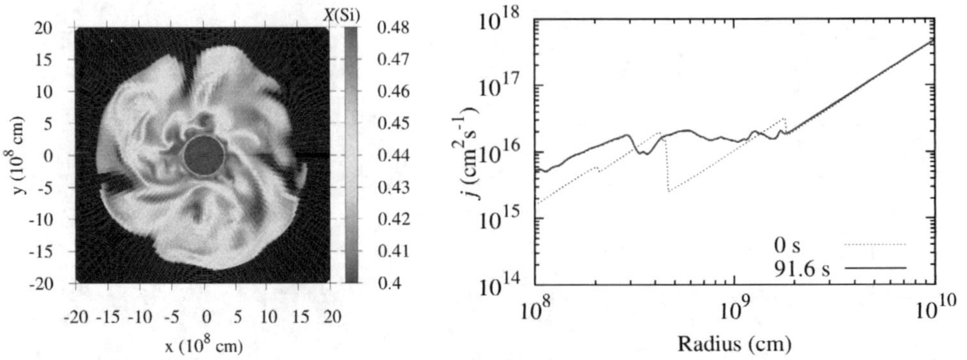

Figure 2. Left panel: Slices in the *x*-*y* plane showing the silicon mass fraction distribution at the last time step (91.6 s) of the fast-rotating 38 $M_\odot$ star. Right panel: Radial profiles of the angle-averaged specific angular momentum at the initial (dotted curve) and last (solid curve) time steps.

Next, we show convection and rotation properties of the oxygen shell burning in the Si/O convective layer of the fast-rotating 38 $M_\odot$ star. This star also shows a large-scale turbulence with high turbulent velocity. The angle-averaged turbulent Mach number becomes $\langle Ma^2 \rangle^{1/2} \sim 0.135$ at the maximum. The power spectrum analysis of the radial turbulent velocity indicates the peak mode of $\ell = 3$ in the range of (5–13) $\times 10^8$ cm. Spiral-arm structures of silicon enhancement in the silicon mass fraction distribution in the equatorial plane are seen as a rotation effect in this layer (left panel of Figure 2). We also obtain dominant low order modes ($m \sim$ 1–3) in the fluctuations of the density and rotational kinetic energy density.

The angular momentum transfer in the Si/O convective layer is investigated for the fast-rotating star. We see that the the monotonically increasing profile of the angle-averaged specific angular momentum in the Si/O layer, corresponding to a rigid rotation, changes to a roughly constant profile in the 3D simulation (right panel of Figure 2). This profile is different from the convective layer considered in 1D spherically symmetrical evolution models. To confirm such kind of angular momentum transfer by convection, we would need to investigate for longer time scale achieving a steady state convection.

References

[1] G. Bazán, D. Arnett, Astrophys. J. **496**, 316 (1994)
[2] C. A. Meakin, D. Arnett, Astrophys. J. Lett. **637**, L53 (2006)
[3] S. M. Couch, C. D. Ott, Astrophys. J. Lett. **778**, L7 (2013)
[4] B. Müller, H.-T. Janka, Mon. Not. R. Astron. Soc. **448**, 2141 (2015)
[5] S. M. Couch, et al., Astrophys. J. Lett. **808**, L21 (2015)
[6] B. Müller et al., Astrophys. J. **833**, 124 (2016)
[7] T. Yoshida et al., Astrophys. J. **881**, 16 (2019)
[8] N. Yadav, et al., Astrophys. J. **890**, 94 (2020)
[9] T. Yoshida, et al., Astrophys. J. **908**, 44 (2021)
[10] D. R. Aguilera-Dena, et al., Astrophys. J. **901**, 114 (2020)
[11] T. Yoshida, et al., Mon. Not. R. Astron. Soc. Lett. **506**, L20 (2021)

Proton emission study as a guide to astrophysical rp process

Pooja Siwach[1,2,*], *P.* Arumugam[1], *S.* Modi[3], *L. S.* Ferreira[4], and *E.* Maglione[4]

[1]Department of Physics, Indian Institute of Technology Roorkee, Roorkee 247667, India
[2]Department of Physics, University of Wisconsin–Madison, Madison, Wisconsin 53706, USA
[3]Department of Physics, BIT Sindri, Dhanbad, Jharkhand-828121, India
[4]Centro de Física e Engenharia de Materiais Avançados CeFEMA, and Instituto Superior Técnico, Universidade de Lisboa, Avenida Rovisco Pais, P1049-001 Lisbon, Portugal

Abstract. Proton emitters play an important role in deciding the path of the astrophysical rapid proton capture (rp) process. The lifetime of these nuclei depends on several factors, like the deformation, angular momentum of the emitted proton, residual interaction between valence proton and neutron (especially in case of odd-odd nuclei) and so on. Therefore, it is worth to investigate the structure of proton emitters to understand the rp process path. However, due to lack of data in this exotic region, the theoretical models should be robust and the dependence on the free parameters should be minimal. In this direction, we have developed the first microscopic approach to study the triaxially deformed odd-odd proton emitters. The application of the developed approach to ^{108}I, a recently observed proton emitter to investigate the end cycle of the rp process, is discussed.

1 Introduction

When the limits of the nuclear landscape are crossed on the proton-rich side, a nucleus becomes incapable to hold more protons resulting into spontaneous proton emission. However, due to the high Coulomb barrier, the proton spends enough time inside the nucleus to be measured before decay. Therefore, this phenomenon has been studied a lot both theoretically and experimentally, and provides a unique tool in unveiling the structural properties of nuclei at extreme limits.

The proton emitting nuclei serve as the decisive elements in the path of astrophysical rapid proton capture (rp) process [1]. By investigating the proton separation energies and half-lives one can get direct or indirect information on the path of rp process. For example, with the help of the proton emission half-life of ^{73}Rb [2], the ^{72}Kr was established as a strong waiting point. Furthermore, the information of proton separation energy of ^{109}I along with the α-decay energy [3] assisted the indirect measurement of proton separation energy of ^{105}Sb to investigate the possible paths in the end-cycle (Sn-Sb-Te cycle) of rp process.

Recently, in an experiment performed at Argonne National Laboratory, a weak proton emission branch was detected in ^{108}I (predominantly an alpha emitter) [4]. This study revealed that the rp process has to deviate its path from ^{106}Sb due to its proton unboundedness.

*e-mail: psiwach@physics.wisc.edu

This study further resolved the ambiguities in the path of rp process end cycle. However, to investigate the structural properties of ^{108}I, we need a theoretical approach which is less reliant on the freely adjustable parameters, and can incorporate the triaxiality and residual neutron-proton (np) interaction in a reliable way. Therefore, we have developed a nonadiabatic quasiparticle approach [5–7] for triaxial odd-odd proton emitters, satisfying the above requirements.

2 Formalism

We treat the odd-odd nuclei as the valence proton and neutron coupled to the even-even core. For the valence proton and neutron, we consider the nuclear mean-field potential in the Woods-Saxon potential form. A very detailed formalism is given elsewhere [6–8]. In the present work, we incorporate the residual np interaction V_{np} in the zero-range interaction form given by [6]

$$V_{np} = -\sqrt{\frac{8\pi^3}{\nu^3}}\,W\left[(1-\alpha)+\alpha\vec{\sigma}_p\cdot\vec{\sigma}_n\right]\delta(\vec{r_p}-\vec{r_n}), \tag{1}$$

where σ_p and σ_n are Pauli spin matrices, α and W are the strength parameters, and ν is the frequency related to the size of the nucleus. The total Hamiltonian of this system is solved to obtain the parent (before decay) wave function. In this microscopic approach, we calculate the overlap of initial (parent) and final (daughter nucleus+proton) wave functions. The partial decay width can be written as

$$\begin{aligned}\Gamma^{II_d}_{l_pj_p} \;=\; & \frac{\hbar^2 k}{\mu}\Big(\frac{2I_d+1}{2I+1}\Big)\Big|\sum_{j_p\Omega_p j_n\Omega_n K_d\Omega K} A^{I_d}_{j_n\Omega_n K_d}\, a^{I}_{\Omega_p\Omega_n\Omega K}\Big(\big\langle I_d K_d j_p\Omega_p|IK\big\rangle \\ & +(-1)^{I_d-K_d}\big\langle I_d-K_d j_p\Omega_p|IK\big\rangle\Big)u_{\Omega_p}N^{\Omega_p}_{l_pj_p}\Big|^2,\end{aligned} \tag{2}$$

where I and I_d are the angular momentum of parent and daughter nucleus, respectively, and K and K_d are their projection on 3-axis. a and A are the mixing coefficients of the parent and daughter wave functions, respectively. j_i and Ω_i $(i = p, n)$ are the quantum numbers of proton and neutron. Square of u_{Ω_p} gives the probability of orbital to be empty in the parent nucleus. The term in angular brackets represents the Clebsch-Gordon coefficients. The asymptotic normalisation constant is given by

$$N^{\Omega_p}_{l_pj_p} = \frac{\phi^{\Omega_p}_{lj}(\mathcal{R})}{G_l(k\mathcal{R})+\iota F_l(k\mathcal{R})}, \tag{3}$$

where ϕ is the proton wave function at distance $\mathcal{R}$. G and F are the irregular and regular parts of the Coulomb wave function.

The total decay width can be obtained by summing over all the states as

$$\Gamma^{II_d} = \sum_{j_p=|I-I_d|}^{I+I_d}\Gamma^{II_d}_{l_pj_p}, \tag{4}$$

which gives the half-life $T_{1/2} = \hbar\ln 2/\Gamma^{II_d}$.

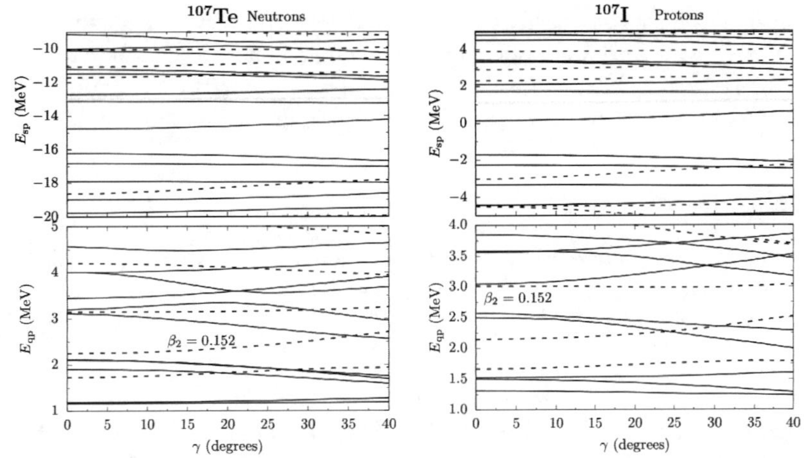

Figure 1. Single-particle (top) and quasiparticle (bottom) energies of neutrons in ^{107}Te (left) and protons in ^{107}I (right) at β_2 = 0.152 as a function of γ. Positive and negative parity levels are represented by the solid and dashed lines, respectively. Green and yellow lines represent the Fermi level and the chemical potential from BCS calculations, respectively.

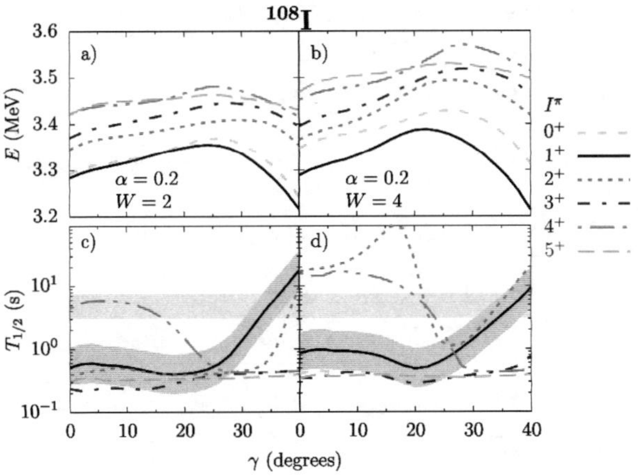

Figure 2. Rotational energies (panels a and b) and proton emission half-lives (panels c and d) at β_2 = 0.152 as a function of γ for different set of parameters α and W of residual np interaction. Gray width in half-life represents the uncertainties due to the one in measured Q_p value.

3 Results and discussion

To calculate the proton emission half-life, we need the information on the ground state of daughter nucleus (^{107}Te), which is found to be $5/2^+$ [7]. Considering the axial deformation ($\gamma = 0°$), the measured data could not be reproduced. Therefore, we incorporate the triaxiality (with non-zero γ), and the resulting single-particle and quasiparticle energies for neutrons in ^{107}Te and protons in ^{107}I are given in Fig. 1. We notice that with increase in γ, the single-particle and quasiparticle energies for neutrons come closer to the Fermi level. Similarly, for protons in ^{107}I, the quasiparticle energies reveal an interesting variation at large γ with one level coming very close to the Fermi level. We see the impact of these variations in the proton emission half-lives as discussed in the forthcoming discussion.

We proceed to investigate the properties of the ^{108}I. We observe from the rotational energies shown in Fig. 2 that the 0^+ and 1^+ are the lowest in energy whereas with increasing the strength of residual np interaction, 1^+ is the more favoured level. The proton emission half-life of this level reproduces the measured one at $\gamma \approx 30^{\circ}$. Other levels crossing the measured half-life at a lower γ are very high in energy. Hence, we conclude that the proton emission occurs from the 1^{+} level in ^{108}I and it is highly triaxial with $\gamma \approx 30^{\circ}$. These conclusions agree with the ones obtained with the residual np interaction in the constant potential form [7].

Acknowledgements

This work was supported in part by the U.S. Department of Energy, Office of Science, Office of High Energy Physics, under Award No. DE-SC0019465.

References

[1] H. Schatz, A. Aprahamian, J. Görres, M. Wiescher, T. Rauscher, J. Rembges, F.K. Thielemann, B. Pfeiffer, P. Moller, K.L. Kratz et al., Phys. Rep. **294**, 167 (1998)

[2] H. Suzuki, L. Sinclair, P.A. Söderström, G. Lorusso, P. Davies, L.S. Ferreira, E. Maglione, R. Wadsworth, J. Wu, Z.Y. Xu et al., Phys. Rev. Lett. **119**, 192503 (2017)

[3] C. Mazzocchi, R. Grzywacz, S.N. Liddick, K.P. Rykaczewski, H. Schatz, J.C. Batchelder, C.R. Bingham, C.J. Gross, J.H. Hamilton, J.K. Hwang et al., Phys. Rev. Lett. **98**, 212501 (2007)

[4] K. Auranen, D. Seweryniak, M. Albers, A. Ayangeakaa, S. Bottoni, M. Carpenter, C. Chiara, P. Copp, H. David, D. Doherty et al., Phys. Lett. B **792**, 187 (2019)

[5] G. Fiorin, E. Maglione, L.S. Ferreira, Phys. Rev. C **67**, 054302 (2003)

[6] P. Siwach, P. Arumugam, S. Modi, L.S. Ferreira, E. Maglione, J. Phys. G: Nucl. Part. Phys. **47**, 125105 (2020)

[7] P. Siwach, P. Arumugam, S. Modi, L.S. Ferreira, E. Maglione, Phys. Rev. C **103**, L031303 (2021)

[8] P. Siwach, Ph.D. thesis, Indian Institute of Technology Roorkee (2021)

Sensitivity Study of Type-I X-ray Bursts to Nuclear Reaction Rates

Irin Sultana[1,3,*], *Alfredo* Estradé[1,3], *Jessica* Borowiak[1], *Jacob* Elliott[1], *Bradley S.* Meyer[4], and *Hendrik* Schatz[2,3]

[1]Department of Physics, Central Michigan University, Mount Pleasant, MI 48859, USA
[2]National Superconducting Cyclotron Laboratory, Michigan State University, East Lansing, MI 48824, USA
[3]JINA Center for the Evolution of the Elements, Michigan State University, East Lansing, MI 48824, USA
[4]Department of Physics and Astronomy, Clemson University, Clemson, SC 29634-0978, USA

Abstract. Type-I X-ray bursts are frequently occurring thermonuclear runaways on the surface of an accreting neutron star. The different nuclear reactions that empower the cataclysmic event play a key role in accurately comparing theoretical models and observations. We investigate the effect of the uncertainties in the nuclear reactions using a ONEZONE model for a set of different compositions and accretion rates that are within the range of the standard observed burst sources. A combination of a full-reaction network and a semi-analytic model is used to obtain the conditions at the time of X-ray burst ignition via simulating the settling process of the accreted material. We then evaluate the sensitivity of the X-ray burst model by varying the proton and alpha-induced reaction rates in JINA REACLIBV2.2 within representative nuclear physics uncertainties.

1 Introduction

Neutron stars in a low mass X-ray binary accreting hydrogen-rich or helium-rich materials on the surface can frequently produce thermonuclear runaways, known as Type-I X-ray bursts (XRBs) [1] [2] [3]. Type-I X-ray burst occurs every few hours to days with a duration of (10 - 100) s depending on the accretion rate and the accreted composition [4]. Data from different X-ray observatory satellites (e.g. EXOSAT, RXTE, BeppoSAX, and Chandra) play an important role in probing the binary system properties with model-observation comparison [5]. Both observations and models show dependence on the accretion rate and composition of the accreted materials [6].

Type-I X-ray bursts are powered by the triple-α process, the αp process, and the rapid proton capture (rp) process [7] [8]. These nuclear processes include many reactions along the proton-rich side of the nuclear chart. Most of the nuclear reactions are determined theoretically due to the absence of experimental data. To understand the phenomena we need to constraints these nuclear reactions experimentally. Sensitivity studies play an important

*e-mail: sulta1c@cmich.edu

role to identify reactions whose reaction rate could be a significant source of uncertainty in the output of the model calculation [4] [9] [10]. These previous studies investigated a limited set of specific models. However, in nature model parameters and the associated burst phenomenology vary widely. Here we use the one-zone model (ONEZONE) [9] to determine the sensitivity of burst observables to nuclear reactions for a broad range of accreted compositions and accretion rates.

2 Method

We study the sensitivity of the XRB model to nuclear reaction rates in two steps. First, we obtain the conditions at burst ignition by combining a semi-analytic model (SETTLE code) [11] with a single-zone reaction network code (NucNet Tools) [12]. Second, we use the one-zone X-ray burst model ONEZONE [9] to follow the time-dependent explosive burning during the burst. Reaction rates are varied in ONEZONE to study their impact on the model predictions. We varied each reaction individually within representative nuclear physics uncertainties.

XRB light curves are dependent on the properties of the binary system along with the nuclear reactions driving the phenomena [9]. Type-I X-ray burst calculations were done for a parameter space based on the observed burst sources [6]. We study different astrophysical conditions with different values of the mass accretion rate and accreted composition. We use the SETTLE code to calculate the temperature and density profile of the surface layers of an accreting neutron star. Figure 1 shows the trajectory obtained from the SETTLE code using an initial accreted hydrogen mass fraction X = 0.9, metallicity Z = 0.02, where Z is the mass fraction of metals (species heavier than helium), and 20% of the Eddington accretion rate ($\dot{m}_{Edd}$). The base heating flux from the interior of the star is assumed to be 0.15 MeV/nucleon [11]. The NucNet Tools were then used to obtain the nucleosynthesis along the trajectory of the surface layers. This single-zone network contains isotopes up to xenon and the reaction data from the JINA reaction database REACLIBV2.2 [13]. We adjust the initial hydrogen and helium mass fraction and scale the solar composition for the metals. We use NucNet Tools to follow the evolution of the nuclear composition of the material until the ignition point of an XRB. Calculations are then continued with the ONEZONE model.

The ONEZONE model used for this study contains a network of 713 nuclei in the proton-rich side of the nuclear chart from hydrogen to xenon. We varied a set of 1422 proton and alpha-induced reactions in our one-zone model individually including their reverse rates from JINA REACLIBV2.2 [13]. For each calculation, we either increased or decreased one reac-

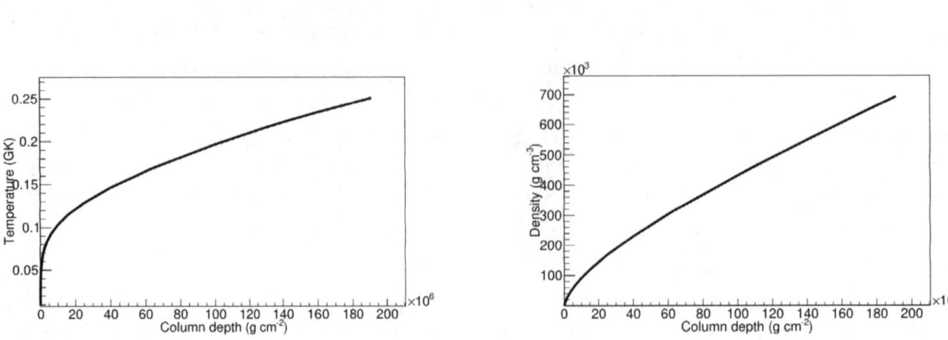

Figure 1. Temperature and density profile of the surface layers for an accreting neutron star using the SETTLE code [11]. See text for initial input. We started our ONEZONE calculation at the column depth of 1.58×10^{8}gcm^{-2}.

 https://doi.org/10.1051/epjconf/202226011040

tion rate by a factor of 100. The large variation factor was used to capture all the sensitive reactions [9].

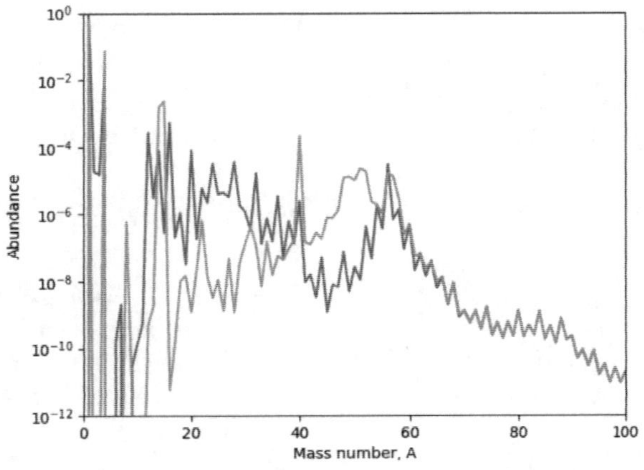

Figure 2. Nuclear composition of the accreted material at the surface (blue line) and at the depth when we change from NucNet Tools to ONEZONE (orange line), after the settling process using the NucNet Tools single-zone reaction network code [12].

3 Results

We present sensitivity results for two astrophysical conditions. Both have the same hydrogen mass fraction X = 0.9 for the accreted material and a mass accretion rate of $0.2\dot{m}_{Edd}$ while we used two metallicities Z = 0.02 and Z = 0.01. We refer to the baseline calculation as

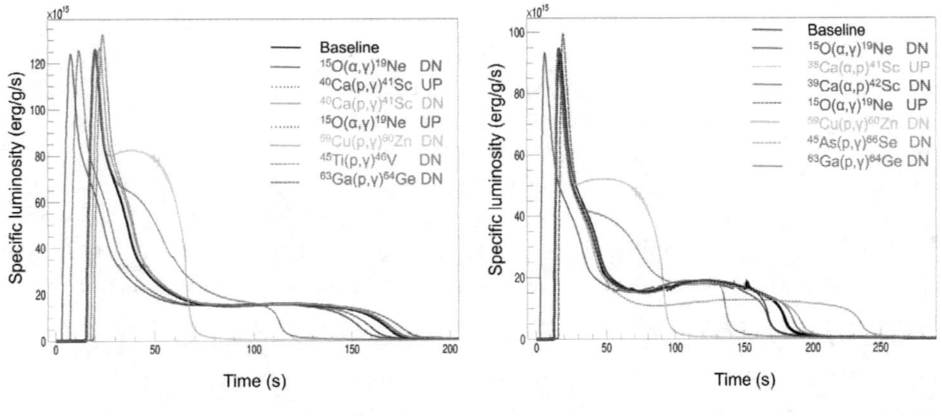

Figure 3. The reactions with higher change in the lightcurve profile as determined through the F_s parameter, for models using two different metallicities for the accreted material. For Z = 0.02 (left), the list of reactions is dominated by the (p, γ) reactions while for Z = 0.01 (right) the list has higher-ranked (α, p) reactions in comparison.

the results obtained with unmodified reaction rates of the JINA REACLIBV2.2 library [13] (black line in figure 3). We define the sensitivity parameter F_s in the following way,

$$F_s = \mid L_c - L_b \mid \quad (1)$$

Here, L_b is the maximum value of the specific luminosity for the baseline calculation, and L_c is the same value for the case with changed rates. We record the high-impact reactions in table 1. Figure 3 shows the light curves for calculations with high-impact reactions with the largest change from the baseline. For a metallicity of 0.02, this is mostly dominated by the (p,γ) reactions while for 0.01 metallicity there are more high-ranked (α, p) reactions in comparison. In both cases, the highest change in the light curve is produced by the $^{15}O(\alpha,\gamma)^{19}Ne$ reaction rate.

Table 1. List of high-impact reactions with impact parameter F_s, for calculations with a hydrogen fraction of 0.9 and 0.2 of Eddington accretion rate, and with two different metallicities (Z) of the accreted material. Calculations, where the reaction rate was increased by a factor of 100, are labeled as U, and those where the rate was decreased by the same factor are labeled as D.

Rank	Reaction (for Z = 0.02)	$F_s[10^{17}$erg/g/s]	Reaction(for Z = 0.01)	$F_s[10^{17}$ erg/g/s]
1	$^{15}O(\alpha,\gamma)^{19}Ne$ (D)	1.18	$^{15}O(\alpha,\gamma)^{19}Ne$ (D)	0.90
2	$^{40}Ca(p,\gamma)^{41}Sc$ (U)	0.95	$^{15}O(\alpha,\gamma)^{19}Ne$ (U)	0.60
3	$^{40}Ca(p,\gamma)^{41}Sc$ (D)	0.83	$^{59}Cu(p,\gamma)^{60}Zn$ (D)	0.35
4	$^{15}O(\alpha,\gamma)^{19}Ne$ (U)	0.74	$^{14}O(\alpha,p)^{17}F$ (U)	0.26
5	$^{59}Cu(p,\gamma)^{60}Zn$ (D)	0.53	$^{63}Ga(p,\gamma)^{64}Ge$ (D)	0.20
6	$^{14}O(\alpha,p)^{17}F$ (U)	0.34	$^{19}Ne(p,\gamma)^{20}Na$ (D)	0.18
7	$^{44}Ti(p,\gamma)^{45}V$ (D)	0.29	$^{38}Ca(\alpha,p)^{41}Sc$ (U)	0.16
8	$^{19}Ne(p,\gamma)^{20}Na$ (D)	0.28	$^{44}Ti(p,\gamma)^{45}V$ (D)	0.14
9	$^{63}Ga(p,\gamma)^{64}Ge$ (D)	0.27	$^{65}As(p,\gamma)^{66}Se$ (D)	0.11
10	$^{45}Ti(p,\gamma)^{46}V$ (D)	0.22	$^{39}Ca(\alpha,p)^{42}Sc$ (D)	0.10

4 Discussions

The different properties of a binary system with an accreting neutron star can lead to X-ray bursts with significantly different characteristics. For metallicity Z = 0.02, the conditions at the burst ignition were a hydrogen mass fraction X = 0.63, helium mass fraction Y = 0.29, temperature T = 0.26 GK, and pressure P = 5.05×10^{22}ergcm^{-3}. For metallicity Z = 0.01, ignition conditions were, hydrogen mass fraction X = 0.65, helium mass fraction Y = 0.28, temperature T = 0.25 GK, and pressure P = 7.19×10^{22}ergcm^{-3}. Our results show that depending on the different ignition conditions different nuclear processes have a stronger influence on the burst light curve.

We have developed a sensitivity study that considers the condition of burst ignition for a range of system parameters. Our preliminary results show the features of the accreted material have a strong impact on the type of important reactions.

5 Acknowledgements

We would like to thank Zach Meisel for useful discussions. This work was supported by U.S. DOE award DE-SC0020406 and by NSF award PHY-1430152 and PHY-1913554.

References

[1] H. Schatz, A. Aprahamian, J. Görres, M. Wiescher, T. Rauscher, J. Rembges, F. Thielemann et al., Physics Reports **294**, 167 (1998)
[2] A. Parikh, J. José, G. Sala, C. Iliadis, Progress in Particle and Nuclear Physics **69**, 225 (2013), 1211.5900
[3] Z. Meisel, A. Deibel, L. Keek, P. Shternin, J. Elfritz, Journal of Physics G Nuclear Physics **45**, 093001 (2018), 1807.01150
[4] A. Parikh, J. José, F. Moreno, C. Iliadis, **52**, 409 (2008), 0806.2975
[5] A. Heger, A. Cumming, D.K. Galloway, S.E. Woosley, **671**, L141 (2007), 0711.1195
[6] D.K. Galloway, A.J. Goodwin, L. Keek, **34**, e019 (2017), 1703.07485
[7] R.K. Wallace, S.E. Woosley, **45**, 389 (1981)
[8] H. Schatz, A. Aprahamian, V. Barnard, L. Bildsten, A. Cumming, M. Ouellette, T. Rauscher, F.K. Thielemann, M. Wiescher, **688**, 150 (2001), nucl-th/0201033
[9] R. Cyburt, A. Amthor, A. Heger, E. Johnson, L. Keek, Z. Meisel, H. Schatz, K. Smith, The Astrophysical Journal **830**, 55 (2016)
[10] H. Schatz, W.J. Ong, **844**, 139 (2017), 1610.07596
[11] A. Cumming, L. Bildsten, The Astrophysical Journal **544**, 453 (2000)
[12] B.S. Meyer, Proceedings of Science **NIC XII**, 096 (2013)
[13] R.H. Cyburt, A.M. Amthor, R. Ferguson, Z. Meisel, K. Smith, S. Warren, A. Heger, R.D. Hoffman, T. Rauscher, A. Sakharuk et al., **189**, 240 (2010)

Self-consistent energy density functional approaches to the crust of neutron stars

Takashi Nakatsukasa[1,2]

[1]Center for Computational Sciences and Faculty of Pure and Applied Sciences, University of Tsukuba Tsukuba 305-8577, Japan
[2]RIKEN Nishina Center, Wako 351-0198, Japan

Abstract. In order to study structure of the crust in neutron stars, we develop a finite-temperature Skyrme-Hartree-Fock method in the full three-dimensional coordinate space using the Fermion operator expansion method. It provides us with a possible order-N approach to non-uniform neutron-star matters.

1 Introduction

The neutron star is a compact star whose radius is about 10 km, which can be regarded as a giant nucleus with macroscopic numbers of hadrons (baryons and mesons). The core part, the central region of the neutron star, is supposed to be uniform nuclear (hadron) matter. However, moving toward the surface, it changes into non-uniform nuclear matter composed of neutron-rich nuclei and neutron matter, called "inner crust". In the vicinity of the boundary between the core and the inner crust, exotic phases, called "nuclear pasta" are predicted. Since a new exotic structure may appear in certain conditions, the calculation without any symmetry restriction is desired.

Structure of these non-uniform exotic nuclear matter have been studied with the Thomas-Fermi (local density) approximation [1] and/or with the Wigner-Seitz approximation [2]. Although these two approximations are complimentary to each other, neither is able to take proper account of transport properties in the non-uniform nuclear matter. Thus, it is highly desirable to go beyond these approximations.

Recently, the three-dimensional (3D) coordinate-space solver of the Hartree-Fock-Bogoliubov (HFB) method has been developed [3]. In order to abstain from diagonalizing the large HFB Hamiltonian matrix, which is computationally demanding, the contour integration in the complex energy plane together with iterative solutions of the shifted linear algebraic equations is adopted to construct densities. The method has been extended to the finite-temperature HFB method, by including the Matsubara frequencies as the imaginary shifts [4]. The method was successfully applied to non-uniform nuclear matter with superfluid neutrons at finite temperature.

In this paper, we present another coordinate-space solver using the Fermion operator expansion (FOE) method [6]. The method was originally developed in the computational condensed matter physics, and known as one of order-N ($O(N)$) methods [5]. The conventional approaches to the density functional theory scales as N^3, where N is the number of particles. This N^3-scaling sets a severe limitation to the system size. In contrast, the $O(N)$ method scales as N^1 in ideal cases. The foundation of the $O(N)$ method is provided by "principle of nearsightedness" [5], namely a consequence of quantum mechanical decoherence among many particles. Calculation of the non-uniform neutron star matter may require us to treat more than thousands of nucleons (baryons) in a large space. In this paper, we demonstrate a test calculation of the FOE method for non-uniform nuclear matter at finite temperature.

2 Fermion operator expansion method

Let us briefly recapitulate the FOE method. $\hat{H}$ is the one-body mean-field Hamiltonian, satisfying $\hat{H}|n\rangle = E_n|n\rangle$. The one-body density at a temperature $T = 1/\beta$ is given by

$$\hat{\rho}_T = \sum_n f(E_n)|n\rangle\langle n| = f(\hat{H})\sum_n |n\rangle\langle n| = f(\hat{H}), \tag{1}$$

where $f(x)$ is the Fermi-Dirac distribution function, $f(x) = (1 + e^{\beta(x-\mu)})^{-1}$. Next, the Fermi-Dirac function is approximated by a series of polynomial functions $\{T_j(x)\}$; $f(x) = \sum_j a_j T_j(x)$. Then, the density, represented by $f(\hat{H})$, can be written as

$$\hat{\rho}_T \approx \sum_{j=0}^{M} a_j T_j(\hat{H}), \tag{2}$$

where $T_j(x)$ is a polynomial of the j-th degree. Here, the summation is truncated at the maximum degree M. In order to obtain a reasonable description of the Fermi-Dirac function, roughly speaking, M should increase at lower T. The density can be constructed by the M times operation of the Hamiltonian. We adopt the Chebyshev polynomials following ref. [6]. They can be determined by the following recursion relations:

$$T_j(x) = 2xT_{j-1}(x) - T_{j-2}(x), \tag{3}$$

with $T_0(x) = 1$ and $T_1(x) = x$. Therefore, the operation of the polynomial of the Hamiltonian on a state $|\psi\rangle$, $|\psi^{(j)}\rangle \equiv T_j(\hat{H})|\psi\rangle$, is calculated as

$$|\psi^{(j)}\rangle = 2\hat{H}|\psi^{(j-1)}\rangle - |\psi^{(j-2)}\rangle, \tag{4}$$

with $|\psi^{(0)}\rangle = |\psi\rangle$ and $|\psi^{(1)}\rangle = \hat{H}|\psi\rangle$. The density is given by

$$\hat{\rho}_T|\psi\rangle = \sum_{j=0}^{M} a_j|\psi^{(j)}\rangle. \tag{5}$$

We need to perform the calculation for, at least, N different states, $\{|\psi_k\rangle, k = 1, \cdots, N\}$. If the Hamiltonian is sparse, the total computational cost is estimated as $O(N^2)$. However, when the "nearsightedness" exists, it can be reduced to $O(N)$ [5].

3 Application of the FOE method

In this paper, we apply the FOE method to the BKN energy density functional [7] which is a simplified Skyrme-like functional without the spin-orbit coupling and assumes the identical densities of protons and neutrons. For simplicity, we neglect the pairing correlation, thus, it is the finite-temperature Hartree-Fock calculation, instead of finite-temperature HFB. The calculation is performed using the 3D coordinate-space representation discretized in the mesh [8]. The number of mesh points is 25^3 with the cubic mesh of $(1\text{ fm})^3$. The baryon density

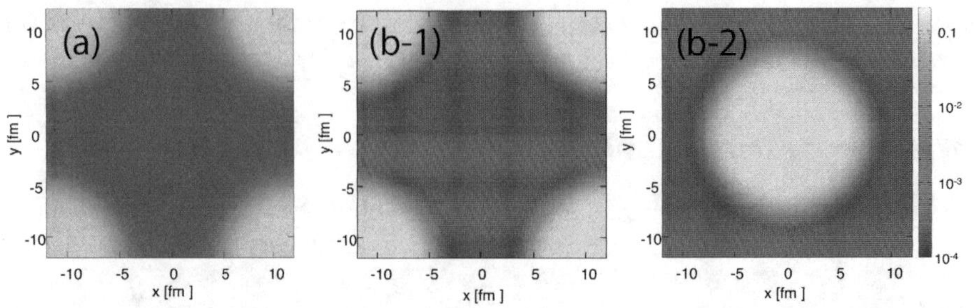

Figure 1. Calculated nucleonic density distribution (log scale) of the bcc phase in the xy plane at $\rho_B = 0.03$ fm^{-3} in the planes $z = |z_{\max}|$ fm (panels (a) and (b-1)), $z = 0$ (b-2). The temperature is $T = 5$ MeV (b) and $T = 10$ MeV (a).

and the proton ratio are fixed at $\rho_B = 0.03$ fm^{-3} and at $Y_p = 0.5$ (symmetric nuclear matter), respectively.

We first set the temperature $T = 5$ MeV, then, perform the self-consistent iterative calculation starting from the body-centered configuration (bcc) as the initial state. The converged state also has a bcc-like structure: In the "sea" of symmetric nuclear matter whose density is about 8.3×10^{-4} fm^{-3}, a "nucleus" (excess density) is embedded at the center and 1/8 of a nucleus is at each corner of the cubic box (Fig. 1(b)). In total, there are two nuclei in the adopted space of volume $V = 25^3$ fm^3. The number of nucleons in the space is $\rho_B \times V = 468.75$, so that, at every iteration, the chemical potential μ is adjusted to produce this particle number. Increasing the temperature (Fig. 1(a)), the nucleons are gradually leaking from "nucleus" to the "sea", approaching to the uniform matter. It should be noted that, since the symmetric matter is assumed, the result is not realistic for neutron star inner crust. However, it is still meaningful for testing performance of the FOE method applied to nuclear matter.

The FOE method is suitable for parallel computing, because calculation of eq. (5) can be performed independently for different $|\psi\rangle$. The FOE method is extremely efficient at high temperature, since the Fermi-Dirac function becomes smooth and is well approximated by a small number of polynomials. The computational cost is roughly proportional to M in eq. (5). M is order of thousands at $T \sim 100$ keV, while it is order ten at $T \sim 10$ MeV.

This work is supported by JSPS KAKENHI Grant No.18H01209 and No.19H05142. This research used computational resources provided by Joint Center for Advanced High Performance Computing (JCAHPC) through the HPCI System Research Project (Project ID: hp200069) and through Multidisciplinary Cooperative Research Program in Center for Computational Sciences, University of Tsukuba.

References

[1] K. Oyamatsu, Nuclear Physics A **561**, 431 (1993)
[2] J. Negele, D. Vautherin, Nuclear Physics A **207**, 298 (1973)
[3] S. Jin, A. Bulgac, K. Roche, G. Wlazłowski, Phys. Rev. C **95**, 044302 (2017)
[4] Y. Kashiwaba, T. Nakatsukasa, Phys. Rev. C **101**, 045804 (2020)
[5] S. Wu, C. Jayanthi, Physics Reports **358**, 1 (2002)
[6] S. Goedecker, M. Teter, Phys. Rev. B **51**, 9455 (1995)
[7] P. Bonche, S. Koonin, J.W. Negele, Phys. Rev. C **13**, 1226 (1976)
[8] T. Nakatsukasa, T. Inakura, K. Yabana, Phys. Rev. C **76**, 024318 (2007)

Massive first star binaries as new tools for Galactic archaeology

Takuma Suda[1,*], *Takayuki R.* Saitoh[2], *Yuki* Moritani[3], *Tadafumi* Matsuno[4], and *Toshikazu* Shigeyama[5]

[1]Department of Liberal Arts, Tokyo University of Technology, Tokyo 144-8535, Japan
[2]Department of Planetology, Graduate School of Science, Kobe University, Hyogo 657-8501, Japan
[3]Kavli Institute for the Physics and Mathematics of the Universe (WPI), The University of Tokyo Institutes of Advanced Study, Tokyo 277-8583, Japan
[4]Kapteyn Astronomical Institute, University of Groningen, 9747AD Groningen, the Netherlands
[5]Research Center for the Early Universe, The University of Tokyo, Tokyo 113-0033, Japan

Abstract. Binary systems are important probes of the origin of stars with peculiar chemical features through the interactions between two stars. We have studied the evolution of a metal-free massive binary after the collision of the supernova ejecta with a low-mass companion. Theoretical models are developed using simulations of binaries after core-collapse supernovae using stellar evolution models, supernova ejecta models, and hydrodynamic simulations of the system consisting of supernova ejecta and companion stars. We find that these first star survivors will be observed as metal-rich halo stars in our Galaxy. In combination with the theoretical research, we looked for observational counterparts in the Galactic halo using the observational database where elemental abundances and kinematic data are available. We have also searched for the progenitor binary systems based on radial velocity monitoring. We report the current status of the search for massive binaries in the solar vicinity. The proposed scenario demands a new channel of star formation in the early universe and is a supplementary scenario for the origin of the known metal-poor stars.

1 Introduction

The search for the first stars in the universe is one of the important goals in astronomy. We proposed a new scenario to identify the survivors of the first stars in the Galactic halo [1]. Our scenario assumes a binary system consisting of massive stars and low-mass stars in a short distance so that the supernova explosion of the primary stars can influence the surface of the companion star at the end of the life of the massive primary. Figure 1 shows the schematic picture of our proposed massive binary scenario.

We performed simulations to quantify the impact of stripping from and accretion onto the surface of companion stars when the ejecta of massive primary collide with the low-mass companion. We also looked for evidence of observational counterparts in the Galaxy through the survey of massive binaries in the solar vicinity using the medium-resolution spectrograph.

*e-mail: sudatkm@stf.teu.ac.jp

In this paper, we review the results of our simulation to provide an example of the survivor of the first stars with the massive binary scenario. The current status of the binary star survey is also reported.

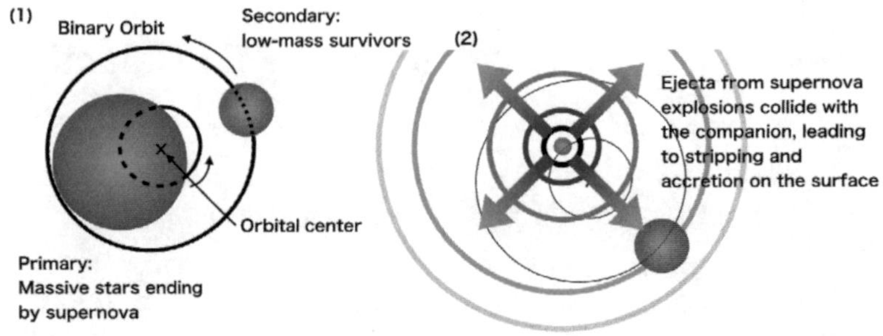

Figure 1. Schematic picture of our massive binary scenario to predict the survivors of the first stars in the Galactic halo.

2 Simulation of the massive binary scenario

Stellar models are taken from a primordial 0.8 $M_{\odot}$ star [2] and the first supernova of a 15 $M_{\odot}$ star [3]. They are mapped on the 3D space to perform SPH simulations using the ASURA code [4]. We followed the evolution of the binary system after the explosion of the primary star by tracing the SPH particles that collide with the low-mass companion. The explosion model is based on the model of SN1987A [5] for simplicity. The detail of the simulation setup and feasibility study is provided in Suda et al. [1].

Figure 2 shows a snapshot of the simulation. We found that the stripping of the envelope by the ejecta does not have a strong impact on the surface chemical composition on the secondary star, while the impact of accretion, mainly caused by slower inner ejecta, is significant. Although depending on the degree of mixing of the ejecta before the collision, it is expected that the secondary star should be metal-rich if the separation is sufficiently small.

We predict that the survivors of the first stars that experienced the accretion of supernova ejecta can be found in the Galactic halo as very metal-rich stars. The comparisons of models with the observations are made with stellar lithium abundances using the SAGA database [6].

3 Search for massive binaries

We looked for metal-rich counterpart of the progenitors of the massive binary scenario because the first massive stars do not exist. We have conducted the survey to identify OB stars with low-mass companion in short-period binaries. The observations have been made by Medium And Low-dispersion Long-slit Spectrograph (MALLS) equipped to the Nayuta telescope and High Dispersion Echelle Spectrograph (HIDES) equipped to the telescope at the Subaru Telescope Okayama Branch Office. Figure 3 shows the ongoing analysis of a binary candidate, HD 93521. The details of the survey is described in Moritani et al. [7] and the result will be reported in a forthcoming paper.

Acknowledgement

This work is supported by JSPS Kakenhi Grant No. 16K05287.

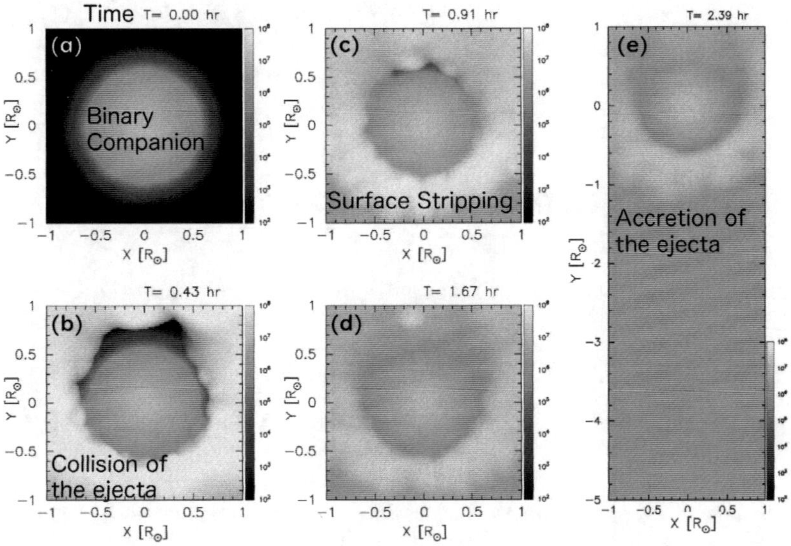

Figure 2. The snapshot of the SPH simulation of the collision between the supernova ejecta of a 15 $M_\odot$ star and a 0.8 $M_\odot$ star with a distance of 0.2 au. The color contour represents the temperature in units of kelvin. Panel (e) shows the simulation box that includes the part of the colliding ejecta.

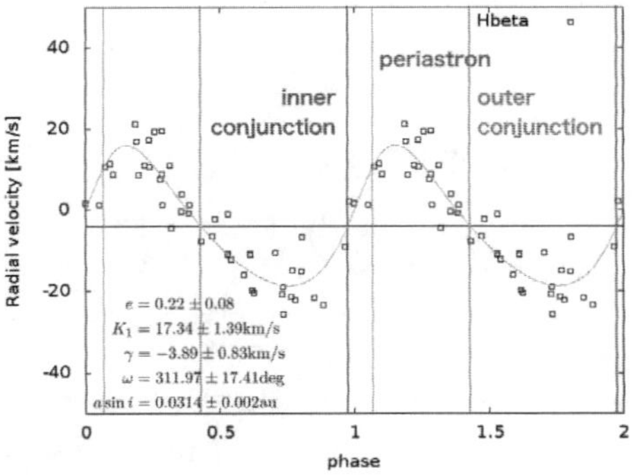

Figure 3. Determination of binary parameters for HD 93521. The derived parameters are given in the bottom left corner. The inner and outer conjuction, together with periastron are shown by the vertical lines.

References

[1] T. Suda, T. R. Saitoh, Y. Moritani, T. Matsuno, & T. Shigeyama, PASJ **73**, 609 (2021)
[2] T. Suda, & M. Y. Fujimoto, MNRAS **405**, 177 (2010)
[3] A. Heger, & S. E. Woosley, ApJ **724**, 341 (2010)
[4] T. R. Saitoh, H. Daisaka, E. Kokubo, J. Makino, T. Okamoto, K. Tomisaka, K. Wada, & N. Yoshida, PASJ, **60**, 667 (2008)
[5] T. Shigeyama, & K. Nomoto, ApJ, **360**, 242 (1990)
[6] T. Suda, Y. Katsuta, S. Yamada, T. Suwa, C. Ishizuka, Y. Komiya, K. Sorai, M. Aikawa, & M. Y. Fujimoto, PASJ, **60**, 1159 (2008)
[7] Y. Moritani, T. Suda, T. Shigeyama, & T. R. Saitoh, Stars and Galaxies **1**, 1 (2018), arXiv e-prints, arXiv:1812.06583

The SHADES neutron detection array

Chemseddine Ananna[1,*], *David* Rapagnani[1], *Andreas* Best[1], *Antonino* Di Leva[1], and *Gianluca* Imbriani[1]

[1]University of Naples "Federico II", Via Vicinale, 80145 Napoli, Italy

Abstract. The detection of neutrons in low energy astrophysics poses two main problems: low cross section measurement requires high detection efficiency and ideally, beam-induced and external neutron backgrounds should be identified through an energy sensitive detector. In recent years, capture-gated neutron spectroscopy has found new opportunities to grow through new materials and better availability of modern data acquisition systems. In this work we will present the design, expected capabilities and preliminary characterization of the hybrid neutron detector array SHADES. Its purpose is the first direct measurement of the reaction ^{22}Ne$(\alpha, \mathrm{n})^{25}$Mg in the Gamow window for s-process nucleosynthesis with LUNA-MV in the deep underground environment of the LNGS.

1 Introduction

The ^{22}Ne$(\alpha, \mathrm{n})^{25}$Mg reaction is not only one of the two main neutron sources feeding the main s-process but it also is the main source of neutrons available for the weak s-process during the AGB phase for massive stars ($M>9M_{\odot}$). It is, therefore, crucial for the study of the stellar nucleosynthesis of the heavy elements in the universe, particularly those with an atomic mass $60<A<90$. As of now, only upper limits are known for this reaction's cross section at the resonance energy of $E_\alpha = 832\,\mathrm{keV}(J^\pi = 2^+)$. Nevertheless, other states do exist below this energy [1–3] but their contribution to the total cross-section hasn't been measured yet because of the low cross section [1, 4, 5].

2 Technical challenges

The ^{22}Ne$(\alpha, \mathrm{n})^{25}$Mg has a negative Q-value of Q = -478 keV, therefore, neutron energies extend down to values that are very low. This represents a first challenge to overcome because the detection and measurement of such low energy neutrons is very difficult.

The measurement of the low cross section requires a high detection efficiency and suppression of external and beam induced background from reactions with impurities like Boron and ^{13}C. In the meanwhile, the external neutron background should be identified and discriminated against through an energy sensitive detector, in addition to a passive shielding.

To overcome the aforementioned challenges, the SHADES project (ERC-StG-2019 #852016) is aiming to perform a high sensitivity measurement, lowering the experimental

*e-mail: chemseddine.ananna@unina.it

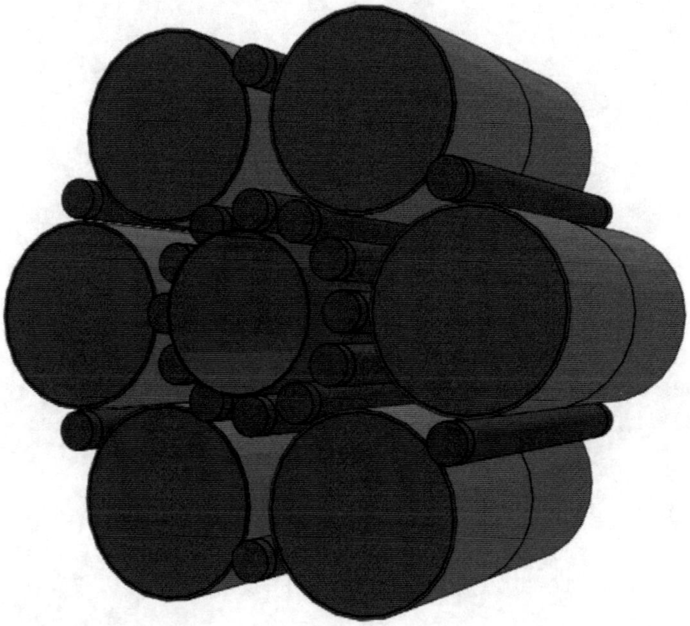

Figure 1. The SHADES detection array

background by going inside the deep underground facility of the Gran Sasso National Laboratory (AQ – Italy) and employing coincidences (and anti-coincidences) between two neutron detector types. The expected background reduction is more than 3 orders of magnitude over the surface lab one.

The new LUNA-MV accelerator [6] will provide a high-current ($\approx 500\,\mu$A), high-stability α beam that will be impinging on a recirculating extended gas target, allowing for a long and interruption-free measurement campaign.

3 Detection setup

The SHADES detector array consists of a 12 EJ-309 liquid scintillators ring in between two ^{3}He counters rings of 12 and 6 units, respectively surrounding a recirculating, windowless gas target (see Fig.1). This combination of multiple detector types will allow a better discrimination against both beam-induced and naatural room background by imposing that only events detected in both are considered and by employing cuts on the neutron energy.

The whole detection array will be placed inside borated polyethylene (see Fig.2 to further reduce the external neutron background.

As a hybrid system, each component of the SHADES detector has a different but complementary task to achieve during the measurement. The liquid scintillators moderate the reaction neutrons and register the recoil energy. When a thermalized neutron enters one of the ^{3}He counters, it will be detected thanks to the high cross section of the reaction ^{3}He(n,p)^{3}T. . For the low-energy neutrons coming from the reaction ^{22}Ne(α, n)^{25}Mg, the ^{3}He detector acts as a counter with a total detection efficiency of about 15-20%. The liquid scintillators are not sensitive to neutrons of these energies but will be very useful as anticoincidence detectors: higher-energy neutrons from beam-induced background reactions like ^{10}B(α,n)^{13}N (Q

Figure 2. Mounting of the Shades detection array inside the borated polyetylene shielding

= 1059 keV), $^{11}B(\alpha,n)^{14}N$ (Q = 158 keV) or $^{13}C(\alpha,n)^{16}O$ (Q = 2216 keV) will be registered in the scintillators and can be identified and removed.

3.1 Characterization

A first characterization measurements of the 12 EJ-309 detectors inside a lead castle have been performed in the LNGS deep-underground from the end of July to the beginning of August 2021. This over a month-long run aimed to measure the intrinsic background activity of the detection array. The gathered data is still under analysis, but the first results seem to be very promising. The Pulse Shape Discrimination figure (see Fig.3) presents a good gamma/alpha separation until an acceptable value towards lower energies.

This first run has also allowed to confirm the low gamma background activity ($\approx$ 45 counts.s^{-1} per detector). A very key feature of the LNGS facility heavily needed in the experiment because of the low cross section of $^{22}Ne(\alpha, n)^{25}Mg$ reaction.

4 Outlook

In the near future, the SHADES team is aiming to better refine the used PSD techniques to get a better alpha/gamma and neutron/gamma separation towards lower energies. The liquid

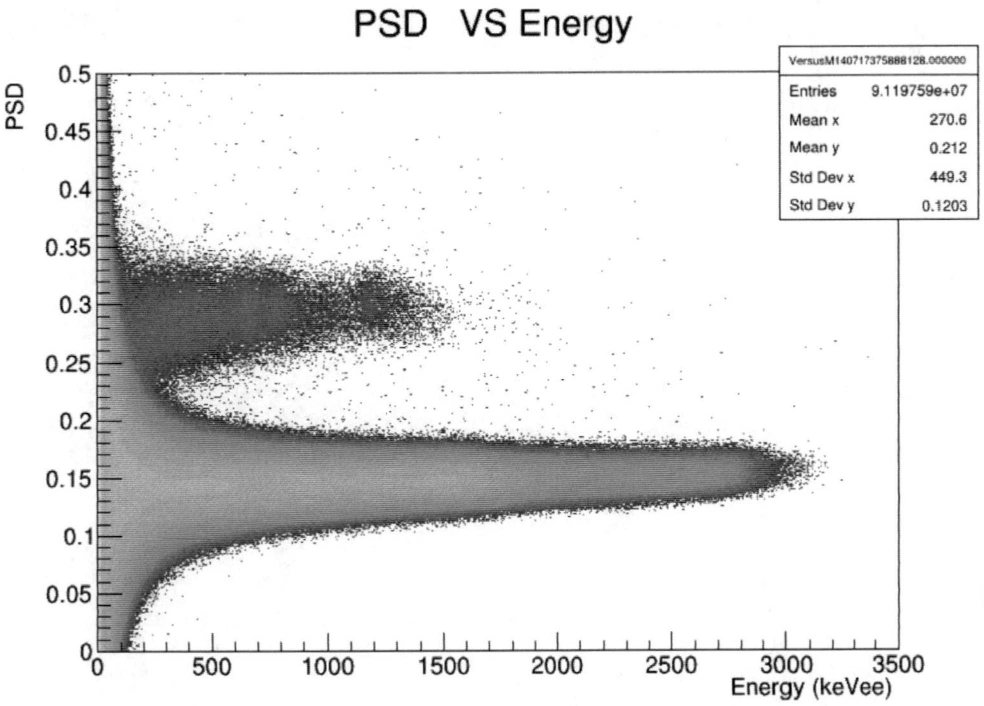

Figure 3. Alpha/Gamma pulse shape discrimination, LNGS, a month-long acquision run using the charge integration method.

scintillator characterization should also be resumed by doing the quenching calibration. The performance of the full array (EJ-309 scintillators + ^{3}He counters) should then be ready to be tested as a whole. The second part of the procedure will consist of the characterization of the ^{22}Ne gas target followed by the first LUNA MV beam trials.

Acknowledgement

The authors acknowledge funding from the European Research Council (ERC-StG 2019 #852016)

References

[1] M. Jaeger, R. Kunz, A. Mayer, J.W. Hammer, G. Staudt, K.L. Kratz, B. Pfeiffer, Phys. Rev. Lett. **87**, 202501 (2001)
[2] V. Harms, K.L. Kratz, M. Wiescher, Phys. Rev. C **43**, 2849 (1991)
[3] H.W. Drotleff, A. Denker, H. Knee, M. Soine, G. Wolf, J.W. Hammer, U. Greife, C. Rolfs, H.P. Trautvetter, Astrophys. J. **414**, 735 (1993)
[4] S. Bisterzo, C. Travaglio, R. Gallino, M. Wiescher, F. Käppeler, Astrophys. J. **787**, 10 (2014)
[5] C. Massimi, P. Koehler, S. Bisterzo, N. Colonna, R. Gallino, F. Gunsing, F. Käppeler, G. Lorusso, A. Mengoni, M. Pignatari et al. (n_TOF Collaboration), Phys. Rev. C **85**, 044615 (2012)

https://doi.org/10.1051/epjconf/202226011043

[6] A. Sen, G. Domínguez-Cañizares, N.C. Podaru, D.J.W. Mous, M. Junker, G. Imbriani, V. Rigato, Nucl. Instrum. Methods Phys. Res. B **450**, 390 (2019)

SECAR: A recoil separator for nuclear astrophysics.

Pelagia Tsintari[1,8,*], *Ruchi* Garg[2,8], *Georg* Berg[3,10], *Jeff* Blackmon[6], *Kelly* Chipps[5], *Manoel* Couder[3,10], *Catherine* Deibel[6], *Nikolaos* Dimitrakopoulos[1,9], *Uwe* Greife[7], *Ashley* Hood[6], *Rahul* Jain[8,2,9], *Caleb* Marshall[4,9], *Zach* Meisel[4,9], *Sara* Miskovich[2,8,9], *Fernando* Montes[2,9], *Georgios* Perdikakis[1,9], *Thomas* Ruland[6], *Hendrik* Schatz[2,8,9], *Kiana* Setoodehnia[2,9], *Michael* Smith[5], and *Louis* Wagner[2,9]

[1]Department of Physics, Central Michigan University, Mt. Pleasant, MI 48859, USA
[2]Facility for Rare Isotope Beams, Michigan State University, East Lansing, MI 48824, USA
[3]Department of Physics, University of Notre Dame, Notre Dame, IN 46556, USA
[4]Department of Physics & Astronomy, Ohio University, Athens, OH 45701, USA
[5]Physics Division, Oak Ridge National Laboratory, Oak Ridge, TN 37831, USA
[6]Department of Physics and Astronomy, Louisiana State University, Baton Rouge, LA 70803, USA
[7]Department of Physics, Colorado School of Mines, Golden, CO 80401, USA
[8]Department of Physics and Astronomy, Michigan State University, East Lansing, MI 48824, USA
[9]The Joint Institute for Nuclear Astrophysics - Center for the Evolution of the Elements, Michigan State University, East Lansing, MI 48824, USA
[10]The Joint Institute for Nuclear Astrophysics - Center for the Evolution of the Elements, University of Notre Dame, Notre Dame, IN 46556, USA

Abstract. Proton- and alpha-capture reactions on unstable proton-rich nuclei power astrophysical explosions like novae and X-ray bursts. Direct measurements of these reactions are crucial for understanding the mechanisms behind these explosions and the nucleosynthesis at such sites. The recoil mass separator, SECAR (SEparator for CApture Reactions) at the National Superconducting Cyclotron Laboratory (NSCL) and the Facility for Rare Isotope Beams (FRIB), has been designed with the required sensitivity to study (p,γ) and (α,γ) reactions, directly at astrophysical energies in inverse kinematics, with radioactive beams of masses up to about A = 65. The complete SECAR system, including two Wien Filters for high mass resolution, has been installed at Michigan State University and is currently being commissioned. The present article introduces the SECAR concept, its scientific goals, and provides an update of the current status of the project.

1 Introduction

Exploding stars are laboratories for extreme physics, involving a complex interplay of high temperature thermonuclear burning and hydrodynamics. One of the interesting aspects of these systems are related to the processes by which the chemical elements are created and dispersed by these catastrophic events.

Nova explosions [1] and X-ray bursts [1] are two examples of such explosive astrophysical scenarios. They take place in the dense and hot atmosphere on the surfaces of accreting

*e-mail: tsint1p@cmich.edu

white dwarfs and neutron stars, respectively. The precise measurements of the reaction cross sections of radiative capture reactions of protons and alpha particles on unstable proton-rich nuclei can potentially address long standing questions associated with the energy generation and the nucleosynthesis occurring in these scenarios.

SECAR (SEparator for CApture Reactions) will allow direct measurements of many of the relevant reactions with proton-rich radioactive beams obtained from ReA3 and FRIB in inverse kinematics (*"i.e."* a heavy, proton-rich radioactive ion beam bombards a hydrogen or helium target). The recoil separation and detection capability enabled by SECAR provides the required sensitivity to detect astrophysical reactions with very low cross sections. SECAR will be used to improve our understanding of the evolution and element creation in these exotic astrophysical sites by enabling direct rate measurements for explosion studies.

2 The SECAR layout

The layout of the SECAR system [3], consisting of a total of 4 sections, is shown in Figure 1. The separator setup begins at the JENSA (Jet Experiments for Nuclear Structure and Astrophysics) chamber [4, 5], where the heavy ion beam enters the high density windowless gas jet target. JENSA is a recirculating gas jet target assembly, providing areal densities up to 10^{19} atoms/cm^2 of hydrogen or helium isotopes in a 4 – 5 mm diameter cylindrical jet.

Section 1 extends from the target to the focal plane 1 (FP1), and is comprised of two dipole magnets (B1–B2), five quadrupoles (Q1–Q5), and one hexapole magnet (Hex1) that allows flexible ion-optical settings and corrections for higher-order aberrations in the beam transport. This section accomplishes the selection of a single charge state, typically the most abundant, using magnetic analysis in B1-B2, and prepares the beam for optimal use of the Wien filter (WF1) in Section 2.

Section 2, between FP1 and FP2, includes two dipole magnets (B3-B4), two quadrupoles (Q6 and Q7), one Wien filter (WF1), two hexapole magnets (Hex2 and Hex3), and one octupole (Oct1). The heavy ion beam and the reaction products have essentially the same magnetic rigidity after the charge state selection and therefore cannot be separated solely by magnetic analysis. A combination of electric and magnetic dipoles is used for the separation of the beam from the reaction products based on their velocity difference. This is achieved with the use of a Wien filter that combines a magnetic and electric field in a ratio chosen such that only the particles with a velocity corresponding to the desired mass follow a straight line, while the rest are deflected off the ion-optical axis.

Following the mass separation of FP2, section 3 consists of two dipole magnets (B5-B6), four quadrupoles (Q8–Q11), and a second Wien filter (WF2), identical to WF1, which provides the required mass resolution at FP3. After a second velocity (mass) separation at FP3, a final momentum analysis is performed in section 4, using two dipole magnets (B7-B8), and four quadrupoles (Q12–Q15). This section enhances the rejection of the background of scattered beam particles and provides a clean environment for the recoil detection at the final focal plane.

Scattered particles ("leaky beam") which are not eliminated by the electromagnetic separation may reach the final focal plane along with the recoils of interest. Two suitable detector systems, are installed at the end of section 4 and at the JENSA chamber, providing a high degree of discrimination between the capture reaction recoils of interest and the background ions. The SECAR detection system at FP4 consists of two microchannel plate detectors (MCP), serving as part of a local time-of-flight (ToF) system, and a hybrid system consisting of a gas ionization chamber (IC) and a double-sided silicon strip detector (DSSD), which provides ΔE-E energy particle identification. The auxiliary detection system at JENSA chamber comprises of a BGO scintillator detector array and two planar silicon detectors. The BGO array is used for the detection of γ-rays from the decay of the compound nucleus formed in the reaction. The two Si detectors, placed symmetrically on both sides of the beam axis, are used for monitoring the beam intensity and the beam-jet overlap by recording scattered target particles.

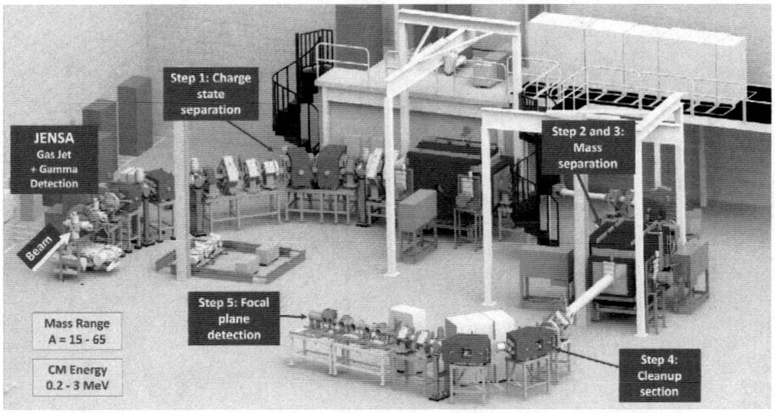

Figure 1. Beam enters SECAR through the JENSA gas jet target system (shown on the left of the picture), where the reaction takes place. γ-Rays, produced from the capture reaction, are detected at the JENSA chamber using an array of BGO detectors. Once a single charge state of the reaction recoils is selected (Step 1), it goes through a Wien Filter (WF1), where the unreacted beam ions are mass separated and eventually stopped using slits (Step 2). The remaining scattered beam particles are separated via the WF2 (Step 3), while the recoils go through a last "clean up" before reaching the detection system (Step 4) at the last focal plane.

Extensive ion-optical calculations and optimizations, using the computer code COSY Infinity [6], were performed to achieve all the design requirements of the SECAR system [3]. Figure 2 shows the ion optics of the SECAR layout that have been optimized and corrected up to the 4th order to be used for (p,γ) and (α,γ) reaction measurements. The upper and lower panel show the characteristic rays in the horizontal and the vertical plane, respectively. The horizontal envelope is relatively large by design, as this is the dispersive plane of the system. The vertical envelope, however, never exceeds $\pm$6 cm to maximize transmission through the system. Also, a large horizontal envelope is required in the Wien filters to achieve a large mass separation. The system has an achromatic focus at FP2 and a zero energy dispersion focus at FP3. A left–right asymmetry of the envelope about the central ray, which is observed in Figure 2, is mainly due to the effect of the second and third order aberrations. Although more symmetric envelopes could be applied, the imposed reduction of the horizontal good-field region invariably resulted in a reduced mass resolution, therefore an asymmetric envelope was deemed more suitable for the maximization of the mass separation.

3 Commissioning results

A first successful direct measurement of the known strength of the 6.360 MeV resonance in the ^{16}O(α,γ)^{20}Ne reaction [7, 8] was recently attained in inverse kinematics using the full SECAR system. The measurement was obtained using an ^{16}O beam at four different energies, namely 1.986, 1.992, 1.996 and 2.003 MeV/u, at a typical beam intensity of 1.8x10^8 pps, provided by ReA3. A helium gas jet target was used, with thicknesses of 3.4 and 5.2 x10^{18} at/cm^2, which corresponded to beam energy losses of 13 and 20 keV/u, respectively. To

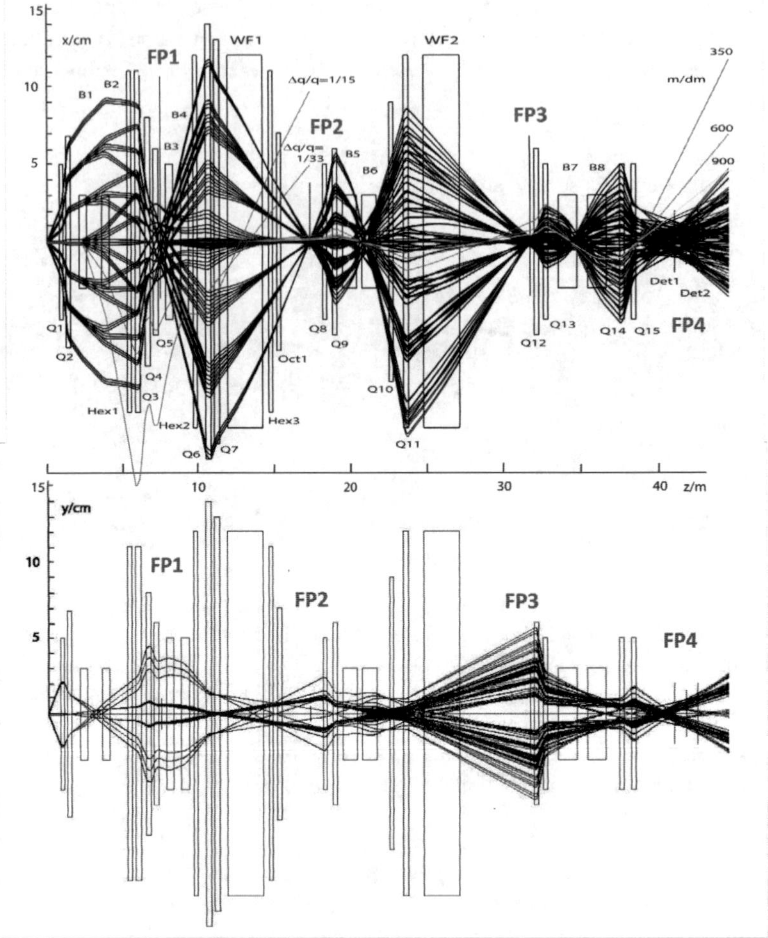

Figure 2. The ion optics of SECAR for the layout shown in Figure 1 is corrected up to the 4th order. The characteristic rays in the horizontal and vertical plane are shown in the upper and lower panel, respectively. The mass resolution in the horizontal plane at FP2 is $m/\Delta m$ = 510 and at FP3, $m/\Delta m$ = 780. The picture was obtained from G.P.A. Berg et al., Nuclear Inst. and Meth. in Phys. Res. A, 877, 87–103 (2018).

ensure charge state equilibrium of the ^{20}Ne recoils of interest, a 54 μg/cm^2 C stripper foil was used immediately after the target.

Figure 3 (left) shows all the ions detected at the IC and DSSD detectors at FP4 with SECAR tuned to the 8^+ charge state of the ^{20}Ne recoils. This particle ID plot shows three distinct groups. The reaction recoils were identified as the top right group, while the other two were confirmed as being "leaky beam" particles. The recoil identification was verified by selecting the ions that had a coincident gamma-ray detection at the target location within a 3.2±0.2 μs window in the BGO - Si Time of Flight spectrum. Figure 3 (right) shows the particle ID plot gated on the BGO - Si ToF peak. The γ-ray energy spectrum of these coincident events is shown in Figure 4, where the three peaks corresponding to the decay cascade of the ^{20}Ne 11.09 MeV level can be clearly seen at 1.633, 2.613, and 6.841 MeV.

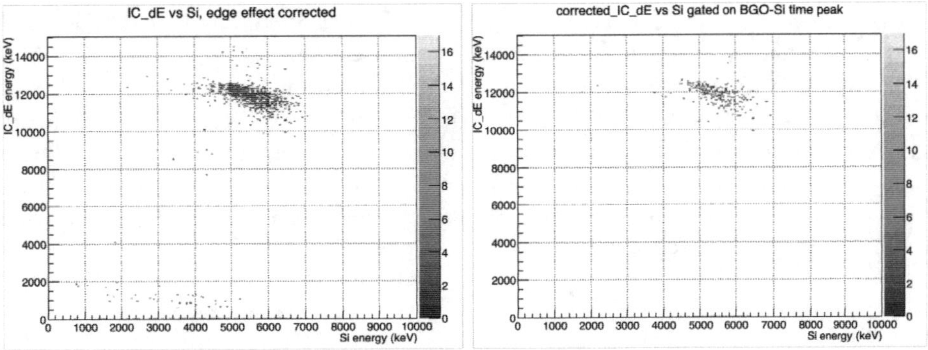

Figure 3. Left: All ions reaching the final focal plane measured with the IC and DSSD detectors (ΔE-E) when SECAR was tuned to the 8^+ charge state of the ^{20}Ne recoils. Three distinctive groups of ions are observed corresponding to recoils and "leaky beam" particles. Right: Same as the left plot, but with the events gated on the BGO - Si time difference peak corresponding to the recoil ToF between the target and the FP4.

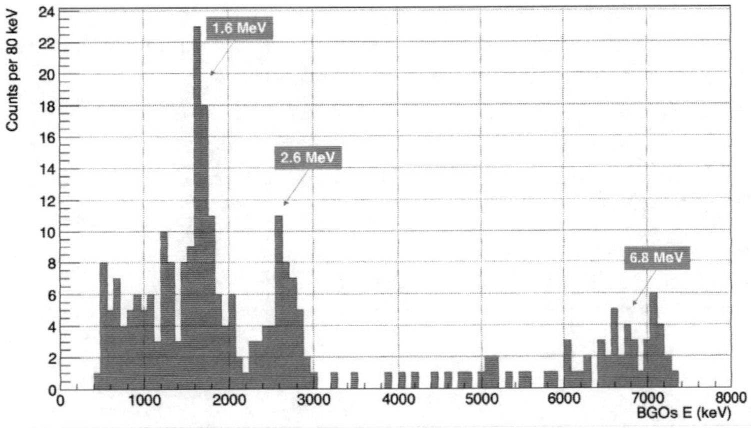

Figure 4. BGO spectrum gated on the events of the BGO - Si time difference peak corresponding to the recoil ToF. The γ-rays of the ^{20}Ne 11.09 MeV level decay cascade can be clearly seen at 1.633, 2.613, and 6.841 MeV.

4 Summary

The new recoil separator, SECAR, at NSCL-FRIB is optimized for measurements of radiative capture reactions (p,γ) and (α,γ) at very low energies of astrophysical interest in inverse kinematics for high intensity short-lived radioactive ion beams, for which such capture reaction measurements in normal kinematics are not feasible. The system is designed to have high transmission of capture reaction recoils and high rejection of scattered unreacted beam particles. With unstable beams from the FRIB facility, it is anticipated that SECAR will enable measurements of thermonuclear reactions critical for our understanding of stellar explosions.

The separation of the beam and reaction products is accomplished by initially selecting one charge state and then transmitting the beam and recoil products having almost identical momenta. Specially designed Wien filters are subsequently used for a high degree (approximately 10-13) of rejection of unreacted scattered beam particles, which is needed to measure the weakest resonances in certain capture reactions. Further background reduction is accomplished in the detector system at the end of the recoil separator.

The successful first direct measurement of the strength of the known 6.360 MeV resonance of the $^{16}O(\alpha,\gamma)^{20}Ne$ reaction, demonstrated that all components of SECAR are operational. Also, the detection of the in-coincidence γ-rays at the target were used to confirm the detection of the recoils of interest, at the end of SECAR system. These results indicate that the construction phase is completed, and that the system is ready for further commissioning and measurements with radioactive beams.

References

[1] J. José, M. Hernanz, C. Iliadis, Nuclear Phys. A, **777**, 550-578 (2006)

[2] H. Schatz, K.E. Rehm, Nuclear Phys. A, **777**, 601-622 (2006)

[3] G.P.A. Berg, M. Couder, M.T. Moran, K. Smith, M. Wiescher, H. Schatz, U. Hager, C. Wrede, F. Montes, G. Perdikakis, X. Wu, A. Zeller, M.S. Smith, D.W. Bardayan, K.A. Chipps, S.D. Pain, J. Blackmon, U. Greife, K.E. Rehm, R.V.F. Janssens, Nucl. Inst. Meth. Phys. Res. A, **877**, 87–103 (2018)

[4] K.A. Chipps, U. Greife, D.W. Bardayan, J.C. Blackmon, A. Kontos, L.E. Linhardt, M. Matos, S.D. Pain, S.T. Pittman, A. Sachs, H. Schatz, K.T. Schmitt, M.S. Smith, P. Thompson, Nucl. Inst. Meth. Phys. Res. A, **763**, 553–564 (2014)

[5] K. Schmidt, K.A. Chipps, S. Ahn, D.W. Bardayan, J. Browne, U. Greife, Z. Meisel, F. Montes, P.D. O'Malley, W-J. Ong, S.D. Pain, H. Schatz, K. Smith, M.S. Smith, P.J. Thompson, Nucl. Inst. Meth. Phys. Res. A, **911**, 1-9 (2018)

[6] M. Berz, COSY Infinity, $http://www.bt.pa.msu.edu/index_files/cosy.html$

[7] D. J. Steck, Phys. Rev. C, **17**, 1034-1050 (1978)

[8] L. K. Fifield, et al., Nuclear Physics A, **334**, 109-126 (1980)

EPJ Web of Conferences **260**, 11045 (2022) https://doi.org/10.1051/epjconf/202226011045
NIC-XVI

Structure of two- and three-alpha systems in cold neutron matter

Hajime Moriya[1,*], *Hiroyuki* Tajima[2,3], *Wataru* Horiuchi[1], *Kei* Iida[3], and *Eiji* Nakano[3]

[1]Department of Physics, Hokkaido University, Sapporo 060-0810, Japan
[2]Department of Physics, Graduate School of Science, The University of Tokyo, Tokyo 113-0033, Japan
[3]Department of Mathematics and Physics, Kochi Univerisity, Kochi 780-8520, Japan

Abstract. We present stability and structure of two- and three-alpha systems embedded in dilute cold neutron matter. By solving a few-alpha Schrödinger equation with quasiparticle properties, i.e., effective mass and induced two- and three-alpha interactions, which are evaluated in terms of Fermi polarons, it is shown that ^{8}Be and the Hoyle state become bound at densities of about 10^{-4} and 10^{-3} of the saturation density, respectively. It is also seen that, under cold neutron matter environment, both systems become smaller than the corresponding systems in vacuum. Our results would affect astrophysical models for stellar collapse and neutron star mergers, as well as relevant reaction rates for nucleosynthesis.

1 Introduction

An alpha (^{4}He nucleus) cluster plays an important role in the nuclear structure of light nuclei and astrophysical reactions relevant for the evolution of chemical elements in a burning star. In particular, the first $J^\pi = 0^+$ excited state of ^{12}C, the so-called Hoyle state, is one of the most famous examples in which three alpha clusters are well developed. The Hoyle state works in such a way as to accelerate the triple-alpha reaction at low energies near the Gamow window [1], where the ^{12}C element is generated through sequential reactions via the resonant ground state of ^{8}Be. It is important to describe such alpha-induced reactions accurately in the astrophysical explosive phenomena, e.g., core-collapse supernovae and neutron star mergers [2].

Recently, quasiparticle properties of an alpha particle in dilute cold neutron matter were discussed [3], where the alpha particle in the neutron matter was described as an impurity in a many-body background like a Fermi polaron in ultracold atomic physics [4]. In the polaron picture, the effective mass of the alpha particle is controlled by the interaction from the background. In this paper we discuss possible changes in the structure of two- and three-alpha systems in dilute neutron matter at zero temperature. By solving a few-alpha Schrödinger equation precisely, we discuss the stability and structure changes of the ground state of ^{8}Be and the Hoyle state in such neutron matter. In addition to the change of the effective mass, we introduce two- and three-alpha interactions induced by the neutron medium. Since all the details of the present evaluation are described in Ref. [5], in this paper, we briefly recapitulate those discussions and present some new results that were not presented there.

*e-mail: moriya@nucl.sci.hokudai.ac.jp

2 Method

We employ a multi-alpha cluster model to describe the two- and three-alpha systems. The effective Hamiltonian for the three-alpha system is given by

$$H = \sum_i \frac{\boldsymbol{p}_i^2}{2M^*} - T_{cm} + \sum_{i>j} \left(V_{\alpha\alpha,ij} + V_{\mathrm{eff},ij}^{(2)} \right) + V_{\alpha\alpha\alpha} + V_{\mathrm{eff}}^{(3)}, \tag{1}$$

where M^* is the effective mass of the alpha particle, $\boldsymbol{p}_i$ is the momentum operator of the ith alpha particle, T_{cm} is the kinetic energy of the center-of-mass motion, $V_{\alpha\alpha}$ and $V_{\alpha\alpha\alpha}$ are two- and three-alpha interactions, respectively. This two-alpha potential is deep enough to have some redundant or forbidden bound states by the Pauli principle between alpha particles. Details of the parameter set are given in Refs. [6, 7]. Physical states should be orthogonal to such forbidden states. These states are eliminated practically by using the projection method [8].

The effective mass of the alpha particle M^* was given as a function of the Fermi momentum k_F in Ref. [3]. Here, medium induced two- and three-alpha interactions ($V_{\mathrm{eff}}^{(2)}$ and $V_{\mathrm{eff}}^{(3)}$) are also introduced by considering the simple Feynman diagrams in which few-alpha clusters exchange momenta. All the details are described in Ref. [5]. The wave function of the few-alpha system is described by a superposition of fully symmetrized correlated Gaussian basis functions whose variational parameters are optimized by the stochastic variational method [9, 10]. All the numerical setups are detailed in Refs. [5, 11].

3 Results and discussion

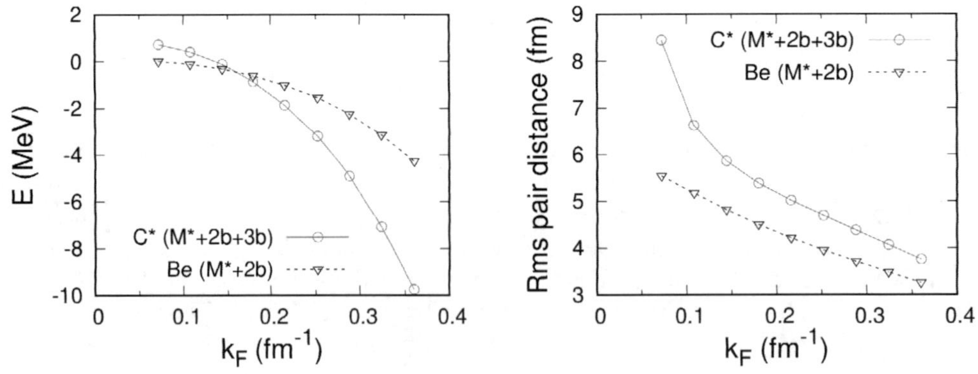

Figure 1. Energies (left) and root-mean-square pair distance (right) of the ground state of ^{8}Be and the Hoyle state C* in neutron matter of the Fermi momentum k_F. M^*, $2b$, and $3b$ denote calculations with the effective mass, induced two- and three-alpha interactions, respectively.

The alpha cluster environment is characterized by the Fermi momentum of background neutrons k_F, which affects the effective mass, induced two- and three-alpha interactions in the Hamiltonian. Here we discuss the energies and structure of the ground state of ^{8}Be and the Hoyle state as a function of k_F. The left panel of Fig. 1 plots the ground state of ^{8}Be and the Hoyle state energies as a function of k_F. The energies of both the ^{8}Be and the Hoyle state become lower for larger k_F due to the short-range attraction of the induced two-alpha interaction. Eventually, the ground state of ^{8}Be becomes a bound state for $k_F \gtrsim 0.08$ fm^{-1}

and the Hoyle state is also bound for $k_F \gtrsim 0.16$ fm^{-1}, which correspond to approximately 10^{-4} and 10^{-3} of the saturation density 0.16 fm^{-3}, respectively. To see the structure of these alpha particle systems, we plot in the right panel of Fig. 1 the root-mean-square (rms) pair distance of the ground state of ^{8}Be and the Hoyle state. Both the ground state of ^{8}Be and the Hoyle state become smaller for larger k_F monotonically. This shrinkage occurs because the internal amplitude of the wave function is strongly modified by the induced interactions mostly by the attractive induced two-alpha interaction. It is interesting to note that the size of the Hoyle state in the cold neutron matter shrinks as is consistent with the study of the finite system, e.g., the $\alpha - \alpha - n$ cluster calculation [12].

4 Conclusion and prospects

We have made precise calculations for few-alpha systems immersed in cold neutron matter. The alpha particle is treated as a polaron, i.e., quasiparticle dressed by excitations of background neutrons. We have found the possibility that the two- and three-alpha systems become bound in cold dilute neutron matter. The ground state of ^{8}Be and the Hoyle state become bound at $\approx 10^{-4}$ and 10^{-3} of the saturation density. The binding of these fundamental light nuclear ingredients would have a significant effect on the modeling of stellar collapse and neutron star mergers, as well as relevant reaction rates for nucleosynthesis. To realize these effects by the presence of the cold neutron matter in experiments, it is interesting to investigate $2\alpha + Xn$ and $3\alpha + Xn$ systems. For this purpose, it is desirable to relate alpha clusters interacting with a few neutrons and with infinite neutron matter. A detailed study along this direction is underway and will be reported elsewhere soon.

5 Acknowledgement

This work was in part supported by JSPS KAKENHI Nos. 17K05445, 18K03635, 18H01211, 18H04569, 18H05406, and 19H05140, and the Collaborative Research Program 2021, Information Initiative Center, Hokkaido University.

References

[1] F. Hoyle, Astrophys. J. Suppl. Ser. **1**, 12 (1954)
[2] M. Oertel, M. Hempel, T. Klähn, and S. Typel, Rev. Mod. Phys. **89**, 015007 (2017)
[3] E. Nakano, K. Iida, and W. Horiuchi, Phys. Rev. C **102**, 055802 (2020)
[4] F. Chevy and C. Mora, Rep. Prog. Phys. **73**, 112401 (2010)
[5] H. Moriya, H. Tajima, W. Horiuchi, K. Iida, and E. Nakano, Phys. Rev. C **104**, 065801 (2021).
[6] C. Kurokawa, K. Katō, Phys. Rev. C **71**, 021301 (2005)
[7] K. Fukatsu and K. Katō, Prog. Theor. Phys. **87**, 151 (1992)
[8] V. I. Kukulin, and V. N. Pomenertsev, Ann. Phys. (N. Y.) **111**, 333 (1978)
[9] K. Varga and Y. Suzuki, Phys. Rev. C **52**, 2885 (1995)
[10] Y. Suzuki and K. Varga, *Stochastic Variational Approach to Quantum-Mechanical Few-Body Problems*, Lecture Notes in Physics, Vol. m54 (Springer, Berlin, 1998)
[11] Lai Hnin Phyu, H. Moriya, W. Horiuchi, K. Iida, K. Noda, and M. T. Yamashita, Prog. Theor. Exp. Phys. **2020**, 093D01 (2020)
[12] M. Lyu, Z. Ren, B. Zhou, Y. Funaki, H. Horiuchi, G. Ropke, P. Schuck, A. Tohsaki, C. Xu, and T. Yamada, Phys. Rev. C **91**, 014313 (2015)

Measuring the $^{15}O(\alpha, \gamma)^{19}Ne$ reaction in Type I X-ray bursts using the GADGET II TPC: Hardware

Tyler Wheeler[1,2,3], *A.* Adams[1,2], *J.* Allmond[4], *H.* Alvarez Pol[5], *E.* Argo[1,2], *Y.* Ayyad[5], *D.* Bardayan[6], *D.* Bazin[1,2], *T.* Budner[1,2], *A.* Chen[7], *K.* Chipps[4], *B.* Davids[8], *J.* Dopfer[1,2], *M.* Friedman[9], *H.* Fynbo[10], *R.* Grzywacz[11], *J.* Jose[12], *J.* Liang[7], *R.* Mahajan[1], *S.* Pain[4], *D.* Pérez-Loureiro[13], *E.* Pollacco[14], *A.* Psaltis[15], *S.* Ravishankar[3], *A.* Rogers[16], *L.* Schaedig[1,2], *L. J.* Sun[1], *J.* Surbrook[1,2], *L.* Weghorn[1,2], and *C.* Wrede[1,2]

[1]Facility for Rare Isotope Beams, Michigan State University, East Lansing, Michigan 48824, USA
[2]Department of Physics and Astronomy, Michigan State University, East Lansing, Michigan 48824, USA
[3]Department of Computational Mathematics, Science, and Engineering, Michigan State University, East Lansing, Michigan 48824, USA
[4]Oak Ridge National Laboratory, Oak Ridge, Tennessee 37830, USA
[5]Instituto Galego de Física de Altas Enerxías (USC), Spain
[6]Department of Physics, University of Notre Dame, Notre Dame, Indiana 46556, USA
[7]McMaster University, Canada
[8]TRIUMF, Vancouver, British Colombia, Canada
[9]The RACAH institute of Physics, Hebrew University of Jerusalem, Israel
[10]Aarhus University, Aarhus C, Denmark
[11]University of Tennessee, Knoxville, USA
[12]Universitat Politècnica de Catalunya, Barcelona, Spain
[13]Canadian Nuclear Laboratories, Canada
[14]IRFU, CEA Saclay, GIF-sur-Yvette, France
[15]Technical University of Darmstadt, Germany
[16]University of Massachusetts Lowell, USA

Abstract. Sensitivity studies have shown that the $^{15}O(\alpha, \gamma)^{19}Ne$ reaction is the most important reaction rate uncertainty affecting the shape of light curves from Type I X-ray bursts. This reaction is dominated by the 4.03 MeV resonance in ^{19}Ne. Previous measurements by our group have shown that this state is populated in the decay sequence of ^{20}Mg. A single $^{20}Mg(\beta p\alpha)^{15}O$ event through the key $^{15}O(\alpha, \gamma)^{19}Ne$ resonance yields a characteristic signature: the emission of a proton and alpha particle. To achieve the granularity necessary for the identification of this signature, we have upgraded the Proton Detector of the Gaseous Detector with Germanium Tagging (GADGET) into a time projection chamber to form the GADGET II detection system. GADGET II has been fully constructed, and is entering the testing phase.

1 Astrophysical Motivation

When a neutron star is orbited by a low-mass population II star it will accrete hydrogen-rich material from this binary companion via a process called Roche lobe overflow. The temperatures and densities on the surface of the neutron star are such that the accreted hydrogen is continuously fused into helium. Eventually the helium is ignited and a thermonuclear runaway proceeds. This can lead to the synthesis of proton-rich nuclides up to mass number 100, and extremely powerful X-ray bursts [1]. These X-ray bursts are measured using space based telescopes, and the shape of the resulting light curves can in principle reveal interesting properties of the neutron star itself (mass, radius, crust elemental abundance). However, there are nuclear reaction rate uncertainties that significantly affect the shapes of simulated light curves [2]. Sensitivity studies have shown that for breakout temperatures of ~0.5 GK, the $^{15}O(\alpha, \gamma)^{19}Ne$ reaction is the most important reaction rate uncertainty that needs to be determined to properly constrain the modelling of Type I X-ray burst light curves [3].

2 Measuring the Reaction

Ideally, we would measure the $^{15}O(\alpha, \gamma)^{19}Ne$ reaction directly, but given that ^{15}O is radioactive ($t_{1/2} \sim 122$ s), beams of sufficient intensity are not currently available. However, the resonance strength, and by extension the reaction rate, can be determined indirectly from the spin, lifetime, and alpha particle branching ratio (Γ_α/Γ). The spin and lifetime are well known, so only Γ_α/Γ needs to be measured [4]. Thankfully, only one dominant resonance determines the $^{15}O(\alpha, \gamma)^{19}Ne$ reaction rate, which corresponds to the 4.03 MeV state in ^{19}Ne. We need only feed this one state to determine Γ_α/Γ [4]. There have been attempts to measure Γ_α/Γ using direct reactions to populate the 4.03 MeV state, but such attempts have only produced a strong upper limit of $(2.9 \pm 2.1) \times 10^{-4}$ [5]. We have proposed a different method for populating the state that will allow us to measure a finite value for Γ_α/Γ. Previous measurements by our group have shown that this state is populated in the $^{20}Mg(\beta p\alpha)$ decay sequence. Additionally, a single $^{20}Mg(\beta p\alpha)^{15}O$ event through the key $^{15}O(\alpha, \gamma)^{19}Ne$ resonance yields a characteristic signature—the simultaneous emission of a proton and alpha particle [6]. This signature is a unique 3D topology that will allow for a finite measurement of Γ_α/Γ and, hence, the thermonuclear rate of the reaction.

To produce the requisite amount of ^{20}Mg for this experiment we will be running at the Facility for Rare Isotope Beams (FRIB). FRIB is the new nuclear facility being built at Michigan State University that will be operational in 2022. Our PAC approved FRIB experiment will use a ^{36}Ar primary beam that will impinge on a ^{12}C target to create a fast beam of ^{20}Mg that will feed the 4.03 MeV ^{19}Ne resonance.

3 GADGET II

To achieve the granularity necessary for the identification of the events of interest, we have upgraded the Proton Detector of the Gaseous Detector with Germanium Tagging (GADGET) into a time projection chamber to form the GADGET II detection system. GADGET II has three distinct elements: a beam-pipe cross with a beam energy degrader and a silicon detector for beam particle identification, the Segmented Germanium Array (SeGA) for γ detection, and the TPC that fits inside SeGA (see figure 1, left panel) [7].

The TPC will allow us to reconstruct 3D images of the decay events inside our detector. Conventional cuts complemented by image reconstruction will enable a background-free measurement of the proton-alpha coincidence events of interest. Our TPC has a compact cylindrical geometry and will be filled with a gas such as P10 (90% Argon, 10% Methane) that acts both as a beam stop for the fast ^{20}Mg beam and the detection medium via ionization electrons. A homogeneous electric field causes the electrons to drift until the charge cluster is collected on a resistive Micromegas. The Micromegas was constructed by the European Organization for Nuclear Research (CERN) and is comprised of over one thousand 2.2 x 2.2 mm^2 pads that will provide the necessary position resolution in 2D, and the difference in drift time for the arriving electrons will provide the third dimension. Once 3D images of decay events are constructed we will use a machine learning algorithm, (a convolutional neural network), to identify the events of interest based on their unique topology.

To accommodate the high density of signals from our TPC we have implemented a new data acquisition system. This new system is the GET System, which stands for Generic Electronics for TPC's, and is a scalable and generic electronics system that was originally designed for gas-filled detector applications in nuclear physics including TPCs [8]. To attach the corresponding front-end electronics to our detector we designed what we call the AsAd box (see figure 1). The AsAd box houses these components in a copper box that acts as a Faraday Cage to prevent EM radiation from inducing noise.

Fig. 1. Left Panel: CAD drawing of the full GADGET II setup, including the beam-cross, SeGA array, and TPC. Middle Panel: Micromegas being installed on GADGET II TPC. Right Panel: Assembled AsAd box attached to the TPC.

GADGET II has been fully constructed and has entered the bench-top testing phase. A radioactive alpha source (^{228}Th) was installed in the GADGET II gas handling system, which allows us to bleed ^{220}Rn into the detector. The alpha signals from the ^{220}Rn alphas have been successfully detected, demonstrating the system's functionality. We are now in the process of comparing detector events to detailed simulations for calibration purposes. GADGET II will be ready to run the ^{15}O(α, γ)^{19}Ne experiment in early 2022.

This work was supported by the U.S. National Science Foundation under Grants No. PHY-1102511, PHY-1565546, PHY-1913554, and PHY-1811855, and the U.S. Department of Energy, Office of Science, under award No. DE-SC0016052

References

[1] H. Schatz *et al.*, Phys. Rev. Lett. **86**, 3471 (2001).
[2] Cyburt et al., Astrophys. J. 830, 55 (2016).
[3] M. Wiescher *et al.*, J. Phys. G **25**, R 133 (1999).
[4] B. Davids *et al.*, Astrophys. J. **735**, 40 (2011).
[5] W. Tan *et al.*, Phys. Rev. Lett. **98**, 242503 (2007).
[6] B. Glassman *et al.* Phys. Rev. C **99**, 065801 (2019).
[7] M. Friedman *et al.*, Nuclear Instrum. Methods in Physics Research A **940**, 93 (2019).
[8] E. Pollacco *et al.*, Nuclear Instrum. Methods in Physics Research A **887**, 81 (2018).

Gamma-ray emission from supermassive black hole binary OJ 287

Vera Yu. Sinitsyna[1,*] and *Vera G.* Sinitsyna[1,**]

[1]P.N. Lebedev Physical Institute, Moscow, Russia

Abstract. OJ 287 is one of the most studied BL Lacs with very long optical measurements which spectrum has been well measured through radio to X-rays. OJ 287 is supposed to be a binary black hole system. Its secondary black hole passes the accretion disk of the primary black hole and produces two impact flashes per 12-year period. Observations of OJ 287 in the GeV – TeV energy range reveal the variable γ-ray flux connected with the flare activity of this object. The spectral energy distributions of BL Lac objects consist of two broad peaks. The Inverse Compton emission of the relativistic electrons in the jet or combined with an external Compton mechanism are considered in the leptonic scenario of second, higher frequency spectrum part generation. The last one supposes the existence of the external to jet photon cloud. Also, the second spectrum part is supposed to be generated due to the acceleration of the cosmic ray hadrons in expanding shock produced by outflow, which then collides with the wind of the primary black hole. The detection of GeV – TeV energy fluxes can help find the configuration parameters of the two-black hole system.

1 Introduction

BL Lacs are a high and rapidly variable subclass of radio-loud, active galactic nuclei (AGN) characterized by non-thermal spectra extending from radio to high and very high energy γ-rays. The short time variability of a day or even minute timescale and the detection at TeV energies points to the origin of emission from the relativistic jets directed to the observer sightline [1]. It suggests that the emission region is located at the parsec scales from the central powering engine. The spectral energy distribution of the BL Lacs is of a two-hump shape with a first peaking at a low-energy in infrared to X-ray range and a second at high energy γ-rays. The production of these emission components and their connection is widely studied. The origin of the high energy emission is a matter of great interest and involves both leptonic and hadronic models. So, the observations of powerful BL Lac-type active galactic nuclei can provide information for the models of described the jet dynamic in AGNi and their emission in the wide range of the electromagnetic spectrum.

2 OJ 287

OJ 287 is one of the most studied blazars of low-frequency peaked BL Lac type which spectrum has been well measured through radio [2], [3] to X-ray [4], [5] and optical bands. The most outstanding characteristic of OJ 287 is its 12 year period, which is discovered throughout the periodic outburst in the optical range [4] and has also been confirmed in the X-ray band. This periodic behavior is supposed to be due to the possible presence of a binary supermassive black hole in which a secondary black hole passes the accretion disk of the primary

*e-mail: sinicinavy@lebedev.ru
**e-mail: sinits@sci.lebedev.ru

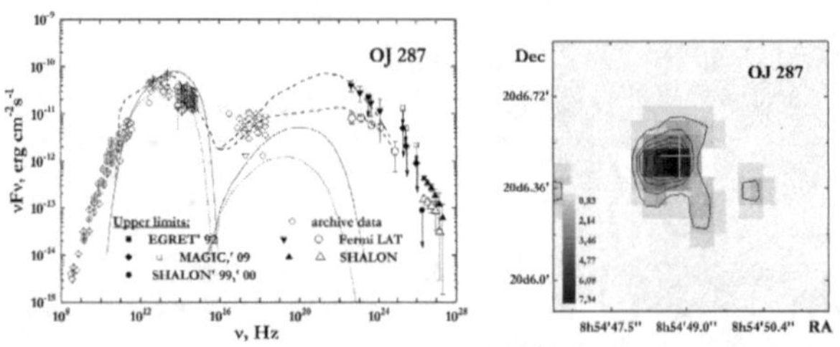

Figure 1. Spectral energy distribution of OJ 287.
△ show the average spectrum;
▲ show the flaring spectra from the SHALON data. Emission map of OJ 287 by SHALON (grey scale).

black hole and produces two impact flashes per period. At high energies, OJ 287 has been detected with GeV emissions by EGRET [10], Fermi-LAT[11], and proposed to be a TeV source.

As mentioned above, the spectral energy distributions of blazars consist of two broad peaks. The first, lower frequency peak is due to the synchrotron emissions of relativistic electrons in the jet. Then in the leptonic scenarios, it is supposed that the second, higher frequency peak is to be due to the Inverse Compton emissions of the same electrons on the surrounding photon fields [6].The inverse Compton scattering of soft photons external to the jet is also considered [7]. The hadronic scenarios invoke proton–synchrotron and proton–photon cascade processes (e.g. [8]), and also cosmic ray accelerated in the shock generated by outflows in the interactions of the parts of the binary black hole system [9].

Under the SHALON program of long-term studies of active galactic nuclei at TeV energies, observations of OJ 287 are being carried out with our experiment. OJ 287 was observed by SHALON in 1999, 2000, 2008 - 2010. The observations of 1999 and 2000 years don't reveal a γ-ray flux from the position of OJ 287, but only an upper limit of $< 1.1 \times 10^{-13}$ph cm^{-2}s^{-1}. In observations of 2008, 2009 and 2010 the weak γ-ray flux was detected [12]. An excess corresponding to a 6.9σ [13] was determined. The observed integral flux at > 800 GeV is $(0.26 \pm 0.07) \times 10^{-12}$ph cm^{-2}s^{-1}. The measured overall energy spectrum of OJ 287 can be fitted by a power-law with an exponential cutoff described by $dN/dE = (1.95 \pm 0.16) \times (E/1\text{TeV})^{-2.37\pm0.20} \times exp(-E/(7.8 \pm 1.4)) \times 10^{-13}$ TeV^{-1}cm^{-2}s^{-1}.

Figure 1 presents the spectral energy distribution of the γ-ray emission from OJ 287 by SHALON (△ and ▲) compared to other experimental data [14], [10], and theoretical predictions from [7], [11] and also [14]. The open triangles at TeV energies on Fig. 1 are the average spectrum of OJ 287 from SHALON experiment; an upper limit at > 0.8 TeV corresponds to SHALON observations in 1999, 2000. The black triangles present the γ-ray spectrum at the increased flux period of 2010. The flux increase over the detected average flux was found on November 14,15, and December 4, 5 of 2010. The average value of increase is $(0.63 \pm 0.15) \times 10^{-12}$ph cm^{-2}s^{-1} (at 6.2σ [13] level), with the softer energy spectrum with a power law with an index of -2.95 ± 0.21. Also, the high flux was detected on November 16, 2009. The duration of the flux increase in December 2010 was 2 days with a peak intensity of $(1.16 \pm 0.51) \times 10^{-12}$ph cm^{-2}s^{-1}. No duration of intervals of flux increase was found in 2009 because the observations were interrupted due to weather conditions. To find correlations of the emissions in various energy ranges, we compared the TeV γ-ray fluxes from OJ287 by SHALON with those at MeV-GeV range by the Fermi-LAT experiment (see Fig. 2) and data in R-band (Suhora Astronomical Observatory) [15].

3 Discussion

It was investigated in [9] a hadronic origin for the X-ray and γ-ray counterparts of one of the major optical flares of OJ 287 occurring after the second lighter black hole passing through the accretion disc of the heavier one. The cosmic ray proton acceleration occurs in the shock generated by outflows interactions with the active galactic nucleus wind of the heavier black hole. A hadronic emission component is compatible with the γ-ray data, but to reproduce the TeV energy spectrum, optical depth for p-p annihilation should be lower.

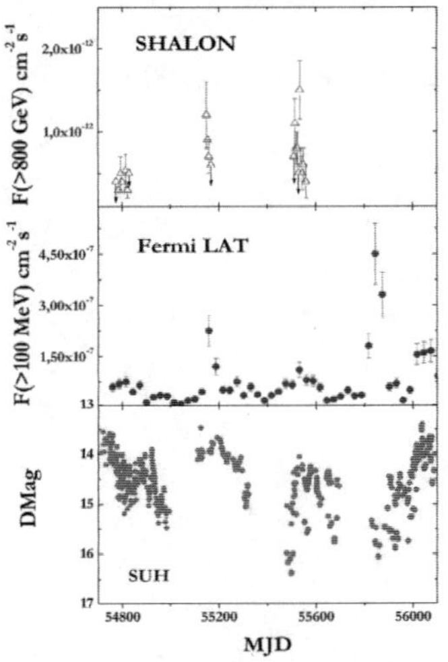

Figure 2. Comparison of OJ 287 light curve by SHALON in the energy range 800 GeV – 35 TeV (upper plot) with data in the MeV-GeV energies by Fermi LAT (middle) and in R-band by Suhora Astronomical Observatory. A slight local flux increase can be seen in the period of November 2010 in Fermi LAT data [16] and at R-band, which corresponds to the above-mentioned γ-ray flux increase observed by SHALON in 2010. The increase of OJ 287 flux in 2009 SHALON data correlates with flares in MeV-GeV energies and R-band.

In [7], the model used the combination of self-Compton and external Compton mechanisms was applied to describe the SED of OJ 287 during a major γ-ray flare of the 2009 year (Fig. 1, dashed lines) also detected at TeV energies. The size of the emission region found to reproduce the observed spectral energy distribution is about 9 pc, consistent with one revealed [17] from the relation of flares in the millimeter, optical, and GeV energy range.

It was found in [17] that the two 0.1–200 GeV flares in OJ 287, as well as two optical flares are related to the two millimeter-wave flares that originated in the second of two >14 pc separated features in the jet. Such multiwavelength behavior can be explained if the γ-rays are produced due to the synchrotron self-Compton scattering of optical photons from the flares. It is proposed that the flare is launched by the interaction of moving plasma blobs with a standing shock. The flare correlations can indicate that the emission observed by SHALON is a TeV particle population also formed the spectrum viewed by Fermi-LAT, which produced by the synchrotron self-Compton process occurs in the jet away from the central core.

4 Conclusions

Experimental data on TeV γ-ray emission from OJ 287 of the different stages, including the flaring activity period of 2008 - 2010 yy are obtained in the SHALON long-term observations and compared with ones at the GeV range by Fermi-LAT and optical data from Suhora Astronomical Observatory. The correlation of the flares in optical, GeV, TeV energies may indicate that the observed spectral energy distribution in TeV energies can be described in terms of the combination of self-Compton and external Compton mechanisms and connected with the interaction of moving plasma blobs launched in the jets with a standing shock.

References

[1] L. Dondi, G. Ghisellini, Mon. Not. R. Astron. Soc. **273**, 583 (1995)
[2] E. Idesawa et al., Publ. Astron. Soc. Jpn. **49**, 631 (1997)
[3] H. Seta et al., Publ. Astron. Soc. Jpn. **61**, 1011 (2009)
[4] E. Valtaoja et al., Astrophys. J. **531**, 744 (2000)
[5] C.E. Tateyama, K.A. Kingham, Astrophys. J. **608**, 149 (2004)
[6] G. Ghisellin et al., Mon. Not. R. Astron. Soc. **301**, 451 (1998)
[7] P. Kushwaha, S. Sahayanathan, K.P. Singh, Mon. Not. R. Astron. Soc. **433**, 2380 (2013)
[8] A. Mücke, R.J. Protheroe, R. Engel et al., Astropart. Phys. **18**, 593 (2003)
[9] J.C. Rodríguez-Ramírez et al., Mon. Not. R. Astron. Soc. **498**, 5424 (2020)
[10] R.C. Hartman et al., ApJS **123**, 79 (1999)
[11] A.A. Abdo et al., Astrophys. J. **700**, 597 (2009); *ibid.* **715**, 429, 132 (2010)

[12] V.G. Sinitsyna et al., Proc. of 32nd ICRC, Beijing, China, 2011; **8/11**, 26 (2011)
[13] T.-P. Li & Y.-Q. Ma, Astrophys. J. **272**, 317 (1983)
[14] M. Hayashida et al., Proc. of 31st ICRC, Lodz, Poland, 2009;
[15] P. Pihajoki, M. Valtonen et al., Astrophys. J. **764**, 5 (2013)
[16] M. Ackermann et al., Astrophys. J. **810**, 14 (2015)
[17] I. Agudo, S.G. Jorstad, A.P. Marscher et al., Astrophys. J. Lett. **726**, L13 (2011)

https://doi.org/10.1051/epjconf/202226011048

Novel experimental techniques for neutron induced charge particle reaction studies in nuclear astrophysics

Chandrabhan Yadav[1,*] and *Moshe* Friedman[1,**]

[1]Racah Institute of Physics, Hebrew University of Jerusalem, 91904 Jerusalem, Israel

Abstract. Neutron induced reaction studies on various unstable and short-lived nuclei play a very important role in understanding the various nucleosynthesis processes occuring in astrophysics scenario. We discuss here our approach to study neutron induced charge particle cross-section for various unstable and short-lived nuclei at effective temperatures of ~1.5-3.5 GK using the ^{7}Li(p, n) reaction as a neutron source with five orders of magnitude higher neutron intensities with respect to currently available neutron time-of-flight facilities, using SARAF accelarator and Liquid Lithium Target (LiLiT) facility.

1 Introduction

Measurements of neutron-induced reaction rates on various unstable and short-lived nuclei are imperative for a complete understanding of the nucleosynthesis occurring in various astrophysical sites and events. The neutron time-of-flight (TOF) method is the most straightforward approach to carry out such experiments. TOF facilities use long flight distance typically few tens of meters and above, resulting in dramatic reduction in neutron intensity. The CERN n_TOF facility EAR2, currently the most intense neutron TOF facility, provide 10^6 neutrons/second at relevant energy range for astrophysics studies [1]. An alternative approach suggested in Ref [2], extensively used for the last three decades. They suggested to use the ^{7}Li(p, n) reaction as a neutron source with a specific proton energy at 1.912 MeV, close to reaction threshold at 1.88 MeV and a thick Li target. At this proton energy, the neutrons are forward-collimated and their energy distribution resembles Maxwellian flux distribution with a temperature of $\approx$ 0.3 GK, allowing nearly direct measurement of the Maxwellian Averaged Cross-Section (MACS) at 0.3 GK, relevant for some s-process sites [3]. An important advantage of this approach is substantial increase in the neutron intensity up to $\approx 10^9$ neutrons/second for 100 μA proton beam. The intense proton beam from superconducting linear RF accelerator SARAF provide beam current of few mA and thus allowing measurements with even higher neutron intensity on the order of 10^{10} neutrons/second [4, 5]. The four order of magnitude increase in neutron intensity allows measurement of smaller cross-section reactions and short-lived isotope. In spite of the successful use of ^{7}Li(p, n) reaction for such experiments, it is only limited for a specific temperature of $\approx$0.3 GK. Various suggestions [6–8] were made to obtain neutron fields at higher temperatures. In this work we discuss our approach to extend the use of varying proton energies in the range of 1.9-3.6 MeV on thick lithium targets to carry out neutron-induced cross sections measurements in

*e-mail: chandrabhan.yadav@mail.huji.ac.il
**e-mail: moshe.friedman@mail.huji.ac.il

the temperatures region of 1.5-3.5 GK. The higher proton energies will result with neutron intensities on the order of 10^{11} neutrons/sec on the sample. Nucleosynthesis in explosive scenarios does not occur in a specific temperature, thus it is important to determine MACS values continuously through-out this temperature region. Our proposed method is based on a series of cross section measurements at different proton energies with proper weighting to obtain the MACS values at any given temperature in this range. The method to produce the neutron field, MACS values at different temperature and required computation tool is described in detail in a recent publication [9]. We will use this method to study (n, p) and (n, α) cross-sections at explosive stellar temperatures, for various nuclei for which experimental data is poor or even non-existent.

2 Experimental plans and details

We will carry out our (n, p) and (n, α) cross-sections measurements for various unstable and short-lived nuclei in two-phases. Figure 1 shows schematic of conceptual design of experimental set-up for Phase-I and Phase-II experiments and respective detection set-up. In Phase-I, experiments with stable and long-lived nuclei will be done at a low intensity proton accelerator. As illustrated in left panel of Fig. 1, a proton beam with varying energies at 1.9 - 3.6 MeV impinges on a thick Li target to produce neutron beam which further impinges on a sample place inside the micromegas based gaseous detector located 2 cm downstream to the neutron source. The reactions product will be detected by the gas filled detector. Phase-II of the project dedicted to experiments with short-lived nuclei will make use of full intensity of SARAF, above 1 mA with Liquid Lithium Target (LiLiT) as a neutron source. As illustrated in right panel of Fig. 1, a proton beam with varying energies impinges on 1.5 mm thick liquid lithium target. The sample is located 1 cm downstream the neutron source. The outgoing charge particles are directed using a quadruple and a dipole magnets to be detected at focal plane nearly 3.5 m from the neutron source. We plan to carry out for example, ^{26}Al(n, p), ^{26}Al(n, α) [$t_{1/2}$=7.2 x 10^5 Years], ^{40}K(n, p), ^{40}K(n, α) [$t_{1/2}$=1.25 x 10^9 Years] and ^{56}Ni(n, p) [$t_{1/2}$= 6.08 Days], ^{75}Se(n, p) [$t_{1/2}$= 119.8 Days] reactions with Phase-I and Phase-II set-up respectively.

We are currently pursuing Phase-I experiments and developing a micromegas based gaseous detector. The detector will be made in cylindrical geometry with cathode, equipotential rings and micro pattern gas amplifier i.e., micromegas structure stretched over the segmented anode pads. The sample for nuclear reaction of interests will be placed in the center of the cathode inside the chamber facing the anode plane. Such design of detection system will provide solid angle of 2π for particle detection. The separation between cathode and anode plane will be 18 cm, and the active diameter of cathode and anode will be around 14 cm. The anode plane will be divided into five circular rings with central pad ring of diameter 3 cm and second, third and fourth rings of diameter 6, 9 and 12 cm respectively, and fifth ring of diameter 14 cm. These outer rings are further segmented into 8 section thus giving total of 33 segmented pads on the anode plane. Left panel of Fig. 2 shows segmented anode pad as designed in Geant4 [10] simulation. The gas mixture of Ne(90%) and CF_4(10%) in the pressure range of 3 – 4 atm will be used as a detection gas. Fig. 2 (right panel) shows electron drift velocity as a function of field for various gas mixtures obtained using [11]. The Ne/CF_4 gas mixture is found to be better choice as detection gas as mixtures containing argon and hydrogen had to omitted in order to avoid background from neutron induced reaction with Ar and H. Geant4 simulation of the detector shows that detector can be operated with proton beam current of 10 μA in a strong γ radiation environment from ^{7}Li$(p, p' \gamma)$ and neutron background.

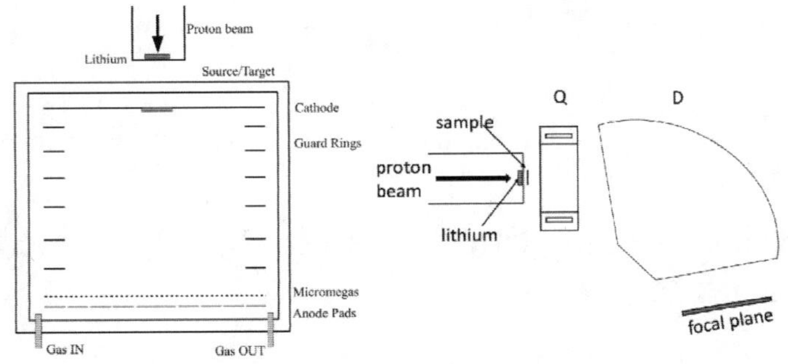

Figure 1. (Left) Schematic of conceptual experimental set-up for Phase-I experiments, and micromegas based gaseous detector. (Right) Schematic of conceptual experimental set-up for Phase-II experiments, and large angle spectrometer.

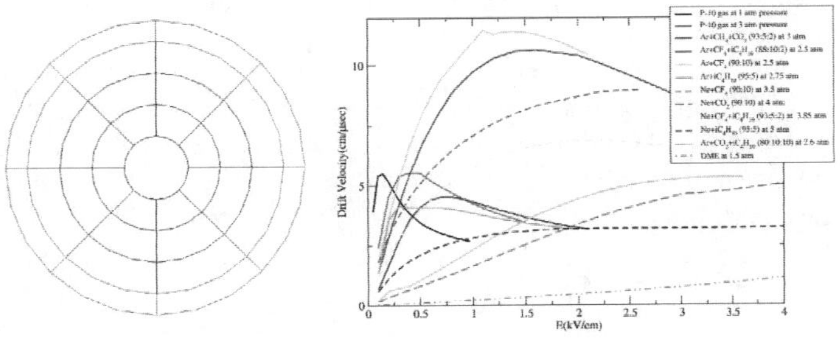

Figure 2. (Left) Segmented anode pad as designed in Geant4 simulation. (Right) Electron drift velocity as a function of field for various gas mixtures obtained using [11].

3 Summary and outlook

Measurements of neutron-induced cross sections on various nuclei of astrophysical interests at explosive stellar temperatures is being plan to study with high-intensity neutron fluxes at the relevant energies. We plan to carry out our first experiments for ^{26}Al(n, p) and ^{26}Al(n, α) reactions with a gas detector being designed and developed very soon.

References

[1] C. Weiß *et al.*, Nucl. Instrum. Method **A 799**, 0168–9002 (2015).
[2] W. Ratynski, and F. Käppeler, Phys. Rev. C **37**, 595 (1988).
[3] F. Käppeler, R. Gallino, S. Bisterzo, W. Aoki, Rev. Mod. Phys.**83**, 157 (2011).
[4] I. Mardor, O. Aviv, M. Avrigeanu *et al.*, Eur. Phys. J. **A 54**, 91 (2018).
[5] M. Paul *et al.*, Eur. Phys. J. A **55**, 44 (2019).
[6] M. Heil *et al.*, Phys. Rev. C **71**, 025803 (2005).
[7] R. Reifarth *et al.*, Eur. Phys. J. Plus **133**, 424 (2018).
[8] R. Reifarth *et al.*, Phys. Rev. C **77**, 015804 (2008).
[9] M. Friedman, Eur. Phys. J. A. **56**, 155 (2020).
[10] S. Agostinelli *et al.*, Nucl. Inst. Methods Phys. Res. A **506**, 250-303 (2003).
[11] Garfield++, https://garfieldpp.web.cern.ch/garfieldpp/

EPJ Web of Conferences **260**, 11049 (2022) https://doi.org/10.1051/epjconf/202226011049
NIC-XVI

β-decay half-lives of neutron-rich N=82,81 isotones by shell-model calculations

Noritaka Shimizu[1,*], *Yutaka* Utsuno[2,1], and *Tomoaki* Togashi[1]

[1]Center for Nuclear Study, The University of Tokyo, Hongo, Bunkyo-ku, 113-0033 Tokyo, Japan
[2]Advanced Science Research Center, Japan Atomic Energy Agency, Tokai, Ibaraki 319-1195, Japan

Abstract. We perform large-scale shell-model calculations and successfully describe the low-lying spectra and half-lives of neutron-rich $N = 82$ and $N = 81$ isotones with $Z = 42 - 49$ in a unified way. Shell-model study shows that their Gamow-Teller strength distribution has a peak in the low-excitation energies, which significantly contributes to the half-lives. This peak is dominated by the $\nu 0g_{7/2} \rightarrow \pi 0g_{9/2}$ transition and enhanced on the proton deficient side of these isotones due to the Pauli blocking effect and the change of the effective single-particle energy of the $\nu 0g_{7/2}$ orbit.

1 Introduction

β-decay half-lives of neutron-rich nuclei around $N = 82$ are key data to understand the second peak of the solar abundance in the *r*-process nucleosynthesis [1]. The *r*-process path is considered to go through the neutron-rich region of the nuclear chart. Where the *r*-process path comes across the magic number $N = 82$, β decay and neutron capture occur alternately and form the waiting points of neutron captures in the r-process. After that, these nuclei cause multiple β-decays and reach the stability line of the nuclear chart, causing the second peak of the solar abundance. The half-lives of $N = 82$ and $N = 81$ isotones are important nuclear data for understanding the *r*-process nucleosynthesis. In the present work, we adopt nuclear shell-model calculations and investigate theoretically the β-decay properties of $N = 82$ and $N = 81$ isotones. This article is based on Ref. [2].

2 Shell-model calculations

We performed large-scale shell-model calculations to discuss β-decay properties of $N = 81$ and $N = 82$ isotones. The model space for the shell-model calculation is taken as $28 < Z < 82$ ($0f_{5/2}$, $1p_{3/2}$, $1p_{1/2}$, $0g_{9/2}$, $0g_{7/2}$, $1d_{5/2}$, $1d_{3/2}$, $2s_{1/2}$, and $0h_{11/2}$) for the proton orbits and $50 < N < 82$ ($0g_{7/2}$, $1d_{5/2}$, $1d_{3/2}$, $2s_{1/2}$, and $0h_{11/2}$) for the neutrons orbits with ^{78}Ni being an inert core. The shell-model interaction consists of the JUN45 interaction [3] for the f_5pg_9 space, the SNBG3 interaction for the neutron $50 < N < 82$ space [4], and the monopole-based universal interaction [5] for the rest part. We truncate the model space by restricting up to 2 proton holes in the pf shell and up to 3-particle 3-hole excitations across the $N = 50$ gap so that the calculation is feasible. Since the M-scheme dimension of the shell-model

*e-mail: shimizu@cns.s.u-tokyo.ac.jp

Hamiltonian matrix reaches 3.1×10^9 and is quite large, efficient usage of a supercomputer is essential. We employ the shell-model code KSHELL [6] for massively parallel computing. The Lawson method is applied to suppress the contamination of center-of-mass motion [7]. In Ref. [2], we demonstrated that the present shell-model interaction successfully reproduces the binding energies and low-lying spectra of $N = 80, 81$, and 82 nuclei, which confirms the validity of the present setup of the shell-model study.

3 Results and discussions

We employ the Lanczos strength function (LSF) method [8] to obtain the Gamow-Teller (GT) strength distribution. In this method, the initial state of the iterations is prepared as $\hat{O}(GT)|\phi_p\rangle$, where $\hat{O}(GT)$ is the Gamow-Teller operator, $\sigma\tau_-$, and $|\phi_p\rangle$ is the wave function of the parent nucleus. We performed 250 Lanczos iterations and evaluate the strength distribution in the subspace spanned by the Lanczos vectors. The quenching factor of the Gamow-Teller transition is taken as 0.7, which is widely used. We do not evaluate the first-forbidden transition since its contribution is expected to be rather unchanged [9] and can be absorbed into the quenching factor.

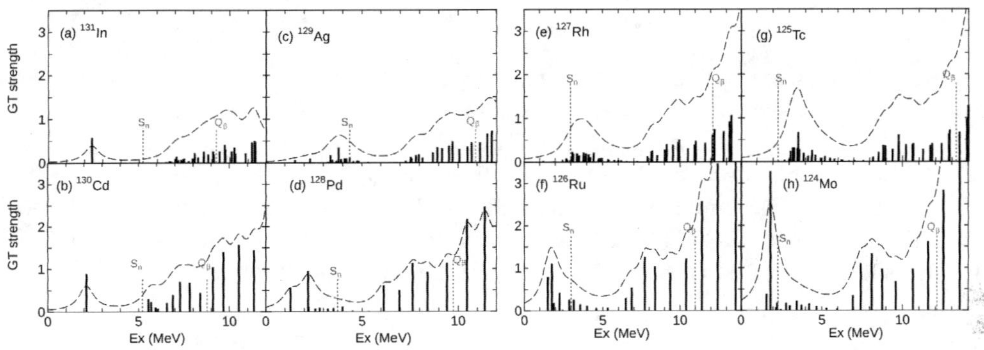

Figure 1. Gamow-Teller strength distribution of $N = 82$ isotones, (a) ^{131}In, (b) ^{130}Cd, (c) ^{129}Ag, (d) ^{128}Pd, (e) ^{127}Rh, (f) ^{126}Ru, (g) ^{125}Tc, and (h) ^{124}Mn, provided by the shell-model calculations. The blue and red dotted lines denote the neutron separation energies and Q-values, respectively. The dashed lines are the folded strength functions by a Lorentzian with 1-MeV width. Revised from Ref. [2].

Figure 1 shows the GT strength distribution of $N = 82$ isotones provided by the LSF method. While the strength of the low-energy peak of ^{131}In ($Z = 49$) at around $E_x = 2$ MeV is small, the strength grows gradually as the proton number decreases. In these isotones, the GT transition from the neutron $0g_{7/2}$ orbit to the proton $0g_{9/2}$ orbit is dominant for the β decay. At $Z = 50$, the proton $0g_{9/2}$ orbit is fully occupied and suppresses the Gamow-Teller transition because of the Pauli blocking. With decreasing the proton number from 50 to 40, the proton $0g_{9/2}$ orbit becomes vacant and the Pauli-blocking effect is weakened. Therefore, the GT strength is gradually enhanced as the proton number decreases till $Z = 40$. In addition, the evolution of the neutron $0g_{7/2}$ orbit contributes coherently to this enhancement as follows.

Figure 2 shows the neutron effective single-particle energies (ESPEs) of $N = 82$ isotones. The proton $0g_{9/2}$ orbit is fully occupied at $Z = 50$ and the ESPE of the neutron $0g_{7/2}$ orbit is the lowest among the five orbits, as indicated by the experimental energy levels of ^{131}Sn [10]. On the other hand, at $Z = 40$ the proton $0g_{9/2}$ orbit is vacant and the neutron $0g_{7/2}$ orbit is raised up and approaches the Fermi level because of the strong attractive monopole force

 https://doi.org/10.1051/epjconf/202226011049

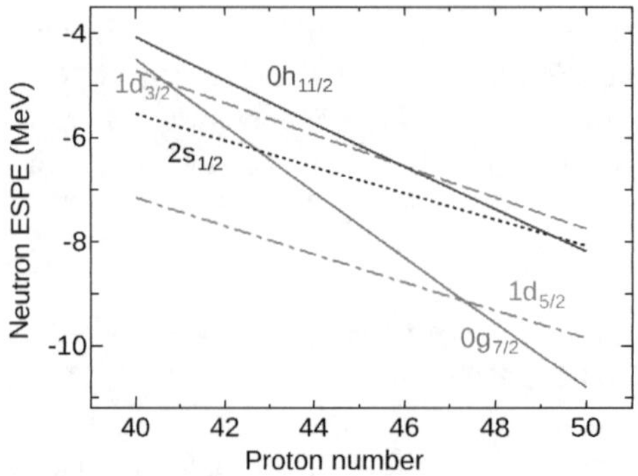

Figure 2. Neutron ESPEs of N=82 isotones. The blue solid, red solid, green dashed-dotted, green dashed, and black dotted lines denote the ESPEs of the neutron $0h_{11/2}, 0g_{7/2}, 1d_{5/2}, 1d_{3/2}$, and $2s_{1/2}$ orbits.

between the $\pi 0g_{9/2}$ and $\nu 0g_{7/2}$ orbits. This strong attractive force is caused by a cooperative attraction of the central and tensor forces [5].

The half-lives are evaluated by accumulating the transition probabilities from the parent ground state to the daughter states whose excitation energies are below the Q_β value. Figure 3 shows the beta-decay half lives of the $N = 81$ and $N = 82$ isotones. The present shell-model results shown as blue solid circles reproduce the experimental values quite well except that of ^{131}In. The half-lives provided by other theoretical models, such as preceding shell-model study [9], relativistic QRPA [11] and the gross theory with the KTUY mass formula [12, 13], are also shown in Fig. 3. These results show a certain overestimation, especially for $N = 81$ isotones.

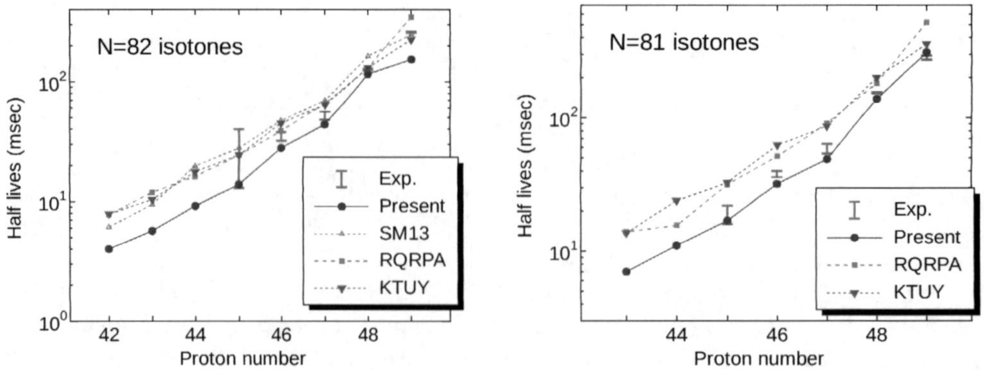

Figure 3. β-decay half-lives of $N = 82$ (left panel) and $N = 81$ (right panel) isotones. The red symbols with error bars denote the experimental values [14]. The blue circles, orange triangles, green squares, and brown inverted triangles denote those of the present shell-model study, the previous shell-model study [9], the relativistic QRPA [11], and the gross theory [13] with the KTUY mass formula [12], respectively.

4 Summary

We present shell-model results of the β-decay half-lives by the GT transition. The experimental half-lives of $N = 82$ and 81 isotones are well produced by the present shell-model study and some predictions are presented. The low-energy peak of the GT strength gradually increases as the proton number decreases from 50 to 40, which is caused by the Pauli-blocking effect of the proton $0g_{9/2}$ orbit and the ascension of the ESPE of the neutron $0g_{7/2}$ orbit coherently.

Acknowledgment

NS acknowledges Hiroyuki Koura for providing us with the KTUY results The authors acknowledge valuable supports by "Program for Promoting Researches on the Supercomputer Fugaku" (JPMXP1020200105), and "Priority Issue on post-K computer" (Elucidation of the Fundamental Laws and Evolution of the Universe) MEXT, Japan. A major part of shell-model calculations was performed on CX400 supercomputer of Nagoya University (hp160146) and Oakforest-PACS supercomputer (hp200130, hp170230, hp160146, xg18i035). We also acknowledge the KAKENHI grants (17K05433, 20K03981), JSPS, Japan.

References

[1] M. Mumpower, J. Cass, G. Passucci, R. Surman, and A. Aprahamian, AIP Advances **4**, 041009 (2014).

[2] N. Shimizu, T. Togashi and Y. Utsuno, Prog. Theor. Exp. Phys. **2021**, 033D01 (2021).

[3] M. Honma, T. Otsuka, T. Mizusaki, and M. Hjorth-Jensen, Phys. Rev. **C 80**, 064323 (2009).

[4] M. Honma *et al.*, RIKEN Accel. Prog. Rep. **45**, 35 (2012).

[5] T. Otsuka, T. Suzuki, M. Honma, *et al.*, Phys. Rev. Lett. **104**, 012501 (2010).

[6] N. Shimizu, T. Mizusaki, Y. Utsuno *et al.*, Comp. Phys. Comm. **244**, 372 (2019).

[7] H. Gloeckner and R. D. Lawson, Phys. Lett. **53B** 313 (1974).

[8] E. Caurier, G. Martinez-Pinedo, F. Nowacki, A. Poves, and A. P. Zuker, Rev. Mod. Phys. **77**, 427, (2005)

[9] Q. Zhi, E. Caurier, J. J. Cuenca-Garcia, K. Langanke, G. Martinez-Pinedo, and Sieja, Phys. Rev. C **87**, 025803 (2013).

[10] NuDat 2.8, http://www.nndc.bnl.gov/nudat2/.

[11] T. Martekin, L. Huther, and G. M.-Pinedo, Phys. Rev. C **93**, 025805 (2016).

[12] H. Koura, T. Tachibana, M. Uno, and M. Yamada, Prog. Theor. Phys. 113 (2005) 305.

[13] T. Tachibana and M. Yamada, Proc. Inc. Conf. on exotic nuclei and atomic masses, Arles, 1995, eds. M. de Saint Simon and O. Sorlin (Editions Frontueres, Gif-sur-Yvette, 1995) p.763. and references therein; T. Yoshida and T. Tachibana, Journal of Nuclear Science and Technology, **37** 491 (2000); T. Tachibana, RIKEN Review, Focused on Models and Theories of the Nuclear Mass, **26** 109 (2000).

[14] G. Lorusso, S. Nishimura, Z. Y. Xu *et al.*, Phys. Rev. Lett. **114**, 192501 (2015).

Cosmic Ray production in SNRs of Pulsar Wind Nebulae type at different ages

Vera G. Sinitsyna[1,*] and *Vera Y.* Sinitsyna[1]

[1]P.N. Lebedev Physical Institute, Moscow, Russia

Abstract. The overall observations of plerions from radio to the very high energy gamma-rays could provide information about the evolution of PWN from the young Crab-like to the older stages. The extended MeV-TeV emission from Geminga middle-aged pulsar of $\sim 3.4 \times 10^5$ year age in SHALON, Milagro, HAWC observations, and Fermi-LAT detection could arise from the PWN associated with the Geminga SNR. The TeV γ-ray emission from 3C 58 PWN whose estimated age varies from ~ 800 to $(5-7) \times 10^3$ years was first detected in year 2011 by SHALON. The overall spectral energy distribution and information about the extension of PWN from radio to GeV-TeV energies from Fermi-LAT and SHALON observations can contribute to particle transport models and also to the understanding of the mechanisms of PWN expanding, which is, in turn, can shed light on the age of 3C 58 and the history of progenitor SN explosion.

1 Introduction

Pulsar wind nebulae (PWNe) are powerful galactic sources of very high energy gamma-rays. PWNe are created by the interaction of pulsar winds with the surrounding media. Such objects contain a pulsar left over the supernova explosion having large electric and magnetic fields and generating a pulsar wind, the outflow of pulsar-accelerated particles [1]. Then the interaction of pulsar winds with the surrounding media generating a shock region where particles are accelerated (e.g. [2]). PWN observations in the wide energy range show that the photon spectra vary with radial distance in some of these objects. So, such observations including ones at high energies can provide information on the energy-dependent and spatially-dependent mechanisms of particle acceleration within the nebula.

2 Geminga

A neutron star in the constellation Gemini is the second brightest source of MeV - GeV γ-rays in the sky, discovered in 1972 by SAS-2 which had no counterpart in other wavelengths. Only in 1991, in X-ray observations by ROSAT satellite, a regular periodicity of 0.237 seconds was detected. This makes Geminga the closest known pulsar to Earth. Geminga is suggested to be a supernova remnant with the expected age of $\sim 3.4 \times 10^5$ yr [3, 4]. Sweeping of the interstellar medium could shape the pulsar wind nebula around Geminga. The morphology of the PWN is unusual, as there is no obvious bow-shock at the front of the nebula, although there is a candidate shock around the pulsar in the optical [5]. The nebula has been extensively studied in X-rays. Also, the very high energy gamma-rays from Geminga [6, 7] were detected by SHALON experiment at 0.8 - 6 TeV (Fig. 1). This weak TeV-emission coincides with the

*e-mail: sinits@sci.lebedev.ru

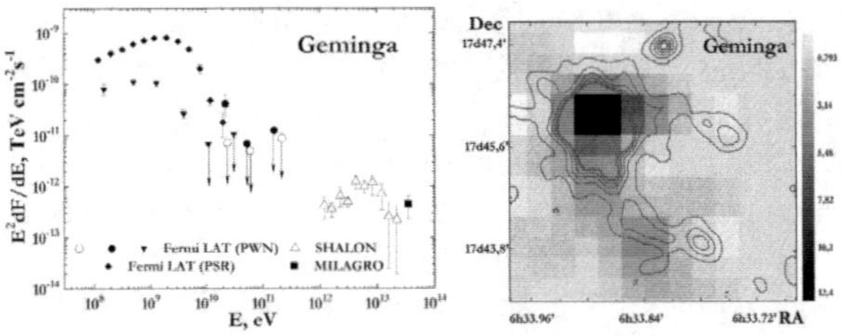

Figure 1. High energy spectral energy distribution of Geminga object.
△ show the average spectrum from the SHALON data.
Emission maps of Geminga by SHALON (grey scale).
Red lines show Chandra data.

outer structures of PWN detected in X-rays by Chandra (Fig. 1, right). The fluxes of MeV-GeV emission from Geminga PWN obtained in observations with Fermi-LAT were estimated in [8]. The ~ 2 degrees extended halo around Geminga was reported first by HAWC [9] and then by [10] presenting the analysis of Fermi-LAT data. Both, TeV gamma-rays and extended MeV-GeV emission viewed by HAWC and extracted from Fermi observations could arise from the PWN associated with the Geminga SNR.

3 3C 58

3C 58 is a supernova remnant (SNR) classified as a PWN due to its flat radio spectrum and very bright core [11]; it is powered by the central pulsar PSR J0205 + 6449 [12]. Radio observations revealed a 6'×9' size filled-centre elliptical morphology of 3C 58 [13, 14]. Similar morphologies have been observed in the infrared (IR) and X-ray bands [15–17]. In X-rays 3C 58 is found to have a non-thermal spectrum from the nebula that varies with radial distance [16, 18] and the nebula size decreases with increasing energy [18].

Gamma-ray emission of MeV-GeV energies from the nebula with a power-law function spectrum with a photon index of 1.61 ± 0.21 has been detected by Fermi-LAT [19, 20].

The source of TeV γ-rays associated with the 3C 58 was detected for the first time by SHALON [21, 22] in 2011 with a statistical significance of 8.1σ. A power law can well describe the energy spectrum of 3C 58, the photon index in the energy range 0.8 - 10 TeV was derived to be 2.36 ± 0.18. Its overall spectral energy distribution is presented in Fig. 2.

It is generally believed that the radio, IR and X-ray part of emission from the 3C 58 nebula is produced by the synchrotron radiation of relativistic electrons within the PWN, but the GeV and TeV γ-rays are generated through inverse Compton scattering-off soft photon fields [23, 24]. The spatially dependent model proposed by [25] was applied to 3C 58 PWN to describe the observed non-thermal emission properties of 3C 58 (Figs. 2, 3).

The radio source 3C 58 was recognized as an SNR at 3.2 kpc [26], and it coincides with the historical supernova SN 1181. But, both the dynamical age of the pulsar 5.4 kyr determined from its spin-down rates and the age of 5.0 ± 2.25 kyr corresponding to the mean expansion velocity of the 3C 58 determined by comparing the VLA radio images in 1973 and

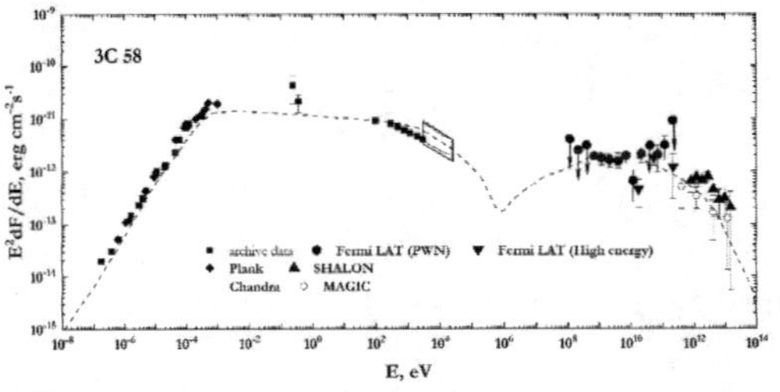

Figure 2. Spectral energy distribution of 3C 58.
▲ show the average spectrum from the SHALON data.
Dashed line presents the modelling results of the photon SED with the parameters found in [25].

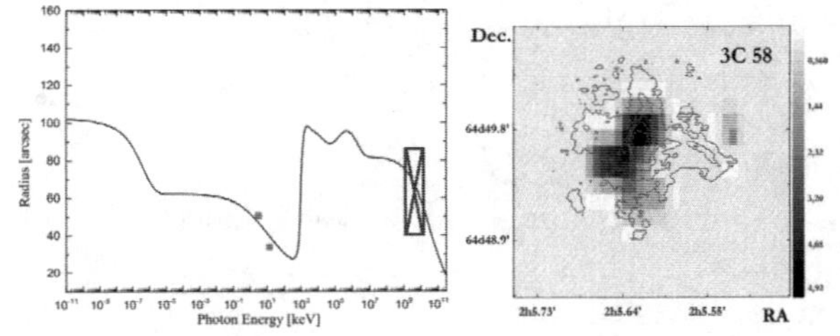

Figure 3. left: Variation of the energy-dependent radial extent of 3C 58. Line is the calculations from [25] model. The observed data at TeV energies is from SHALON observations. **right:** Emission maps of 3C 58 by SHALON (grey scale). Red lines show Chandra data.

1998 are incompatible with the age of the remnant of SN 1181. The system age estimation from the PWN evolution and energetics gave the value of ~ 2.5 kyr [27], also incompatible with the supernova explosion of 1181. Different studies derive ages are from 0.8 to 7 kyr.

New HI absorption study towards 3C 58 PWN 3C 58 in combination with a newly determined distance-velocity relation and by relating this PWN to a nearby H II region and SNR complex of well-known distance revealed the distance value of 2 kpc [28], which then points to the younger age of this object. Also, approximately the type of the progenitor star of 3C 58 were determined in these studies that favour to a point to a late O-type star. Comparison of the result of the model (with parameters including the distance of 2 kpc) with the observed photon spectral energy distribution of 3C 58 is presented in Figs. 2, 3 shows modeled in [25] nebula radial extent in the radio, IR, GeV and TeV bands. The observed extension in TeV energies is consistent with this model.

4 Conclusions

Experimental data on TeV γ-ray emission from Pulsar Wind Nebulae of the different evolution stages are obtained in the SHALON long-term observations. The extended emission at energies > 800 GeV from Geminga was detected and compared with one from the X-ray observations with Chandra telescope. It could arise from the PWN associated with the middle-aged Geminga SN remnant. The Geminga PWN structures viewed in TeV gamma-rays by SHALON and X-rays by Chandra together with the presence of extended halo detected both in MeV-GeV by Fermi-LAT and TeV by HAWC is notably different from fluxes and morphology of 3C 58 detected across the electromagnetic spectrum. Gamma-ray emission from 3C 58 PWN which may associated with historical SN 1181 was detected with SHALON in the range of 0.8 - 10 TeV. 3C 58 γ-ray emission structure was obtained and overlaps with one visible in X-rays by Chandra. Spatially dependent particle transport model [25] was applied to 3C 58 to compare the observed γ-ray emission spectra and radial extent at TeV energies with the model results. The observed extension and spectral energy distribution in TeV energies are consistent with this model, which points to the variation of the 3C 58 nebula extent with energy and may impact the study of the injection particle distribution and the particle acceleration within the PWN.

References

[1] M.J. Rees & J.E. Gunn, MNRAS **167**, 1 (1974)
[2] S.P. Reynolds & R.A. Chevalier, ApJ **278**, 630 (1984)
[3] M. Salvati & B. Sacco, A&A **485**, 527 (2008)
[4] M. Salvati, A&A **513**, A28 (2010)
[5] Y.A. Shibanov et al., A&A **448**, 313 (2006)
[6] V.G. Sinitsyna, AIP (Conf. Proc.) **515**, 205, 293 (2000)
[7] V.G. Sinitsyna et al., Nucl. Phys. B (Proc.Suppl.) **196**, 437 (2009)
[8] M. Ackermann et al, ApJ **726**, 35 (2011)
[9] R. López-Coto, F. Salesa, H. Zhou (for the HAWC collaboration), AIP (Conf. Proc.) **1792**, 040014 (2017)
[10] M. Di Mauro, S. Manconi, F. Donato, Phys. Rev. D **100**, 419 (2019)
[11] K.W. Weiler, N. Panagia, A&A **70**, 419 (1978)
[12] S.S. Murray, P.O. Slane, F.D. Seward et al., ApJ **568**, 226 (2002)
[13] S.P. Reynolds, H.D. Aller, ApJ **327**, 845 (1988)
[14] M.F. Bietenholz, ApJ **645**, 1180 (2006)
[15] D.A. Green, P.A.G. Scheuer, MNRAS **258**, 833 (1992)
[16] P. Slane, D.J. Helfand, E. van der Swaluw, S/S. Murray, ApJ **616**, 403 (2004)

[17] P. Slane, D.J. Helfand, S.P. Reynolds et al., ApJ **676**, L33 (2008)
[18] H. An, ApJ **876**, 150 (2019)
[19] A.A. Abdo et al., ApJ **699**, L102 (2009)
[20] M. Ackermann et al., ApJS **209**, 34 (2013)
[21] V.G. Sinitsyna & V.Yu. Sinitsyna, J. Phys.: Conf. Ser. **409**, 012110 (2013)
[22] V.G. Sinitsyna & V.Yu. Sinitsyna, *"Astroparticle, Particle, Space Physics and Detectors Physics Applications"* ed S Giani, C Leroy, P-G Rancoita, L Price and R Ruchti (Singapore: World Scientific, 2014) vol 8 p 135
[23] S.J. Tanaka & F. Takahara, MNRAS **429**, 2945 (2013)
[24] F.-W. Lu, Q.-G. Gao, B.-T. Zhu, L. Zhang, MNRAS **472**, 2926 (2017a)
[25] F.-W. Lu, Q.-G. Gao, B.-T. Zhu, L. Zhang, ApJ **834**, 43 (2017b)
[26] D.A. Roberts, W.M. Goss, P.M. Kalberla et al., A&A **274**, 427 (1993)
[27] R.A. Chevalier, ApJ **619**, 839 (2005)
[28] R. Kothes, A&A **560**, A18 (2013)

Unresolved nature source TeV J2032+4130

V. G. Sinitsyna[1,*], *K. A.* Balygin[1,2], *S. S.* Borisov[1], *A. I.* Klimov[1,2], *R. M.* Mirzafatikhov[1], *N. I.* Moseiko[2], *I. E.* Ostashev[1,2], and *V. Y.* Sinitsyna[1]

[1]P.N. Lebedev Physical Institute, Moscow, Russia
[2]NRC "Kurchatov Institute", Moscow, Russia

Abstract. Cygnus Region contains many objects that are bright in all wavelengths, including one of the most powerful active star formation regions: Cyg OB2, pulsars, and supernova remnants. Some of the sources have been detected at high and very high energies. One of them was discovered due to the proximity to well-known microquasar Cyg X-3 is object TeV J2032+4130. This object is still of unresolved nature and is being intensively studied in the different energy ranges. The numerous X-ray point sources and diffuse X-ray emission regions were found within the TeV J2032+4130 region by Chandra and Suzaku. Intensities detected in X-rays from these regions may favor a scenario with the dominantly nucleonic, not electronic origin of TeV emission. The results of the twenty-year observation of TeV J2032+4130 object by the SHALON experiment are presented. The collected experimental data on fluxes, spectrum shape, and morphology of TeV J2032+413 can help determine an object type and shed light on the source nature.

1 Introduction

Cygnus Region is one of the brightest areas of the sky in all wavelengths. It is known to contain a number of potential GeV and TeV gamma-ray sources like active star formation regions, supernova remnants (SNRs), relativistic objects. Objects connected to this region include star association Cygnus OB2, the microquasar Cygnus X-3, γCygni SNR and TeV J2032+4130 object. Some of the sources listed above have been detected at high energies with Fermi LAT experiment [1–3] and earlier with EGRET telescope [4, 5]. Also, it was detected at very high energies with Milagro air shower array detector [6] and by imaging atmospheric Cherenkov telescopes Whipple [7], HEGRA [8], SHALON [9–11], VERITAS [12] and MAGIC[13]. TeV J2032+4130 was detected by the HEGRA telescopic system [8] at energies >1 TeV during observations made in 1999 – 2001 yy. This object was the first TeV gamma-ray detection having no bright counterpart at any other wavelength as radio and X-ray, and was also the first extended source to be discovered in the very high energy range.

2 TeV J2032+4130 at very high energies

In accordance with the program on long-term studies of microquasar Cygnus X-3 [9–11] at very high energies, observations of Cygnus Region and its member objects, including TeV J2032+4130 are being carried out with the high-mountain SHALON imaging atmospheric Cherenkov telescopes since 1995.

*e-mail: sinits@sci.lebedev.ru

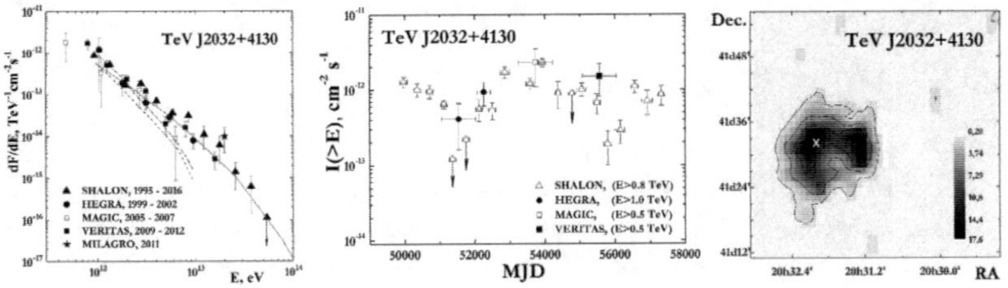

Figure 1. TeV J2032+4130 characteristics: **from left to right:** Differential spectrum of object by SHALON (▲) in comparison with HEGRA, VERITAS, MAGIC and MILAGRO experiments; The light curve at TeV-energies obtained in the SHALON observations (△) in comparison with the HEGRA, MAGIC, and VERITAS data; Emission map of TeV J2032+4130 at > 0.8 TeV by SHALON.

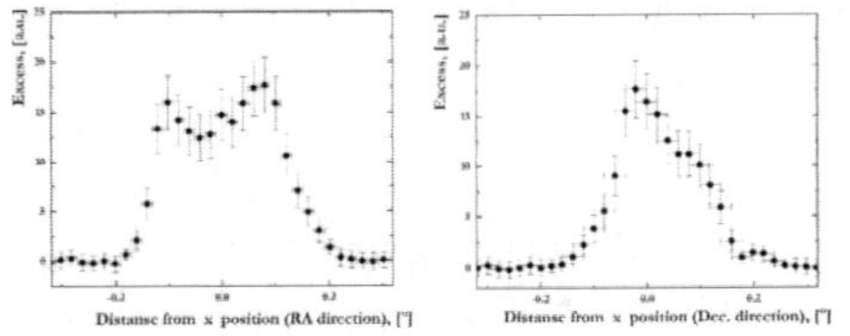

Figure 2.
The profiles of the TeV J2032+4130 emission map along the RA and Dec direction crossing through the point (white ×) of source position (Fig. 1, right).

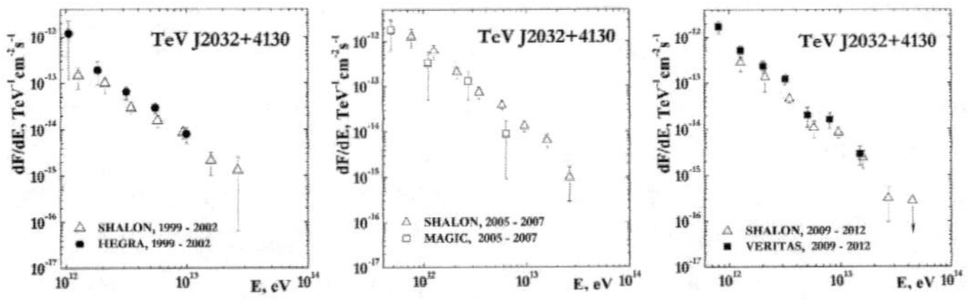

Figure 3.
Differential spectra of TeV J2032+4130 in different observation periods from the SHALON data compared with the HEGRA (1999 - 2002), MAGIC (2005 - 2007), and VERITAS (2009 - 2012) data.

As a result, TeV J2032+4130 was detected above 0.8 TeV by SHALON with the average integral flux of $(0,84 \pm 0,05) \times 10^{-12}$ cm^{-2}s^{-1} at level of 31σ determined by method [15]. The differential γ-ray energy spectrum of TeV J2032+4130 in the energy range from 0,8 to 50 TeV obtained by SHALON telescope is fitted well to a hard power law with an exponential cutoff: $dN/dE = (7,1 \pm 1,1) \times 10^{-13} \times (E_\gamma/1TeV)^{-1,50\pm0,11} \times exp(-E_\gamma/(16 \pm 3)$ TeV) with the $\chi^2/Dof = 1,31$ where degree of freedom $Dof = 8$. It is presented in Fig. 1 with ▲ together with light-curve of this object at the period 1995 - 2015 yy. Also, the source image in TeV-energies by SHALON are shown in Figure 1. The profiles of the TeV J2032+4130 emission map along the RA and Dec direction crossing through the × point of source position are shown in Fig. 2. Analysis of the γ-shower direction showed that the emission source is an extended area of $\sim 10'$ size, coinciding with the experimental data of HEGRA [8], VERITAS [12] and MAGIC [13].

Very high energy data on the TeV J2032+4130 energy spectrum were obtained in the different years with HEGRA [8], VERITAS [12] and MAGIC [13] ground-based telescopes and with the MILAGRO [6] detector. Figure 3 shows the comparison of differential energy spectra of TeV J2032+4130 by SHALON (1995-2016), HEGRA(1999 - 2002), MAGIC (2005 - 2007), VERITAS (2009 - 2012) and data of MILAGRO (2011).

The twenty-year-long SHALON observations of TeV J2032+4130 revealed flux variations at energies > 0.8 TeV. The average integral flux for TeV J2032+4130 at energies > 0.8 TeV from the SHALON data was $(1.01 \pm 0.15) \times 10^{-12}$ cm^{-2}s^{-1} in the periods of observations 1995 – 1997, 2003 – 2006, and 2013 – 2015; $(0.58 \pm 0.12) \times 10^{-12}$cm^{-2}s^{-1} in 1998, 2001, 2002, 2010, and 2016; and a low flux, $(0.12 \pm 0.09) \times 10^{-12}$ cm^{-2}s^{-1} , was recorded in 1999, 2000, 2011, and 2012. Figure 1 presents the light curve of TeV J2032+4130 obtained in the SHALON experiment over the entire period of observations from 1995 to 2015 in comparison with the HEGRA, MAGIC, and VERITAS data. The white triangles indicate the integral fluxes from the SHALON data averaged over each year of observation.

3 TeV J2032+413 nature

The nature of extended object TeV J2032+413 still remains unresolved. Since the discovery of TeV J2032+4130, several region observations have been made by X-ray telescopes Chandra, Suzaku, XMM-Newton, which operate in the energy range of 2 – 10 keV, 0.1 - 10 keV, and 0.2 - 12 keV, respectively. Multiple point sources were detected in the region of TeV J2032+413 by Chandra, XMM-Newton telescopes (see Fig. 4, left), and then after the known point sources were subtracted, XMM-Newton detected an extended X-ray emission region (see [12, 19]). Also, Suzaku detected two structures of diffuse X-ray emission in the TeV J2032+413 region [16]. One of them coincides with the GeV old γ-ray pulsar PSR J2032+4127 detected by Fermi LAT. And overall diffuse emission of the studied region may be attributed to the IC gamma-ray origin in old PWN forming by the pulsar. Another possibility is connecting with long-period binary system forming by has mentioned gamma-ray pulsar PSR J2032+4127 and B0Ve star, MT91 213 having circumstellar disk of varying radius. This object lies on the edge of TeV emission area. In [17] spectra gamma-ray emission of the IC origin from the system are calculated, shown in the Fig. 1 with dash and dash-dot lines. In [18] studies related to the TeV J2032+413 part of Cyg OB2 massive OB association in the energy bands from radio to X-ray and simulations of the broad multifrequency demonstrated that TeV gamma rays are dominantly of a nucleonic, rather than an electronic, origin (see Fig. 1 with red line). On the other hand, radio observations of the region have been made using the GMRT, yielding at least three non-thermal emission sources and an extended non-thermal diffuse emission. Moreover, VLA observations revealed a shell-like structure of weak radio emission [19] consistent with the dimensions of the TeV source reported by HEGRA, VERITAS, MAGIC, and SHALON (see Fig. 4, right).

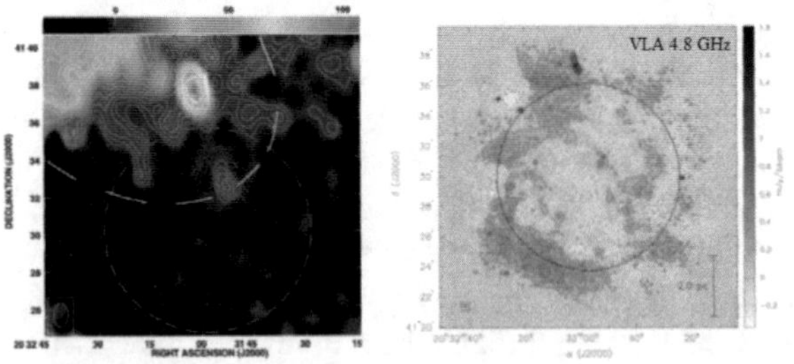

Figure 4. left: Multiple point Chandra X-ray sources (red crosses) were detected in the region of TeV J2032+413 marked with red dashed circle line;
right: Shell-like structure of weak radio emission observed by VLA.(pictures adopted from [19]) TeV J2032+413 source is indicated by the overlaid black line circle)

4 Conclusions

The results of long-term observations of TeV J2032+4130 object by imaging atmospheric Cherenkov telescope SHALON are presented. Characteristics of TeV J2032+4130 in the energy range from 1 up to 50 TeV were obtained with SHALON telescope for the first time. The collected experimental data and refining of fluxes, spectrum shape, and morphology of TeV J2032+413 favor the nucleonic origin of the observed emission and can help in the future determine an object type on the source and shed light on its nature.

References

[1] A.A. Abdo, B. Allen, D. Berley et al., Astrophys. J. **658**, L33 (2007)
[2] A.A. Abdo, M. Ackermann, M. Ajello et al., Science **326**, 1512 (2009)
[3] A.A. Abdo, M. Ackermann, M. Ajell et al., Astrophys. J. **187**, 460 (2010)
[4] D.J. Thompson et al., Astrophys. J. Suppl. Ser. **101**, 259 (1995)
[5] J.A. Esposito, S.D. Hunter, G. Kanbach, P. Sreekumar, Astrophys. J. **461**, 820 (1996)

[6] A.A. Abdo, B. Allen, T. Aune et al., Astrophys. J. **734**, 28 (2011)
[7] J.H. Buckley et al., Astron.& Astrophys. **329**, 639 (1998)
[8] F. Aharonian, A. Akhperjanian, M. Beilicke et al., Astron.& Astrophys. **431**, 197 (2005)
[9] V.G. Sinitsyna, in *Proc. 16th European Cosmic Ray Symposium*, ed. J. Medina, (Alcala de Henares: Universidad de Alcala, Spain, 1998), 383
[10] V.G. Sinitsyna & V.Y. Sinitsyna, Bulletin of the Lebedev Physics Institute, (New York: Allerton Press, Inc.), **40(5)**,113 (2013)
[11] V.G. Sinitsyna & V.Y. Sinitsyna, EPJ Web of Conferences **145** 04002, (2017)
[12] E. Aliu, T. Aune, B. Behera et al., Astrophys. J. **783**, 16 (2014)
[13] J. Albert, E. Aliu, H. Anderhub et al., Astrophys. J. **675**, L25 (2008)
[14] V.G. Sinitsyna, Nuovo Cimento **19C(6)**, 965 (1996)
[15] T.-P. Li & Y.-Q. Ma, Astrophys. J. **272**, 317 (1983)
[16] H. Murakami & S. Kitamoto, Publ. Astron. Soc. Japan **63**, S873 (2011)
[17] W. Bednarek, P. Banasiński, J. Sitarek, JPhG **45**, 015201 (2018)
[18] Y.M. Butt, P. Benaglia, J.A. Combi et al.,Astrophys. J. **597**, 494 (2003)
[19] Y.M. Butt, J.A. Combi, J. Drake et al., Mon. Not. Roy. Astron. Soc. **385**, 1764 (2008)

Symposium Summary

Weiping Liu
Chair of NIC-XVI
China Institute of Atomic Energy (CIAE), P. O. Box 275(1), Beijing 102413, China
February 14, 2022, Beijing

It has been a great pleasure for us to organize the 16th International Symposium on Nuclei in the Cosmos (NIC-XVI) in September 21-25, 2021, which is the first time for this important series of symposium to be held in China and in the form of on-line meeting. Total 388 participants including 170 research institutes and/or universities from 27 countries and regions joined together in this NIC-XVI. Owing to the strong support from participants all over the world, the symposium has been a success in the sense of providing a platform to summarize the exciting new advances of nuclear astrophysics and related fields. I feel highly honored to have hosted this symposium as the Chair of NIC-XVI.

NIC-XVI was co-hosted by China Institute of Atomic Energy (CIAE) and Institute of Modern Physics (IMP). The symposium is financially supported by the International Union of Pure and Applied Physics (IUPAP), National Natural Science Foundation of China (NSFC), Asia-Pacific Center for Theoretical Physics (APCTP), Asian Nuclear Physics Association (ANPhA), and other institutes and associations. Since the first NIC symposium in Austria in 1990, every two years we have such a meeting which combines many science fields of astronomy, astrophysics, cosmology, elementary particle physics, earth and planetary science, and meteorite science as well as nuclear physics. Past meetings had been held in Gran Sasso - Italy (2018), Niigata- Japan (2016), Debrecen-Hungary (2014), Cairns-Australia (2012), Heidelberg-Germany (2010), Mackinac Island-USA (2008), Geneva-Switzerland (2006), Vancouver-Canada (2004), Fuji Yoshida-Japan (2002), Aarhus-Denmark (2000), Volos-Greece (1998), Notre Dame-USA (1996), Gran Sasso-Italy (1994), Karlsruhe-Germany (1992), Baden bei Wien-Austria (1990).

The symposium topics include: Nuclear reaction rates and stellar abundances, The s process, Nuclear properties for astrophysics, High density matter, Novae and XRB, SNIa and the p process, Core collapse SN, mergers and the r process, Underground nuclear astrophysics, The early universe, Galaxy evolution, Radioactivity and meteorites, Inputs for Astrophysics. The whole scientific program consists of 23 invited talks, 62 oral talks, and 87 poster presentations of theory, experiment and observation. The selection of the fields and presenters reflected the recent advances, like interplay between mergers and r process.

New insights were added in this symposium owing to the exciting progress of gravitational wave and underground nuclear astrophysics. Substantial arguments were made on the cosmic, galactic, and stellar evolution based on foremost understanding of the origin and production mechanism of atomic nuclides under the free and generous atmosphere. A special emphasis was put in NIC-XIV on opening new scientific domains by cooperation and the synergy effect among researchers in interdisciplinary fields. In recent years, the observational and experimental efforts to detect the signals from celestial phenomena like supernovae or mergers of compact objects were highly rewarded for the development of new fields of astrophysics, i.e. neutrino astronomy and gravitational wave astronomy. Neutrino astronomy is based on technology of laboratory nuclear physics, and provides valuable knowledge in astronomy and astrophysics on how the atomic nuclides are created in stellar core collapse and explosions, which was one of the scientific highlights intensively discussed in NIC-XVI.

NIC school took place online from Sept. 13 to 20, 2021 with the help of JINA. The school covered a wide variety of topics such as gamma-ray astronomy, stellar grain, stellar modeling and nuclear experiments. By watching video lectures and discussing online with other participants and lecturers, about 20 students from 10 countries learned the basic knowledge and the most recent progress at the frontiers of nuclear astrophysics, were better prepared for the forthcoming NIC conferences. In the poster session, 5 students were awarded by ANPhA as the best posters based on the evaluation of senior researchers in the field.

Due to the COVID-19 pandemic situation globally, this symposium has to be made in hybrid mode. Except a few nearby Chinese participants who were able to attend the meeting in real location to ensure the atmosphere, whereas all foreign and most Chinese participants joined the symposium online. Special considerations were taken to make each presenter assigned a suitable time slot to present his/her talk. Thanks to the support and joining efforts of all participants, the symposium has been surprisingly smooth and did not encounter any interruption. This hybrid symposium with largest participants will give a unique memory in NIC history. I wish to take the chance to thank LoC members and young students very much for their skills and tireless efforts to make hardware and network working smoothly during the symposium.

Finally, I report that the NIC-XVI International Advisory Committee decided that the next NIC-XVII will be hosted by IBS in Korea in 2023. On behalf of NIC-XVI Organizing Committee, I hope that the 16th International Symposium on Nuclei in the Cosmos served for all participants to exchange expertise and deeper knowledge on Nuclei in the Cosmos, and for the young generations in particular to raise international competitiveness and to get motivation on establishing new international collaborations and the leadership in the studies of nuclear astrophysics and related interdisciplinary science in the future.

Wish to meet you all in summer 2023 in Daejeon Korea!

AUTHOR INDEX

AUTHOR INDEX

AUTHOR INDEX

AUTHOR INDEX

AUTHOR INDEX

AUTHOR INDEX

AUTHOR INDEX